D1295599

DICTIONARY OF ECOLOGY AND THE ENVIRONMENT

3rd Edition

Titles in the series

American Business Dictionary	ISBN 0-948549-11-4
Dictionary of Accounting	ISBN 0-948549-27-0
Dictionary of Agriculture, 2nd ed	ISBN 0-948549-78-5
Dictionary of Automobile Engineering	ISBN 0-948549-66-1
Dictionary of Banking & Finance	ISBN 0-948549-12-2
Dictionary of Business, 2nd ed.	ISBN 0-948549-51-3
Dictionary of Computing, 2nd ed.	ISBN 0-948549-44-0
Dictionary of Ecology & Environment, 3rd ed	ISBN 0-948549-74-2
Dictionary of Government and Politics	ISBN 0-948549-05-X
Dictionary of Hotels, Tourism, Catering	ISBN 0-948549-40-8
Dictionary of Information Technology	ISBN 0-948549-03-3
Dictionary of Law, 2nd ed.	ISBN 0-948549-33-5
Dictionary of Library & Information Mgmnt	ISBN 0-948549-68-8
Dictionary of Marketing	ISBN 0-948549-08-4
Dictionary of Medicine, 2nd ed.	ISBN 0-948549-36-X
Dictionary of Multimedia	ISBN 0-948549-69-6
Dictionary of Personnel Management, 2nd ed	ISBN 0-948549-79-3
Dictionary of Printing and Publishing	ISBN 0-948549-09-2
Dictionary of Science & Technology	ISBN 0-948549-67-X

(see back of this book for full title list and information request form)

Also Available

Workbooks for teachers and students of specialist English:

Test your

Vocabulary for Computing	ISBN 0-948549-58-0
Vocabulary for Medicine	ISBN 0-948549-59-9
Vocabulary for Business	ISBN 0-948549-72-6
Vocabulary for Hotels & Tourism	ISBN 0-948549-75-0

DICTIONARY OF ECOLOGY AND THE ENVIRONMENT

3rd Edition

P.H. Collin

PETER COLLIN PUBLISHING

First published in Great Britain 1985
Second Edition 1992
Third Edition 1995

Published by Peter Collin Publishing Ltd
1 Cambridge Road, Teddington, Middlesex, TW11 8DT

British Library Cataloguing in Publication Data

A Catalogue record for this book is available from the British Library

ISBN 0-948549-74-2

Text computer typeset by Barbers Ltd, Wrotham, Kent
Printed and bound in Finland by WSOY

Cover design by Gary Weston

PREFACE TO THE FIRST EDITION

Problems concerning the environment and man's contribution to its destruction or conservation have become more and more urgently discussed over the past years, and are the subject of both academic and general study.

This dictionary provides the user with the basic vocabulary used in studies relating to ecology and the environment. Pollution, climatology, farming practice, waste disposal and many other topics are all covered.

The terms are defined in simple English in a way which makes them accessible to the layman; examples of usage are given, in particular extracts from newspapers and magazines published both in Great Britain and the United States. More extended coverage of some topics is provided by the encyclopaedic comments. At the back of the book the reader will find a useful supplement of information in the form of tables.

We are particularly grateful to Margaret Jull Costa who read the text in proof and to Dr Roger Panaman who both read the text and contributed many scientific points.

PREFACE TO THE SECOND EDITION

There have been so many developments in the fields covered by this dictionary since the first edition, that a new edition has been made, incorporating a very large number of new items to the text and updating existing items. We are again grateful to Margaret Jull Costa for updating material in the Supplement; and in particular we would like to thank Hazel Curties for her work in checking and revising the whole text and David Curties for his advice on meteorology and scientific nomenclature in general.

PREFACE TO THE THIRD EDITION

Further developments in the fields of ecology and environmental studies have been included in this third edition; we are grateful to Liz Greasby for her research work on the revision of the text, and in particular the supplement.

In this edition, we have also added phonetic pronunciation for all the headwords within the dictionary.

Pronunciation

The following symbols have been used to show the pronunciation of the main words in the dictionary:

Stress has been indicated by a main stress mark ('), but this is only a guide, as the stress of the word changes according to its position in the sentence.

Vowels

æ	b**a**ck
ɑː	h**ar**d
ɒ	f**o**g
aɪ	fl**y**
aʊ	pl**ough**
aɪə	f**ire**
aʊə	sh**ower**
ɔː	c**oar**se
ɔɪ	n**oi**se
e	h**ea**d
eə	f**air**
eɪ	m**ai**n
ə	**a**bsorb
əʊ	n**o**de
əʊə	l**ower**
ɜː	b**ir**d
iː	s**ee**p
ɪ	f**i**t
ɪə	cl**ear**
uː	p**oo**l
ʊ	w**oo**d
ʌ	n**u**t

Consonants

b	**b**ud
d	**d**itch
ð	wea**th**er
dʒ	**j**et
f	**f**arm
g	**g**old
h	**h**ead
j	**y**east
k	**c**oke
l	**l**eaf
m	**m**ixed
n	**n**est
ŋ	spri**ng**
p	**p**ond
r	**r**ust
s	**s**cale
ʃ	**sh**ell
t	**t**eak
tʃ	**ch**ain
θ	**th**aw
v	**v**alue
w	**w**ork
z	**z**one
ʒ	fu**s**ion

Aa

Vitamin A ['vɪtəmɪn 'eɪ] *noun (= retinol)* vitamin which is soluble in fat and can be formed in the body, but which is mainly found in food, such as liver, vegetables, egg yolks and cod liver oil

> COMMENT: lack of Vitamin A affects the body's growth and resistance to disease and can cause night blindness

abate [ə'beɪt] *verb* to become less strong

◇ **abatement** [ə'beɪtmənt] *noun* reduction *or* becoming less strong; **pollution abatement** = reduction of pollution; **water pollution abatement** = reduction of pollution in rivers, lakes, etc.; **Noise Abatement Society** = association formed to try to influence people to reduce noise pollution

abiotic [eɪbaɪ'ɒtɪk] *adjective* not biological *or* not relating to living organisms

ablation [ə'bleɪʃn] *noun* removal of the top layer of something; removal of snow *or* ice from the surface of a glacier by melting *or* by the action of the wind; **ablation zone** = section of a glacier's movement downstream during which it loses snow *or* ice by ablation

ABO system [eɪbiː'əʊˌsɪstəm] *noun* system of classifying blood groups

> COMMENT: blood is classified in various ways. The most common classifications are by the agglutinogens in red blood corpuscles (factors A and B) and by the Rhesus factor. Blood can therefore have either factor (Group A and Group B), or both factors (Group AB) or neither (Group O), and each of these groups can be Rhesus negative or positive

abrasion [ə'breɪʒn] *noun* wearing away of rock

abscission [æb'sɪʃn] *noun* shedding of a leaf *or* fruit due to the formation of an abscission layer of cells between the leaf *or* fruit and the rest of the plant (it occurs in autumn, or at any time of the year in diseased parts of a plant)

absolute ['æbsəluːt] *adjective* (i) complete; (ii) terminal point (not compared with anything else); **absolute humidity** = vapour concentration *or* mass of water vapour in a given quantity of air

absorb [əb'zɔːb] *verb* to swallow up *or* consume; to take something up by chemical action; *(of a solid)* to take up a liquid; ***salt absorbs moisture in the air***

◇ **absorbent** [əb'zɔːbənt] **1** *adjective* which absorbs; **oxygen absorbent** = able to take up oxygen **2** *noun* substance *or* part of organism (e.g. root tip) which can take up moisture, nutrient, etc.

◇ **absorption** [əb'zɔːpʃən] *noun* action of taking a liquid into a solid; **absorption plant** = part of a petroleum processing plant, where oil is extracted from natural gas; **sound absorption factor** = number indicating the amount of sound energy absorbed by a surface

◇ **absorptive capacity** [əb'zɔːptɪv kə'pæsɪti] *noun* ability to take up moisture, nutrient, etc.

abstraction [æb'strækʃn] *noun* removal; **abstraction licence** = a licence issued by a Water Board to allow abstraction of water from a river or lake for domestic or commercial use (it is needed for irrigation)

> QUOTE: a statutory limit on water abstraction from the Thames was set in 1911, in essence to maintain the quality of water in the river and provide for navigational needs
> **London Environmental Bulletin**

abundance [ə'bʌndns] *noun* large amount *or* number of something; **relative abundance** = number of individual specimens of an animal *or* plant seen over a certain period of time in a certain place

◇ **abundant** [ə'bʌndnt] *adjective* occurring in large numbers; ***the ocean has an abundant supply of krill***

abyss [ə'bɪs] *noun* (i) very deep hole; (ii) very deep part of the sea

◇ **abyssal** [ə'bɪsəl] *adjective* referring to the deepest part of the sea; **abyssal plain** = flat part of the seabed at the deepest level, approximately 4,000m below sea level; **abyssal zone** = deepest and darkest part of the sea below the euphotic zone (about 4,000 metres deep) where light cannot reach and plant and animal life is rare

◇ **abyssobenthic** [əbɪsəʊ'benθɪk] *adjective* (organism) living on the floor of the deepest part of the sea *or* of a lake

◇ **abyssopelagic** [əbɪsəʊpə'lædʒɪk] *adjective* (organism) living in the deepest water in the sea *or* of a lake, at depths greater than 3,000m

Ac *chemical symbol for* ACTINIUM

acarid ['ækərɪd] *noun* mite *or* tick, a small insect which feeds on plants or animals by piercing the outer skin and sucking juices

◇ **acaricide** *or* **acaridicide** [ə'kærɪsaɪd *or* ækə'rɪdɪsaɪd] *noun* poison used to kill mites and ticks

◇ **Acarida** *or* **Acarina** [ə'kærɪdə *or* əkə'ri:nə] *noun* scientific name for the order of animals including mites and ticks

acceptable daily intake (ADI) [ək'septəbəl,deɪlɪ'ɪnteɪk] *noun* quantity of a substance (nutrient, vitamin, additive, pollutant, etc.) which a person *or* animal can safely consume in his *or* its life

access ['ækses] *noun* **right of access** = (i) right of someone to be able to get to land by passing over someone else's property; (ii) right of the public to walk in areas of the countryside, providing they do not harm crops or farm animals; **access order** = court order which gives the public the right to go on private land; **access road** = road giving access only to the properties on it

acclimatize [ə'klaɪmətaɪz] *verb* (i) to make something become used to a different sort of environment, usually a change in climate; (ii) to become used to a different sort of environment; ***plants take some time to become acclimatized to tropical conditions***

◇ **acclimatization** *or* **acclimation** [əklaɪmətaɪ'zeɪʃn *or* əklaɪ'meɪʃn] *noun* action of becoming acclimatized

> COMMENT: when an organism such as a plant or animal is acclimatizing, it is adapting physically to different environmental conditions, such as changes in food supply, temperature or altitude

accretion [ə'kri:ʃn] *noun* (i) growth of inorganic objects by the attachment of material to their surface; (ii) accumulation of sediments

accumulate [ə'kju:mjəleɪt] *verb* to make *or* become greater in size or quantity over a period of time; ***sediment and debris accumulate at the bottom of a lake;*** **accumulated temperature** = temperature (the number of hours and degrees) above a certain point, usually taken in the UK to be the number of hours above 6°C, which is the minimum temperature necessary for growing crops

◇ **accumulation** [əkju:mjə'leɪʃn] *noun* becoming greater in size or quantity over a period of time; ***the risk of accumulation of toxins in the food chain;*** **accumulation zone** = section of a glacier's movement downstream during which it increases in mass

◇ **accumulative** [ə'kju:mjələtɪv] *adjective* produced by accumulation; ***the accumulative effect of these toxins is considerable***

◇ **accumulator** [ə'kju:mjəleɪtə] *noun* rechargeable electric cell; **heat** *or* **thermal accumulator** = vessel for storing hot liquid

Acer ['eɪsə] *noun* scientific name for the maple and sycamore

acetylene **(C_2H_2)** [ə'setɪli:n] *noun* colourless, flammable gas used in the production of chemicals and in welding

acid ['æsɪd] *noun* chemical compound containing hydrogen, which dissolves in water and forms hydrogen, or reacts with an alkali to form a salt and water, and turns litmus paper red; ***hydrochloric acid is secreted in the stomach and forms part of the gastric juices;*** **inorganic acids** = acids which are derived from minerals, such as hydrochloric acid and sulphuric acid; **organic acids** = weak acids which contain carbon, some of which are pesticides; **acid mine drainage** = water containing acids, which drains from mine workings or from heaps of mine refuse, and enters the drinking water supply; **acid mine water** = water in mine workings which contains acid from rocks; **acid-neutralizing capacity (ANC)** = ability of water (shown by the amount of bicarbonate it contains) to neutralize acids entering from runoff or acid rain

◇ **acidic** [ə'sɪdɪk] *adjective* referring to acids; ***soil and vegetation in high altitude forests are directly exposed to an extremely acidic cloud base;*** **acidic properties** = properties associated with acids; **acidic rocks** = rocks which contain a high proportion of silica; **acidic water** = water which contains acid

◇ **acidification** [əsɪdɪfɪ'keɪʃn] *noun* process of becoming acid *or* of making a substance more acid; ***acidification of the soil leads to the destruction of some living organisms***

◇ **acidify** [ə'sɪdɪfaɪ] *verb* to make a substance more acid; ***fallout causes acidified lakes with no fish population***

◇ **acidity** [ə'sɪdɪti] *noun* level of acid in a solution; ***the alkaline solution may help to reduce acidity***

◇ **acid-proof** ['æsɪd'pru:f] *adjective* able to resist the effect of acid

COMMENT: acidity and alkalinity are measured according to the pH scale. pH7 is neutral; numbers above show alkalinity, while pH6 and below is acid

acid rain *or* **acid deposition** *or* **acid precipitation** [æsɪd 'reɪn *or* æsɪd depə'zɪʃn *or* æsɪd presɪpɪ'teɪʃn] *noun* rain (or snow) which contains a higher level of acid than normal

◇ **acid soot** [æsɪd 'sʊt] *noun* acid carbon particles which fall from smoke from chimneys

COMMENT: acid rain is mainly caused by sulphur dioxide, nitrogen oxide and other pollutants which are released into the atmosphere when fossil fuels (such as oil or coal) containing sulphur are burnt. Acid rain rarely falls near the source of the pollution, because the smoke from chimneys can be carried by air currents for many kilometres before it finally falls as rain. So Scandinavia receives acid rain which is caused by pollution from British and German factories; Canada receives acid rain from factories in the US. Acid soot, on the other hand, can fall relatively close to the source of pollution. It is caused when carbon combines with sulphur trioxide from sulphur-rich fuel to form particles of an acid substance which can damage the surfaces it falls on (such as stone buildings). The effects of acid rain are primarily felt by wildlife. The water in lakes becomes very clear as fish and microscopic animal life are killed. It is believed that it is acid rain that kills trees, especially conifers, which gradually lose their leaves and die

acorn ['eɪkɔ:n] *noun* fruit of the oak tree

acoustics [ə'ku:stɪks] *noun* study of sound, especially noise levels in buildings

◇ **acoustician** [əku:s'tɪʃn] *noun* person (such as an architect) who specializes in the study of noise

acquired [ə'kwaɪəd] *adjective* (condition) which is neither congenital nor hereditary, and which develops in reaction to the environment; **acquired immunity** = immunity which a body acquires and which is not congenital

acre ['eɪkə] *noun* unit of measurement of land area, equal to 4,840 square yards, or 0.4047 hectares

◇ **acreage** ['eɪkərɪdʒ] *noun* area of land measured in acres; **cultivable acreage** = number of acres on which crops can be grown; **acreage reduction program (ARP)** = American federal programme under which farmers are only eligible for subsidies if they reduce the acreage of certain crops planted (NOTE: the British equivalent is **set-aside)**

acrid ['ækrɪd] *adjective* (smoke, fumes, flavour, etc.) having a strong, bitter smell or taste

acrolein [ə'krəʊlɪn] *noun* poisonous, strong-smelling liquid used in the production of resins and medicines

act [ækt] *verb* to take action *or* to have an effect on something; ***the emergency services acted quickly to contain the oil spill; pesticides act on an animal's nervous system***

actinide ['æktɪnaɪd] *noun* one of the radioactive elements (actinium to lawrencium) which are in the same category as uranium in the periodic table, with atomic numbers from 89 to 104

◇ **actinium** ['æktɪniəm] *noun* natural radioactive element, produced from the decay of uranium-235 (NOTE: chemical symbol is **Ac;** atomic number is **89)**

COMMENT: actinides are waste products from nuclear fission. They pose problems for disposal as some of them have very long half-lives. They can be reduced to more disposable forms by burning in fast reactors

action ['ækʃn] *noun* thing which is done; effect; **to take action against something** *or* **to stop something** = to work to prevent something happening; ***the government is taking action to stop the spread of pollution***

◇ **activate** ['æktɪveɪt] *verb* to make something start to work, especially a chemical reaction; **activated carbon** *or* **activated charcoal** = form of carbon to which gases can stick, used in gas masks or as a filter to control pollution or added to water as it is being treated before domestic consumption; **activated sludge** = solid sewage containing active microorganisms and air, which is used to mix with untreated sewage to speed up the purification process

◇ **activation** [æktɪ'veɪʃn] *noun* making (something) start to work; **activation of sludge** = mixing of microorganisms and air into sewage to speed up the purification process

◇ **activator** ['æktɪveɪtə] *noun* substance which activates; **compost activator** = chemical added to a compost heap to speed up the decomposition of decaying plant matter

◇ **active** ['æktɪv] *adjective* working; **active ingredient (AI)** = substance which works to produce the desired effect (as in a medicine, etc.); **active margin** = area at the edge of a continental mass, where volcanic activity is frequent; **active organic matter (AOM)** =

organic matter in the process of being broken down by bacteria; **active volcano** = volcano which is erupting *or* likely to erupt

◇ **activity** [ək'tɪvɪti] *noun* action *or* movement; **volcanic activity** = earthquakes, eruptions, lava flows, smoke emissions, etc. which show that a volcano is not extinct

acute [ə'kju:t] *adjective* (i) (problem *or* situation) which has rapidly become very serious; (ii) (disease) which comes on rapidly and can be dangerous; ***the region is suffering from an acute shortage of medical supplies; after the acute stage of the illness had passed, he felt very weak*** *compare* CHRONIC

adapt [ə'dæpt] *verb* to change to fit a new situation; ***the animals have gradually adapted to the change in climate; people adapt to the reduced amounts of oxygen available at high altitudes***

◇ **adaptability** [ədæptə'bɪlɪti] *noun* ability (of an organism) to change to fit a new situation; **degree of adaptability** = extent to which an organism can change to fit a new situation

◇ **adaptation** [ædæp'teɪʃn] *noun* change in an organism so that it is better able to survive *or* reproduce, thereby contributing to its fitness

◇ **adaptive radiation** [ə'dæptɪv reɪdɪ'eɪʃn] *noun* development of a species from a single ancestor in such a way that different forms evolve to fit different environmental conditions

additive ['ædɪtɪv] *noun* chemical substance which is added, especially one which is added to food to improve its appearance, smell or taste, or to prevent it going bad; ***the orange juice contains a number of additives; allergic reactions to additives are frequently found in workers in food processing factories; these animal foodstuffs are free from all additives;*** **food additive** = chemical substance added to food, especially one which is added to food to improve its appearance or to prevent it going bad; **fuel additive** *or* **lead-based additive** = substance (such as tetraethyl lead) which is added to petrol to prevent knocking

COMMENT: colour additives are added to food to improve its appearance. Some are natural organic substances like saffron, carrot juice or caramel, but other colour additives are synthetic. Other substances are added to food to prevent decay or to keep the food in the right form: these can be emulsifiers, which bind different foods together as mixtures in sauces, for example, and stabilizers, which can keep a sauce semi-liquid and prevent it from separating into solids and liquids. The European Community allows certain additives to be added to food and these are given E numbers

ADI [eɪdi:'aɪ] = ACCEPTABLE DAILY INTAKE

adiabatic [ædɪə'bætɪk] *adjective* (process) where no heat leaves or enters the system; **adiabatic lapse rate** = rate of temperature change in rising air (10°C per thousand metres for dry air and 5.8°C per thousand metres for damp air)

◇ **adiabatically** [ædɪə'bætɪkli] *adverb* without losing or gaining heat; ***the air mass rises adiabatically through the atmosphere***

COMMENT: a parcel of rising air expands, because the surrounding pressure falls. Also its temperature falls because no heat can enter or leave it. If a parcel of air descends, the opposite happens and the air temperature rises. See also INVERSION

adipose tissue ['ædɪpəʊz'tɪʃu:] *noun* body fat *or* tissue where the cells contain fat which replaces the normal fibrous tissue when too much food is eaten

adjust [ə'dʒʌst] *verb* to change to fit in with new circumstances; ***plants take some time to adjust to a new climatic system***

◇ **adjustment** [ə'dʒʌstmənt] *noun* process of physical change in response to external environmental changes

adjuvant ['ædʒu:vənt] *noun* something which helps *or* assists

adobe [ə'dəʊbi] *noun* (i) fine clay from which bricks can be made; (ii) bricks made from fine clay, dried in the sun

adrenaline [ə'drenəli:n] *noun* hormone secreted by the adrenal glands

◇ **adrenal glands** [ə'dri:nəl 'glændz] *plural noun* two endocrine glands at the top of the kidneys which produce adrenaline and other hormones

COMMENT: adrenaline is produced when a person *or* animal is experiencing surprise *or* shock *or* fear *or* excitement: it speeds up the heartbeat and raises the blood pressure

adsorb [æd'zɔ:b] *verb (of a solid)* to bond with a gas *or* vapour which touches its surface

◇ **adsorbable** [əd'zɔ:bəbl] *adjective* (solid) which is able to be bonded with a gas *or* vapour which touches its surface

◇ **adsorbent** [əd'zɔ:bənt] **1** *adjective* capable of adsorption **2** *noun* solid which is able to be bonded with a gas *or* vapour which touches its surface

◇ **adsorption** [əd'zɔ:pʃən] *noun* bonding of a solid with a gas *or* vapour which touches its surface

advanced gas-cooled reactor (AGR) [əd'vɑ:nstgæsku:ldrɪ'æktə] *noun* type of nuclear reactor, in which carbon dioxide is used as the coolant and is passed into water tanks to create the steam which will drive the turbines

advection [æd'vekʃn] *noun* movement of air in a horizontal direction (as opposed to convection, where the air rises); **advection fog** = fog which forms when a warmer, moist air mass moves over a colder surface (land *or* sea)

adventitious [ædven'tɪʃəs] *adjective* (root) which develops from a plant's stem and not from another root

adverse ['ædvɜːs] *adjective* **(a)** not favourable; ***the plant will not prosper in these adverse conditions; adverse weather conditions delayed planting*** **(b)** moving in the opposite direction; ***adverse winds***

advisory [æd'vaɪzəri] *adjective* giving advice; **advisory board** = group of specialists who can give advice

AEC [eiː'siː] *US* = ATOMIC ENERGY COMMISSION

aeolian [iː'əʊlɪən] *adjective* caused by wind; **aeolian deposits** = sediments which are blown by the wind; *see also* LOESS

aerate [eə'reɪt] *verb* to put air into a substance (as to replace stagnant air in soil with fresh air); ***worms are useful because they aerate the soil***

◇ **aeration** [eə'reɪʃn] *noun* putting air into a substance

◇ **aerator** [eə'reɪtə] *noun* device for putting air into a substance

> COMMENT: the process of aeration of soil is mainly brought about by the movement of water into and out of the soil; rainwater drives out the air, and then as the water drains away or is used by plants, fresh air is drawn into the soil to fill the spaces. The aeration process is also assisted by changes in temperature, good drainage, cultivation and open soil structure. Sandy soils are usually well aerated; clay soils are poorly aerated

aerial ['eərɪəl] *adjective* which exists in the air; **aerial root** = root of certain plants, which hangs in the air or clings to other plants and takes up moisture from the air; **aerial spraying** = spraying of crops by plane or helicopter

> COMMENT: aerial spraying involves both pesticides and fertilizers. It is subject to various safety precautions and care must be taken to avoid drifting of harmful chemicals onto land next to the land being sprayed

aerobic [eə'rəʊbɪk] *adjective* needing oxygen for its existence; **aerobic digestion** = processing waste, especially organic waste such as manure, in the presence of oxygen (NOTE: opposite is **anaerobic**)

◇ **aerobiosis** [eərəʊbaɪ'əʊsɪs] *noun* biological activity which occurs in the presence of oxygen

aerogenerator [eərəʊ'dʒenəreɪtə] *noun* windmill with fast-moving sails used to generate mechanical power or electricity

aerosol ['eərəsɒl] *noun* **(a)** tiny particles of liquid *or* powder which stay suspended in the atmosphere (such as mist); **aerosol dispenser** = container *or* device from which liquid *or* powder can be sprayed in tiny particles; **aerosol propellant** = gas used in an aerosol can to make the spray of liquid come out; **condensation aerosol** = droplets of moisture which form in warm damp air as it cools, producing mist; **dispersion aerosol** = droplets of moisture which are blown into the air (as spray) **(b)** can of liquid with a propellant gas under pressure, which is used to spray the liquid (such as an insecticide *or* medicinal liquid) in the form of tiny drops

> COMMENT: aerosols in the atmosphere may be formed of liquid (as in the case of mist) or solid particles (as in the case of dust storms). Aerosols in the atmosphere are the form in which pollutants such as smoke are dispersed. Commercial aerosols (that is, the metal containers) may use CFCs as propellants, but these are believed to be responsible for the destruction of ozone in the upper atmosphere and are gradually being replaced by less destructive agents

aestivation [estɪ'veɪʃn] *noun* state in some animals when they become dormant during the hot summer months (not to be confused with hibernation)

aetiology *or US* **etiology** [iːti'ɒlədʒi] *noun* (study of) the cause *or* origin of a disease

◇ **aetiological agent** [iːtiə'lɒdʒɪkl 'eɪdʒənt] *noun* agent which causes a disease

affect [ə'fekt] *verb* to make something change *or* have a result on; ***plant growth has been affected by the change in climate; acid rain affects all freshwater lakes in the area***

affluent ['æfluənt] **1** *adjective* **(a)** wealthy **(b)** (water) which is flowing freely **2** *noun* stream which flows into a larger river

afforest [ə'fɒrɪst] *verb* to plant (an area) with trees

◇ **afforestation** [əfɒrɪs'teɪʃn] *noun* (i) growing trees as a crop; (ii) planting trees on land previously used for other purposes; ***there is likely to be an increase in afforestation of upland areas if the scheme is introduced***

aflatoxin [æflə'tɒksɪn] *noun* very toxic substance formed by a fungus *Aspergillus* *flavus,* which grows on seeds and nuts and affects stored grain

afrormosia [æfrɔː'məʊzɪə] *noun* hardwood from West Africa, now becoming scarce

aftershock ['ɑːftəʃɒk] *noun* weaker shock which follows the main shock of an earthquake

Ag *chemical symbol for* SILVER

agar *or* **agar agar** ['eɪgə] *noun* jelly made from seaweed, used to cultivate bacterial cultures in laboratories

agent ['eɪdʒənt] *noun* **(a)** person who acts for another, usually in another country; ***she is the local agent for the relief organization*** **(b)** chemical substance which makes another substance react; substance *or* organism which causes a disease *or* condition; **Agent Orange** = extremely poisonous herbicide used as a defoliant (it was used by US forces in the Vietnam War); **bleaching agent** = something which removes colour, such as a chemical or the action of sunlight; **carcinogenic agent** = substance which causes cancer; **dispersing agent** = chemical substance sprayed onto an oil slick to try and break up the oil into smaller particles; **polluting agent** = substance which causes pollution; **surface-active agent** = substance which reduces surface tension; **thickening agent** = substance which causes a liquid to become thicker

◇ **agency** ['eɪdʒənsi] *noun* **(a)** force that causes something to happen; ***the disease develops through the agency of certain bacteria present in the bloodstream*** **(b)** office *or* organization which provides help; ***a famine relief agency***

◇ **Agency for International Development (AID)** *noun* the main organization in the US dealing with aid to developing countries

Agenda 21 [ə'dʒendətwentɪ'wʌn] *noun* lengthy document agreed at the UN Conference on Environment and Development summit in Rio outlining the extent of global environmental problems and the measures needed to ensure sustainable development; *see* EARTH SUMMIT

> COMMENT: Agenda 21, a 720-page document, was a principal outcome of the UNCED summit held in Rio de Janeiro during June 1992. It covered in some detail the measures needed to ensure sustainable development. Described as a 'blueprint for action', its recommendations were not, however, binding. Its core chapters covered in some detail the establishment of a framework for implementing UNCED's resolutions. The document outlined broad strategies for combating problems such as over-consumption of non-renewable resources; loss of biodiversity; air, land and marine pollution; and toxic wastes

agglomerate [ə'glɒmərət] *noun* rock made up of fragments of lava fused by heat

agglutinogen [əgluː'tɪnədʒn] *noun* substance in red blood cells which reacts with a substance in serum

aggravate ['ægrəveɪt] *verb* to make worse; ***the effects of acid rain on the soil have been aggravated by chemical runoff***

aggregate ['ægrəgət] *noun* crushed stones used to make concrete or road surfaces

◇ **aggregation** [ægre'geɪʃn] *noun* dispersal of plants *or* animals, where the individuals remain quite close together

AGR [eɪdʒiː'ɒː] = ADVANCED GAS-COOLED REACTOR

agri- *or* **agro-** ['ægrɪ *or* 'ægrəʊ] *prefix* referring to agriculture *or* cultivation of land

◇ **agribusiness** [ægrɪ'bɪznəs] *noun* the business of farming and making products and equipment used by farmers

> COMMENT: the term is used to refer to large-scale farming businesses run along the lines of a conventional company, often involving the growing, processing, packaging and sale of farm products

agriculture ['ægrɪkʌltʃə] *noun* farming, the cultivation of land, including horticulture, fruit growing, crop and seed growing, dairy farming and livestock breeding

◇ **agricultural** [ægrɪ'kʌltʃərl] *adjective* referring to farming; **agricultural engineer** = (i) person trained in applying the principles of science to farming; (ii) person who designs, manufactures or repairs farm machinery and equipment; **agricultural waste** = waste matter produced on a farm such as manure from animals *or* excess fertilizers and pesticides which run off the fields

◇ **agriculturalist** [ægrɪ'kʌltʃərəlɪst] *noun* person trained in applying the principles of science to farming

agrochemicals [ægrəʊ'kemikəlz] *noun* pesticides and fertilizers developed artificially for agricultural use; **the agrochemical industry** = the branch of industry which produces pesticides and fertilizers used on farms

> QUOTE: more efficient agrochemicals applied at lower rates and which target pests weeds and diseases are reflected in the British Agrochemicals Association figures for 1988. Since 1983, the amount of active pesticide ingredients applied to UK soils and tcrops fell from 33,5[illegible] The area increased from 15.1m hectares to 19.6m hectares
>
> **Farmers Weekly**

agroclimatology [ægrəʊklaɪmə'tɒlədʒi] *noun* study of climate and its effect on agriculture

◇ **agroecology** [ægrəʊɪ'kɒlədʒi] *noun* ecology of a crop-producing area

◇ **agroecosystem** [ægrəʊ'iːkəʊsɪstəm] *noun* community of organisms in a crop-producing area

◇ **agroforestry** [ægrəʊ'fɒrɪstri] *noun* growing farm crops and trees together as a farming unit

◇ **agroindustry** [ægrəʊ'ɪndəstri] *noun* industry dealing with the supply, processing and distribution of farm products

◇ **agronomist** [ə'grɒnəmɪst] *noun* person who studies crop cultivation; **pasture agronomist** = person who specializes in the study of types of grass grown in pastures

◇ **agronomy** [ə'grɒnəmi] *noun* scientific study of the cultivation of crops

COMMENT: the use of land to raise crops for eating first started about 10,000 years ago. All plants grown for food have been developed over many centuries from wild plants, which have been progressively bred to give the best yields in different types of environment. Genes from wild plants tend to be more hardy and resistant to disease and are still kept in gene banks to strengthen new cultivated varieties

AI [eɪ'aɪ] = ARTIFICIAL INSEMINATION

aid [eɪd] **1** *noun* **(a)** help; **first aid** = temporary help given rapidly to someone who is injured *or* ill until full-scale medical treatment can be given; **medical aid** = treatment of someone who is ill *or* injured, given by a doctor; medical supplies and experts sent to a country after a disaster **(b)** help in the form of people, food, medicines, equipment, etc., given to a developing country; ***food aid to the Third World; the country spends 7% of its gross national product on aid to Third World countries*** **(c)** machine *or* tool *or* drug which helps someone do something; ***crop sprayers are useful aids in combating insect-borne disease*** **2** *verb* to help *or* give support to; ***changing to lead-free petrol will aid the fight against air pollution***

◇ **aid agency** ['eɪd 'eɪdʒənsi] *noun* organization (such as Oxfam, Save The Children Fund) which specializes in sending help to countries which need it

◇ **aider** *or* **aid worker** ['eɪdə *or* 'eɪd wɜːkə] *noun* person who helps; **first aider** = person who offers first aid to someone who is suddenly injured *or* ill

QUOTE: Since 1960, about $1.4 trillion (in 1988 dollars) of aid has been transferred from rich countries to poor ones

Economist

AID [eɪaɪ'diː] = AGENCY FOR INTERNATIONAL DEVELOPMENT

aim [eɪm] *verb* to intend to do something; ***the government aims to bring in stricter pollution controls***

air [eə] *noun* mixture of gases (mainly oxygen and nitrogen) which cannot be seen, but which exists all around us and which is breathed by animals; ***go for a walk in the mountains and get some fresh air; children playing in the streets are breathing polluted air;*** **air cleaner** = filter which removes unwanted substances from the air; **air conditioner** = device which controls the temperature, ventilation and humidity in a building *or* vehicle; **air conditioning** = system which controls the temperature, ventilation and humidity in a building *or* vehicle; **air current** = flow of air; **air frost** = condition where the air temperature above ground level is below 0°C; **air mass** = very large mass of air in the atmosphere in which the temperature is almost constant, divided from another mass by a front; **air pollutant** = substance (such as gas *or* smoke) which pollutes the air; **air pollution** = polluting of the air by gas *or* smoke, etc.; **air pressure** = normal pressure of the air on the surface of the earth; **air purification** = removal of unwanted substances from the air; **air quality** = way of assessing how 'good' air is, that is, if it is clean, fresh, unpolluted, etc. (NOTE: where referring to a distinct mass of air, the word **parcel** is used)

◇ **airborne** ['eəbɔːn] *adjective* carried by the air; ***there is a high risk of airborne pollution near the factory;*** **airborne combustion product** = gas, vapour *or* solid produced by burning and which is transported through the air; **airborne pollutant** = substance which causes pollution and which is carried along by the air

◇ **air-cooled** ['eəkuːld] *adjective* cooled by means of a current of air

◇ **air-cooling** [eə'kuːlɪŋ] *noun* cooling by means of a current of air

◇ **airstream** ['eəstriːm] *noun* flow of air in a certain direction; ***a westerly airstream is flowing across the country***

COMMENT: the composition of air in the lower atmosphere is: nitrogen (78%), oxygen (21%), argon (less than 1%) and trace quantities of carbon dioxide, helium, hydrogen, krypton, neon, ozone and xenon. Air pollution can be caused by human action such as industrial processes or smoking tobacco, but also by natural disasters such as volcanic eruptions, forest fires, etc.

Al *chemical symbol for* aluminium

Alaska [ə'læskə] *noun US* unit of measurement of the quantity of oil produced in Alaskan oilfields, needed for a certain use; ***energy use in commercial buildings has fallen by almost 0.5 Alaska; heat conservation has eliminated fuel needs equivalent to almost 2 Alaskas***

albedo [æl'biːdəʊ] *noun* measurement of the ability of a surface to reflect light, shown as the proportion of solar energy which strikes the earth and is reflected back by a particular surface

COMMENT: albedo is highest on light shiny surfaces such as snow. It is lowest on dark uneven surfaces, such as masses of leaves which absorb solar energy

QUOTE: if cloud cover increases, the Earth's albedo can increase - clouds over the oceans contribute significantly to the overall reflectivity of the earth
Nature

albino [æl'biːnəʊ] *noun* animal which is deficient in the colouring pigment melanin, with little or no pigmentation in skin, hair or eyes

aldehyde ['ældɪhaɪd] *noun* one of several compound hydrocarbons, found in smog

alcohol (C_2H_5OH) ['ælkəhɒl] *noun* colourless inflammable liquid, produced by the fermentation of sugars and used as an ingredient of organic chemicals, intoxicating drinks and medicines

alder ['ɔːldə] *noun* fine European hardwood *(Alnus glutinosa)* which is waterproof

aldosterone [æl'dɒstərəʊn] *noun* hormone secreted by the adrenal gland and which regulates the balance of sodium and potassium in the body and the amount of body fluid

aldrin ['ɔːldrɪn] *noun* one of the organochlorine insecticides, used to control cabbage root fly, wireworms and leatherjackets. There is an agreement to restrict the use of this insecticide

alert [ə'lɜːt] *noun* warning *or* alarm; **pollution alert** = warning that pollution levels are or will be high

aleurone ['æluːrəʊn] *noun* protein found in the outer skin of seeds

alfalfa [æl'fælfə] *noun* lucerne, a plant of the Leguminosae family, grown to use as fodder

algae ['ældʒiː] *plural noun* tiny plants living in water or in moist conditions, which contain chlorophyll and have no stems or roots or leaves; **blue-green algae** = Cyanophyta, algae found mainly in fresh water; **brown algae** = Phaeophyta *or* brown seaweed; **green algae** = Chlorophyta *or* green plants living in water; **red algae** = Rhodophyta *or* type of very small algae *or* phytoplankton, mainly found on the seabed and which cause the phenomenon called red tide

◊ **algaecide** ['ældʒɪsaɪd] *noun* substance used to kill algae

◊ **algal** ['ælgəl] *adjective* referring to algae; ***algal populations increase rapidly when phosphates are present;*** **algal bloom** = mass of algae which develops rapidly in a lake due to eutrophication; **algal control** = keeping in check the growth of algae; **algal culture** = algae grown in a laboratory; **algal culture installation** = place where algae is grown commercially *or* for scientific purposes

◊ **algoculture** ['ælgəkʌltʃə] *noun* growing of algae commercially *or* for scientific purposes

COMMENT: algae grow rapidly in water which is rich in phosphates. When the phosphate level increases, as when fertilizer runoff enters the water, the algae multiply greatly to form huge floating mats (or blooms), blocking out the light and inhibiting the growth of other organisms. When the algae die, they combine with all the oxygen in the water so that other organisms suffocate

alien ['eɪliən] *noun & adjective* (species) which is not native to an area, but which has been introduced by people; ***a fifth of the area of the national park is under alien conifers; alien species, introduced by settlers as domestic animals, have brought about the extinction of some endemic species***

alimentation [ælɪmən'teɪʃn] *noun* feeding *or* taking in food

◊ **alimentary canal** [ælɪ'mentəri kə'næl] *noun* tube in the body going from the mouth to the anus and including the throat, stomach, intestine, etc., through which food passes and is digested

◊ **alimentary system** [æli'mentəri 'sɪstem] *noun* arrangement of tubes and organs, including the alimentary canal, salivary glands, liver, etc., through which food passes and is digested

aliphatic hydrocarbons [ælɪ'fætɪk haɪdrəʊ'kɑːbənz] *noun* paraffins, acetylenes and olefins, hydrocarbon compounds that do not contain benzene; *compare* AROMATIC

alkali ['ælkəlaɪ] *noun* one of many substances which neutralize acids and form salts (NOTE: British English plural is **alkalis,** but US English is **alkalies)**

◊ **alkaline** ['ælkəlaɪn] *adjective* containing more alkali than acid

◊ **alkalinity** [ælkə'lɪnɪti] *noun* amount of alkali in something such as soil *or* water

COMMENT: alkalinity and acidity are measured according to the pH scale. pH7 is neutral, and pH8 and upwards are alkaline. One of the commonest alkalis is caustic soda, used to clear blocked drains

alkaloid ['ælkəlɔɪd] **1** *adjective* similar to an alkali **2** *noun* one of many poisonous substances (such as atropine *or* morphine *or* quinine) found in plants which use them as defence against herbivores; they are also useful as medicines

alkyl benzene sulphonate ['ælkɪl benziːn 'sʌlfəneɪt] *noun* surface-acting agent used in detergent, which is not biodegradable and creates large amounts of foam in sewers and rivers

allele [ə'liːl] *noun* one of two or more alternative forms of a gene, which can imitate each other's form: they are situated in the same area of a pair of chromosomes and produce different characteristics

◇ **allelopathy** [ælɪ'lɒpəθi] *noun* harm caused by one plant to another plant, usually by producing a chemical substance

allergen ['ælədʒen] *noun* substance which produces hypersensitivity; **food allergen** = substance in food which produces an allergy

◇ **allergenic** [ælə'dʒenɪk] *adjective* which produces an allergy; ***the allergenic properties of fungal spores;*** **allergenic agent** = substance which produces an allergy

◇ **allergy** ['ælədʒi] *noun* sensitivity to certain substances such as pollen *or* dust, which cause a physical reaction; ***she has an allergy to household dust;*** **food allergy** = reaction caused by sensitivity to certain foods (the commonest being strawberries, chocolate, milk, eggs, oranges)

◇ **allergic** [ə'lɜːdʒɪk] *adjective* suffering from an allergy; ***she is allergic to pollen; he showed an allergic reaction to rotting hay;*** **allergic agent** = substance which produces an allergic reaction; **allergic reaction** = effect (such as a skin rash or sneezing) produced by a substance to which a person has an allergy; **allergic person** = person who has an allergy to something

COMMENT: allergens are usually proteins, and include foods, dust, animal hair, as well as pollen from flowers. Treatment of allergies depends on correctly identifying the allergen to which the patient is sensitive. This is done by patch tests, in which drops of different allergens are placed on scratches in the skin. Food allergens discovered in this way can be avoided, but it is hard to avoid other common allergens like dust and pollen, and these have to be treated by a course of desensitizing injections

alleviate [ə'liːvɪeɪt] *verb* to reduce difficulties (usually only temporarily); ***they are trying to alleviate the water shortage in the south of the country***

◇ **alleviation** [əliːvi'eɪʃn] *noun* reducing difficulties; **flood alleviation** = helping to reduce the possibility of flooding by controlling the flow of water in rivers

allo- ['æləʊ] *prefix* meaning different

◇ **allogamy** [ə'lɒgəmi] *noun* fertilization by pollen from different flowers *or* from flowers of genetically different plants of the same species

COMMENT: some fruit trees are self-fertile, that is, they fertilize themselves with their own pollen. Others need pollinators, usually different cultivars of the same species

allograft ['æləʊgrɑːft] *noun* homograft, graft of tissue from one specimen to another of the same species

◇ **allopatric** [ælə'pætrɪk] *adjective* (plants of the same species) which grow in different parts of the world and so do not cross-breed

alloy ['ælɔɪ] *noun* metal (such as brass) made from a compound of two or more metallic elements

COMMENT: many metals have practical disadvantages in the pure state and alloys have been developed to make the best use of their advantages

alluvium [ə'luːviəm] *noun* silt deposited by rivers *or* by a lake

◇ **alluvial** [ə'luːviəl] *adjective* referring to alluvium; **alluvial deposits** *or* **alluvial soils** = deposits of silt on the bed of a river *or* lake; **alluvial fan** = cone-shaped deposit of sediment built up by a river where the slope of the bed is sharply reduced; **alluvial flat** *or* **alluvial plain** = flat area along a river where silt is deposited when the river floods; **alluvial mining** = mining to get minerals from alluvial deposits (such as panning for gold); **alluvial silt** = = ALLUVIUM

alpha ['ælfə] *noun* first letter of the Greek alphabet; **alpha particle** = nucleus of the same composition as a helium atom, which is emitted by the nuclei of some radioactive elements, such as radon and which, when emitted, will pass through gas but not through solids; **alpha radiation** = radiation by alpha particles from radioactive nuclei; **alpha ray** = stream of alpha particles; **alpha waste** = radioactive waste emitting alpha particles

alpine ['ælpaɪn] *adjective* referring to the Alps *or* to other high mountains; **alpine plant** = plant which grows on *or* comes originally from high mountains; ***alpine vegetation grows above the treeline***

alternative [ɔːl'tɜnətɪv] *adjective* different *or* not following the usual way; **alternative energy** = energy produced by alternative technology (such as tidal power, wind power, etc.); **alternative medicine** = treating diseases and disorders by means which are not normally used by traditionally trained doctors (such as herbal medicines, acupuncture, etc.); **alternative source of energy** = means of providing energy by tidal power, wind power, solar power, etc., rather than by using fossil fuels or nuclear power; **alternative technology** = using methods to produce energy which are different and less polluting than the usual ways (i.e. wind power, tidal power, solar power, as opposed to traditional *or* nuclear power)

alti- ['æltɪ] *prefix* meaning height

◇ **altimeter** ['æltɪmiːtə] *noun* instrument which records altitude

◇ **altitude** ['æltɪtjuːd] *noun* height of an object, especially above sea-level; **altitude sickness** = condition where a person suffers from oxygen deficiency at a high altitude (as on a mountain) where the level of oxygen in the air is low

◇ **altocumulus** [æltəʊ'kjuːmjələs] *noun* small white cumulus clouds which form as a layer high in the atmosphere, above 3,000m; *compare* STRATOCUMULUS

◇ **altostratus** [æltəʊ'strɑːtəs] *noun* high thin uniform cloud (above 3,000m), usually seen as a front is approaching

aluminium [ælju'mɪniəm] *or US* **aluminum** [ə'luːmɪnəm] *noun* metallic element, extracted from the ore bauxite; **aluminium oxide (Al_2O_3)** = white *or* colourless powder used in the manufacture of glass, ceramics and abrasives (NOTE: chemical symbol is **Al**; atomic number is **13**)

◇ **alumina (Al_2O_3)** [ə'luːmɪnə] *noun* = ALUMINIUM OXIDE; **high-alumina cement** *or* **aluminous cement** = cement made of bauxite and limestone, used because it resists heat

Am *chemical symbol for* americium

amber ['æmbə] *noun* yellow translucent fossil conifer resin, sometimes containing fossilized insects

ambient ['æmbjənt] *adjective* surrounding an organism *or* an object; ***deaths from ambient carbon monoxide poisoning are increasing;*** **ambient noise** = general noise which surrounds an organism (such as traffic noise, waterfalls, etc.); **ambient quality standards** = levels of acceptable clean air which a national body tries to enforce; **ambient temperature** = temperature of the air in which an organism lives

amelioration [əmiːljə'reɪʃn] *noun* improvement *or* making better; ***there has been some amelioration in pollution levels***

amenity [ə'miːnəti] *noun* **(a)** pleasantness of the surroundings **(b)** something which makes the surroundings more pleasant, such as a park, swimming pool, sports centre, etc.; **amenity society** = group of people dedicated to the protection and improvement of their local surroundings; **natural amenities** = things which make the surroundings more pleasant and are not man-made, such as rivers, lakes, heathland, etc.; **public amenities** = things which make the surroundings more pleasant and are open to the general public

americium [æmə'rɪsiəm] *noun* artificial radioactive element (NOTE: chemical symbol is **Am**; Atomic number is **95**)

amino acid [ə'miːnəʊ 'æsɪd] *noun* chemical compound which is broken down from proteins in the digestive system and then used by the body to form its own protein; ***proteins are first broken down into amino acids;*** **essential amino acids** = eight amino acids which are essential for growth, but which cannot be synthesized and so must be obtained from food or medicinal substances

> COMMENT: amino acids all contain carbon, hydrogen, nitrogen and oxygen, as well as other elements. Some amino acids are produced in the body itself, but others have to be absorbed from food. The eight essential amino acids are: isoleucine, leucine, lysine, methionine, phenylalanine, threonine, tryptophan and valine

ammonia (NH_3) [ə'məʊniə] *noun* gas with a strong smell, a compound of nitrogen and hydrogen, which is a normal product of organic metabolism and is used in compounds to make artificial fertilizers *or* in liquid form as a refrigerant; **ammonia water (NH_4OH)** = solution of ammonia in water

> COMMENT: ammonia is released into the atmosphere from animal dung. It has the effect of neutralizing acid rain but in combination with sulphur dioxide it forms ammonium sulphate which damages the green leaves of plants

◇ **ammonification** [əməʊnɪfɪ'keɪʃn] *noun* treatment *or* impregnation with ammonia

ammonium [ə'məʊniəm] *noun* ion formed from ammonia; **ammonium fixation** = the absorption of ammonium ions by the soil; **ammonium nitrate** = a popular fertilizer used as top dressing. It is available in a special prilled or granular form, and can be used both as a straight fertilizer and in compounds; **ammonium phosphate** = fertilizer which can be used straight, but is more often used in compounds. A certain amount of care is needed in its use, as applications may increase the acidity of the soil; **ammonium sulphate** *see* SULPHATE OF AMMONIA

amoeba [ə'mi:bə] *noun* single cell organism, characterized by a constant changing of its shape (NOTE: plural is **amoebae.** Note also the US spelling **ameba)**

◇ **amoebiasis** [əmɪ'baɪəsɪs] *noun* infection caused by amoeba, which can result in amoebic dysentery in the large intestine (intestinal amoebiasis) and can sometimes infect the lungs (pulmonary amoebiasis)

◇ **amoebic** [ə'mi:bɪk] *adjective* referring to an amoeba; **amoebic dysentery** = mainly tropical form of dysentery which is caused by *Entamoeba histolytica* which enters the body through contaminated water or unwashed food

◇ **amoebicide** [ə'mi:bɪsaɪd] *noun* substance which kills amoebae

amorphous [ə'mɔ:fəs] *adjective* with no regular shape; (mineral) which does not form crystals

amount [ə'maunt] **1** *noun* quantity; ***the amount of pollutants that the factory is allowed to discharge into the atmosphere is carefully controlled*** **2** *verb* to be equal (to); ***rainfall in some areas only amounts to a few millimetres per annum; the government's attitude amounts to a dismissal of the acid rain problem***

ampere ['æmpeə] *noun* base SI unit of electrical current, the current flowing through an impedance of one ohm which has a voltage of one volt across it

Amphibia [æm'fɪbiə] *noun* scientific name for a class of egg-laying animals which live partly in water and partly on land (including frogs, toads, etc.), where the larvae live in water and the adults can live both in water and on land

◇ **amphibian** [æm'fɪbiən] **1** *adjective* (organism) which lives both in water and on land **2** *noun* animal which lives both in water and on land (such as a frog)

◇ **amphibious** [æm'fɪbiəs] *adjective* (animal) which lives both in water and on land; (vehicle) which can travel both on water and on land

AMU = ATOMIC MASS UNIT

anabatic wind [ænə'bætɪk 'wɪnd] *noun* wind which blows up the slope of a mountain as it is heated by the ground (NOTE: opposite is **katabatic)**

anabolism [ə'næbəlɪzm] *noun* building up of simple chemicals into complex ones in an organism

anadromy [ə'nædrəmi] *noun* form of migration of fish (such as salmon) which are hatched in fresh water, migrate to the sea and then return to fresh water to spawn

◇ **anadromous** [ə'nædrəməs] *adjective* (species of fish) which hatches in fresh water and becomes adult in salt water; *see also* CATADROMY

anaerobic [ænə'rəʊbɪk] *adjective* not needing oxygen for existence; **anaerobic decomposition** = breakdown of organic material by microorganisms without the presence of oxygen; **anaerobic digestion** = breakdown of organic material without the presence of oxygen, a process which permanently removes the unpleasant odour of many organic wastes so that they can be used on agricultural land (NOTE: opposite is **aerobic)**

◇ **anaerobiosis** [æneərəʊbaɪ'əʊsɪs] *noun* biological activity which occurs without the presence of oxygen

> COMMENT: anaerobic digesters for pig, cattle and poultry waste feed the waste into a tank where it breaks down biologically to give off large amounts of methane. This gas is then used to generate electricity. The remaining slurry can be applied directly to the land

analyse ['ænəlaɪz] *verb* (i) to examine something in detail; (ii) to separate something into its component parts; ***the laboratory is analysing the soil samples; when the water sample was analysed it was found to contain traces of bacteria***

◇ **analysis** [æ'nælɪsɪs] *noun* examination of a substance to find out what it is made of (NOTE: plural is **analyses)**

◇ **analyst** ['ænəlɪst] *noun* person who examines samples of substances *or* tissue to find out what they are made of

> COMMENT: chemical and electrical methods are used in soil analysis to determine the pH and lime requirements of a soil. Portable testing equipment, using colour charts, is sometimes used to test for pH

> QUOTE: the ozone layer is thinning in the northern as well as the southern hemisphere according to a new and exhaustive analysis of data
> **New Scientist**

ANC [eɪen'si:] = ACID-NEUTRALIZING CAPACITY

anemograph [ə'neməgrɑ:f] *noun* instrument which records wind speed on a roll of paper

anemometer [ænɪ'mɒmɪtə] *noun* instrument for measuring wind speed

> COMMENT: an anemometer is formed of four cups at the ends of the arms of a cross-piece, which is mounted horizontally on a pivot and turns round as the wind

blows; it can be linked to an anemograph, which records the wind speed on a roll of paper

aneroid barometer ['ænərɔɪd bə'rɒmɪtə] *noun* barometer with a vacuum to which a diaphragm is attached, which moves as the atmospheric pressure changes

angledozer ['æŋgəldəʊzə] *noun* piece of earth-moving equipment: a bulldozer with the blade set at an angle

animal ['ænɪməl] **1** *adjective* of *or* or concerning living organisms which can feel sensation and move voluntarily **2** *noun* living organism which can feel sensation and move voluntarily; **animal husbandry** rearing and tending farm animals

anion ['ænaɪən] *noun* ion with a negative electric charge

annoyance [ə'nɔɪəns] *noun* something that causes a nuisance, which is harmful or offensive to the community; harm *or* offence caused; **annoyance scale** = system of classifying the amount of nuisance *or* harm *or* offence caused by something; **annoyance sound level** = point at which a noise begins to cause a nuisance *or* become offensive

annual ['ænjuəl] **1** *adjective* which happens each year *or* once a year; **annual ring** = ring of new wood formed each year in the trunk of a tree and which can easily be seen when the tree is cut down; *see also* DENDROCHRONOLOGY **2** *noun* plant whose life cycle (germination, flowering, fruiting) takes place within the period of a year

COMMENT: as a tree grows, the wood formed in the spring has more open cells than that formed in later summer. The difference in texture forms the visible rings. Note that in tropical countries, trees grow all the year round and so do not form rings

anomaly [ə'nɒməli] *noun* something which is different; **magnetic anomaly** = way in which the local magnetic field differs from the normal magnetic field in a certain area

anopheles [ə'nɒfɪli:z] *noun* mosquito which carries the malaria parasite

anoxia [æ'nɒksiə] *noun* lack of oxygen (as in tissues *or* water)

◇ **anoxic** [æ'nɒksɪk] *adjective* (water) which lacks oxygen

◇ **anoxybiosis** [ænɒksɪbaɪ'əʊsɪs] *noun* biological activity occurring where there is a lack of oxygen

antacid [ænt'æsɪd] *adjective & noun* (substance, such as a medicine) that stops too much acid forming *or* alters the amount of acid (as in the stomach)

Antarctic [ænt'ɑ:ktɪk] **1** *noun* **the Antarctic** = continent at the South Pole, largely covered with snow and ice **2** *adjective* referring to the Antarctic; **Antarctic air** = mass of cold air which is permanently over the Antarctic region; **the Antarctic Circle** = parallel running round the earth at latitude 66°32S, to the south of which lies the Antarctic region

◇ **Antarctica** [ænt'ɑ:ktɪkə] *noun* area of land around the South Pole

QUOTE: about 90 per cent of the world's fresh water is held in Antarctica and, if it all melted, sea levels could rise as much as 60 metres

Times

anther ['ænθə] *noun* part of a stamen which produces pollen

anthracite ['ænθrəsaɪt] *noun* type of shiny hard black coal which burns well and does not produce much smoke

◇ **anthracosis** [ænθrə'kəʊsɪs] *noun* disease of the lungs, caused by inhaling coal dust

anthropogenic [ænθrɒpə'dʒi:nɪk] *adjective* caused by or resulting from man's activities

anti- ['ænti] *prefix* meaning against; **anticondensation paint** = paint which prevents the formation of condensation; **anti-pollution legislation** = laws banning pollution; **antipredator nets** = nets which keep out predators

antibacterial [æntɪbæk'tɪəriəl] *adjective* which destroys bacteria

antibiotic [æntɪbaɪ'ɒtɪk] **1** *adjective* which stops the spread of bacteria **2** *noun* drug (such as penicillin) which is developed from living substances and which kills or stops the spread of microorganisms; ***he was given a course of antibiotics; antibiotics are no use against viral diseases;*** **broad-spectrum antibiotic** = antibiotic used to control many types of bacteria

COMMENT: penicillin is one of the commonest antibiotics, together with streptomycin, tetracycline, erythromycin and many others. Although antibiotics are widely and successfully used, new forms of bacteria have developed which are resistant to them

QUOTE: some antibiotics, such as penicillin and cloxacillin, are effective against staphs and streps, but have no effect on coliform bacteria. Others, like streptomycin, are effective against streptococci. A third group, the broad-spectrum antibiotics, which includes tetracycline, are effective against the common causes of mastitis

Practical Farmer

antibody ['æntɪbɒdi] *noun* substance which is naturally present in the body and which attacks foreign substances (such as bacteria)

◊ **anticaking additive** [æntɪ'keɪkɪŋ 'ædɪtɪv] *noun* additive added to food to prevent it becoming solid (E numbers E530 - 578)

◊ **anticline** ['æntɪklaɪn] *noun (in rock formations)* fold where the newest layers of rock are on the surface; *compare* SYNCLINE

◊ **anticyclone** [æntɪ'saɪkləʊn] *noun* area of high atmospheric pressure, usually associated with fine dry weather in summer and fog in winter; ***winds circulate round an anticyclone clockwise in the Northern Hemisphere and anticlockwise in the Southern***

◊ **anticyclonic** [æntɪsaɪ'klɒnɪk] *adjective* (i) referring to anticyclones; (ii) referring to the opposite direction to the rotation of the earth; **anticyclonic gloom** = darkness during the daytime, when low stratocumulus clouds form at the approach of an anticyclone

◊ **anticyclonically** [æntɪsaɪ'klɒnɪkli] *adverb* in the opposite direction to the rotation of the earth

◊ **antidote** ['æntɪdəʊt] *noun* substance which counteracts the effect of a poison; ***there is no satisfactory antidote to cyanide***

◊ **antifoam** *or* **anti-foaming agent** ['æntɪfəʊm *or* æntɪ'fəʊmɪŋ 'eɪdʒənt] *noun* chemical substance added to a detergent *or* to sewage to prevent foam from forming

◊ **antifouling paint** [æntɪ'faʊlɪŋ 'peɪnt] *noun* special pesticide painted onto the bottom of a ship to prevent organisms growing on the hull and which may be toxic enough to pollute sea water; *see also* TBT

◊ **antifungal** [æntɪ'fʌŋgəl] *adjective* (substance) which kills *or* controls fungi

◊ **antigen** ['æntɪdʒən] *noun* substance (such as a virus *or* germ) in the body which makes the body produce antibodies to attack it

◊ **antihistamine (drug)** [æntɪ'hɪstəmiːn] *noun* drug used to control the effects of an allergy which releases histamine

◊ **antiknock additive** [æntɪ'nɒk 'ædɪtɪv] *noun* substance (such as tetraethyl lead) which is added to petrol to prevent knocking

> COMMENT: antiknock additives increase the power of petrol but create dangerous lead pollution through exhaust fumes

antimalarial [æntɪmə'leəriəl] *adjective & noun* (drug) used to treat malaria

◊ **antioxidant** [æntɪ'ɒksɪdənt] *noun* substance which prevents oxidation, used to prevent materials such as rubber from deteriorating; also added to processed food to prevent oil going bad (in the EU, antioxidant food additives have numbers E300 - 321)

◊ **antipodes** [æn'tɪpədiːz] *plural noun* two points on opposite sides of the earth

◊ **antiseptic** [æntɪ'septɪk] **1** *adjective* which prevents germs spreading; ***she gargled with an antiseptic mouthwash*** **2** *noun* substance which prevents germs growing or spreading; ***the nurse painted the wound with antiseptic***

◊ **antiserum** [æntɪ'siːərəm] *noun* serum taken from an animal which has developed antibodies to bacteria and used to give temporary immunity to a disease (NOTE: plural is **antisera**)

◊ **antitoxin** [æntɪ'tɒksɪn] *noun* antibody produced by the body to counteract a poison in the body

◊ **antivenene** *or* **antivenom (serum)** [æntɪ'veniːn *or* æntɪ'venəm] *noun* serum which is used to counteract the poison from snake *or* insect bites

anvil ['ænvɪl] *noun* cloud formation in a dark thundercloud, with a flat top and a point similar in shape to a blacksmith's anvil

AOM = ACTIVE ORGANIC MATTER

AONB = AREA OF OUTSTANDING NATURAL BEAUTY

aphid ['eɪfɪd] *noun* small insect (popularly called blackfly *or* greenfly) which sucks sap from plants and can multiply very rapidly; ***analysis of the effects on the aphid population showed a 19% increase in the rate of production***

> COMMENT: cereal aphids are various species of greenfly. Winged females are found feeding on cereal crops in May and June. The grain aphid causes empty or small grain by puncturing the grain in the milk ripe stage, letting the grain contents seep out. Aphids can carry virus diseases from infected plants to clean ones

Aphis ['eɪfɪs] *noun* Latin name for various species of aphid

aphotic zone [eɪ'fɒtɪk 'zəʊn] *noun* water in the sea or a lake below about 1,500m, so deep that sunlight cannot penetrate it

API [eɪpiː'aɪ] *noun* American Petroleum Institute; **API scale** = scale of gravity of crude oil (the heaviest oils have the lowest numbers on the scale)

aposematic [eɪpəʊse'mætɪk] *adjective* (markings on an animal) which are very conspicuous and serve to discourage potential predators

apparatus [æpə'reɪtəs] *noun* equipment used (in scientific experiments *or* in a laboratory) (NOTE: no plural: for one item say **a piece of apparatus; some new apparatus)**

appliance [ə'plaɪəns] *noun* device *or* instrument especially an electrical one used in the home, such as a vacuum cleaner, washing machine, iron, etc.

apply [ə'plaɪ] *verb* **(a)** to ask for a job; ***she applied for a job in an environmental health department*** **(b)** to refer to; ***this order applies to all producers of the pesticide; the rule applies to visitors only*** **(c)** to put (a substance) on; ***the fungicide should be applied in early spring***

◇ **application** ['æplɪkeɪʃn] *noun* **(a)** asking for a job (usually in writing); ***if you are applying for the job, you must fill in an application form*** **(b)** putting a substance on; ***two applications of the pesticide should be sufficient to keep most pests off***

> QUOTE: nitrogen application is only part of an overall system of management aimed at getting the best out of the crop
>
> **Farmers Weekly**

approach [ə'prəʊtʃ] **1** *noun* way of dealing with a problem; ***the government has adopted a radical approach to the problem of desertification*** **2** *verb* to go *or* come nearer; ***do not approach the deer in the mating season; the sand dunes are approaching the forest area***

appropriate [ə'prəʊprɪət] *verb* to take control of something for one's own use; ***the town council has appropriated the site for a car park***

appropriate technology [ə'prəʊprɪət tek'nɒlədʒi] *noun* technology that is suited to the local environment, usually involving skills or materials that are easily available locally; ***biomethanation seems an eminently appropriate technology for use in rural areas***

> COMMENT: in many parts of world, devices to help the local population cultivate the land can be made out of simple pipes or pieces of metal. Expensive tractors may not only be unsuitable for the terrain involved, but also use fuel which costs more than the crops produced

aqu- ['ækw] *prefix* meaning water

◇ **aquaculture** *or* **aquiculture** *or* **aquafarming** ['ækwəkʌltʃə *or* 'ækwɪkʌltʃə *or* 'ækwəfɑːmɪŋ] *noun* breeding and rearing fish, shellfish, etc. *or* growing plants for food in special ponds

◇ **aquarium** [ə'kweəriəm] *noun* container with water and a display of fish and other animals or plants that live in water; *see also* -ARIUM

◇ **aquatic** [ə'kwætɪk] *adjective* referring to water; **aquatic animals** = animals which live in water

◇ **aquiclude** ['ækwɪkluːd] *noun* rock *or* soil through which water passes very slowly (such as clay)

◇ **aquifer** ['ækwɪfə] *noun* porous rock *or* soil through which water passes and in which water gathers to supply wells; **confined aquifer** = aquifer which has a layer of rock *or* soil above it; **unconfined aquifer** = aquifer whose upper surface is at ground level

Ar *chemical symbol for* argon

arable ['ærəbl] *adjective* (land) on which crops are grown; **arable farming** = growing crops (as opposed to dairy farming, cattle farming, etc.); **arable soil** = soil which is able to be tilled for the cultivation of crops

arachidonic acid ['ærəkɪdɒnɪk 'æsɪd] *noun* essential fatty acid

Arachnida [ə'ræknɪdə] *noun* class of animals with eight legs (such as spiders, mites, etc.)

> COMMENT: Arachnida have pincers on the first pair of legs. Their bodies are divided into two parts and they have no antennae

arbor- ['ɑːbɔː] *prefix* referring to trees

◇ **arboreal** [ɑː'bɔːriəl] *adjective* referring to trees; **arboreal animals** = animals which live in trees

◇ **arboretum** [ɑːbə'reɪtəm] *noun* collection of trees from different parts of the world, grown for scientific study

◇ **arboricide** [ɑː'bɒrɪsaɪd] *noun* chemical substance, such as 2,4,5-T, which kills trees

◇ **arboriculture** [ɑː'bɒrɪkʌltʃə] *noun* study of the cultivation of trees

◇ **arborist** ['ɑːbərɪst] *noun* person who studies the cultivation of trees

arbovirus ['ɑːbəvaɪrəs] *noun* virus transmitted by blood-sucking insects

archipelago [ɑːkɪ'peləgəʊ] *noun* group of islands

Arctic ['ɑːktɪk] **1** *noun* **the Arctic** = area of ice and snow around the North Pole, north of the Arctic Circle **2** *adjective* referring to the Arctic; **Arctic air** = mass of cold air which forms over the Arctic region and then moves south; **the Arctic Circle** = parallel running round the earth at latitude 66°32N, to the north of which lies the Arctic region

◇ **Arctogea** [ɑːktə'dʒeɪə] *noun* one of the main biogeographical regions, formed of the Palaearctic, Nearctic, Oriental and Ethiopian regions; *see also* NEOGEA, NOTOGEA

area ['eəriə] *noun* **(a)** measurement of the space occupied by something; ***to measure the area of the field you must multiply the length by the width; the desert covers an area of***

2,500 square miles **(b)** region of land; ***the whole area has been contaminated by waste from the power station;*** **Area of Outstanding Natural Beauty (AONB)** = region in England and Wales which is not a National Park but which is considered sufficiently attractive to be preserved from overdevelopment; **distribution area** = number of places in which a species is found; **mining subsidence area** = region in which the ground has subsided because of mine workings; **pedestrian area** = part of a town where people can only go on foot and where motor vehicles are not allowed; **urban area** = town *or* city, area which is completely built up

arete [ə'reɪt] *noun (of mountain)* sharp ridge between two valleys

argon ['ɑːgɒn] *noun* inert gas, which occurs in air and of which isotopes form in the cooling systems of reactors. It is used in electric light bulbs (NOTE: chemical symbol is **Ar**; atomic number is **18**)

arid ['ærɪd] *adjective* (soil) which is very dry; (area of land) which has very little rain; **arid zone** = area in the tropics (between about 15° and 30° north and south) which is very dry and covered with deserts

◇ **aridity** [ə'rɪdɪtɪ] *noun* state of being extremely dry

-arium ['eəriəm] *suffix* referring to a display, usually involving water; **dolphinarium** = display of dolphins; **herbarium** = display of plants in a glass container; **oceanarium** = display of animals which live in the sea

aromatic [ærə'mætɪk] **1** *adjective* having a pleasant smell; **aromatic compound** = compound like benzene, with a ring of carbon atoms with single and double bonds **2** *noun* substance *or* plant *or* drug which has a pleasant smell

ARP = ACREAGE REDUCTION PROGRAM

array [ə'reɪ] *noun* set of numbers shown in a display, such as a table

arrester [ə'restə] *noun* device which stops something happening; **lightning arrester** = device which prevents surges of the electrical current which are caused when lightning strikes a building and which can damage equipment

arroyo [ə'rɒɪəʊ] *noun* gully with a stream at the bottom, found in desert regions of America; *compare* WADI

arsenic ['ɑːsənɪk] *noun* chemical element which forms poisonous compounds, such as arsenic trioxide, and is used to kill rodents (NOTE; symbol is **As**; atomic number is **33**)

◇ **arsenical** [ɑː'senɪkl] *noun* drug *or* insecticide, one of the group of poisonous oxides of arsenic

artefact *or* **artifact** ['ɑːtɪfækt] *noun* man-made object; something which has arisen as a result of the process of observation *or* investigation

artesian well [ɑː'tiːziən 'wel] *noun* well which has been bored into a confined aquifer; the hydrostatic pressure is usually strong enough to force the water to the surface

artificial [ɑːtɪ'fɪʃl] *adjective* which is made by man *or* which does not exist naturally; **artificial community** = plant community kept by man (as in a garden); **artificial insemination (AI)** = way of breeding livestock by injecting sperm from specially selected males into the female; **artificial rain** = rain which is made by scattering crystals of salt and other substances into clouds

As *chemical symbol for* arsenic

asbestos [æz'bestəs] *noun* fibrous mineral substance, used as a shield against fire and as an insulating material in many industrial and construction processes; **asbestos cement** = mixture of asbestos and cement, used to make pipes, tiles and other small items used in construction; **asbestos fibre** = threads of asbestos which can be woven into rope or tape, etc.

◇ **asbestosis** [æsbes'təʊsɪs] *noun* disease of the lungs caused by inhaling asbestos dust

> COMMENT: asbestos was formerly widely used in cement and cladding and other types of fireproof construction materials; it is now recognized that asbestos dust can cause many lung diseases, leading in some cases to forms of cancer, with the result that constructions containing asbestos are being demolished or rebuilt with alternative materials. Blue asbestos is extremely toxic and is banned in many countries; white and brown asbestos can be safely used in some forms

ascorbic acid [ə'skɔːbɪk 'æsɪd] *noun* vitamin C

> COMMENT: ascorbic acid is found in fresh fruit (especially oranges and lemons) and in vegetables. Lack of Vitamin C can cause anaemia and scurvy

asepsis [eɪ'sepsɪs] *noun* state of having no infection

◇ **aseptic** [eɪ'septɪk] *adjective* referring to asepsis; ***it is important that aseptic techniques should be used in microbiological experiments;*** **aseptic surgery** = surgery using sterilized equipment, rather than relying on killing germs with antiseptic drugs; *compare* ANTISEPTIC

asexual [eɪ'sekʃjuəl] *adjective* not sexual *or* not involving sexual intercourse; **asexual reproduction** = reproduction by taking cuttings of plants *or* by cloning

ash [æʃ] *noun* **(a)** European hardwood tree *(Fraxinus excelsior)* **(b)** grey *or* black powder formed of minerals left after an organic substance has been burnt; **fly ash** = fine ash which is carried in smoke and fumes from burning processes (which can be collected and used to make bricks); **volcanic ash** = ash and small pieces of lava and rock which are thrown up by an erupting volcano

◇ **ash bin** *or* **ash can** ['æʃ bɪn *or* 'æʃ kæn] *noun* container for waste, especially for ashes from a fire or boiler

aspen ['æspən] *noun* North American hardwood

asphalt ['æsfɔːlt] *noun* black substance formed from bitumen

COMMENT: asphalt is found naturally in tar sands, but is also manufactured as a by-product of petroleum distillation. It is used, when melted, to paint on roofs to make them waterproof *or* to mix with aggregate to make hard road surfaces

assessment [ə'sesmənt] *noun* evaluation; **impact assessment** = evaluation of the effect upon the environment of a large construction programme, draining of marshes, etc.

assimilate [ə'sɪməleɪt] *verb* to take food substances which have been absorbed into the blood *or* into the tissues of an organism

◇ **assimilation** [əsɪmɪ'leɪʃn] *noun* action of assimilating food substances (such as photosynthesis, by which plants convert light into energy and tissue)

associate [ə'səʊsiəeɪt] *verb* to be related to *or* to be connected with; ***the government is closely associated with the project***

◇ **association** [əsəʊsi'eɪʃn] *noun* (i) group of plants living together in a large area, forming a stable community; (ii) group of people with similar interests; **biological association** = group of associated organisms

astatine ['æstətiːn] *noun* natural radioactive element (NOTE: chemical symbol is **At;** atomic number is **85)**

aster ['æstə] *noun* structure shaped like a star, seen around the centrosome during cell division

asthenosphere [æs'θenəsfɪə] *noun* part of the interior of the earth, formed of molten matter below the lithosphere

asulam ['æsjuləm] *noun* powerful herbicide, used to remove tenacious plants such as bracken

At *chemical symbol for* astatine

Atlantic [ət'læntɪk] *adjective & noun* **the Atlantic (Ocean)** = ocean to the north of the Antarctic, south of the Arctic, west of Europe and Africa and east of North and South America

atmosphere ['ætməsfɪə] *noun* **(a)** gaseous zone which surrounds the earth **(b)** unit of measurement of pressure

◇ **atmospheric** [ætməs'ferɪk] *adjective* referring to the atmosphere; **atmospheric pressure** = normal pressure of the air on the surface of the earth; **atmospheric pressure zones** = bands of high and low pressure running round the earth: high pressure near the poles, low pressure between 40 and 70 degrees latitude north and south, high pressure along the 30 degrees latitude line in the subtropics, then low pressure again around the equator

COMMENT: the atmosphere surrounds the earth to a height of several thousand kilometres, but is concentrated in the 20 kilometres immediately above the planet's surface. Its pressure decreases with height. The atmosphere is divided into various layers or zones: troposphere, stratosphere, mesosphere and thermosphere. See also comment at AIR

atoll ['ætɒl] *noun* island in warm seas, shaped like a ring and made of coral

atom ['ætəm] *noun* fundamental unit of a chemical element

◇ **atomic** [ə'tɒmɪk] *adjective* referring to atoms; **atomic bomb** = bomb whose destructive power is produced by nuclear fission *or* fusion; **atomic energy** = energy created during a nuclear reaction, either fission *or* fusion, which, in a nuclear reactor, produces heat which warms water and forms steam which runs a turbine to generate electricity; **atomic fission** = splitting of the nucleus of an atom (such as uranium-235) into several small nuclei which then releases energy and neutrons; **atomic fusion** = joining together of several nuclei to form a single large nucleus, creating energy (as in a hydrogen bomb); **atomic mass** = mass of an atom measured in atomic mass units; **atomic mass unit (AMU)** = unit of measurement of mass, used to express the weight of an atom *or* molecule and equal to one twelfth of the mass of an atom of carbon-12; **atomic number** = number of positive electric charges round the nucleus of an atom (equal to the number of protons in the atom), giving the element its place in the periodic table; **atomic pile** = nuclear reactor; **atomic power** = power generated by a nuclear reactor; electricity generated by a nuclear power station; **atomic power station** = power station in which nuclear reactions are used to provide energy to run turbines which generate electricity; **atomic-powered** = operated by nuclear power; **atomic warfare** = war using atomic weapons; **atomic waste** = radioactive waste from a nuclear reactor (including spent fuel rods and coolant); **atomic weapon** = bomb *or* missile whose destructive power is produced by nuclear fission *or* fusion

◇ **Atomic Energy Authority (UKAEA)** *noun* government agency responsible for nuclear energy in the UK

◇ **Atomic Energy Commission** *noun* *US* agency responsible for nuclear energy in the US

COMMENT: each atom consists of protons and neutrons grouped together to form a nucleus, and electrons which surround the nucleus. Atomic energy is produced from the fission of atoms of uranium-235. One of the problems associated with the production of atomic energy is the radioactive waste produced by nuclear reactors. This takes various forms: it can be gas (such as krypton or xenon),spent fuel rods or water from cooling processes

atomize ['ætəmaɪz] *verb* to reduce a liquid to very small particles

◇ **atomizer** ['ætəmaɪzə] *noun* instrument which sprays liquid in the form of very small drops like mist

atrazine ['ætrəzi:n] *noun* a residual herbicide which acts on the soil

atropine ['ætrəpi:n] *noun* alkaloid drug found in deadly nightshade

attenuation [ətenju'eɪʃn] *noun* lessening of a property or quantity, as in the reduction of pollutants

attract [ə'trækt] *verb* to make something come nearer; ***the solid attracts the gas to its surface***

◇ **attractant** [ə'træktənt] *noun* chemical which attracts an organism; **sexual attractant** = chemical produced by an insect which attracts other insects of the same species

◇ **attraction** [ə'trækʃn] *noun* act *or* power of attracting

COMMENT: artificially produced attractants can be used to attract insects which can then be killed

Au *chemical symbol for* gold

audio- ['ɔ:diəʊ] *prefix* referring to hearing *or* sound

◇ **audiogram** ['ɔ:diəgræm] *noun* graph drawn by an audiometer

◇ **audiometer** [ɔ:di'ɒmɪtə] *noun* instrument for testing hearing *or* for testing the range of sounds that the human ear can detect

Audubon Society ['ɔ:dəbɒn sə'saɪti] *noun* society in the US, whose aims are the conservation of wildlife, especially birds

aureole ['ɔ:riəʊl] *noun* glow visible round the sun, when seen through thin mist

auriferous [ɔ:'rɪfərəs] *adjective* (deposit) which bears gold

Aurora Australis *or* **Aurora Borealis** [ə'rɔ:rə ɒ'strɑ:lɪs *or* 'ə'rɔ:rə bɔ:rɪ'eɪlɪs] *noun* spectacular illumination of the sky caused by ionized particles striking the atmosphere. It is called Aurora Borealis or Northern Lights in the Northern Hemisphere and Aurora Australis or Southern Lights in the Southern Hemisphere

Australasian Region ['ɒstrə'leɪʒiən 'ri:dʒn] *noun* Notogea *or* one of the distinct biogeographical regions into which the earth is divided, covering Australia, New Zealand and most of the islands in the Pacific Ocean

autecology [ɔ:tek'ɒlədʒi] *noun* study of an individual species in its environment; *compare* SYNECOLOGY

authority [ɔ:'θɒrəti] *noun* **(a)** power to act; **to abuse one's authority** = to use powers in an illegal *or* harmful way **(b)** official body which controls an area *or* region; ***you will have to apply to the local planning authority***

auto- ['ɔ:təʊ] *prefix* meaning self

◇ **autoecology** [ɔ:təʊe'kɒlədʒi] *noun* = AUTECOLOGY

◇ **autogamy** [ɔ:'tɒgəmi] *noun* pollination with pollen from the same flower

◇ **autolysis** [ɔ:'tɒlɪsɪs] *noun* action of cells destroying themselves with their own enzymes

◇ **autonomic nervous system** [ɔ:tə'nɒmɪk 'nɜ:vəs sɪstəm] *noun* nervous system formed of ganglia linked to the spinal column, which regulates the automatic functioning of the main organs of the body, such as the heart and lungs

◇ **autotroph** *or* **autotrophic organism** ['ɔ:tətrɒf *or* ɔ:tə'trɒfɪk] *noun* organism (such as a green plant) which takes its energy from the sun; *compare* HETEROTROPH

autumn ['ɔ:təm] *noun* season of the year, following summer and before winter, when days become shorter and the weather progressively colder

◇ **autumnal** [ɔ:'tʌmnəl] *adjective* referring to the autumn; **autumnal equinox** = about September 22nd, one of the two occasions in the year when the sun crosses the celestial equator and night and day are each twelve hours long

auxin ['ɔ:kzɪn] *noun* plant hormone which encourages tissue growth

COMMENT: some herbicides act as synthetic auxins by upsetting the balance of the plant's growth

avalanche ['ævəlɑ:nʃ] *noun* large mass of snow which becomes detached and falls down the side of a mountain; large mass of rock *or* mud which falls down a mountainside in heavy rain; **avalanche wind** = very strong wind caused by an avalanche

average ['ævrɪdʒ] **1** *noun* **(a)** usual amount, size, speed, etc.; ***the temperature has been above (the) average for the time of year*** **(b)** value calculated by adding together several quantities and then dividing the total by the number of quantities; **peak average** = average of all the highest points on a graph *or* in a series of figures **2** *adjective* **(a)** usual, ordinary; ***their son is of above average intelligence*** **(b)** calculated by adding together several quantities and then dividing the total by the number of quantities; ***his average speed was 30 miles per hour***

Aves ['eɪvi:z] *noun* birds, the class of egg-laying feathered animals which are adapted to fly

◇ **avicide** ['ævɪsaɪd] *noun* substance which kills birds

◇ **avifauna** ['ævɪfɔ:nə] *noun* all the birds which live naturally in a certain area (NOTE: plural is **avifauna** *or* **avifaunas**)

COMMENT: birds are closely related to reptiles and have scales on their legs. Their forelimbs have developed into wings

axial-flow turbine ['æksiəlfləʊ 'tɜ:baɪn] *noun* turbine with blades like those on a ship's propeller, rotating horizontally

azobacter ['eɪzəʊbæktə] *noun* nitrogen-fixing bacteria present in the soil

azo dyes ['eɪzəʊ 'daɪz] *noun* artificial colouring additives derived from coal tar, added to food to give it colour

COMMENT: many of the azo dyes (such as tartrazine) provoke allergic reactions; some are believed to be carcinogenic

Bb

B *chemical symbol for* boron

Vitamin B ['vɪtəmɪn 'bi:] *noun* **Vitamin B complex** = group of vitamins which are soluble in water, including folic acid, pyridoxine, riboflavine and many others; **Vitamin B_1** *or* **thiamine** = vitamin found in yeast, liver, cereals and pork; **Vitamin B_2** *or* **riboflavine** = vitamin found in eggs, liver, green vegetables, milk and yeast; **Vitamin B_6** *or* **pyridoxine** = vitamin found in meat, cereals and molasses; **Vitamin B_{12}** *or* **cyanocobalamin** = vitamin found in liver and kidney, but not present in vegetables

COMMENT: lack of vitamins from the B complex can have different results: lack of thiamine causes beriberi; lack of riboflavine affects a child's growth and can cause anaemia and inflammation of the tongue and mouth; lack of pyridoxine causes convulsions and vomiting in babies; lack of vitamin B_{12} causes anaemia

Ba *chemical symbol for* barium

bacillus [bə'sɪləs] *noun* bacterium shaped like a rod (NOTE: plural is **bacilli**)

◇ **bacillary** [bə'sɪləri] *adjective* referring to bacillus; **bacillary dysentery** = dysentery caused by the bacillus *Shigella* in contaminated food

back [bæk] *verb (of wind)* to change direction, anticlockwise in the Northern Hemisphere and clockwise in the Southern Hemisphere (NOTE: opposite is **veer**)

background ['bækgraʊnd] *noun* conditions which are always present in the environment, but are less obvious (or less important) than others; **background carboxyhaemoglobin level** = level of carboxyhaemoglobin in the blood of a person living a normal existence without exposure to particularly high levels of carbon dioxide; **background concentration** = (i) BACKGROUND POLLUTION; (ii) BACKGROUND RADIATION; **background level** = general level of noise, pollution, etc. which is always there; **background noise** = (i); *(in the environment)* general level of noise which is always there (ii); *(in an electronic instrument)* unwanted interference noise; **background pollution** = general level of air pollution in an area, disregarding any specifically local factors, such as the presence of a coal-fired power station; **background radiation** = radiation which comes from natural sources like rocks *or* the earth *or* the atmosphere and not from a single man-made source

> COMMENT: background radiation can depend on the geological structure of the area. Places above granite are particularly subject to high levels of radiation. Other sources of background radiation are: cosmic rays from outer space, radiation from waste products of nuclear power plants which has escaped into the environment, radiation from TV and computer screens

backscatter ['bækskætə] **1** *noun* sending back of radiation; ***backscatter contributes to an increase in albedo*** **2** *verb* to send back (radiation); ***a proportion of incoming solar radiation is backscattered by air in the atmosphere***

backshore ['bækʃɔː] *noun* part of a beach between the foreshore and where permanent vegetation grows

backwash ['bækwɒʃ] *noun* flow of seawater down a beach; *compare* SWASH

backwater ['bækwɔːtə] *noun* (i) stagnant water connected to a river or stream; (ii) water behind a dam or tide

bacteria [bək'tɪəriə] *plural noun* submicroscopic organisms, which help in the decomposition of organic matter, some of which are permanently present in the intestines of animals and can break down food tissue; many of them can cause disease; **bacteria bed** = filter bed of rough stone, forming the last stage in the treatment of sewage; **oil-eating bacteria** = bacteria which can consume and destroy oil (NOTE: the singular is **bacterium**)

> COMMENT: bacteria can be shaped like rods (bacilli), like balls (cocci) or have a spiral form (such as spirochaetes). Bacteria, especially bacilli and spirochaetes, can move and reproduce very rapidly

◇ **bacterial** [bæk'tɪəriəl] *adjective* referring to bacteria *or* caused by bacteria; **bacterial contamination** = state of something (such as water *or* food) which has been contaminated by bacteria; **bacterial strain** = distinct variety of bacteria

◇ **bactericidal** [bæktɪərɪ'saɪdəl] *adjective* (substance) which destroys bacteria

◇ **bactericide** [bæk'tɪərɪsaɪd] *noun* substance which destroys bacteria

◇ **bacteriological warfare** [bæktɪərɪ'lɒdʒɪkl 'wɔːfeə] *noun* germ warfare, war where one side tries to kill *or* affect the people of the enemy side by infecting them with bacteria

◇ **bacteriology** [bæktɪəi'ɒlədʒi] *noun* scientific study of bacteria

◇ **bacteriophage** [bæktɪəriəʊ'feɪdʒ] *noun* virus which affects bacteria

badlands ['bædlænz] *noun (in North and South America)* areas of land which is (or has become) unsuitable for agriculture

bag [bæg] *noun* filter made of cloth or artifical fibre, used to remove particles of matter from waste gas from industrial processes

balance ['bæləns] *noun* state where two sides are equal; **the balance of nature** *or* **ecological balance** = situation where different organisms live in a stable state in the same ecosystem; **to disturb the balance of nature** = to make a change to the environment which has the effect of putting some organisms at a disadvantage against others; **energy balance** = measurements showing the movement of energy between organisms and their environment; **heat balance** = state in which the earth loses as much heat by radiation and reflection as it gains from the sun, making the earth's temperature constant from year to year; **water balance** = (i) state where the water lost in an area by evaporation *or* by runoff is replaced by water received in the form of rain; (ii) state where the water lost by the body (in urine *or* perspiration, etc.) is balanced by water absorbed from food and drink; **water-salt balance** = state where the water in the soil balances the amount of salts in the soil

◇ **balanced diet** ['bælənsd 'daɪət] *noun* diet which provides the animal with all the nutrients it needs in the correct proportions

baleen [bə'liːn] *noun* series of plates like a comb, which hang down from the upper jaw of some whales and act like a sieve; **baleen whales** = Mysticeti, including blue whales, humpbacks, etc.

> COMMENT: the baleen whales (which are the larger of the two groups of whales, the others being the toothed whales or Odontoceti) live on plankton and other tiny marine animals. They eat by sucking in huge quantities of water and then forcing it out with their tongues through the baleen, which catches the plankton for the whales to eat

ballast ['bæləst] *noun* material (such as stones) carried in the hull of a ship to make it heavier and so less likely to roll; **solid-state ballast lamp** = fluorescent light bulb containing a ballast which reduces the current needed to keep the lamp bright

balsa wood ['bɒːlsə 'wʊd] *noun* very light tropical wood, used for making light-weight models

ban [bæn] **1** *noun* official statement forbidding something *or* saying that something should not be done; **nuclear test ban** = ban on testing nuclear weapons **2** *verb* to forbid, to say that something should not be done; ***the US has banned the use of CFCs in aerosols; smoking is banned in most restaurants***

band [bænd] *noun* **(a)** layer (of rock) **(b)** range of frequencies between two limits; ***the telephone voice band is between 300 and 3400 Hz***

◇ **banded** ['bændɪd] *adjective* (rock) arranged in layers

◇ **band sprayer** ['bænd 'spreɪə] *noun* crop sprayer that applies chemicals in narrow strips, mostly used with precision seeders

bank [bæŋk] **1** *noun* **(a)** land at the side of water, such as a river or lake **(b)** long heap of sand *or* snow, such as a sandbank in shallow water, either in a river or the sea, or a snowbank along the side of a road **(c)** place where blood *or* organs from donors can be stored until needed; *see also* BLOOD BANK, EYE BANK, GENE BANK, SEED BANK, SPERM BANK **(d)** place where waste materials can be collected for recycling; *see also* BOTTLE BANK, CAN BANK **2** *verb* to cover a fire *or* furnace with fine solid fuel and so prevent it burning too fast

bar [bɑ:] *noun* **(a)** long bank of sand submerged at high tide (at the entrance to a harbour, river or bay); **longshore bar** = bank of sand submerged at high tide and running parallel with the coast **(b)** unit of measurement of atmospheric pressure, divided into millibars (the atmospheric pressure at sea level is generally calculated as being 1,013 millibars)

bar- [bær] *prefix* referring to atmospheric pressure; *see also* BAROGRAPH, BAROMETER

barium ['beəriəm] *noun* chemical element, forming poisonous compounds, used as a contrast when taking X-ray photographs of soft tissue; **barium concrete** = concrete with barium added to it, used to absorb radiation; **barium meal** *or* **barium solution** = liquid solution containing barium sulphate which a patient drinks so that an X-ray can be taken of his alimentary tract; **barium sulphate ($BaSO_4$)** = salt of barium, not soluble in water and which shows as opaque in X-ray photographs (NOTE: chemical symbol is **Ba;** atomic number is **56)**

bark [bɑ:k] *noun* **(a)** hard outer layer of a tree, formed of dead tissue **(b)** cry of an animal of the dog family (such as a wolf *or* fox)

barley ['bɑ:li] *noun* common cereal crop *(Hordeum sativum),* grown in temperate areas

COMMENT: barley is widely grown in northern temperate countries, with the largest production in Germany and France; it is a very important arable crop in the UK. The grain is mainly used for livestock feeding and for malting for use in producing alcoholic drinks. It is rarely used for making flour

baro- ['bærəʊ] *prefix* meaning weight *or* pressure

◇ **barograph** ['bærəgrɑ:f] *noun* instrument which records changes in atmospheric pressure, made by attaching a pen to a barometer and recording fluctuations in pressure on a roll of paper

◇ **barometer** [bə'rɒmɪtə] *noun* instrument which measures changes in atmospheric pressure and is, therefore, used to forecast changes in the weather; **aneroid barometer =** barometer with a vacuum to which a diaphragm is attached, which moves as the atmospheric pressure changes; **mercury barometer =** barometer made of a glass tube containing mercury: one end of the tube is sealed, the other is open, resting in a bowl of mercury; as the atmospheric pressure changes, so the column of mercury in the tube rises or falls

◇ **barometric** [bærə'metrɪk] *adjective* referring to a barometer; **barometric corrections =** corrections made to the reading on a mercury thermometer to allow for altitude and outside temperature; **barometric pressure** = atmospheric pressure indicated by a barometer

◇ **barometrically** [bærə'metrɪkli] *adverb* measured using a barometer

barrage ['bərɑ:ʒ] *noun* construction to prevent *or* regulate the flow of tides, used either to prevent flooding (as in the Thames Barrage) or to harness tidal power

COMMENT: the advantages of building barrages to harness tidal power are that they are economical to run and use no fuel. They do pose particular environmental problems (as do all dams) in that they may change the ecosystem of the surrounding countryside. In the case of a barrage, it would have an effect on the estuary marshlands, on the river behind the barrage, and possibly on the movement of coastal silt by the tides

QUOTE: the barrage would cost more to build than a conventional or nuclear plant, but it could run for a century or more with only minor engineering attention

Environment Now

barrel ['bærəl] *noun* large round container for liquids (such as beer, wine, oil); the amount of a liquid contained in a standard barrel (42 US gallons), used as a measure of the quantity of crude oil produced

barren ['bærən] *adjective* (land) which cannot produce vegetation; (river, lake, etc.) which cannot support plant *or* animal life; (plant) which does not produce fruit; (animal) which cannot bear young

barrier ['bæriə] *noun* wall which closes *or* which prevents something going through; **biotic barrier** = conditions which prevent members of a species moving to other regions, and so prevent the species from expanding; **sound barrier** = air resistance encountered by objects moving at speeds near the speed of sound; **barrier beach** *or* **barrier island** = bank of sand *or* strip of land lying along the shore and separated from it by a lagoon; **barrier reef** = long coral reef lying along the shore and enclosing a lagoon

basal ['beɪsəl] *adjective* extremely important *or* which affects a base; **basal area** = area of woodland actually covered by the trunks of trees; **basal metabolism** *or* **basal metabolic rate (BMR)** = amount of energy used by a body in exchanging oxygen and carbon dioxide when at rest (i.e. energy needed to keep the body functioning and the temperature normal)

basalt ['bæsɔːlt] *noun* fine dark volcanic rock

◇ **basaltic** [bə'sɔːltɪk] *adjective* referring to basalt; ***most volcanic lava is basaltic***

base [beɪs] **1** *noun* **(a)** bottom part; ***the pituitary gland hangs down from the base of the brain;*** **base unit** = one of the seven SI units (ampere, candela, kelvin, kilogram, metre, mole, second) on which other units are based **(b)** main ingredient of something (such as a paint *or* an ointment) **(c)** substance which reacts with an acid to form a salt **2** *verb* to use something as a basis; ***statistics, based on information gathered from several monitoring stations***

◇ **base level** ['beɪs 'levl] *noun* **(a)** lowest level of something, from which other levels are calculated **(b)** the depth below which erosion would be unable to occur

◇ **basement** ['beɪsmənt] *noun* lowest level of rocks, which have been covered by sediment

◇ **base metal** ['beɪs 'metl] *noun* common metal such as copper, lead, tin, etc. (as opposed to the noble metals)

basic ['beɪsɪk] *adjective* **(a)** very simple *or* from which everything else comes; ***you should know basic maths if you want to work in a shop;*** **basic law** = fundamental rule **(b) basic rock** = rock, such as basalt, which contains little silica; **basic slag** = calcium phosphate, waste from blast furnaces, used as a fertilizer because of its phosphate content **(c)** (chemical substance) which reacts with an acid to form a salt; **basic salt** = chemical compound formed when an acid reacts with a base

basin ['beɪsɪn] *noun* large, low-lying area of land, drained by a large river system; large area of land surrounding an ocean; ***thousands of tributaries drain into the Amazon basin; a ring of volcanoes around the edge of the Pacific basin;*** **catchment basin** *or* **drainage basin** = area of land which collects and drains the rainwater which falls on it (such as the area around a lake or river); **coal basin** = part of the earth's surface containing layers of coal; **ground-water basin** = area of land where water stays in the top layers of soil or in porous rocks and can collect pollution; **river basin** = large, low-lying area of land, drained by a river; **sedimentation basin** = area of land where the rocks have been formed from matter carried there by wind and water; **settling basin** = tank in which sewage is allowed to stand so that solid particles can sink to the bottom

Batesian mimicry ['beɪtsiən 'mɪmɪkri] *noun* form of mimicry where an animal mimics another animal which is poisonous, so as to avoid being eaten

bathy- ['bæθi] *prefix* referring to the part of the seabed between 1,000 and 3,000m deep

◇ **bathyal** ['bæθiəl] *adjective* referring to the deep part of the sea

◇ **bathylimnetic** [bæθɪlɪm'netɪk] *adjective* referring to the deepest part of a lake

battery ['bætri] *noun* **(a) electric battery** = small device for storing electric energy; ***the calculator runs on batteries; a battery-powered car;*** **battery-operated** = powered *or* run by an electric battery **(b)** series of small cages in which thousands of chickens are kept; **battery farming** = system of keeping thousands of chickens in a series of small cages; **battery hen** = chicken which spends its life confined in a small cage

> COMMENT: a method of egg production which is very energy-efficient. It is criticized, however, because of the quality of the eggs, the possibility of disease and the polluting substances produced, and also on grounds of cruelty because of the stress caused to the birds. See also FREE-RANGE EGGS

bauxite ['bɔːksaɪt] *noun* mineral which contains aluminium ore

bay [beɪ] *noun* wide curved coastline, partly enclosing an area of sea; **bay bar** *or* **bay barrier** = bank of sand *or* strip of land lying along the coastline of a bay

Be *chemical symbol for* beryllium

beach [biːtʃ] *noun* area of sand or small stones by the side of the sea or a lake or river; **pebble beach** = beach covered with small stones; **sand(y) beach** beach covered with sand; **beach sediment** = stone, sand, mud and shells deposited on a beach by the sea, by a river or by erosion of the cliffs

◇ **beachcomber** ['bi:tʃkəʊmə] *noun* (i) person who collects debris on beaches; (ii) tall, long wave rolling onto a beach

Beaufort scale ['bəʊfət 'skeɪl] *noun* scale (from 0 to 12) used to refer to the strength of wind; ***the meteorological office has issued a warning of force 12 winds*** (NOTE: used in weather forecasts as an indication of the force of the wind: **'gales force 8, northerly'**)

COMMENT: the Beaufort scale was devised in the 18th century by a British admiral. The descriptions of the winds and their speeds in knots are: 0: calm (0 knots); 1: light air (2 knots); 2: light breeze (5 knots); 3: gentle breeze (9 knots); 4: moderate breeze (13 knots); 5: fresh breeze (19 knots); 6: strong breeze (24 knots); 7: near gale (30 knots); 8: gale (37 knots); 9: strong gale (44 knots); 10: storm (52 knots); 11: violent storm (60 knots); 12: hurricane (above 60 knots)

becquerel ['bekərel] *noun* SI unit of measurement of radiation: 1 becquerel is the amount of radioactivity in a substance where one nucleus decays per second (NOTE: now used in place of the **curie.** See also **rad. Becquerel** is written **bq** with figures: **200bq)**

bed [bed] *noun* **(a)** bottom of a river *or* lake *or* the sea; ***a fish which feeds on the seabed; the river bed is choked with weeds*** **(b)** layer of sediment in rock; ***the cliffs show clearly several beds of sandstone***

◇ **bedding** ['bedɪŋ] *noun* different layers of sediment

◇ **bedrock** ['bedrɒk] *noun* rock which is found under a layer of ore *or* coal

beech [bi:tʃ] *noun* common temperate hardwood tree

beetle ['bi:tl] *noun* insect of the order Coleoptera, with hard covers on the wings; **black beetle** = cockroach, insect of the order Dictyoptera, a common household pest

behaviour [bɪ'heɪvjə] *noun* way in which a living organism responds to a stimulus

◇ **behavioural** [bɪ'heɪvjərəl] *adjective* referring to behaviour; **behavioural scientist** = person who specializes in the study of behaviour

◇ **behaviourism** [bɪ'heɪvjərɪzm] *noun* psychological theory that only the patient's behaviour should be studied to discover his psychological problems

◇ **behaviourist** [bɪ'heɪvjərɪst] *noun* psychologist who follows behaviourism

below [bɪ'ləʊ] *preposition* **(a)** lower in position than; ***below sea level; the beach below the cliffs*** **(b)** less in number *or* quantity than; **below freezing point** = less than *or* lower than freezing point **(c)** downstream from; **the Thames below London** = part of the River Thames lying downstream from London

belt [belt] *noun see* GREEN BELT

benthos ['benθɒs] *noun* bottom of the sea *or* of a lake

◇ **benthic** ['benθɪk] *adjective* living on the bottom of the sea *or* of a lake; **benthic fauna** = animals living on the bottom of the sea *or* of a lake

◇ **benthon** *or* **benthic organism** ['benθɒn *or* 'bentθɪk 'ɔ:gənɪzm] *noun* organism living on the bottom of the sea *or* of a lake

benzene ['benzi:n] *noun* simple aromatic hydrocarbon, produced from coal tar, and very carcinogenic; **benzene hexachloride (BHC) ($C_6H_6Cl_6$)** = white or yellow powder containing lindane, used as an insecticide as a dust or spray against pea and bean weevil, and as a seed dressing against wireworm

◇ **benzpyrene** [benz'paɪri:n] *noun* inflammable carcinogenic substance found in coal tar, produced in the exhaust fumes from petrol engines, from coal- and oil-burning appliances and from smoking tobacco

QUOTE: benzene and other aromatics in car exhaust fumes react with sunlight and encourage the formation of traffic smogs which aggravate asthma and other respiratory conditions

Environment Times

bergshrund ['beəgʃrʌnt] *noun* deep, wide crevasse found between a cirque glacier and its back wall

beriberi [beri'beri] *noun* disease of the nervous system caused by lack of vitamin B_1

COMMENT: beriberi is prevalent in tropical countries where the diet is mainly white rice which is deficient in thiamine

berry ['beri] *noun* small fleshy seed-bearing fruit of a bush; there are usually many seeds in the same fruit, and the seeds are enclosed in a pulp (as in a tomato or gooseberry)

beryllium [bə'rɪliəm] *noun* chemical element, a metal used in making various alloys (NOTE: chemical symbol is **Be;** atomic number is **4)**

◇ **berylliosis** [berɪli'əʊsɪs] *noun* poisoning caused by breathing in particles of beryllium oxide

Bessemer process ['besəmə 'prəʊses] *noun* method of making steel

COMMENT: the process involves heating molten metal and blowing air into it at the same time; this is done in a type of furnace called a Bessemer converter. The process is used to remove phosphorus and carbon from pig iron. The air forms iron oxide, which removes impurities from the molten metal, including carbon monoxide which burns off. Finally, manganese is added to the metal to remove the iron oxide

best-before date ['bestbɪ'fɔː deɪt] *noun* date stamped on foodstuffs sold in supermarkets, which is the last date when the food is guaranteed to be in good quality; *similar to* SELL-BY DATE

beta ['biːtə] *noun* second letter of the Greek alphabet; **beta decay** = process of disintegration of a radioactive substance during which electrons are emitted; **beta particle** = electron which will pass through thin substances such as metal and can harm living tissue; **beta radiation** = radiation formed of beta particles

BHC = BENZENE HEXACHLORIDE

Bhopal [bəʊ'pæl] *noun* town in India, scene of an accident in 1984, where lethal methyl isocyanate gas escaped from a chemical plant and caused many casualties

BHT = BUTYLATED HYDROXYTOLUENE

bicarbonate (HCO_3) [baɪ'kɑːbəneɪt] *noun* any acid salt of carbonic acid

biennial [baɪ'eniəl] **1** *adjective* happening every two years; (plant) which takes two years to complete its life cycle **2** *noun* plant which completes its life cycle (germination, flowering, fruiting) over a period of two years

bight [baɪt] *noun* wide curve in a shoreline

bile [baɪl] *noun* thick bitter brownish yellow fluid produced by the liver and used to digest fatty substances

Bilharzia [bɪl'hɑːtsiə] *noun* Schistosoma, genus of fluke which enters the patient's bloodstream and causes bilharziasis

◇ **bilharziasis** [bɪlhɑːts'aɪəsɪs] *noun* schistosomiasis, a tropical disease caused by flukes in the intestine *or* bladder (NOTE: although strictly speaking, **Bilharzia** is the name of the fluke, it is also generally used for the name of the disease: **bilharzia patients; six cases of bilharzia**)

COMMENT: the larvae of the fluke enter the skin through the feet and lodge in the walls of the intestine or bladder. They are passed out of the body in stools or urine and return to water, where they lodge and develop in the water snail, the secondary host, before going back into humans. Patients suffer from fever and anaemia

bilirubin [bɪli'ruːbɪn] *noun* red pigment in bile

billion ['bɪljən] *noun* one thousand million (NOTE: in the US it has always meant one thousand million, but in GB it formerly meant one million million, and it is still sometimes used with this meaning. With figures it is usually written **bn: $5bn** say 'five billion dollars')

bin [bɪn] *noun* container for keeping something in until it is needed; ***the grain is kept in large storage bins***

binary ['baɪnəri] *adjective* (i) made of two parts; (ii) (compound) made of two elements; **binary fission** = common method of asexual reproduction by which one cell divides into two similar or identical cells

binding agent ['baɪndɪŋ 'eɪdʒənt] *noun* substance which causes two or more other substances to stick together *or* combine

binomial classification [baɪ'nəʊmiəl klæsɪfɪ'keɪʃn] *noun* scientific system of naming organisms devised by the Swedish scientist, Carolus Linnaeus (1707-1778)

COMMENT: the Linnaean system (or binomial classification) gives each organism a name made up of two Latin words. The first is a generic name referring to the genus to which the organism belongs, and the second is a specific name referring to the particular species. Organisms are usually identified by using both their generic and specific names, e.g. *Homo sapiens* (man), *Felix catus* (domestic cat) and *Sequoia sempervirens* (redwood). A third name can be added to give a subspecies. The generic name is written or printed with a capital letter. Both names are usually given in italics or are underlined if written or typed

bio- ['baɪəʊ] *prefix* referring to living organisms

◇ **bioaccumulation** [baɪəʊəkjuːmjuː'leɪʃn] *noun* accumulation of substances such as toxic chemicals up the food chain

◇ **bioaeration** [baɪəʊeə'reɪʃn] *noun* treatment of sewage by pumping it with activated sludge

◇ **bioassay** [baɪəʊə'seɪ] *noun* test of a substance by examining the effect it has on living organisms

◇ **biocenosis** [baɪəʊsə'nəʊsɪs] *noun* = BIOCOENOSIS

◇ **biochemical** [baɪəʊ'kemɪkl] *adjective* referring to biochemistry; **biochemical oxygen demand (BOD)** = amount of pollution in water, shown as the amount of oxygen which will be needed to oxidize the polluting substances; ***the main aim of sewage treatment is to reduce the BOD of the liquid***

COMMENT: diluted sewage passed into rivers contains dissolved oxygen which is absorbed by bacteria as they oxidize the pollutants in the sewage. The oxygen is replaced by oxygen from the air. Diluted sewage should not absorb more than 20ppm of dissolved oxygen

biochemist [baɪəʊ'kemɪst] *noun* scientist who specializes in biochemistry

◇ **biochemistry** [baɪəʊ'kemɪstri] *noun* chemistry of living tissues

◇ **biocide** ['baɪəʊsaɪd] *noun* substance which kills living organisms; ***biocides used in agriculture run off into lakes and rivers; the biological effect of biocides in surface waters can be very severe;*** **biocide pollution** = pollution of lakes and rivers caused by the runoff from biocides

◇ **bioclimatology** [baɪəʊklaɪmə'tɒlədʒi] *noun* study of the effect which the climate has on living organisms

◇ **biocoenosis** *or* **biocenosis** [baɪəʊsə'nəʊsɪs] *noun* varied community of organisms living in the same small area e.g. in the bark of a tree; relationship between these organisms; **marine biocoenosis** *or* **marine biocenosis** = varied community of organisms living in the sea

◇ **biocontrol** [baɪəʊkən'trəʊl] *noun* = BIOLOGICAL CONTROL

◇ **bioconversion** [baɪəʊkən'vɜːʃn] *noun* changing organic waste into a source of energy, such as the production of methane gas from the decomposition of organic matter

biodegradable [baɪəʊdɪ'greɪdəbl] *adjective* (substance) which can be easily decomposed by organisms such as bacteria or by the effect of sunlight, the sea, etc.; ***organochlorines are not biodegradable and enter the food chain easily;*** **biodegradable packaging** = boxes, cartons, bottles, etc. which can be decomposed by organisms such as bacteria or by the effect of sunlight, the sea, etc.

◇ **biodegradability** [baɪəʊdɪgreɪdə'bɪlɪti] *noun* degree to which a material, packaging, etc. can be decomposed by organisms

◇ **biodegradation** [baɪəʊdegrə'deɪʃn] *noun* breaking down of a substance by bacteria (as in the case of activated sludge)

COMMENT: manufacturers are trying to produce more biodegradable products, as the effect of non-biodegradable substances (such as detergents) on the environment can be serious

QUOTE: human sewage is a totally biodegradable product, and sea and sunlight will break it down through the natural process of oxidation
Environment Now

biodetergent [baɪəʊdɪ'tɜːdʒənt] *noun* biological detergent, a detergent with added enzymes which make the washing process more effective

biodiversity [baɪəʊdaɪ'vɜːsɪti] *noun* richness of the number of species

QUOTE: but as more hardwoods are cut down, the biodiversity of the area is lost and the delicate balance of the ecosystem destroyed
Green Magazine

biodynamics [baɪəʊdaɪ'næmɪks] *noun* study of living organisms and the production of energy

bioecology [baɪəʊi'kɒlədʒi] *noun* study of the relationships among organisms and the relationship between them and their physical environment, with particular emphasis upon the effect of humans on the environment environment

bioenergetics [baɪəʊenə'dʒetɪks] *noun* study of energy with reference to living organisms

bioengineering [baɪəʊendʒɪ'niːərɪŋ] *noun* use of biochemical processes on an industrial scale to produce drugs and foodstuffs, to recycle waste, etc.

biofuel ['baɪəʊfjuːəl] *noun* fuel from organic domestic waste and other sources which does not use up fossil fuels; ***coppiced wood can be grown as biofuel***

◇ **biogas** ['baɪəʊgæs] *noun* gas (partly methane and partly carbon dioxide) which is produced from fermenting waste, such as animal refuse; ***farm biogas systems may be uneconomic unless there is a constant demand for heat; the use of biogas systems in rural areas of developing countries is increasing***

biogeochemical [baɪəʊdʒiːəʊ'kemɪkl] *adjective* referring to biogeochemistry; **biogeochemical cycle** = process in which nutrients from living organisms are transferred into the physical environment and back to the organisms: this process is essential for organic life to continue

◇ **biogeochemistry** [baɪəʊdʒiːəʊ'kemɪstri] *noun* study of living organisms and their relationship to the chemical composition of the earth, its soil, rocks, minerals, etc.

◇ **biogeographer** [baɪəʊdʒiː'ɒgrəfə] *noun* scientist who studies regions with distinct fauna and flora

◇ **biogeographical region** [baɪəʊdʒiːəʊ'græfɪkl 'riːdʒən] *noun* large region of the earth with distinct fauna and flora

◇ **biogeosphere** [baɪəʊ'dʒiːəʊsfiːə] *noun* top layer of the lithosphere (the earth's crust) which contains living organisms

◇ **bioinsecticide** [baɪəʊɪn'sektɪsaɪd] *noun* insecticide developed from natural toxins

biology [baɪ'ɒlədʒi] *noun* study of living organisms

◇ **biological** [baɪə'lɒdʒɪkl] *adjective* referring to biology; **biological clock** *or* **circadian rhythm** = rhythm of daily activities and bodily processes (eating *or* defecating *or* sleeping, etc.) frequently controlled by hormones, which repeats every twenty-four hours and is found both in plants and animals; **biological detergent** *or* **biological washing powder** = detergent with added enzymes which make the washing process more effective; **biological half-life** = time taken for half of an amount of radioactive material to be eliminated naturally from a living organism; **biological indicator** = organism which is used to show changes in the environment; **biological magnification** = way in which a pollutant increases in concentration at each level of the food chain; **biological monitoring** = checking the changes which take place in a habitat; **biological oxygen demand (BOD)** = amount of pollution in water, shown as the amount of oxygen which will be needed to oxidize the polluting substances; **biological warfare** = war where one side tries to kill *or* affect the people of the enemy side by infecting them with living organisms *or* poison derived from living organisms

◇ **biological control** *or* **biocontrol** [baɪə'lɒdʒɪkl kən'trəʊl *or* baɪəʊkən'trəʊl] *noun* control of pests by using predators to eat them; **biological diversity** = richness of the number of species; **Biological Diversity Convention** = one of two binding treaties agreed at the Earth Summit, requiring states to take steps to preserve ecologically valuable areas and species; **biological weathering** = changing the state of soil *or* rock through the action of living organisms, such as burrowing insects or the movement of tree roots

◇ **biologist** [baɪ'ɒlədʒɪst] *noun* scientist who specializes in biology

> COMMENT: biological control of insects involves using bacteria, viruses, parasites and predators to destroy the insects. Plants can be controlled by herbivorous animals such as cattle

biomagnification [baɪəʊmægnɪfɪ'keɪʃn] *noun* increase in the concentration of substances up the food chain

biomass ['baɪəʊmæs] *noun* **(a)** all living organisms in a given area or at a given trophic level (usually expressed in terms of living or dry weight) **(b)** organic matter used to produce energy; **biomass fuel** = fuel produced from biomass

> COMMENT: the use of biofuels for domestic heating may be a way of disposing of refuse, but it carries the risk of adding to air pollution, since the biomass itself may pollute

> QUOTE: from one layer of the food chain to the next, the biomass decreases by approximately a factor of ten
> **Natural Resources Forum**

biome ['baɪəʊm] *noun* large ecological region characterized by similar vegetation and climate (such as the deserts, the tundra, etc.) and all the living organisms in it

> COMMENT: the ten principal biomes are: mountain and polar regions, tropical rainforest, grasslands, deserts, temperate forests, monsoon forests, deciduous forests, coniferous forests and evergreen shrub forests

biometeorology [baɪəʊmiːtjə'rɒlədʒi] *noun* scientific study of the weather and its effect on organisms

biomethanation [baɪəʊmeθə'neɪʃn] *noun* system of producing biogas for use as fuel or light

> QUOTE: biomethanation is attractive for use in rural areas for several reasons: it is an anaerobic digestion process, which is the simplest, safest way that has been found for treating human excreta and animal manure
> **Appropriate Technology**

bion ['baɪɒn] *noun* single living organism in an ecosystem

biopesticide [baɪəʊ'pestɪsaɪd] *noun* pesticide made from biological sources, that is from toxins which occur naturally

> COMMENT: biopesticides have the advantage that they do not harm the environment as they are easily inactivated and broken down by sunlight. This is, however, a practical disadvantage for the farmer who uses them, since they may not be as efficient in controlling pests as artificial chemical pesticides (which are persistent and difficult to control)

biophyte ['baɪəʊfaɪt] *noun* plant (such as a sundew) which obtains nutrients from the decomposing bodies of insects which it traps and kills

◇ **biosphere** ['baɪəʊsfiːə] *noun* part of the earth and its atmosphere where living organisms exist (including parts of the lithosphere, the hydrosphere and the atmosphere)

◇ **biosynthesis** [baɪəʊ'sɪnθesɪs] *noun* production of chemical compounds by a living organism

◇ **biota** [baɪ'əʊtə] *noun* flora and fauna of a region

◇ **biotechnology** [baɪəʊtek'nɒlədʒi] *noun* use of technology to manipulate and combine different genetic materials to produce living organisms with particular characteristics; ***a biotechnology company is developing a range of new pesticides based on naturally occurring toxins; artificial insemination of cattle was one of the first examples of biotechnology***

> COMMENT: biotechnology offers great potential to increase farm production and food processing efficiency, to lower food costs, to enhance food quality and safety and to increase international competitiveness

biotic [baɪ'ɒtɪk] *adjective* referring to living organisms (as opposed to the chemical constituents of an environment); **biotic barrier** = conditions which prevent members of a species moving to other regions, and so prevent the species from expanding; **biotic carrier potential** = assessment of the maximum increase in the number of individuals in a species, disregarding the effect of natural selection; **biotic climax** = stable biotic community; **biotic community** = community of organisms in a certain area; **biotic factors** = different organisms in an area, and the way in which they affect the plants in the area; **biotic index** = scale for showing the quality of an environment (e.g. how clean a river is) by indicating the types of organisms present in it; **biotic pyramid** = ecological pyramid *or* graphical representation of the structure of an ecosystem in terms of who eats what (the base is composed of producer organisms, usually plants, then herbivores, then carnivores. It may be measured in terms of number, biomass or energy); **biotic succession** = changes which take place in a group of organisms under the influence of their changing environment

biotope ['baɪətəʊp] *noun* small area with uniform biological conditions (climate, soil, altitude, etc.)

◇ **biotype** ['baɪəʊtaɪp] *noun* group of similar individuals within a species

biphenyl ($C_6H_5C_6H_5$) [baɪ'fenɪl] *noun* white *or* colourless crystalline substance used as a fungicide, in the production of dyes and as a preservative (E number E230) on the skins of citrus fruit

birch [bɜːtʃ] *noun* common hardwood tree *(Betula pendula)* found in northern temperate zones

bird [bɜːd] *noun* animal which lays eggs and has wings; **flightless bird** = bird (such as the penguin *or* ostrich) which has wings but cannot fly; **bird of prey** = carnivorous bird which kills and eats animals; **bird sanctuary** = place where birds can breed and live in a protected environment

◇ **birdsong** ['bɜːdsɒŋ] *noun* singing calls made by birds to communicate with each other

◇ **birdwatcher** ['bɜːdwɒtʃə] *noun* ornithologist, a person who observes and studies birds and takes notes of their behaviour

◇ **birdwatching** ['bɜːdwɒtʃɪŋ] *noun* observing and studying birds

> COMMENT: all birds are members of the class Aves. They have feathers and the forelimbs have developed into wings, though not all birds are now able to fly. Birds are closely related to reptiles, and have scales on their legs. Birds (rooks, pigeons, pheasants) can cause very serious damage to crops: various controls can be used such as shooting, bird-scarers and destruction of nests. Birds also destroy many pests, such as wireworms, leatherjackets and caterpillars. In general, birds are more helpful than harmful, though this will depend on the type of farming undertaken

birth [bɜːθ] *noun* being born; **birth control** = restricting the number of children born by using contraception; **birth defect** = malformation which exists in a person's *or* animal's body from birth; **birth rate** = number of births per year, shown per thousand of the population; ***a birth rate of 15 per thousand; there has been a severe decline in the birth rate***

bitumen ['bɪtʃʊmən] *noun* solid hydrocarbon, contained in coal

◇ **bituminous coal** [bɪ'tjuːmɪnəs 'kəʊl] *noun* coal containing a high percentage of tar, which is harder than lignite but not as hard as anthracite, and which gives off smoke

◇ **bituminous sand** *or* **bituminous shale** [bi'tjuːmɪnəs 'sænd *or* 'ʃeɪl] *noun* geological formation of sedimentary rocks containing bitumen which can be extracted and processed to give oil

black [blæk] *adjective* having no colour due to the absorption of all *or* nearly all light; having the very darkest colour; ***thick black smoke poured out of the factory chimneys;*** **black bean** = type of very hard tropical wood, resistant to termites; **black beetle** = cockroach, insect of the order Dictyoptera a common household pest; **black carbon** = fine particles of carbon which rise in the smoke produced by the burning of coal, wood, oil, etc.; **Black Death** = violent form of bubonic plague, a pandemic during the Middle Ages; **black earth** = chernozem, a dark fertile soil, rich in organic matter, found in the temperate, grass-covered plains of Russia and North and South America; **black frost** = condition when the air is dry and the ground temperature is lower than the air temperature, but no white frost forms; **black ice** = ice which is clear, not white and opaque as frost, and which forms on the surface of roads, etc.; **black spot** = fungus which attacks plants, causing black spots to appear on the leaves

◇ **blackwood** ['blækwʊd] *noun* Australian hardwood tree

blade [bleɪd] *noun* thin flat leaf (such as a leaf of grass)

blanket ['blæŋkɪt] **1** *noun* thick covering; ***a blanket of snow covered the fields; the town was covered in a blanket of smog;*** **blanket bog** = bog covering a wide area **2** *verb* to cover; ***thick fog blanketed the airport***

blast [blɑːst] *noun* impact from an explosion; ***thousands of people would be killed by the blast;*** **blast effect** = result of the impact from an explosion, such as damage caused; **blast furnace** = heating device for producing iron or copper from ore, in which the ore, coke and limestone are heated together, air is blown through the mixture and the molten metal is drawn off into moulds. The waste matter from this process is known as slag

blaze [bleɪz] **1** *noun* mark put on a tree to show that it needs to be felled *or* to indicate a path through a forest **2** *verb* **to blaze a trail** = to show where a path goes by making marks on trees or rocks

bleaching ['bliːtʃɪŋ] *noun* removing of colour by the action of a chemical *or* sunlight; **bleaching agent** = something which removes colour, such as a chemical *or* the action of sunlight

blight [blaɪt] **1** *noun* **(a)** fungus disease of plants; ***they are taking steps to eradicate the epidemic of potato blight*** **(b)** **(urban) blight** = unattractive, dirty and dilapidated area in a city *or* town **2** *verb* to ruin *or* to spoil the environment; ***the landscape was blighted by open-cast mining***

blizzard ['blɪzəd] *noun* heavy snowstorm with wind; lying snow which is blown by strong wind; ***there were blizzards on the highlands during the weekend***

blood [blʌd] *noun* red liquid in an animal's body; **blood bank** = section of a hospital where blood given by donors is stored for use in transfusions; **blood cell** *or* **blood corpuscle** = cell (red blood cell *or* white blood cell) which is one of the components of blood; **red blood cell** = blood cell which contains haemoglobin and carries oxygen; **white blood cell** blood cell which contains a nucleus, is formed in bone marrow and creates antibodies; **blood chemistry** = (i) record of the changes which take place in blood during disease and treatment; (ii) substances which make up blood, which can be analysed in blood tests, the results of which are useful in diagnosing disease; **blood count** = test to count the number of different blood cells in a certain quantity of blood; number of blood cells counted; **blood plasma** = watery liquid which forms the greatest part of blood; **blood platelet** = small blood cell which releases thromboplastin and which multiplies rapidly after an injury to cause the blood to clot; **blood serum** = watery liquid which separates from coagulated blood; **blood sugar level** = amount of glucose in the blood; **blood test** = laboratory test to find the chemical composition of a patient's blood

◇ **blood group** ['blʌd 'gruːp] *noun* one of the different types of blood by which groups of people are identified

> COMMENT: blood is classified in various ways. The most common classifications are by the agglutinogens in red blood corpuscles (factors A and B) and by the Rhesus factor. Blood can therefore have either factor (Group A and Group B), or both factors (Group AB) or neither (Group O), and each of these groups can be Rhesus negative or positive

◇ **bloodstream** ['blʌdstriːm] *noun* blood as it passes round the body

> COMMENT: blood is formed of red and white corpuscles, platelets and plasma. It circulates round the body, going from the heart and lungs along arteries and returns to the heart through the veins. As it moves round the body it takes oxygen to the tissues and removes waste material from them. Waste material is removed from the blood by the kidneys or exhaled through the lungs. It also carries hormones produced by glands to the various organs which need them

bloom [bluːm] **1** *noun* **(a)** flower; ***the blooms on the orchids have been ruined by frost*** **(b)** **algal bloom** = mass of algae which develops rapidly in a lake due to eutrophication **2** *verb* to flower; ***the plant blooms at night; some cacti only bloom once every seven years***

blow [bləʊ] **1** *noun* sudden movement of air *or* gas; ***the blow of air through a Bessemer converter*** **2** *verb* **(a)** *(of wind)* to be moving; ***it was blowing hard on the top of the mountain*** **(b)** to move something (by the wind); ***topsoil was blown away by the strong winds***

◇ **blow-by** ['bləʊbaɪ] *noun* unburnt fuel mixed with air and other gases produced by an internal combustion engine which escapes past the piston rings

> COMMENT: in many parts of the world emission of gases from pistons is controlled by law, since these gases contribute significantly to atmospheric pollution

blowhole ['bləʊhəʊl] *noun* **(a)** hole in the roof of a cave by the sea, leading to the surface of the cliff above, through which air and water are sent out by the pressure of the waves breaking beneath **(b)** nostril of a whale, situated towards the back of the skull, through which the whale spouts air and water

◇ **blowout** ['bləʊaʊt] *noun* sudden rush of oil *or* gas from an oil well

blubber ['blʌbə] *noun* thick layer of fat under the skin of marine animals such as whales

> COMMENT: fat-soluble pollutants like PCBs tend to collect in the blubber of seals and other marine animals

blue [blu:] *adjective & noun* colour such as that of a clear unclouded sky in the daytime; *see COMMENT at* SKY; **blue baby =** baby suffering from congenital cyanosis, born either with a congenital heart defect or with a collapsed lung which prevents an adequate supply of oxygen reaching the tissues, giving the baby's skin a bluish colour; **blue whale =** large whale with no teeth, which feeds by straining large amounts of water through its baleen and so retaining plankton and other food

BMR = BASAL METABOLIC RATE

BNFL = BRITISH NUCLEAR FUELS LTD company which makes fuel rods for nuclear power stations and reprocesses spent fuel

board [bɔ:d] *noun* group of people, such as directors, councillors, examiners, etc., who administer an organization; **advisory board =** group of specialists who can give advice; *US* **Board of Health =** government department dealing with matters of public health; **board of inquiry =** group of people who carry out a formal investigation when there has been an accident *or* something has gone wrong

BOD = BIOCHEMICAL OXYGEN DEMAND

body ['bɒdi] *noun* **(a)** group of people *or* organization carrying out specific duties; **consultative body =** group of people who can give their advice and opinion on a subject but who do not have the power to make laws; **legislative body =** group of people who make laws **(b) body of water =** separate and distinct mass of water

bog [bɒg] *noun* soft wet land, usually with moss growing on it, which does not decompose, but forms a thick layer of acid peat; ***bog mosses live on nutrients which fall in rain;*** **blanket bog =** wide area of bog

◇ **boggy** ['bɒgi] *adjective* soft and wet (soil)

◇ **bogland** ['bɒglænd] *noun* area of bog

boil [bɔɪl] *verb* to bring a liquid to the temperature at which it changes into gas; ***water is boiled to form steam; tourists are advised that ordinary tap water should be boiled before drinking it;*** **boiling point =** temperature at which a liquid changes into gas; ***the boiling point of water is 100°C;*** **boiling water reactor (BWR) =** nuclear reactor fuelled by uranium, in which light water is heated to form steam which drives the turbines

◇ **boiler** ['bɔɪlə] *noun* container in which water is boiled to make steam to drive turbines, etc.

bole [bəʊl] *noun* base of a tree trunk

bomb [bɒm] *noun* large piece of molten lava ejected from an erupting volcano

bond [bɒnd] **1** *noun* close link between two substances *or* two atoms; **chemical bond =** force which links atoms to form molecules **2** *verb (of two substances)* to link together; ***adsorption is the bonding of a gas to a solid surface***

bone [bəʊn] *noun* **(a)** one of the calcified pieces of connective tissue which make a skeleton; ***there are several small bones in the human ear; fossil bones are often found in limestone deposits*** **(b)** hard substance which forms a bone; **bone conduction =** osteophony, the conduction of sound waves to the inner ear through the bones of the skull (as opposed to air conduction); **bone structure =** (i) system of jointed bones as it forms a body; (ii) the arrangement of the various components of a bone

> COMMENT: bones are formed of a hard outer layer (compact bone) which is made up of a series of layers of tissue and a softer inner part which contains bone marrow

◇ **bonemeal** ['bəʊnmi:l] *noun* fertilizer made of ground bones or horns, reduced to a fine powder. It is also used as a fine meal for growing animals, used for the calcium, phosphorus and magnesium it contains

boom [bu:m] *noun* loud noise; **sonic boom =** loud noise made by the shock waves produced by any object (such as an aircraft *or* a bullet) travelling through the air at *or* faster than the speed of sound

> COMMENT: the shock waves can cause objects to resonate so violently that they are damaged. Supersonic aircraft generally fly at speeds greater than the speed of sound

only over the sea, to avoid noise nuisance and damage to property

Bordeaux mixture [bɔː'dəʊ 'mɪkstʃə] *noun* mixture of copper sulphate, lime and water, used to spray on plants to prevent attacks by blight

bore [bɔː] **1** *noun* **(a)** tidal wave which rushes up the estuary of a river at high tide **(b)** measurement across the inside of a pipe *or* hole; ***the central heating uses small-bore copper piping; the well has a 2-metre bore*** **2** *verb* to make a round hole (in the ground); ***they have bored six test holes to try to find water***

◇ **borehole** ['bɔːhəʊl] *noun* hole bored in the ground; ***the borehole is intended to test the geology of the site to see if it is suitable for burying nuclear waste; boreholes supply water of excellent quality***

boreal ['bɔːriəl] *adjective* referring to the north; (climate in the Northern Hemisphere between 60° and 40°N) with short hot summers and longer cold winters

◇ **borealis** [bɔːri'eɪlɪs] *see* AURORA

QUOTE: unless levels of greenhouse gases in the atmosphere are quickly stablilized, global warming is likely to reduce between 50 and 90 per cent of the world's existing boreal forests to patchy open woodland or grassland within the next 30-50 years
Ecologist

-borne [bɔːn] *suffix* meaning 'carried by'; **airborne pollutants** = pollutants which are carried in the air; **waterborne diseases** = diseases which are spread by animals living in water

boron ['bɔːrɒn] *noun* chemical element from which control rods in reactors are made. It is present in borax, and essential for healthy plant growth as a trace element in soils (NOTE: chemical symbol is **B**; atomic number is **5**)

botany ['bɒtəni] *noun* scientific study of plants

◇ **botanical** [bɒ'tænɪkl] *adjective* referring to botany; **botanical garden** = place where plants are grown for showing to the public and for scientific study; **botanical insecticide** = insecticide made from a substance extracted from plants (the best-known botanical insecticides are pyrethrum, derived from the chrysanthemum, and nicotine, derived from the tobacco plant); **botanical specimens** = plants gathered for study

◇ **botanist** ['bɒtənɪst] *noun* scientist who studies plants

bottle bank ['bɒtl 'bæŋk] *noun* large container into which people put empty glass bottles and jars which can then be recycled into new glass

COMMENT: because recycled glass does not require extremely high temperatures to melt, it is more energy-efficient than using the usual raw materials. Also one tonne of cullet (broken glass) will produce the same amount of finished glass as ten tonnes of raw materials (mainly sand, limestone and sodium carbonate)

bottled gas ['bɒtəld 'gæs] *noun* gas (liquefied petroleum gas) produced from refining crude oil and sold in pressurized metal containers for use as fuel for domestic heating, camp stoves, etc.

bottom ['bɒtəm] *noun* (i) seabed *or* floor of a lake *or* river; (ii) flat area along a river where silt is deposited when the river floods; **bottom feeder** = organism (such as a fish) which collects food on the bottom *or* in the deepest water of the sea *or* of a lake *or* of a river; **bottom water** = water in the deepest part of the sea *or* of a lake *or* of a river

botulism ['bɒtjuːlɪzm] *noun* type of food poisoning, caused by a toxin of *Clostridium botulinum* in badly canned or preserved food

COMMENT: the symptoms include paralysis of the muscles, vomiting and hallucinations. Botulism is often fatal

boulder ['bəʊldə] *noun* large rounded piece of rock; **boulder clay** = clay soil mixed with rocks of different sizes, found in glacial deposits

boundary layer ['baʊndəri 'leɪə] *noun* altitude at which airflow is affected by the ground

Bovidae ['bɒvɪdiː] *noun* largest class of even-toed ungulates, including cattle, antelopes, gazelles, sheep and goats

◇ **bovine** ['bəʊvaɪn] *adjective* referring to cattle; **bovine somatotropin (BST)** = natural hormone found in cows, which has been produced artificially by genetic engineering; it is used to increase milk yields and is said to increase them by between 12% and 20%. It is not licensed for use in the UK, but is being used in trials; **bovine spongiform encephalopathy (BSE)** = fatal disease of cattle, affecting the nervous system; *see also* MAD COW DISEASE

COMMENT: ruminant-based additives are now banned from cattle feed

QUOTE: BSE first appeared on English dairy farms in 1987. By December, 1988, 1,677 cattle had been slaughtered after contracting the infection. BSE is a new addition to a group of animal viruses known for about 200 years. The similarity between BSE and scrapie suggests that scrapie has been transmitted from sheep to cattle. Processed sheep carcasses, offal and heads are commonly fed to cattle.
Guardian

QUOTE: the USA plans to approve the use of BST. The EU is uncertain and in making its decision will need to consider 1) milk surpluses; 2) that BST use will favour large-scale dairy operations over the small farmers; 3) possible consumer opposition; 4) US trade retaliation; 5) BST's effects on animal health

New Scientist

bq *abbreviation for* becquerel

Br *chemical symbol for* bromine

bracken ['brækən] *noun* fern which grows widely on acid soils

COMMENT: bracken is hard to eradicate as it sprouts easily after being burnt. The spores of bracken contain a carcinogenic substance

brackish ['brækɪʃ] *adjective* (water) which contains salt (though not as much as sea water) and is not good to drink

QUOTE: the Baltic Sea is the world's largest stretch of brackish water, a poorly mixed cocktail of freshwater from rivers that drain into it, pollution and saltwater from the North Sea. The salt content of the North Sea is around 3.5% but within the Baltic it falls fast with distance from the open sea, to below 0.4% in places

New Scientist

bract [brækt] *noun* small green leaf at the base of a flower

braiding ['breɪdɪŋ] *noun* phenomenon which occurs when a river becomes divided into several channels with small islands or bars between them

bran [bræn] *noun* outside covering of the wheat seed, removed to make white flour, but an important source of roughage, hence used in muesli and other breakfast cereals

branch [brɑːnʃ] *noun* **(a)** *(of a tree)* woody stem growing out from the main trunk **(b)** *(of a river)* smaller stream **(c)** *(of a science, company, etc.)* subdivision of something larger; ***he knows nothing about this branch of engineering; they have just opened a branch office in Paris***

brass [brɑːs] *noun* a metal, an alloy of copper and zinc

breakbone fever ['breɪkbəʊn 'fiːvə] *noun* tropical disease caused by an arbovirus, transmitted by mosquitoes, where the patient suffers a high fever, pains in the joints, headache and a rash

breakdown ['breɪkdaʊn] *noun* (of machine, communications, etc.) failure; (of system, organization, etc.) disintegration; (of chemical substance, etc.) separation into elements; (of organic matter, etc.) decomposition

breakthrough ['breɪkθruː] *noun* situation where sewage or other pollutants get into the main domestic water supply

breakwater ['breɪkwɔːtə] *noun* strong wall *or* fence which is built from the shore into the sea in order to block the force of waves and so prevent erosion

breccia ['bretʃɪə] *noun* type of rough rock made of sharp fragments of other rocks fused together

breed [briːd] **1** *noun* organisms of a certain species, which have been developed by people over a period of time to stress certain characteristics; ***a hardy breed of sheep; two new breeds of rice have been developed*** **2** *verb* **(a)** *(of organisms)* to produce young; ***rabbits breed very rapidly*** **(b)** to encourage to develop; ***insanitary conditions help to breed disease*** **(c)** to raise a certain type of animal *or* plant by crossing one variety with another to produce a new variety where the desired characteristics are strongest; ***farmers have bred new hardy forms of sheep***

QUOTE: Milk Marketing Board AI figures show a swing away from traditional British beef breeds. Between April and December 1988, Limousin accounted for 13.8%, Charolais 10.8%, Hereford 6.8%, Belgian Blue 5.4%, Simmental 4.4% of all inseminations

Farmers Weekly

QUOTE: the large birds bred in the poorly drained wetlands of central Illinois

Birder's World

breeder ['briːdə] *noun* **(a)** person who breeds new forms of animals *or* plants; ***cat breeder; cattle breeder; rose breeder; plant breeder*** **(b) breeder reactor** = nuclear reactor which produces more fissile material than it consumes; **fast breeder** *or* **fast breeder reactor (FBR)** = nuclear reactor which produces more fissile material than it consumes, using fast-moving neutrons and making plutonium-239 from uranium-238, thereby increasing the reactor's efficiency

COMMENT: uranium-238 is a natural uranium isotope and is fertile, i.e. it can be used to produce the fissile plutonium-239. In a breeder reactor, uranium-238 is used as a blanket round the plutonium fuel and when the plutonium is fissioned high-speed neutrons are produced which change on contact with the uranium-238 and eventually produce a slightly greater quantity of plutonium-239 than that originally used as fuel. The excess plutonium can be used as a fuel in another breeder reactor or in an ordinary burner reactor

breeding ['briːdɪŋ] *noun* raising a certain type of animal *or* plant by crossing one variety with another to produce a new variety where the desired characteristics are strongest

breeze [bri:z] *noun* **(a)** light wind; **land breeze** = light wind which blows from the land to the sea, for example, during the day when the land is warm; **sea breeze** = light wind which blows from the sea towards the land, for example, in the evening when the land cools; **stiff breeze** = quite a strong wind; *see also* BEAUFORT SCALE **(b)** solid waste of burnt coal, etc. from a furnace; **breeze block** = large building brick made from breeze and cement

◇ **breezy** ['bri:zi] *adjective* fresh and windy (weather)

British thermal unit (Btu *or* **BTU)** ['brɪtɪʃ 'θɜ:məl 'ju:nɪt] *noun* unit of measurement of heat, the amount of heat needed to heat one pound of water one degree Fahrenheit

COMMENT: a non-metric unit, used in the US rather more than in the UK

broad [brɔ:d] *adjective* wide; ***broad paths are left through the forest plantations to act as firebreaks***

◇ **broadleaf (tree)** *or* **broad-leaved tree** ['brɔ:dli:f *or* 'brɔ:dli:vd 'tri:] *noun* deciduous tree (such as beech, oak, etc.) which has wide leaves, as opposed to the needles on conifers; ***the plan is to plant 12,000 hectares of agricultural land a year, one third of which will be broadleaves;*** **broad-leaved evergreen** = evergreen tree with large leaves, such as rhododendron, tulip tree, etc.

◇ **broad-spectrum** [brɔ:d'spektrəm] *adjective* (antibiotic) which destroys *or* controls many types of bacteria; (pesticide) which kills many types of pest

bromine ['brəʊmi:n] *noun* chemical element used in various industrial processes and in antiknock additives for petrol (NOTE: chemical symbol is **Br;** atomic number is **35)**

bronch- [brɒŋk] *prefix* referring to the windpipe

◇ **bronchi** ['brɒŋki:] *plural noun* air passages leading from the windpipe into the lungs (NOTE: singular is **bronchus)**

◇ **bronchial** ['brɒŋkiəl] *adjective* referring to the bronchi; **bronchial asthma** = type of asthma mainly caused by an allergen *or* exertion; **bronchial pneumonia** = inflammation of the bronchioles which may lead to general infection of the lungs; **bronchial tubes** = bronchi *or* air tubes leading from the windpipe into the lungs

◇ **bronchiole** ['brɒŋkiəʊl] *noun* tiny air passage at the end of a bronchus

◇ **bronchitis** [brɒŋ'kaɪtɪs] *noun* inflammation of the mucous membrane of the bronchi; **acute bronchitis** = attack of bronchitis caused by a virus *or* exposure to cold and wet; **chronic bronchitis** = long-lasting form of bronchial inflammation

◇ **bronchus** ['brɒŋkəs] *noun* air passage leading from the windpipe into the lung and branching out into many bronchioles (NOTE: plural is **bronchi)**

brood [bru:d] *noun* group of offspring produced at the same time, especially group of young birds; ***the territory provides enough food for two adults and a brood of six or eight young***

◇ **brooding time** ['bru:dɪŋ 'taɪm] *noun* length of time a bird sits on its eggs to hatch them out

brook [brʊk] *noun* little stream

brown [braʊn] *adjective* of a colour like the colour of earth or wood; **brown bread** = bread made with flour which has not been bleached *or* refined; ***not all brown bread is made with wholemeal flour;*** **brown coal** = lignite *or* type of soft coal which is not as efficient a fuel as anthracite and produces smoke when it burns; **brown earth** *or* **brown forest soil** = good fertile soil, slightly acid, containing humus; **brown fat (tissue)** = animal fat which can easily be converted to energy and is believed to offset the effects of ordinary white fat; **brown flour** = wheat flour (such as wholemeal flour) which contains some bran and has not been bleached; **brown fumes** *or* **brown smoke** = fumes *or* smoke from tarry substances produced by coal burning at low temperatures; **brown podzolic soil** = brown earth from which humus particles have been leached by rain; **brown rice** = rice grain that has had the husk removed, but has not been milled and polished to remove the bran

Brucella [bru:'selə] *noun* type of rod-shaped bacterium

◇ **brucellosis** [bru:sɪ'ləʊsɪs] *noun* disease which can be caught from cattle or goats or from drinking infected milk, spread by a species of the bacterium *Brucella*

COMMENT: symptoms include tiredness, arthritis, headache, sweating, irritability and swelling of the spleen

brush(wood) [brʌʃ *or* 'brʌʃwʊd] *noun* undergrowth, thicket; **brush killer** = powerful herbicide which destroys the undergrowth

BSE = BOVINE SPONGIFORM ENCEPHALOPATHY

BST = BOVINE SOMATOTROPIN

Btu *or* **BTU** *or* **BThU** = BRITISH THERMAL UNIT

bubonic plague [bju:'bɒnɪk 'pleɪg] *noun* fatal disease caused by *Pasteurella pestis* in the lymph system transmitted to humans by fleas from rats

COMMENT: bubonic plague was the Black Death of the Middle Ages; its symptoms are fever, delirium, vomiting and swelling of the lymph nodes

buckwheat ['bʌkwi:t] *noun* grain crop *(Fagopyrum esculentum)* which is not a member of the grass family; it can be grown on the poorest of soils. In the USA, when buckwheat is ground into flour, it is used to make grits

bud [bʌd] *noun* young shoot on a plant, which will later become a leaf or flower

◇ **budding** ['bʌdɪŋ] *noun* way of propagating plants, where a bud from one plant is grafted onto another plant (used, for example, to propagate roses)

budget ['bʌdʒɪt] *see* ENERGY BUDGET

buffalo ['bʌfələʊ] *noun* **(a)** common domestic animal in tropical countries, used for milk and also as a draught animal **(b)** *US* type of large wild cattle

QUOTE: Egypt's main milk producing animal is still the water buffalo. For milk production, female buffalo calves growing from 200kg to the proper mating weight of 350kg are reared totally on forages, with minimum concentrates, berseem and rice straw in winter, followed by berseem hay, rice straw and green maize in summer
Middle East Agribusiness

buffer ['bʌfə] **1** *noun* (i) substance that keeps a constant balance between acid and alkali; (ii) solution where the pH is not changed by adding acid or alkali; **buffer action** = balancing between acid and alkali; **buffer land** *or* **buffer zone** = land between a protected area, for example, a nature reserve, and the surrounding countryside or town **2** *verb* to prevent a solution from becoming acid; ***if a lake is well buffered, it will not have a low pH factor, even if acid rain falls into it; bicarbonate is the main buffering factor in fresh water***

◇ **buffer stock** ['bʌfə 'stɒk] *noun* stock of a commodity (such as coffee) held by an international organization and used to control movements of the price on international commodity markets

bug [bʌg] *noun informal* common name for any little insect; name of certain winged insects belonging to the class of animals called Hemiptera

build [bɪld] *verb* to make a construction; ***the developer is planning to build 2,500 new houses on the greenfield site; the female birds build nests of straw in holes in trees;*** **the built environment** = built-up areas seen as the environment in which humans live

◇ **build up** ['bɪld 'ʌp] *verb* to form by accumulation; ***the traces of pesticide gradually built up in the food chain***

◇ **build-up** ['bɪldʌp] *noun* gradual accumulation; ***a build-up of DDT in the food chain***

◇ **built-up area** ['bɪltʌp 'eəriə] *noun* area which is full of houses, shops, offices and other buildings, with very little open space

building ['bɪldɪŋ] *noun* (i) construction, such as a house, shop, office, etc.; (ii) process of constructing; **building area** *or* **building zone** = area of land on which building may take place *or* is taking place; **building certificate** *or* **building permit** = official document allowing a person *or* company to build a property on empty land; **building density** = number of buildings allowed on a certain area of land, for example, fifteen houses to the acre; **building development** = area of land on which building is taking *or* has taken place; **building ground** *or* **building land** = area of land suitable for building on; **building industry** = business *or* trade of constructing houses, offices, etc.; **building site** = area of land on which building is taking place; ***a cement mixer has been stolen from the building site***

bulb [bʌlb] *noun* fleshy stem like an onion, formed of layers of tissue, which can be planted and which will produce flowers and seed

bulk [bʌlk] *noun* large quantity; **bulk collection** = collection of something in large quantities, for example, milk, which can be collected from farms by tankers; **bulk plant** = factory which produces something in large quantities

◇ **bulking agent** ['bʌlkɪŋ 'eɪdʒənt] *noun* additive which causes a substance to stick together in coagulated masses; **bulking of sludge** = causing sludge to stick together in coagulated masses

bulldozer ['bʊldəʊzə] *noun* piece of earth-moving equipment, like a tractor with a blade in front, used for pushing soil *or* rocks and for levelling surfaces

bund [bʌnd] *noun* soil wall built across a slope to retain water *or* to hold waste in a sloping landfill site

QUOTE: farmers have traditionally relied on a system of bunding to grow sorghum in the area's semi-arid soil
New Scientist

bunt [bʌnt] *noun* disease of wheat caused by the smut fungus

buoyancy ['bɔɪənsi] *noun* ability to float on a liquid or air

bureaucracy [bjuː'rɒkrəsi] *noun (often used as criticism)* **(a)** group of civil servants *or* officials of central or local government **(b)** slow and complicated way of working; ***the distribution of aid is held up by bureaucracy; the group has complained about the bureaucracy involved in getting planning permission***

◇ **bureaucratic** [bjuːrə'krætɪk] *adjective* referring to bureaucracy; ***relief workers are continually fighting against bureaucratic muddle***

burial ['beriəl] *noun* action of putting something in a hole in the ground and covering it with earth; **burial site** = place where nuclear waste is buried; ***the group is opposed to the burial of nuclear waste on the site***

burn [bɜːn] **1** *noun* **(a)** *(in Scotland)* small stream **(b)** injury to skin and tissue caused by light *or* heat *or* radiation *or* electricity *or* chemical **2** *verb* to destroy *or* damage by fire; ***several hundred hectares of forest were burnt in the fire*** *see also* SLASH AND BURN

◇ **burner** ['bɜːnə] *noun* device for burning something such as fuel *or* waste; **open burner** = outdoor site where waste, such as automobile tyres, rags, etc., are destroyed by fire, thereby causing atmospheric pollution

◇ **burner reactor** ['bɜːnə riː'æktə] *noun* normal type of nuclear reactor in which fuel, such as uranium-239, is used to generate heat by fission

◇ **burning** ['bɜːnɪŋ] *noun* destroying *or* damaging by fire; **open burning** = destroying waste, such as automobile tyres, rags, etc., by setting fire to it out of doors, and thereby causing atmospheric pollution

◇ **burn-up** ['bɜːnʌp] *noun* amount of fuel burnt in a nuclear reactor, shown as a proportion of the fuel originally used

bush [bʊʃ] *noun* **(a)** low shrub *or* small tree; ***a coffee bush*** **(b)** *(in semi-arid regions)* **the bush** = wild land covered with bushes and small trees; **bush-fallow** = subsistence type of agriculture in which land is cultivated for a few years until its natural fertility is exhausted, then allowed to rest for a considerable period during which the natural vegetation regenerates itself, after which the land is cleared and cultivated again

butane **(C_4H_{10})** ['bjuːteɪn] *noun* gas produced during petroleum distillation, used domestically for heating and sold in special containers as bottled gas

butylated hydroxytoluene (BHT) ['bjuːtɪleɪtɪd haɪdrɒksɪ'tɒljuiːn] *noun* common antioxidant additive (E321) used in processed foods containing fat, probably carcinogenic

Buys Ballot's Law ['baɪz 'bælət 'lɔː] *noun* rule for identifying low pressure areas (based on the Coriolis effect) that in the Northern Hemisphere, if the wind is blowing from behind you, then the low pressure area is to the left, while in the Southern Hemisphere it is to your right

BWR = BOILING WATER REACTOR

by-pass ['baɪpɑːs] **1** *noun* road built around a town, to relieve traffic congestion; ***since the by-pass was built, traffic in the town has been reduced by half*** **2** *verb* to avoid (something) by going around; ***they by-passed the normal complaints procedure and went straight to the director of the chemical company; the plans are to build a new main road, by-passing the town***

by-product ['baɪprɒdʌkt] *noun* something additional produced during a process

byssinosis [bɪsɪ'nəʊsɪs] *noun* lung disease (a form of pneumoconiosis) caused by inhaling cotton dust

Cc

C **1** *abbreviation for* Celsius **2** *chemical symbol for* carbon

Vitamin C ['vɪtəmɪn 'siː] *noun* ascorbic acid, vitamin which is soluble in water and is found in fresh fruit (especially oranges and lemons), raw vegetables, liver and milk

COMMENT: lack of Vitamin C can cause anaemia and scurvy

Ca *chemical symbol for* calcium

cactus ['kæktəs] *noun* succulent plant with a fleshy stem (often protected by spines), found in the deserts of North and Central America (NOTE: plural is **cacti** or **cactuses**)

cadmium ['kædmiəm] *noun* metallic element which is naturally present in soil and rock in association with zinc; **cadmium sulphide (CdS)** = orange *or* yellow solid used as a colouring in paints, etc. (NOTE: chemical symbol is **Cd**; atomic number is **48**)

COMMENT: cadmium is used for making rods for nuclear reactors. It is also present in tobacco smoke and is found in fish and shellfish such as oysters

caducous [kə'dju:kəs] *adjective* (part of plant *or* animal) which becomes detached during the organism's life

caesium ['si:ziəm] *noun* metallic alkali element which is one of the main radioactive pollutants taken up by fish (NOTE: chemical symbol is **Cs;** atomic number is **55)**

caffeine ['kæfi:n] *noun* alkaloid found in coffee, chocolate and tea which acts as a stimulant

COMMENT: apart from acting as a stimulant, caffeine also helps in the production of urine. It can be addictive and exists in both tea and coffee in about the same percentages as well as in chocolate and other drinks

cal *abbreviation for* calorie

Cal *abbreviation for* Calorie *or* kilocalorie

calcareous [kæl'keəriəs] *adjective* (soil) which is chalky and contains calcium; (rock such as chalk or limestone) which contains calcium

calcicole *or* **calciphile** *or* **calcicolous plant** ['kælsɪkəʊl *or* 'kælsɪfaɪl *or* kæl'sɪkələs plɑ:nt] *noun* plant which grows well on chalky *or* alkaline soils

calcification [kælsɪfɪ'keɪʃn] *noun* hardening by forming deposits of calcium salts

◇ **calcified** ['kælsɪfaɪd] *adjective* made hard

◇ **calcifuge** *or* **calciphobe** ['kælsɪfju:dʒ *or* 'kælsɪfəʊb] *noun* plant which prefers acid soils and cannot exist on chalky *or* alkali soils

◇ **calcimorphic soil** [kælsɪ'mɔ:fɪk 'sɔɪl] *noun* soil which is rich in lime

◇ **calcination** [kælsɪ'neɪʃn] *noun* heating at high temperature as, for example, in the production of metal oxides

◇ **calciphile** ['kælsɪfaɪl] *see* CALCICOLE

◇ **calciphobe** ['kælsɪfəʊb] *see* CALCIFUGE

calcium ['kælsiəm] *noun* metallic chemical element which is naturally present in limestone, chalk and is essential to biological life. (It is a major component of bones and teeth); **calcium carbonate ($CaCO_3$)** = chalk, a white mineral found widely in many parts of the world, formed from animal organisms; **calcium cycle** = cycle of events by which calcium in the soil is taken up into plants, passed to animals which eat the plants and then passed back to the soil again when the animals die and decompose; **calcium deficiency** = lack of calcium in an animal's bloodstream; **calcium hydroxide ($Ca(OH)_2$)** = slaked lime, a mixture of calcium oxide and water, used on soils to improve their quality; **calcium oxide (CaO)** = quicklime, a chemical used in many industrial processes and also spread on soil to reduce acidity; **calcium phosphate ($Ca_3(Po_4)_2$)** = main constituent of bones and bone ash fertilizer; **calcium supplement** = addition of calcium to the diet, or as injections, to improve the level of calcium in the bloodstream (NOTE: chemical symbol is **Ca;** atomic number is **20)**

COMMENT: calcium is essential for various bodily processes such as blood clotting and is an important element in a balanced diet. Milk, cheese, eggs and certain vegetables are its main sources. In birds, calcium is responsible for the formation of strong eggshells. Water which passes through limestone contains a high level of calcium and is called 'hard'

calm [kɑ:m] *noun* period when there is no wind at all; *see also* BEAUFORT SCALE

calomel ['kæləmel] *noun* mercurous chloride, poisonous substance used to kill moss on lawns and to treat pinworms in the intestine

calor ['kælə] *noun* heat

◇ **caloric** [kə'lɒrɪk] *adjective* referring to calories; **caloric energy** = amount of energy shown as a number of calories; **caloric requirement** = amount of energy (shown in calories) which an animal such as a human needs each day

◇ **calorie** *or* **gram calorie** ['kæləri *or* 'græm 'kæləri] *noun* unit of measurement of heat *or* energy (NOTE: the **joule** is now more usual; also written **cal** after figures: **2,500 cal)**

◇ **Calorie** *or* **large calorie** ['kæləri] *noun* kilocalorie *or* 1,000 calories (NOTE: spelt with a capital; also written **Cal** after figures: **250 Cal)**

◇ **calorific value** [kælə'rɪfɪk 'vælju:] *noun* heat value of a substance *or* number of calories which a certain amount of a substance (such as a certain food) contains; ***the tin of beans has 250 calories*** *or* ***has a calorific value of 250 calories***

COMMENT: one calorie is the amount of heat needed to raise the temperature of one gram of water by one degree Celsius. A Calorie or kilocalorie is the amount of heat needed to raise the temperature of a kilogram of water by one degree Celsius. The Calorie is also used as a measurement of the energy content of food and to show the caloric requirement or amount of energy needed by an average person. The average adult in an office job requires about 3,000 Calories per day, supplied by carbohydrates and fats to give energy and proteins to replace tissue. More strenuous physical work needs more Calories. If a person eats more than the number of Calories needed by his energy output or for his growth, the extra Calories are stored in the body as fat

calyx ['keɪlɪks] *noun* part of a flower shaped like a cup, made up of the green sepals which cover the flower when it is in bud (NOTE: the plural is **calyces**)

camouflage ['kæməflɑ:ʒ] **1** *noun* hiding of an animal's shape by colours or patterns; ***the stripes on the zebra are a form of camouflage which makes the animal less easy to see in long grass*** **2** *verb* to hide the shape (of an animal) by using colours or patterns on the skin

can [kæn] *noun* metal container for food *or* drink, made of iron with a lining of tin *or* made entirely of aluminium; **can bank** = large container into which people put empty cans which are then recycled

COMMENT: aluminium cans are particularly valuable and can be collected separately. Other cans are made of ferrous metals and can be separated by testing with a magnet

Canadian deuterium-uranium reactor (CANDU) *noun* nuclear reactor using uranium oxide as a fuel and heavy water containing deuterium as a moderator and coolant. (There is a reactor of this type at Pickering, Ontario in Canada)

canal [kə'næl] *noun* (i) waterway made by people for ships to travel along; (ii) waterway made by people to take water to irrigate land

cancer ['kænsə] *noun* **(a) Tropic of Cancer** = parallel running round the earth at latitude 23°28N **(b)** malignant growth *or* tumour which develops in tissue and destroys it, which can spread by metastasis to other parts of the body and cannot be controlled by the body itself; ***he has been diagnosed as having lung cancer*** *or* ***as having cancer of the lung***

◇ **cancerous** ['kænsərəs] *adjective* referring to cancer; ***the X-ray revealed a cancerous growth in the breast***

COMMENT: cancers can be divided into cancers of the skin (carcinomas) or cancers of connective tissue, such as bone or muscle (sarcomas). Cancer can be caused by tobacco, radiation, certain diets and many other factors. There is evidence that constant exposure to the sun can cause cancer (or melanoma) of white skin. Depletion of the ozone layer in the atmosphere may increase the incidence of skin cancer. Many cancers are curable by surgery, by chemotherapy or by radiation, especially if they are detected early

candela [kæn'delə] *noun* base SI unit of measurement of light intensity (NOTE: usually written **cd** with figures)

candida ['kændɪdə] *noun* type of fungus which is normally present in the mouth and throat without causing any illness, but which can cause thrush

CANDU *noun* = CANADIAN DEUTERIUM-URANIUM REACTOR

canker ['kæŋkə] *noun* disease causing lesions on the skin, in the mouth, etc. or on the stem, etc., of a plant

◇ **cankered** ['kæŋkəd] *adjective* (skin *or* plant) having lesions

cannabis ['kænəbɪs] *noun* Indian hemp plant *(Cannabis sativa)* which produces an addictive drug (marijuana) (NOTE: also called **marijuana**)

canopy ['kænəpi] *noun* leaves of trees which act as an umbrella over the ground beneath; **canopy cover** = percentage of the surface of the ground which is covered by the leaves of trees

QUOTE: trees that grow to form the tallest part of the canopy suffer more damage than the slower growing trees forming the understorey
Guardian

canyon ['kænjən] *noun* deep valley with steep sides (usually in North America)

CAP = COMMON AGRICULTURAL POLICY

capacity [kə'pæsɪti] *noun* (i) ability to hold *or* absorb; (ii) amount which can be held *or* absorbed; ***what is the capacity of the grain storage bins?;*** **engine capacity** = total volume of a reciprocating engine's cylinders, expressed in litres, cubic centimetres or cubic inches; **water holding capacity** = ability of a material *or* substance, such as paper, peat, etc. to retain water

capillary [kə'pɪləri] *noun* tiny tube carrying a liquid in an organism

◇ **capillary action** *or* **capillary flow** [kə'pɪləri 'ækʃn *or* fləʊ] *noun* movement of a liquid upwards inside a narrow tube *or* upwards through the soil

◇ **capillarity** [kæpɪ'lærəti] *noun* = CAPILLARY ACTION

COMMENT: capillary flow is important in water in soil, as it does not drain away. It moves through the soil by capillary action, i.e., by the surface tension between the water and the walls of the fine tubes or capillaries. It is a very slow movement, and may not be fast enough to supply plant roots in a soil which is drying out

capital expenditure ['kæpɪtəl ɪk'spendɪtʃə] *noun* money spent on buying something *or* on improving something already owned; **capital goods** = items such as machines and equipment used to produce *or* manufacture goods (NOTE: the opposite of **capital goods** is **consumer goods**)

Capricorn ['kæprɪkɔːn] *noun* **Tropic of Capricorn** = parallel running round the earth at latitude 23°28S

capsid bug ['kæpsɪd bʌg] *noun* tiny insect which sucks the sap of plants

capsular ['kæpsjulə] *adjective* referring to a capsule

◇ **capsule** ['kæpsjuːl] *noun* **(a)** membrane round an organ *or* egg, etc. **(b)** dry seed case which bursts open to allow the seeds to shoot out

capture ['kæptʃə] *noun see* RIVER CAPTURE

caramel ['kærəmel] *noun* burnt sugar, a colouring additive (E150) which is believed to be carcinogenic

carbamate ['kɑːbəmeɪt] *noun* type of pesticide *or* insecticide developed to replace organochlorines such as DDT

> COMMENT: the advantage of carbamates over DDT is that they are not persistent and do not enter the human food chain

carbohydrate [kɑːbəʊ'haɪdreɪt] *noun* organic compound which derives from sugar and which is the main ingredient of many types of food

> COMMENT: carbohydrates are compounds of carbon, hydrogen and oxygen. They are found in particular in sugar and starch from plants, and provide the body with energy. Plants build up valuable organic substances from simple materials. The most important part of this process, which is called photosynthesis, is the production of carbohydrates such as sugars, starches and cellulose. They form the largest part of food of animals

carbon ['kɑːbən] *noun* one of the common non-metallic elements, an essential component of living matter and organic chemical compounds; **carbon cycle** *or* **circulation of carbon** = carbon atoms from carbon dioxide are incorporated into organic compounds in plants during photosynthesis. They are then oxidized into carbon dioxide again during respiration by the plants or by herbivores which eat them and by carnivores which eat the herbivores, thus releasing carbon to go round the cycle again; **carbon dating** *or* **carbon-14 dating** = process of finding out how old something is by analysing the amount of carbon dioxide in it which has decayed; **carbon fibre** = black thread of carbon which can be combined with other materials, such as metals and resins, to make them stronger; **carbon sink** = part of the biosphere (such as a tropical forest) which absorbs carbon, as opposed to animals which release carbon into the atmosphere in the form of carbon dioxide; **carbon tax** = amount of money added by a government to the price of fuel in an attempt to reduce greenhouse gas emissions; **carbon tetrachloride** (**CCl_4**) = liquid used a cleaning agent and insecticide (NOTE: chemical symbol is **C**; atomic number is **6**)

◇ **carbonaceous** [kɑːbə'neɪʃs] *adjective* (rock, such as coal) which is rich in hydrocarbons

◇ **carbonate** ['kɑːbəneɪt] **1** *noun* compound formed from a base and carbonic acid; **calcium carbonate** (**$CaCO_3$**) = chalk, a white mineral found widely in many parts of the world, formed from animal organisms **2** *verb* to add carbon dioxide to a drink to make it fizzy

◇ **carbonation** [kɑːbə'neɪʃn] *noun* adding carbon dioxide to a drink to make it fizzy

◇ **carbonization** [kɑːbənaɪ'zeɪʃn] *noun* process by which fossil plants have become carbon

carbon dioxide (**CO_2**) ['kɑːbən daɪ'ɒksaɪd] *noun* colourless gas produced when carbon is burnt with oxygen.

> COMMENT: Carbon dioxide exists naturally in air and is produced by burning or rotting organic matter. In animals, the body's metabolism makes the tissues burn carbon, which is then breathed out by the lungs as waste carbon dioxide. Carbon dioxide is removed from the atmosphere by plants when it is split by chlorophyll in photosynthesis to form carbon and oxygen. It is also dissolved from the atmosphere in sea water. The increasing release of carbon dioxide into the atmosphere, especially from burning fossil fuels, contributes to the greenhouse effect. Carbon dioxide is used in solid form as a means of keeping food cold. It is also used in fizzy drinks and has the E number 290. It is used as a coolant in some nuclear reactors

carbon monoxide (**CO**) ['kɑːbən mən'ɒksaɪd] *noun* poisonous gas found in fumes from car engines, from burning gas and cigarette smoke; **carbon monoxide poisoning** = poisoning caused by breathing carbon monoxide which starves the body of oxygen

> COMMENT: Carbon monoxide combines very easily with haemoglobin in blood to form carboxyhaemoglobin, which has the effect of starving the tissues of oxygen. It exists in tobacco smoke (see PASSIVE SMOKING) and in car exhaust fumes

carboxyhaemoglobin [kɑ:bɒksɪhi:mə'gləʊbɪn] *noun* compound of carbon monoxide and haemoglobin formed when a person breathes in carbon monoxide from car fumes or from ordinary cigarette smoke; **background carboxyhaemoglobin level** = level of carboxyhaemoglobin in the blood of a person living a normal existence without exposure to particularly high levels of carbon monoxide

carcin- ['kɑ:sɪn] *prefix* referring to cancer

◇ **carcinogen** [kɑ:'sɪnədʒən] *noun* substance which produces cancer

◇ **carcinogenesis** [kɑ:sɪnə'dʒenəsɪs] *noun* formation of cancer in tissue; **radiation carcinogenesis** = formation of cancer in tissue caused by exposure to a radioactive substance

◇ **carcinogenic** [kɑ:sɪnə'dʒenɪk] *adjective* which produces cancer

◇ **carcinoma** [kɑ:sɪ'nəʊmə] *noun* cancer of the skin *or* glands

COMMENT: carcinogens are found in pesticides such as DDT, in asbestos, aromatic compounds such as benzene, and radioactive substances

cardinal points ['kɑ:dɪnəl 'pɔɪnts] *plural noun* the four points on the horizon which are used to show direction (north, south, east and west)

carnivore *or* **carnivorous animal** ['kɑ:nɪvɔ: *or* kɑ:'nɪvərəs] *noun* animal which eats meat; **social carnivores** = meat-eating animals which live and hunt in groups, such as lions, wolves, etc.; *compare* HERBIVORE, OMNIVORE

◇ **Carnivora** [kɑ:'nɪvərə] *noun* order of meat-eating animals, including dogs, cats, bears and seals

◇ **carnivorous** [kɑ:'nɪvərəs] *adjective* which eats meat; **carnivorous plant** = plant which attracts insects, traps them and then digests their bodies

carpel ['kɑ:pəl] *noun* female part of a plant, formed of an ovary and stigma

carr [kɑ:] *noun* type of fen which supports some trees

carrageen ['kærəgi:n] *noun* Irish moss, a purplish-brown seaweed

◇ **carrageenan** [kærə'gi:nən] *noun* an extract of seaweed (E407) used as an emulsifier, but possibly carcinogenic

carrier ['kæriə] *noun* **(a)** person who carries bacteria of a disease in his body and who can transmit the disease to others without showing any sign of it himself **(b)** insect which carries disease and infects humans **(c)** healthy person who carries the chromosome defect of a hereditary disease (such as haemophilia)

◇ **carrier gas** *or* **carrier solvent** ['kæriə 'gæs *or* 'sɒlvənt] *noun* gas used in an aerosol can to make the spray come out

◇ **carrying capacity** ['kæriɪŋ kə'pæsɪti] *noun* maximum number of individuals of a species that can be supported in a given area

cascade [kæs'keɪd] **1** *noun* **(a)** small waterfall **(b)** system for purifying substances, where the substance passes through a series of identical processes, each stage increasing the level of purity **2** *verb (of liquid)* to fall down like a waterfall

case [keɪs] *noun* **(a)** hard outer cover; **case-hardened** = (soft steel) which has been given a hard outside layer by heating with carbon **(b)** (i) single occurrence of a disease; (ii) person who has a disease *or* who is undergoing treatment; ***there were two hundred cases of cholera in the recent outbreak;*** **case history** = details of what has happened to a patient undergoing treatment

cash crop ['kæʃ krɒp] *noun* crop which is grown to be sold rather than eaten by the person who grows it

QUOTE: in the Philippines as a whole, more than 30% of the total cultivated land area was given over to cash crop production for export in 1980, mainly bananas, pineapples and sugar cane
The Ecologist

caste [kɑ:st] *noun* social position *or* position in society; **caste system** = system where animals occupy different positions in a society (as in the case of bees, where there are three different groups - the queen, the workers (sterile females) and the drones (males)

catabolism [kə'tæbəlɪzm] *noun* breaking down of complex chemicals into simple chemicals

◇ **catabolic** [kætə'bɒlɪk] *adjective* referring to catabolism

catadromous [kə'tædrəməs] *adjective* (fish) which lives in fresh water and goes into the sea to spawn

◇ **catadromy** [kə'tædrəmi] *noun* migration of fish (such as eels) from fresh water to the sea for spawning; *see also* ANADROMY, ANADROMOUS

catalysis [kə'tælɪsɪs] *noun* process where a chemical reaction is helped by a substance (the catalyst) which does not change during the process

◇ **catalyst** ['kætəlɪst] *noun* substance which produces *or* helps a chemical process without itself changing; ***the sun acts as a catalyst in forming ozone; an enzyme which acts as a catalyst in the digestive process***

◇ **catalytic** [kætə'lɪtɪk] *adjective* referring to catalysis; **catalytic converter** *or also US*

catalytic muffler = device attached to the exhaust pipe of a motor vehicle which reduces the emission of carbon monoxide; **catalytic reaction** = chemical reaction which is caused by a catalyst which does not change during the reaction

◇ **catalyze** ['kætəlaɪz] *verb* to act as a catalyst *or* to help make a chemical process take place

COMMENT: a catalytic converter is a box filled with a catalyst, such as platinum. Converters can only be used on motor vehicles burning unleaded petrol, as the lead compounds in leaded petrol rapidly coat the catalyst in the converter and prevent it functioning

catch [kætʃ] **1** *noun* amount of fish caught; ***regulations to limit the herring catch;*** **total allowable catch (TAC)** = maximum amount of fish permitted to be caught **2** *verb* to hunt and take animals (usually fish)

QUOTE: more than 10,000 fishermen here used to produce about 11 per cent of the fish catch of the entire country

Guardian

catchment (area) ['kætʃmənt] *noun* **(a)** area of land which collects and drains the water which falls on it (such as the area round a lake or the basin of a river) **(b)** area around a school, hospital, shopping centre, etc., from which pupils, patients, customers, etc., come

catchwater drain ['kætʃwɔːtə 'dreɪn] *noun* type of drain designed to take rainwater from sloping ground

catena [kə'tiːnə] *noun* diagram showing the differences in soil caused by drainage

cation ['kætaɪən] *noun* ion with a positive electric charge; **cation exchange** = exchange which takes place when the ions of calcium, magnesium and other metals found in soil replace the hydrogen ions in acid

catkin ['kætkɪn] *noun* long structure consisting of many flowers (found on certain temperate trees such as hazel, birch and oak)

cattle ['kætl] *noun* domestic farm animals of the class Bovidae, raised for their milk and meat; ***the herdsmen drive their herds of cattle across the plain; thousands of cattle have died in the drought***

caustic ['kɒstɪk] *adjective* acid (substance); **caustic soda (NaOH)** = sodium hydroxide, a compound of sodium and water used to make soap and to clear blocked drains; **caustic lime (CaO)** = quicklime *or* calcium oxide

cave [keɪv] *noun* large hole under the ground, usually in rock

◇ **cavern** ['kævən] *noun* very large cave, formed by water which has dissolved limestone or other calcareous rock

cavity ['kævɪti] *noun* hole inside a solid substance; **cavity insulation** = insulation of cavity walls by putting insulating material inside them; **cavity wall** = wall made of two ranges of bricks, with a space in between them, giving greater insulation

Cd *chemical symbol for* cadmium

cd = CANDELA

cedar ['siːdə] *noun* American red wood *(Thuya plicata),* soft, but resistant to water and now becoming scarce. It is used mainly for outdoor construction work

ceiling ['siːlɪŋ] *noun* upper layer *or* level; **cloud ceiling** = height of cloud cover

-cele *or* **-coele** [siːl] *suffix* referring to a hollow

celestial [sɪ'lestiəl] *adjective* referring to the sky; **celestial equator** = imaginary line in the sky above the earth's equator

celi- *US* = coeli-

cell [sel] *noun* tiny unit of matter which is the base of all plant and animal tissue; **blood cell** = cell, a blood corpuscle (red blood cell *or* white blood cell) which is one of the components of blood; **red blood cell** = blood cell which contains haemoglobin and carries oxygen; **white blood cell** = blood cell which contains a nucleus, is formed in bone marrow and creates antibodies; **daughter cell** = one of the cells which develop by mitosis from a single parent cell; **mast cell** = large cell in connective tissue which carries histamine and reacts to allergens; **mother cell** *or* **parent cell** = original cell which splits into daughter cells during mitosis; **receptor cell** = cell which senses a change in the surrounding environment *or* in the body (such as cold, heat, pressure, pain) and reacts to it by sending an impulse through the nervous system to the brain; **cell division** = way in which a cell reproduces itself by mitosis; **cell membrane** = membrane enclosing the cytoplasm of a cell; **cell wall** = outside wall of a plant cell, formed of cellulose

◇ **cellular** ['seljʊlə] *adjective* **(a)** referring to cells *or* formed of cells; **cellular plant** = plant with no distinct stem, leaves, etc. **(b)** made of many similar parts connected together; **cellular tissue** = form of connective tissue with large spaces

COMMENT: the cell is a unit which can reproduce itself. It is made up of a jelly-like substance (cytoplasm) which surrounds a nucleus and contains many other small organisms which vary according to the type of cell. Cells reproduce by division (mitosis). Some microorganisms are formed of a single cell, but most organisms are formed by the division and reproduction of many millions of cells

cellulose ['seljuləuz] *noun* **(a)** carbohydrate which makes up a large percentage of plant matter, especially cell walls **(b)** chemical substance processed from wood, used for making paper, film and artificial fibres

COMMENT: cellulose is not digestible, and is passed through the digestive system as roughage

celom *US* = COELOM

Celsius ['selsiəs] *noun* scale of temperature where the freezing and boiling points of water are 0° and 100° (NOTE: used in many countries, but not in the USA, where the Fahrenheit system is still preferred. Normally written as a **C** after the degree sign: **52°C** (say: 'fifty-two degrees Celsius'). Was formerly called **centigrade**)

COMMENT: to convert Celsius temperatures to Fahrenheit, multiply by 1.8 and add 32. So 20°C is equal to 68°F

cement [sɪ'ment] *noun* **(a)** material which binds things together, such as that which binds minerals together to form sedimentary rocks **(b)** powder, which if mixed with water and then dried, sets hard like stone (used in building); **high-alumina cement** *or* **aluminous cement** = cement made of bauxite and limestone, used because it resists heat

COMMENT: various types of cement are used in construction. The commonest is made of burnt lime with clay and other mineral compounds. High-alumina cement is used in stressed concrete and is made from ground bauxite

centigrade ['sentɪgreɪd] *noun* scale of temperature where the freezing and boiling points of water are 0° and 100°; *see note at* CELSIUS

centimetre *or US* **centimeter** ['sentɪmi:tə] *noun* unit of measurement of length (one hundredth of a metre) (NOTE: with figures **centimetre** is usually written **cm: 10cm**)

central ['sentrəl] *adjective* referring to the centre; **central heating** = system for heating a building where hot water *or* hot air is circulated round from a single source of heat, usually a boiler; **central nervous system (CNS)** = the brain and spinal cord which link together all the nerves

◊ **centre** *or US* **center** ['sentə] *noun* **(a)** middle point *or* main part; ***the centre of the hurricane passed over the city; a park is to be built in the centre of the town*** **(b)** large building; **research centre** = place where scientific research is carried out

centrifugal [sentrɪ'fju:gəl] *adjective* which goes away from the centre; **centrifugal force** = force which pulls a body away from the centre of a curved path

◊ **centrifugation** *or* **centrifuging** [sentrɪfju'geɪʃn *or* 'sentrɪfju:dʒɪŋ] *noun* separating the components of a liquid in a centrifuge

◊ **centrifuge** ['sentrɪfju:dʒ] *noun* device to separate the components of a liquid

centriole ['sentriəʊl] *noun* small structure found in the cytoplasm of a cell, which forms asters during cell division

centripetal [sen'trɪpətəl] *adjective* which goes towards the centre; **centripetal force** = force which attracts a body towards the centre of a curved path

centrosome ['sentrəsəʊm] *noun* structure of the cytoplasm in a cell, near the nucleus, and containing the centrioles

cephal- [se'fæl] *prefix* referring to the head

◊ **cephalic** [sə'fælɪk] *adjective* referring to the head; **cephalic index** = measurement of the shape of the skull

◊ **cephalopods** *or* **Cephalopoda** ['sefələpɒds *or* sefəl'ɒpədə] *noun* class of mollusc (such as an octopus *or* squid) which has a large head and a shell inside the soft body

cereals ['sɪəriəlz] *noun* specialized types of grasses that are cultivated for their large seeds or grains, especially to make flour for breadmaking or for animal feed; ***the European Union grows large quantities of cereals***

COMMENT: cereal plants are all members of the Graminales family or grasses. The commonest are oats, wheat, barley, maize and rye in colder temperate areas, and rice and sorghum in warmer regions. Other cereals of local importance are millet, teff in Ethiopia and adlay in India. Cereal production has considerably expanded and improved with the introduction of better methods of sowing, combine harvesters, driers, bulk handling and chemical aids such as herbicides, fungicides, insecticides and growth regulators

cerebellar [serə'belə] *adjective* referring to the cerebellum

◊ **cerebellum** [serə'beləm] *noun* section of the brain located at the back of the head, beneath the back part of the cerebrum

COMMENT: the cerebellum is formed of two hemispheres, with the vermis in the centre. The cerebellum is the part of the brain where voluntary movements are coordinated and is associated with the sense of balance

cerebr- *or* **cerebro-** ['serəb *or* 'serəbrəʊ] *prefix* referring to the cerebrum

◇ **cerebral** ['serəbrəl] *adjective* referring to the cerebrum *or* to the brain in general; **cerebral hemisphere** = one of the two halves of the cerebrum

cerebrospinal [serəbrəʊ'spaɪnəl] *adjective* referring to the brain and the spinal cord; **cerebrospinal fluid (CSF)** = fluid which surrounds the brain and the spinal cord

COMMENT: CSF is found in the space between the arachnoid mater and pia mater of the brain, between the ventricles of the brain and in the central canal of the spinal cord. CSF consists mainly of water, with some sugar and sodium chloride. Its function is to cushion the brain and spinal cord

cerebrum ['serəbrəm] *noun* main part of the brain

COMMENT: the cerebrum is the largest part of the brain, formed of two sections (the cerebral hemispheres) which run along the length of the head. The cerebrum controls the main mental processes, including the memory

certificate [sə'tɪfɪkət] *noun* official paper which states something; **birth certificate** = paper giving details of a person's date and place of birth and parents; **death certificate** = paper signed by a doctor, stating that a person has died, giving details of the person and the cause of death

◇ **certify** ['sɜːtɪfaɪ] *verb* to make an official statement in writing

cesspool *or* **cesspit** ['sespuːl *or* 'sespɪt] *noun* tank for household sewage, constructed in the ground near a house which is not connected to the main drainage system, and in which the waste is stored, then pumped out for disposal somewhere else

Cetacea *or* **cetaceans** [sɪ'teɪʃə *or* sɪ'teɪʃnz] *noun* order of large mammals, such as dolphins, porpoises and whales, which live in the sea

CFC = CHLOROFLUOROCARBON

CFM = CHLOROFLUOROMETHANE

chain [tʃeɪn] *noun* (i) number of metal rings attached together to make a line; (ii) number of components linked together *or* number of connected events; **food chain** = series of organisms which pass energy from one to another as each provides food for the next. The first organism in the food chain is the producer and the rest are consumers; **chain reaction** = (i) nuclear reaction where a neutron hits a nucleus, makes it split, and so releases further neutrons; (ii) chemical reaction where each stage is started by a chemical substance which reacts with another, producing further substances which can continue to react

chalk [tʃɔːk] *noun* fine white limestone rock formed of calcium carbonate

◇ **chalkpit** ['tʃɔːkpɪt] *noun* hole in the ground to extract chalk from

◇ **chalky** ['tʃɔːki] *adjective* (soil) which is full of chalk

COMMENT: chalk is found widely in many parts of Northern Europe. Formed from animal organisms it is also used as an additive (E170) in white flour

channel ['tʃænəl] **1** *noun* **(a)** deep part of a harbour or sea passage where ships can pass; stretch of water between two seas; **the English Channel** = stretch of sea lying between England and France **(b)** bed of a river *or* the ground across which a river or stream flows; **drainage channel** = small ditch made to remove rainwater from the soil surface; **storm channel** = large drain for taking away heavy storm water **2** *verb* to send (water) in a particular direction (NOTE: UK English is **channelled, channelling** while US English is **channeled, channeling**)

◇ **channelization** [tʃænəlaɪ'zeɪʃn] *noun* process of straightening a stream which has many bends, so as to make the water flow faster

◇ **channelize** ['tʃænəlaɪz] *verb* to straighten a stream which has many bends, so as to make the water flow faster

character ['kærəktə] *noun* way in which a person thinks and behaves; **acquired character** = character which develops in reaction to the environment; **hereditary character** = character which is transmitted from parents to offspring

◇ **characteristic** [kærəktə'rɪstɪk] **1** *adjective* typical *or* special; **characteristic species** = species typical of and only occurring in a certain region **2** *noun* quality *or* trait that distinguishes something; ***one of the characteristics of the octopus is that it can send out a cloud of ink when attacked***

◇ **characterize** ['kærəktəraɪz] *verb* to be a characteristic of; ***deserts are characterized by little rainfall, arid soil and very little vegetation***

charcoal ['tʃɑːkəʊl] *noun* impure form of carbon, formed when wood is burnt in the absence of oxygen; **activated charcoal** = form of carbon to which gases can stick, used in gas masks or as a filter to control pollution or added to water as it is being treated before domestic consumption

charge [tʃɑːdʒ] *noun* cost, price; **effluent charge** = fee paid by a company to be allowed to discharge waste into the sea *or* a river; **noise charge** = fee paid by a company to be allowed to make a certain amount of noise in the course of its business; **pollution charges** = cost of repairing *or* stopping environmental pollution

chart [tʃɑːt] **1** *noun* **(a)** map of an area of water, such as the sea *or* a large lake **(b)** diagram *or* record of information shown as a series of lines *or* points on graph paper; ***a chart showing the increase in cases of lung cancer; this chart shows the extent of forest areas affected by acid rain;*** **temperature chart** = chart showing changes in temperature over a period of time; **weather chart** = chart showing the state of the weather at a particular moment *or* changes which are expected to happen in the weather in the near future **2** *verb* to make a chart

chem- [kem] *prefix* referring to chemistry *or* chemicals

◇ **chemical** ['kemɪkl] **1** *adjective* referring to chemistry; **chemical agent** = substance which makes another substance react; **chemical closet** *or* **chemical toilet** = toilet closet used without running water, in which faeces are covered with chemicals; **chemical compound** = substance formed from two or more chemical elements, in which the proportions of the elements are always the same; **chemical elements** = the 105 different substances which exist independently and cannot be broken down to simpler substances; *see* PERIODIC TABLE; **chemical engineering** = branch of engineering which deals with the design, construction and repair of the machines and equipment used in industrial chemical processes; **chemical factory** = building where chemicals are manufactured; **chemical mutagen** = chemical substance which causes mutation; **chemical name** = technical name for a substance; ***sodium hydroxide is the chemical name for caustic soda;*** **chemical oxygen demand (COD)** = amount of oxygen taken up by organic matter in water (used as a measurement of the amount of organic matter in sewage); **chemical residue** = waste left after a chemical process has taken place; **chemical symbol** = letter (or letters) used to indicate an element. (Used especially in formulae: C for carbon, Co for cobalt, etc.); **chemical warfare** = war using chemical weapons; **chemical warhead** = front part of a missile containing a poisonous chemical; **chemical weapon** = missile, etc. containing a poisonous chemical, such as a nerve gas, defoliant, etc. **2** *noun* substance produced by a chemical process *or* formed of chemical elements; ***the widespread use of chemicals in agriculture;*** **process chemical** = chemical which is manufactured by industrial process

◇ **chemist** ['kemɪst] *noun* **(a)** scientist who specializes in the study of chemistry **(b)** **dispensing chemist** = pharmacist who prepares and sells drugs according to doctors' prescriptions

◇ **chemistry** ['kemɪstri] *noun* **(a)** study of substances, elements and compounds and their reactions with each other; **blood chemistry** *or* **chemistry of the blood** = (i) record of the changes which take place in blood during disease and treatment; (ii) substances which make up blood, which can be analysed in blood tests, the results of which are useful in diagnosing disease **(b)** chemical substances existing together; ***human action has radically altered the chemistry of the atmosphere***

chemo- ['keməʊ *or* 'kiːməʊ] *prefix* referring to chemistry

◇ **chemoautotrophic bacteria** [keməʊ'ɔːtəʊtrɒfɪk bæk'tiːəriə] *plural noun* bacteria which make protoplasm using energy from chemical (as opposed to phototrophic) reactions

◇ **chemolithotrophic** [keməʊ'lɪθəʊtrɒfɪk] *adjective* (organisms, such as bacteria) which obtain energy from inorganic substances

◇ **chemo-organotrophic** [keməʊ'ɔːgənəʊtrɒfɪk] *adjective* (organism) which obtains its energy from organic sources. (All animals are chemo-organotrophic)

◇ **chemoreceptor** [keməʊrɪ'septə] *noun* cell which responds to the presence of a chemical compound by activating a sensory nerve (such as a taste bud reacting to food); *see also* EXTEROCEPTOR, INTEROCEPTOR, RECEPTOR

◇ **chemosphere** ['keməʊsfiːə] *noun* zone in the earth's atmosphere, above the upper part of the troposphere and within the stratosphere, where chemical changes take place under the influence of the sun's radiation

◇ **chemosterilant** [keməʊ'sterɪlənt] *noun* chemical substance which sterilizes (by killing microbes *or* bacteria)

◇ **chemosynthesis** [kiːməʊ'sɪnθesɪs] *noun* production by bacteria of organic material using chemical reaction

◇ **chemosynthetic bacteria** [kiːməʊsɪn'θetɪk bæk'tiːəriə] *plural noun* = CHEMOAUTOTROPHIC BACTERIA

◇ **chemotaxis** [kiːməʊ'tæksɪs] *noun* movement of a cell which is attracted to *or* repelled by a chemical substance

◇ **chemotherapy** [kiːməʊ'θerəpi] *noun* using chemical drugs (such as antiobiotics *or* painkillers *or* antiseptic lotions) to fight a disease, especially using toxic chemicals to destroy rapidly developing cancer cells

◇ **chemotrophic** [kiːməʊ'trɒfɪk] *adjective* (organism) which takes energy not from light, but from other sources such as organic matter. (Almost all animals are chemotrophic, while plants are phototrophic)

◇ **chemotropism** [kiːməʊ'trɒpɪzm] *noun* growth of a plant caused by stimulation from a chemical

Chernobyl [tʃɜː'nɒbəl] *noun* name of a large nuclear power station in the Ukraine, the scene of a disastrous fire in 1986

chernozem ['tʃɜːnəʊzem] *noun* dark fertile soil, full of organic matter, above a lighter lime soil

COMMENT: chernozem is found in the temperate grass-covered plains of Russia and North and South America

chestnut ['tʃesnʌt] *noun* **sweet chestnut** = European hardwood tree *(Castanea sativa)* important for its nuts and timber

chill [tʃɪl] *noun* coldness; **wind chill factor** = way of calculating the risk of exposure in cold weather by adding the speed of the wind to the number of degrees of temperature below zero

◇ **chillshelter** ['tʃɪlʃeltə] *noun* form of feedlot, where a small area of land for the intensive fattening of cattle is surrounded by a high embankment to protect the cattle against the cold

chimney ['tʃɪmni] *noun* tall tube, either inside a building or separate from it, which takes smoke and fumes away from a fire; ***smoke poured out of the factory chimneys; chimneys were built very tall, so that the polluting smoke would be carried a long way away by the wind;*** **chimney stack** = very tall chimney, as in a factory, usually containing several flues

china clay ['tʃaɪnə 'kleɪ] *noun* kaolin, a fine white clay used for making china and also for coating shiny paper; ***spoil heaps from china clay workings are bright white***

China syndrome ['tʃaɪnə 'sɪndrəʊm] *noun* imaginary potential disaster where the core of a nuclear reactor overheats

COMMENT: according to the China syndrome, the core of the reactor melts its surroundings, and sinks into the earth with such energy that it would supposedly burn its way through to China (the antipode of the USA). In reality it would not burn its way so far, but would heavily contaminate the environment of the reactor

chinook [tʃɪ'nuːk] *noun* warm wind which blows from the Rocky Mountains down onto the Canadian plains in winter

chlorella [klɔː'relə] *noun* very small green freshwater algae, occurring naturally *or* grown commercially for use in human food

chlor(o)- ['klɔːrəʊ] *prefix* (i) referring to chlorine; (ii) green

chlordane *or* **chlordan** **($C_{10}H_6Cl_8$)** ['klɔːdeɪn] *noun* poisonous substance used as an insecticide, especially to kill earthworms in turf. (Its use in the UK is now limited because of its harmful side effects)

chloride ['klɔːraɪd] *noun* salt which is a compound of chlorine; **sodium chloride (NaCl)** = common salt

chlorinate ['klɒrɪneɪt] *verb* to treat something with chlorine, especially, to sterilize drinking water *or* water in a swimming pool by adding chlorine

◇ **chlorinated** ['klɒrɪneɪtɪd] *adjective* treated with chlorine; **chlorinated hydrocarbon insecticide** = organochlorine *or* type of insecticide made synthetically as a compound of chlorine

COMMENT: chlorinated hydrocarbon insecticides include DDT, aldrin and lindane. These types of insecticide are very persistent, with a long half-life of up to 15 years, while organic phosphorous insecticides have a much shorter life. Chlorinated hydrocarbon insecticides not only kill insects, but also enter the food chain and kill small animals and birds which feed on the insects

chlorination [klɒrɪ'neɪʃn] *noun* sterilizing by adding chlorine

◇ **chlorinator** ['klɒrɪneɪtə] *noun* apparatus for adding chlorine to water

COMMENT: chlorination is used to kill bacteria in drinking water, in swimming pools and sewage farms, and has many industrial applications such as sterilization in food processing

chlorine ['klɔːriːn] *noun* powerful greenish gas, used to sterilize water and to bleach; **chlorine demand** = amount of chlorine needed to kill bacteria in a given quantity of sewage; **chlorine monoxide** = substance which forms in the stratosphere and destroys ozone (NOTE: chemical symbol is **Cl;** atomic number is **17)**

QUOTE: chlorine monoxide, at the concentrations found over the Antarctic in spring, destroys ozone at a rate of about 2 per cent per day
New Scientist

chlorofluorocarbon (CFC) *or* **chlorofluoromethane (CFM)** [klɔːrəʊfluːərəʊ'kɑːbən *or* klɔːrəʊfluːərəʊ'miːθeɪn] *noun* compound of fluorine and chlorine, used as a propellant in aerosol cans, in the manufacture of plastic foam boxes for takeaway food, as a refrigerant in refrigerators and air conditioners and as cleaners of circuit boards for computers

COMMENT: CFCs are classified by numbers: CFC-10 is used in aerosols; CFC-11 is used to make plastic foam; CFC-12 is a coolant for refrigerators; CFC-13 is the cleaning substance used in the electronics industry. When CFCs are released into the atmosphere, they rise slowly taking about seven years to reach the stratosphere. But once they are there, under the influence of the sun's ultraviolet light they break down into chlorine atoms which destroy the ozone layer. This allows harmful solar UV radiation to pass through to the earth's surface. Because it takes so long for the CFCs to reach the stratosphere, any reduction in their use on earth does not have an immediate effect on the concentrations in the stratosphere. Replacements for CFCs are being developed, which should reduce the problem eventually. It is a pity that CFCs have this unfortunate effect on the ozone layer, since in other ways they are ideal gases, being stable and non-toxic

QUOTE: he plans to publish lists of aerosols detailing their use of CFCs. As products become CFC-free, they will be removed from the list
Independent

QUOTE: the electronics industry accounts for some 45% of the use of CFC-13, which is used mainly to clean printed circuit boards
New Scientist

QUOTE: it remains to be seen if alternatives will prove satisfactory for use as solvents in electronics manufacture or as blowing agents for making cushioning foams - two of the principal uses of CFCs at present. CFCs are valuable in these areas because they are relatively non-toxic, non-flammable and stable
Environment Now

QUOTE: no other chemicals in the home are as harmless as CFCs, and few so useful in their role as the working fluids of refrigerators
New Scientist

chloromethane [klɔːrəʊ'miːθeɪn] *noun* gas (compound of carbon and chlorine) formed by fungi as they rot wood, which acts in a similar way to CFCs in depleting the ozone layer

chlorophyll ['klɒrəfɪl] *noun* green pigment in plants

◇ **Chlorophyta** [klɒr'ɒfɪtə] *noun* genus of green algae, the largest class of algae

◇ **chlorosis** [klɒ'rəʊsɪs] *noun* reduction of chlorophyll in plants, making the leaves turn yellow

COMMENT: chlorophyll absorbs light energy from the sun and supplies plants with the energy to enable them to carry out photosynthesis. It is also used as a colouring (E140) in processed food

choke [tʃəʊk] *verb* to kill a living organism by cutting off air or other life-giving substances; ***the reservoirs became choked with rapidly-growing weeds; the centre of the town is choked with traffic***

cholera ['kɒlərə] *noun* serious bacterial disease spread through food *or* water which has been infected by *Vibrio cholerae*; ***a cholera epidemic broke out after the flood***

COMMENT: the infected person suffers diarrhoea, cramp in the intestines and dehydration. The disease is often fatal and vaccination is only effective for a relatively short period

cholesterol [kə'lestərɒl] *noun* fatty substance found in fats and oils, also produced by the liver, and forming an essential part of all cells

COMMENT: cholesterol is formed by the body and is found in brain cells, the adrenal glands, liver and bile acids. High levels of cholesterol in the blood are found in diabetes. High blood cholesterol levels are associated with diets rich in animal fat (such as those found in butter and fat meat). Excess cholesterol can be deposited in the walls of arteries, causing atherosclerosis

CHP = COMBINED HEAT AND POWER

◇ **CHP plant** *or* **station** [siːeɪtʃ'piː 'plɑːnt] *noun* power station which produces both electricity and hot water for the local population. A CHP plant may operate on almost any fuel, including refuse

chrom- [krɒm] *prefix* (i) indicating colour; (ii) indicating chromium

chromatid ['krəʊmətɪd] *noun* one of two parallel filaments making up a chromosome

◇ **chromatin** ['krəʊmətɪn] *noun* network which forms the nucleus of a cell and can be stained with basic dyes; **sex chromatin** = chromatin which is only found in female cells and which can be used to identify the sex of a baby before birth

chromatography [krəʊmə'tɒgrəfi] *noun* scientific method for separating and analysing chemicals through a porous medium; **column chromatography** = chromatography using a column of powder as the porous medium; **gas chromatography** = scientific method for analysing a mixture of volatile substances; **paper chromatography** = chromatography using a strip of paper as the porous medium

chromatophore [krəʊ'mætəfɔː] *noun* **(a)** any cell which contains pigment, such as the cells in the eyes, hair and skin, or the cells by which an animal such as the chameleon can change colour **(b)** cell in plants *or* algae which contains chlorophyll

chromium ['krəʊmiəm] *noun* metallic trace element, used to make alloys (NOTE: chemical symbol is **Cr;** atomic number is **24)**

chromo- ['krəʊməʊ] *prefix* (i) indicating colour; (ii) indicating chromium

chromosome ['krəʊməsəʊm] *noun* rod-shaped structure in the nucleus of a cell, formed of DNA which carries the genes

◇ **chromosomal** [krəʊmə'səʊməl] *adjective* referring to chromosomes

COMMENT: each species has its own chromosomal make-up. Cattle have 60 chromosomes, horses 64, and chickens 78. Plants have fewer chromosomes than animals: carrots have 18, maize 20, rice 24, etc. The human cell has 46 chromosomes, 23 inherited from each parent. Among these are the X and Y chromosomes, which are responsible for sexual differences. The female has one pair of X chromosomes and the male one X and one Y chromosome. Sperm carry either an X or a Y chromosome. If a sperm with a Y chromosome fertilizes a female's ovum the resulting offspring will be a male (XY); otherwise it will be a female (XX)

chronic ['krɒnɪk] *adjective* (disease *or* condition) which lasts for a long time; ***the forest was suffering from chronic soil acidification*** *compare* ACUTE

chrysalis ['krɪsəlɪs] *noun* stage in the development of a butterfly *or* moth, the hard case in which the pupa is protected

Chrysophyta [krɪsə'fiːtə] *noun* golden-brown algae, which store oil in their cell walls

chute [ʃuːt] *noun* **(a)** sloping channel along which water, rubbish, etc. may pass **(b)** waterfall

Ci = CURIE

-cide [saɪd] *suffix* referring to killing; **herbicide** = substance which kills plants; **insecticide** substance which kills insects

cilia ['sɪliə] *see* CILIUM

◇ **ciliary** ['sɪliəri] *adjective* referring to cilia; **ciliary feeder** = mollusc which feeds by sucking water and the organisms it contains through cilia

◇ **ciliated epithelium** [sɪli'eɪtɪd epɪ'θeliəm] *noun* simple epithelium where the cells have tiny hairs *or* cilia

◇ **cilium** ['sɪliəm] *noun* one of many tiny hair-like processes which line cells in passages in the body and, by moving backwards and forwards, drive particles or fluid along the passage (NOTE: plural is **cilia)**

cinders ['sɪndəz] *plural noun* material left when the flames from a burning substance have gone out; small pieces of lava and rock thrown up in a volcanic eruption

circadian rhythm [sɜː'keɪdiən 'rɪðm] *noun* rhythm of daily activities and bodily processes (eating *or* defecating *or* sleeping, etc.) frequently controlled by hormones, which repeats every twenty-four hours and is found both in plants and animals

circulate ['sɜːkjuleɪt] *verb (of air or fluid)* to move around in a circle; ***blood circulates around the body***

◇ **circulation** [sɜːkjuː'leɪʃn] *noun* (i) movement of air in a circle; (ii) movement of blood around the body from the heart through the arteries to the capillaries and back to the heart through the veins; ***a circulation of cold air across the British Isles;*** **circulation of carbon** = carbon atoms from carbon dioxide are incorporated into organic compounds in plants during photosynthesis. They are then oxidized into carbon dioxide again during respiration by the plants or by herbivores which eat them and by carnivores which eat the herbivores, so releasing carbon to go round the cycle again

◇ **circulatory** [sɜːkjuː'leɪtəri] *adjective* referring to the circulation of the blood; **circulatory system** = system of arteries and veins which, together with the heart, makes the blood circulate around the body

COMMENT: blood circulates around the body, carrying oxygen from the lungs and nutrients from the liver through the arteries and capillaries to the tissues. The capillaries exchange the oxygen for waste matter such as carbon dioxide which is taken back to the lungs to be expelled. At the same time the blood obtains more oxygen in the lungs to be taken to the tissues

circum- ['sɜːkəm] *prefix* meaning around

◇ **circumpolar** [sɜːkəm'pəʊlə] *adjective* around the North *or* South Pole; **circumpolar vortex** = circular movement of air around one of the poles

cirque [sɜːk] *noun* scoop-shaped hollow formed in mountains or high plateaux by small, separate glaciers (NOTE: also called **corrie, cwm)**

cirrocumulus [sɪrəʊ'kjuːmjʊləs] *noun* form of high cloud, occurring above 5,000m, like altocumulus with little clouds

◊ **cirrostratus** [sɪrəʊ'strɑːtəs] *noun* high layer of cloud veil

◊ **cirrus** ['sɪrəs] *noun* high cloud, occurring above 5,000m, forming a mass of separate clouds which look as if they are made of fibres, but which are, in fact, formed of ice crystals

CITES = CONVENTION ON INTERNATIONAL TRADE IN ENDANGERED SPECIES OF WILD FAUNA AND FLORA

citric acid ['sɪtrɪk 'æsɪd] *noun* acid found in fruit such as oranges, lemons and grapefruit

◊ **citric acid cycle** ['sɪtrɪk 'æsɪd saɪkl] *noun* Krebs cycle, an important series of reactions in which the intermediate products of fats, carbohydrates and amino acid metabolism are converted to carbon dioxide and water in the mitochondria

Cl *chemical symbol for* chlorine

cladding ['klædɪŋ] *noun* material which is put round something to protect it, such as material put on the outside of buildings, or round pipes or tanks to prevent heat loss; ***the hot water tank needs to be insulated with at least 10cm of cladding; cladding is used in nuclear reactors to separate the nuclear fuel rods from the coolant***

clarification [klærɪfɪ'keɪʃn] *noun* process of removing solid waste matter from sewage; **clarification basin** = tank in which solid waste matter is removed from sewage

class [klɑːs] *noun* one of the divisions into which organisms are categorized; ***groups of families are classified into orders and groups of orders into classes***

◊ **classify** ['klæsɪfaɪ] *verb* **(a)** to arrange in classes; ***the area has been classified as an Area of Outstanding Natural Beauty*** **(b)** to put things into order so as to be able to refer to them again and identify them easily; ***the pollution records are classified under the name of the site which has been polluted; blood groups are classified according to the ABO system*** **(c)** to make information secret; ***the reports on the accident are classified and may not be consulted by the public***

◊ **classification** [klæsɪfɪ'keɪʃn] *noun* system by which things are put into order so as to be able to refer to them again and identify them easily; ***the Linnaean classification of plants and animals; the ABO classification of blood***

clay [kleɪ] *noun* type of heavy, non-porous soil made of fine particles of silicate; **clay land** = area of land where the soil has a lot of clay in it

◊ **clayey** ['kleɪi] *adjective* containing clay; ***these plants do best in clayey soils***

◊ **claypan** ['kleɪpæn] *noun* hollow on the surface of clay land where rain collects

clean [kliːn] **1** *adjective* not dirty; **clean air** = air which does not contain impurities *or* air which is free from pollution; **clean power station** = power station which gives off little pollution *or* radiation; **clean technology** = industrial process which causes little or no pollution **2** *verb* to make clean by taking away dirt *or* impurities; **gas cleaning** = removing pollutants from gas, especially from emissions from factories and power stations

◊ **cleanse** [klenz] *verb* to make very clean

◊ **cleanser** ['klenzə] *noun* powder *or* liquid which cleanses

◊ **clean up** ['kliːn 'ʌp] *verb* to remove refuse *or* waste substances *or* pollutants from a place; ***they are working to clean up the beaches after the oil spill***

◊ **cleaning up** *or* **clean-up** ['kliːnɪŋ 'ʌp *or* 'kliːnʌp] *noun* action of removing refuse *or* waste substances *or* pollutants; ***the local authorities have organized a large-scale clean-up of polluted beaches;*** **clean-up costs** = amount of money which has to be spent to remove pollutants as, for example, after an environmental disaster such as an oil spill

clear [klɪə] **1** *adjective* **(a)** easily understood; ***there's no clear way of explaining the problem*** **(b)** which is not cloudy, which you can easily see through; ***acid rain turns lakes crystal clear;*** **clear air** = air with no mist or smoke; **clear sky** = sky with no clouds **(c)** **clear of** = free from; ***the area is now clear of pollution*** **2** *verb* to remove something which is in the way in order to make space for something else; ***they cleared hectares of jungle to make a new road to the capital; we are clearing rainforest at a faster rate than before***

◊ **clearance** ['klɪərəns] *noun* act of removing something which is in the way in order to make space for something else

◊ **clearcut** ['klɪəkʌt] **1** *noun* cutting down of all the trees in an area **2** *verb* to clear an area of forest by cutting down all the trees; *see also* CUT

◊ **clearcutting** *or* **clearfelling** [klɪə'kʌtɪŋ *or* klɪə'felɪŋ] *noun* cutting down all the trees in an area at the same time; ***the greatest threat to wildlife is the destruction of habitats by clearfelling the forest for paper pulp***

> COMMENT: clearcutting can be a way of managing a forest: once the felled timber has been removed, the land is cleared of stumps and roots and then sown with new tree seed. But clearcutting can cause environmental damage if the felled trees are not replaced

◇ **clearfell** [klɪə'fel] *verb* to clear an area of forest by cutting down all the trees

cliff [klɪf] *noun* high wall of rock by the sea *or* along a canyon; ***many types of birds nest on ledges in the cliffs***

climate ['klaɪmət] *noun* general weather of a certain place; **continental climate** = type of climate found in the centre of a large continent away from the sea, with long dry summers, very cold winters and not much precipitation; **temperate climate** = type of climate which is neither very hot in summer nor very cold in winter

◇ **climatic** [klaɪ'mætɪk] *adjective* referring to climate; **climatic climax** = climax which is controlled by climatic factors; **climatic factors** = conditions of climate which affect the organisms living in a certain area

◇ **climatic zone** [klaɪ'mætɪk 'zəʊn] *noun* one of the eight areas of the earth which have distinct climates

COMMENT: the climatic zones are: the Arctic (covering the two polar regions); the boreal in the Northern Hemisphere, south of the Arctic; two temperate zones in the Northern and Southern Hemispheres; two subtropical zones, including the deserts; and the equatorial zone which has a damp tropical climate

climatological [klaɪmətə'lɒdʒɪkl] *adjective* referring to climatology; ***scientists have gathered climatological statistics from weather stations all round the world;*** **climatological data** = information, statistics, etc. about climate; **climatological station** = scientific research centre where the climate is studied

◇ **climatologist** [klaɪmə'tɒlədʒɪst] *noun* scientist who specializes in the study of the climate

◇ **climatology** [klaɪmə'tɒlədʒi] *noun* scientific study of the climate

COMMENT: climate is influenced by many factors, in particular the latitude of the region, whether it is near the sea or far inland, the altitude, etc.

climax ['klaɪmæks] *noun* final stable state in the development of an ecosystem; **edaphic climax** = climax caused by the type of soil in a certain area; **climax community** = plant community which has been stable for many years and which is unlikely to change unless the climate changes or there is human interference

cline [klaɪn] *noun* changes which take place in a species according to the environment in which it lives; *see also* ECOCLINE, GEOCLINE

clinker ['klɪŋkə] *noun* lumps of ash and hard residue from furnaces, used to make road surfaces or breeze blocks

clisere ['kliːsɜː] *noun* succession of communities influenced by the climate of an area

clone [kləʊn] **1** *noun* (i) group of cells derived from a single cell by asexual reproduction and so identical to the first cell; (ii) organism produced asexually (as by taking cuttings from a plant) (NOTE: the word originally referred to a group of cells derived from the same source, but is now used in the singular to refer to a single organism) **2** *verb* to reproduce an individual organism by asexual means

◇ **cloning** ['kləʊnɪŋ] *noun* reproduction of an individual organism by asexual means

Clostridium [klɒst'rɪdiəm] *noun* type of bacterium

COMMENT: species of Clostridium cause botulism, tetanus and gas gangrene, but also increase the nitrogen content of soil

cloud [klaʊd] *noun* **(a)** mass of water vapour *or* ice particles in the sky which can produce rain; **storm cloud** = dark-coloured cloud in which vigorous activity produces heavy precipitation and sometimes other pronounced meteorological effects such as squalls of wind **(b)** mass of particles suspended in the air; ***clouds of smoke poured out of the factory chimney; the eruption sent a cloud of ash high into the atmosphere; dust clouds swept across the plains*** (NOTE: as a plural, **clouds** means several separate clouds; otherwise the singular **cloud** can be used to refer to a large continuous mass: **there is a mass of cloud over the southern part of the country**)

◇ **cloudbank** ['klaʊdbæŋk] *noun* mass of low clouds

◇ **cloudbase** ['klaʊdbeɪs] *noun* bottom part of a layer of cloud

◇ **cloudburst** ['klaʊdbɜːst] *noun* sudden rainstorm

◇ **cloud chamber** ['klaʊd 'tʃeɪmbə] *noun* piece of laboratory equipment in which clouds can be formed for the study of ionization

◇ **cloud cover** ['klaʊd 'kʌvə] *noun* amount of sky which is covered by clouds; *see also* OKTA

◇ **cloudlayer** ['klaʊdleɪə] *noun* mass of clouds at a certain height above the land

◇ **cloudless** ['klaʊdləs] *adjective* (sky) which has no clouds

◇ **cloudy** ['klaʊdi] *adjective* (sky) which is covered with clouds

COMMENT: clouds are formed as humid air rises and then cools, causing the water in it to condense. They are classified by meteorologists into ten categories: high clouds (cirrus, cirrocumulus, cirrostratus); middle clouds (altocumulus and altostratus); low clouds (stratocumulus, nimbostratus, cumulus, cumulonimbus and stratus)

clump [klʌmp] **1** *noun* group of particles which stick together **2** *verb (of particles)* to stick together in groups

cm = CENTIMETRE

CNS = CENTRAL NERVOUS SYSTEM

Co *symbol for* cobalt

coagulate [kəʊ'ægju:leɪt] *verb (of a liquid)* to become semi-solid as suspended particles clump together; ***blood coagulates in contact with air***

◇ **coagulant** [kəʊ'ægjʊlənt] *noun* substance which can make blood coagulate

◇ **coagulation** [kəʊægjʊ'leɪʃn] *noun (of a liquid)* becoming semi-solid

coal [kəʊl] *noun* solid black organic substance found in layers underground in most parts of the world, burnt to provide heat or power; **brown coal** = lignite, a type of soft coal which is not as efficient a fuel as anthracite and produces more smoke when it burns; **gas coal** = coal used for making coal gas; **high-sulphur coal** = coal with a lot of sulphur in it, therefore producing more sulphur dioxide when it is burnt; **low-sulphur coal** = coal with little sulphur in it, therefore producing less sulphur dioxide when it is burnt; **soft coal** = coal which is not as efficient a fuel as anthracite and produces more smoke when it burns; **coal basin** = part of the earth's surface containing layers of coal; **coal-fired power station** = power station which burns coal, as opposed to a gas-fired station or nuclear power station; **coal gas** = gas produced by processing coal, giving coke as a residue; **coal gasification** = process of converting coal into gas to be used as fuel in gas-fired power stations; **coal mine** = hole dug in the ground to extract coal; **coal seam** = layer of coal in the rocks beneath the earth's surface; **coal tar** = any of several liquids formed by distillation of coal, used in the pharmaceutical industry and as a wood preservative

COMMENT: coal was formed many millions of years ago from organic refuse deposited in swamps. Decaying plants formed peat, which was pressed by other layers of deposits to form first lignite (or brown coal) and then coal itself. Coal is composed mainly of carbon. It can be classified into various grades. Lignite (or brown coal) is soft and not a very efficient fuel. Bituminous coals are harder and contain more carbon. Anthracite is the hardest coal, being almost pure carbon, and is the most efficient producer of heat. Coal is used commercially to fuel power stations, to burn to produce gas and to make coke, which burns at a higher temperature and is used in metal refining. It is also processed to make various forms of smokeless fuels used in domestic heating appliances. Coal affects the environment in two ways: (a) in the mining process, waste matter can create ugly slag heaps and can contaminate rivers; old mines create subsidence as they fill in; (b) burning coal emits toxic smoke, especially sulphur oxides and fly ash, together with carcinogenic substances such as benzpyrene. These emissions cause smog at low levels and can rise into the atmosphere and contribute to the greenhouse effect

QUOTE: on the evidence available, coal-fired electricity's generation, which provides more than 40% of the world's power, contributes about 8% to the calculated greenhouse effect
Green Magazine

coalescence [kəʊə'lesəns] *noun* process by which the size of water droplets in clouds increases as larger drops collide with smaller drops

coarse [kɔ:s] *adjective* (particle *or* grain of sand) which is larger than others; ***coarse sand fell to the bottom of the liquid as sediment, while the fine grains remained suspended;*** **coarse fishing** = catching freshwater fish other than salmon and trout

coast [kəʊst] *noun* land along by the sea

◇ **coastal** ['kəʊstəl] *adjective* by the sea; ***a plant found in coastal waters;*** **coastal fog** = type of advection fog which forms along the coast *or* over sea; **coastal protection** = protecting the coast from being eroded by the action of the sea

◇ **coastline** ['kəʊstlaɪn] *noun* the outline of a coast

cobalt ['kəʊbɔ:lt] *noun* metallic element, used to make alloys; **cobalt-60** = radioactive isotope which is used in radiotherapy to treat cancer (NOTE: chemical symbol is **Co;** atomic number is **27)**

cobble [kɒbl] *noun* large round stone used; *see also* WENTWORTH-UDDEN SCALE

coccidioidomycosis [kɒksɪdɪɔəʊaɪdəʊmaɪ'kəʊsɪs] *noun* lung disease caused by inhaling spores of the fungus *Coccidioides immitis*

coccus ['kɒkəs] *noun* bacterium shaped like a ball (NOTE: plural is **cocci)**

COMMENT: cocci grow together in groups: either in clumps (Staphylococci) or in long chains (Streptococci)

cockroach ['kɒkrəʊtʃ] *noun* black beetle, insect of the order Dictyoptera, a common household pest

coconut (palm *or* **tree)** ['kəʊkənʌt] *noun* tropical palm tree *(Cocos nucifera)* with a large hard fruit containing a white edible pulp; the outer husk is used to make a rough fibre called coir; **coconut oil** = oil extracted from dried coconut flesh or copra

COMMENT: the dried flesh of the coconut is called 'copra'; it is the main form in which coconuts are traded on the world markets; it is used for the extraction of coconut oil, which is used in making margarine, soap and other products. The Philippines, Malaysia and Mexico are all important producers

COD = CHEMICAL OXYGEN DEMAND

code [kəʊd] **1** *noun* signs which have a hidden meaning; **genetic code** = information which determines the synthesis of a cell, is held in the DNA of a cell and is passed on when the cell divides **2** *verb* to give a meaning; ***genes are sequences of DNA that code for specific proteins***

codistillation [kəʊdɪstɪ'leɪʃn] *noun* process by which molecules of toxic substances can be evaporated into clouds over land and then fall back into the sea as rain

cod liver oil [kɒdlɪvə'ɔɪl] *noun* oil from the liver of codfish, which is rich in calories and vitamins A and D

codominant [kəʊ'dɒmɪnənt] *noun & adjective* (species) which is dominant with another species

-coele = -CELE

coeli- *US* **celi-** ['si:lɪ] *prefix* referring to a hollow, usually the abdomen

◇ **coeliac** ['si:liæk] *adjective* referring to the abdomen

coelom ['si:ləm] *noun* body cavity in an embryo, which divides to form the thorax and abdomen

cogeneration [kəʊdʒenə'reɪʃn] *noun* production of heat and power as in a combined heat and power installation

coir [kɔɪə] *noun* rough fibre from the outer husk of coconuts

coke [kəʊk] *noun* fuel manufactured by heating coal to high temperatures without the presence of air; **gas coke** = coke resulting from the processing of coal for gas; **oil coke** = coke resulting from the processing of oil; **coke burner** *or* **coke oven** *or* **coking oven** = device in which coal is heated to produce coke

COMMENT: coke is produced in coking ovens, where coal is heated to white heat without air. This removes most of the tar and the resulting fuel burns at a much higher temperature than coal and produces very little smoke or ash. It is used in blast furnaces

col [kɒl] *noun* (i) high pass between two mountains; (ii) low pressure area between two anticyclones

cold [kəʊld] **1** *adjective* not warm *or* not hot; ***the weather is colder than last week and they say it will be even colder tomorrow; many old people suffer from hypothermia in very cold weather;*** **cold front** = edge of an advancing mass of cold air, associated with an area of low pressure, bringing clouds and rain as it meets the warmer air which it displaces; **cold hardening** = process by which a conifer prepares for the winter by reducing the amount of water contained in its needles; **cold trough** = long area of low pressure with cold air in it, leading away from the centre of a depression **2** *noun* cold weather; ***sheep can withstand the cold***

◇ **cold-blooded** ['kəʊldblʌdɪd] *adjective* (animal, such as a reptile *or* fish) which has cold blood; *see also* POIKILOTHERM

COMMENT: the body temperature of cold-blooded animals changes with the outside temperature

coliform ['kɒlɪfɔ:m] *adjective* (bacteria) which are similar to *Escherichia coli*

COMMENT: if *Escherichia coli* is found in water, it indicates that the water has been polluted by faeces, although it may not cause any illness

collagen ['kɒlədʒen] *noun* bundles of protein fibres forming connective tissue, bone and cartilage; **collagen disease** = any of several diseases of the connective tissue

◇ **collagenous** [kɒ'lædʒənəs] *adjective* (i) containing collagen; (ii) referring to collagen disease

COMMENT: collagen diseases include rheumatic fever and rheumatoid arthritis. Collagen diseases can be treated with cortisone

collection [kə'lekʃn] *noun* picking up *or* gathering together and taking away (of refuse, waste material, etc.) from houses, factories, etc. (as carried out, for example, by municipal refuse collectors); **separate collection** *or* **selective collection** = collection of different types of waste material at different times; ***we have a separate collection of glass on Thursdays***

collector panel *or* **solar collector** [kə'lektə] *noun* device with a dark surface which absorbs the sun's radiation and uses it to heat water

colloid ['kɒlɔɪd] *noun* substance with very small particles which form a solution when in suspension in a liquid

◇ **colloidal** [kə'lɔɪdl] *adjective* (very fine particle) which does not settle but remains in suspension in a liquid; **colloidally dispersed particles** = particles which remain in suspension in a liquid

colony ['kɒləni] *noun* group of animals *or* plants *or* microorganisms living together in a certain place; ***a colony of ants***

◇ **colonial** [kə'ləʊniəl] *adjective* referring to a colony; **colonial animals** = animals (such as ants *or* some seabirds *or* polyps) which usually live in colonies

◇ **colonization** [kɒlənaɪ'zeɪʃn] *noun* act of colonizing a place; ***islands are particularly subject to colonization by species of plants or animals introduced by man***

◇ **colonize** ['kɒlənaɪz] *verb* to begin to live in a place (as a group); ***derelict city sites rapidly become colonized by plants; rats have colonized the river banks***

◇ **colonizer** ['kɒlənaɪzə] *noun* animal *or* plant which colonizes an area

Colorado beetle [kɒlə'rɑːdəʊ 'biːtl] *noun* beetle *(Leptinotarsa decemlineata)* with black and yellow stripes which eats and destroys potato plants

coloration [kɒlə'reɪʃn] *noun* colours *or* patterns of an animal; **cryptic coloration** = pattern of colouring which makes an animal less easy to see (such as the stripes on a zebra); **protective coloration** = pattern of colouring which protects an animal from attack

colostrum [kə'lɒstrəm] *noun* fluid secreted by an animal's mammary glands at the birth of young and before the true milk starts to flow

colour *or US* **color** ['kʌlə] **1** *noun* differing wavelengths of light (red *or* blue *or* yellow, etc.) that are reflected from objects and sensed by the eyes; ***flowers use colour to attract insects*** **2** *verb* to give colour to; ***the chemical coloured the water blue***

◇ **colouring (matter)** ['kʌlərɪŋ 'mætə] *noun* substance which colours a processed food

◇ **colourless** ['kʌlələs] *adjective* with no colour; ***water is a colourless liquid***

COMMENT: colouring additives have E numbers 100 to 180. Some are natural pigments, such as riboflavine (E101), carrot juice (E160) or chlorophyll (E140) and are safe. Others, such as tartrazine (E102) and other azo dyes are suspected of being carcinogenic. Also suspect is caramel (E150), which is the most widely used colouring substance

column ['kɒləm] *noun* usually circular mass standing upright like a tree; ***basalt rocks form columns in certain parts of the world***

◇ **columnar** [kə'lʌmnə] *adjective* shaped like a column; ***igneous rocks may have a columnar structure***

combat ['kɒmbæt] *verb* to fight against; ***the medical team is combating an outbreak of diphtheria; what can we do to combat the spread of the disease?***

combine [kəm'baɪn] *verb* to join together; **combine drill** = drill which sows grain and fertilizer at the same time

◇ **combine harvester** ['kɒmbaɪn 'hɑːvɪstə] *noun* machine used to harvest a vast range of crops: cereals, grass, clover, peas, oilseed rape, and other arable crops. The combine cuts the crop, passes it to the threshing mechanism, then sorts the grain or seed from the straw or chaff

◇ **combined heat and power (CHP) plant** *or* **station** [kəm'baɪnd 'hiːt ənd 'paʊə] *noun* power station which produces both electricity and hot water for the local population. A CHP plant may operate on almost any fuel, including refuse

◇ **combination** [kɒmbɪ'neɪʃn] *noun* act of joining together

combust [kəm'bʌst] *verb* to burn; ***Denmark combusts 75% of its refuse for heat reclamation***

◇ **combustion** [kəm'bʌstʃən] *noun* burning of a substance with oxygen; **internal combustion engine** = type of engine used in motor vehicles, where the fuel is a mixture of petrol and air burnt in a closed chamber (the combustion chamber) to give energy to the pistons; **combustion residue** = material left after combustion has taken place

commensal [kə'mensəl] *noun & adjective* (plant *or* animal) which lives on another plant *or* animal but does not harm it or influence it in any way, though one of the two may benefit from the other (NOTE: if it causes harm it is a **parasite**; if the two organisms are dependent on each other they exist in **symbiosis**)

◇ **commensalism** [kə'mensəlɪzm] *noun* existing together as commensals

comminution [kɒmɪ'njuːʃn] *noun* crushing *or* grinding of rock, ore *or* sewage into small particles

◇ **comminutor** [kɒmɪ'njuːtə] *noun* crushing machine which reduces particles to a smaller size, as in the first treatment given to raw sewage, where the sewage is ground into small particles

committee [kə'mɪti:] *noun* group of people dealing with a particular subject *or* problem; **advisory committee** = group of people who can give advice

common ['kɒmən] **1** *noun* land on which anyone can walk and where anyone may have the right to graze animals, gather wood, etc.; **right of common** = right to walk on someone else's land, graze animals on it, gather wood, etc. (NOTE: now usually used in place names such as **Clapham Common**) **2** *adjective* **(a)** which happens very often; ***noise pollution is a common problem in large cities*** **(b)** referring to *or* belonging to several different people or to everyone; *(in the EU)* **Common Agricultural Policy (CAP)** = agreement between members of the EU to protect farmers by paying subsidies to fix prices of farm produce; **common land** = land on which anyone can walk and where anyone may have the right to graze animals, gather wood, etc.; **common nuisance** = criminal act which causes harm *or* danger to members of the public in general *or* to their rights

community [kə'mju:nɪti] *noun* **(a)** group of different organisms which live together in an area, and which are usually dependent on each other for existence; **climax community** = plant community which has been stable for many years and which is unlikely to change unless the climate changes or there is human interference **(b)** group of people who live and work in a district; ***the health services serve the local community; community care is an important part of primary health care;*** **community architecture** = way of designing new housing projects *or* adapting old buildings, in which the people living in the area as well as specialists are involved in the planning; **community heating** = heating of houses and shops in an area from a central source; **community medicine** = medical practice which examines groups of people and the health of the community, including factors such as housing, pollution and the environment; **community physician** = doctor who specializes in community medicine; **community services** = nursing services which are available to the community; **community transport** = bus *or* rail service which is available to the community **(c)** = EUROPEAN ECONOMIC COMMUNITY

commuter [kə'mju:tə] *noun* person who travels regularly to and from his place of work, especially in a city; **commuter train** = train which carries mainly commuters

compact [kəm'pækt] *verb* to compress the ground and make it hard, as by driving over it with heavy machinery

◇ **compacting** *or* **compaction** [kəm'pæktɪŋ *or* kəm'pækʃn] *noun* compressing the ground and making it hard, as by driving over it with heavy machinery

◇ **compactor** [kəm'pæktə] *noun* machine which compresses the ground and makes it hard

companion plants [kəm'pænjən 'plɑ:nts] *noun* (i) plants which grow best with other plants; (ii) plants which are grown with others by horticulturists because they encourage growth *or* because they reduce pest infestation

> COMMENT: some plants grow better when planted near others. Beans and peas help root plants such as carrots and beetroot. Most herbs (except fennel) are helpful to other plants. Marigolds help reduce aphids if they are planted near plants such as broad beans or roses which are subject to aphid infestation. The strong smell of onions is disliked by the carrot fly, so planting onions near carrots makes sense. On the other hand, most other plants (and especially peas and beans) dislike onions and will not grow well near them

compartment [kəm'pɑ:tmənt] *noun* section of a managed plantation of trees

compensation [kɒmpən'seɪʃn] *noun* **compensation depth** = point in a lake *or* sea at which the rate of creation of organic matter by photosynthesis is the same as the rate of loss of matter by respiration; **compensation point** = point at which the rate of creation of organic matter in photosynthesis is the same as the rate of loss of matter by respiration (i.e. the point at which carbon dioxide used by plants equals the oxygen they release)

comply with [kəm'plaɪ wɪθ] *verb* to obey (a regulation); ***the factory was forced to comply with the new regulations on emission of toxic gases***

◇ **compliance** [kəm'plaɪəns] *noun* act of obeying a regulation; **cost of compliance** = money required *or* spent in order to obey a regulation; ***the cost of compliance with the new regulations on emission of toxic gases will be enormous***

> QUOTE: some cities have tried to bring their ozone levels into compliance with federal safety standards, but failed
>
> **Environmental Action**

compose [kəm'pəʊz] *verb* to make up; ***air is composed of a mixture of gases: 75% nitrogen, 23% oxygen, 1% argon and very small quantities of several other gases***

◇ **composition** [kɒmpə'zɪʃn] *noun* way in which a compound is formed; **chemical composition** = the chemicals which make up a substance

Compositae [kəm'pɒzɪtaɪ] *noun* common and very large family of plants, where the flowers are made of many florets arranged in heads (as in dandelions)

compost ['kɒmpɒst] **1** *noun* **(a)** rotted vegetation, used as fertilizer or mulch; **compost heap** = pile of rotting vegetation in a garden, which when fully rotted can be spread on the soil as a fertilizer or mulch **(b)** prepared soil *or* peat mixture in which plants are grown in horticulture; **mushroom compost** = special growing medium for commercial production of mushrooms **2** *verb* to encourage the breaking down of organic waste, as in a compost heap

◇ **composting** ['kɒmpɒstɪŋ] *noun* action of encouraging the breaking down of organic waste, as in a compost heap; **composting drum** = cylindrical container in which organic waste is rotted down for use as a fertilizer or mulch; **sludge composting** = rotting down of sewage for use as a fertilizer or mulch

> QUOTE: at least 20 per cent and as much as 40 per cent of the City's waste stream is organic matter that could be readily composted
> **Ecologist**

compound ['kɒmpaʊnd] *noun* substance formed from two or more chemical elements; **chemical compound** = substance formed from two or more chemical elements, in which the proportions of the elements are always the same; **nitrogen compound** = substance, such as a fertilizer, containing mostly nitrogen with other elements

> COMMENT: compounds are stable (i.e. the proportions of the elements in them are always the same) but can be split into their basic elements by chemical reactions

compulsory [kəm'pʌlsəri] *adjective* which is forced or ordered by an authority; ***the compulsory slaughter of infected animals***

concentrate ['kɒnsəntreɪt] **1** *noun* **(a)** strength of a solution; quantity of a substance in a certain volume **(b)** strong solution which is to be diluted **2** *verb* **(a) to concentrate on** = to examine something in particular **(b)** to reduce a solution and increase its strength by evaporation (NOTE: opposite is **dilute**)

◇ **concentration** [kɒnsən'treɪʃn] *noun* **(a)** amount of a substance in a solution *or* in a certain volume; **background concentration** = (i) general level of air pollution in an area, disregarding any specifically local factors, such as the presence of a coal- fired power station; (ii) amount of radiation which comes from natural sources like rocks *or* the earth *or* the atmosphere and not from a single man-made source; **ground-level concentration** = amount of a pollutant measured at the height of the ground *or* just above it *or* just below it; **maximum allowable concentration (MAC)** = largest amount of a pollutant which workers are allowed to be in contact with in their work environment **(b)** (large) amount gathered together in one place; **population concentration** = (large) number of people in one place; **traffic concentration** = (large) number of motor vehicles in one place

> QUOTE: the concentration of atmospheric carbon dioxide is now 23 per cent higher
> **Scientific American**

> QUOTE: flights into the Antarctic ozone hole revealed very high concentrations of chlorine monoxide
> **New Scientist**

concrete ['kɒŋkri:t] *noun* hard stone-like substance made by mixing cement, sand and water and letting it dry; **reinforced concrete** = blocks of concrete with steel rods inside to give extra strength, used in the construction of large buildings; **concrete shield** = protective cover made of concrete, as over nuclear waste or around a nuclear reactor

◇ **concretion** [kən'kri:ʃn] *noun* formation of a solid mass of stone, made of pieces of stones and other sedimentary materials

◇ **concretionary** [kən'kri:ʃənəri] *adjective* (stone) which is formed by concretion

condense [kən'dens] *verb* **(a)** (i) to make a vapour become liquid; (ii); *(of a vapour)* to become liquid **(b)** to make compact *or* to make more dense

◇ **condensate** ['kɒndənseɪt] *noun* substance formed by condensation, such as a liquid formed from a vapour; substance which is a gas when occurring naturally underground but becomes liquid when brought to the surface

◇ **condensation** [kɒndən'seɪʃn] *noun* **(a)** action of making vapour into liquid; **condensation aerosol** = droplets of moisture which form in warm damp air as it cools, producing mist; **condensation nucleus** = particle on which moisture condenses, forming a raindrop; **condensation trail** *or* **vapour trail** = white streak in the sky left by an aircraft flying at high altitude and caused by condensation and freezing of components of its exhaust gases, mainly water **(b)** water which forms on walls *or* windows, etc. when warm damp air meets a cold surface

◇ **condensed** [kən'denst] *adjective* made compact *or* more dense

◇ **condenser** [kən'densə] *noun* device which cools steam from a power station, converts it back into water, which is then sent back to the boilers for reheating

> COMMENT: condensers are also used in domestic clothes driers (where the warm air is recycled through the machine and is used again to save energy), and in

refrigerators (where heat is transferred to the cooling liquid)

condition [kən'dɪʃn] *noun* state (of health *or* of cleanliness); ***the road system is in very good condition; half the population is suffering from malnutrition and their condition is getting worse; conditions in the urban areas are very bad***

◇ **conditioned reflex** [kən'dɪʃnd 'ri:fleks] *noun* automatic reaction by an animal to a stimulus; normal reaction to a normal stimulus learned from past experience

◇ **conditioning** [kən'dɪʃənɪŋ] *noun* improving the quality of something; **radioactive waste conditioning** = processing of radioactive waste to use it again *or* to make it safe for disposal

conduction [kən'dʌkʃn] *noun* passing of heat *or* sound from one part to another; **air conduction** = conduction of sounds through the channel from the outside to the inner ear; **bone conduction** *or* **osteophony** = conduction of sound waves to the inner ear through the bones of the skull

◇ **conductive** [kən'dʌktɪv] *adjective* referring to conduction; (material) which is able to pass heat *or* sound from one part to another

◇ **conductivity** [kɒndʌ'tɪvɪti] *noun* ability of a material to pass heat *or* sound from one part to another

conduit ['kɒndjuɪt] *noun* channel *or* passage along which a fluid flows

cone [kəʊn] *noun* **(a)** circular shape like a tube with a wide bottom, narrowing to a point **(b)** one of two types of cell in the retina of the eye which is sensitive to light. (Cones are sensitive to bright light and colours and do not function in bad light) **(c)** hard case containing the seeds of a conifer

◇ **coning** ['kəʊnɪŋ] *noun* effect of a plume of smoke from a chimney which widens into the shape of a cone as it leaves the chimney

confluence ['kɒnfluəns] *noun* (i) place where two rivers join; (ii) place where two streams of air join

congenital [kən'dʒenɪtl] *adjective* which exists at *or* before birth; **congenital defect** = defect which exists in a baby from birth

COMMENT: a congenital condition is not always inherited from a parent through the genes. It may be due to abnormalities which develop in the foetus because of factors such as a disease which the mother has (as in the case of German measles) or a drug which she has taken

◇ **congenitally** [kən'dʒenɪtəli] *adverb* in a way which exists at *or* before birth; ***she was born congenitally deaf***

congestion [kən'dʒestʃn] *noun* blocking of a tube *or* a passage; **traffic congestion** = blocking of the streets in a town by motor vehicles

◇ **congested** [kən'dʒestɪd] *adjective* blocked (passage); ***it is hoped that building the by-pass will relieve the congested inner city streets***

QUOTE: in the 39 metropolitan areas in the US with a population of one million or more, roughly one-third of all vehicular travel takes place under congested conditions in which speed averages half of its free- flow value
American Scientist

conglomerate [kən'glɒmərɪt] *noun* sedimentary layer formed of small round stones

conifer ['kɒnɪfə] *noun* evergreen tree with long thin leaves (called needles) and which produces fruit in the form of cones; ***afforestation with conifers has had a noticeable effect on the landscape of the upland region***

◇ **coniferous** [kə'nɪfərəs] *adjective* (tree) which has cones

COMMENT: Conifers are members of the order Coniferales and include pines, firs, spruce, etc. They are natives of the cooler temperate regions, are softwoods and grow very fast. They exude resin, and this, together with the hardness of the needles makes it impossible for plants to grow on the forest floor beneath them. If conifers are planted close together they will grow straight and tall, making them ideal for cheap construction material. Because of this, they are frequently used in afforestation plantations in places where they would not have grown naturally, and where, unfortunately, they hide the landscape and reduce wildlife

QUOTE: scientists have discovered evidence about insect life in conifer plantations which undermines the popular conception that they are less species-rich than broadleaf woodlands
New Scientist

coning ['kəʊnɪŋ] *see* CONE

connect [kə'nekt] *verb* to join; ***a sandbar connects the island to the mainland at high tide***

◇ **connection** [kə'nekʃn] *noun* something which joins

◇ **connective tissue** [kə'nektɪv 'tɪʃu:] *noun* tissue which forms the main part of bones and cartilage, ligaments and tendons, in which a large proportion of fibrous material surrounds the tissue cells

conserve [kən'sɜ:v] *verb* to keep *or* not to waste; to look after and keep in the same state

◇ **conservancy** [kən'sɜːvənsi] *noun* official body which protects a part of the environment, such as a river, moorland, etc.; **Nature Conservancy Council (NCC)** = official body in the UK, established in 1973, which takes responsibility for the conservation of fauna and flora. Since April 1991 the branch of the Council dealing with England has also been called English Nature

◇ **conservation** [kɒnsə'veɪʃn] *noun* **(a)** keeping *or* not wasting; **conservation of energy** = making consumption of energy more efficient; preventing loss *or* waste of energy (such as the loss of heat from buildings); **conservation of resources** = managing resources, such as fossil fuels and other natural materials, so as not to waste them, damage them or use them too quickly; **conservation of soil** *or* **soil conservation** = using methods such as irrigation, mulching, etc., to prevent soil from being eroded *or* overcultivated; **forest conservation** = keeping forests by not cutting down trees **(b)** maintenance of environmental quality and resources by the use of ecological knowledge and principles; **nature conservation** = active management of the earth's natural resources and environment to ensure their quality is maintained and that they are wisely used; **conservation body** = group of people *or* organization which promotes the preservation of the countryside and the careful management of natural resources; **conservation measures** = ways in which environmental quality can be maintained

◇ **Conservation Area** [kɒnsə'veɪʃn 'eəriə] *noun* area of a town where the buildings are of special architectural and historical interest

◇ **conservationist** [kɒnsə'veɪʃnɪst] *noun* person who promotes the preservation of the countryside and the careful management of natural resources

◇ **Conservation Reserve Program (CRP)** *US* federal programme which pays farmers to let land lie fallow

> QUOTE: the genus Martes comprises eight species, all of which have solved the problem of conserving heat in the long northern winters by growing dense water-repellent coats
>
> **BBC Wildlife**

> QUOTE: farmers will be paid cash to help conserve threatened wildlife on the Somerset Levels
>
> **Green Magazine**

> QUOTE: if all cost-effective conservation measures were taken and the U.S. became as energy-efficient as Japan, the country would consume half as much energy as it does today
>
> **Scientific American**

constant ['kɒnstənt] *adjective* **(a)** continuous *or* not stopping; ***the wardens complained about the constant stream of visitors to the reservation*** **(b)** level *or* unchanging; ***rainfall remains fairly constant throughout the year;*** **constant level chart** = chart showing the isobars of a certain pressure at a certain level in the upper atmosphere; **constant pressure chart** = chart showing the different heights at which a pressure reading is obtained

constituent [kɒn'stɪtjuənt] **1** *adjective* forming part of a whole; **the constituent elements of air** = various substances which make up air **2** *noun* substance *or* component which forms part of a whole

construction [kən'strʌkʃn] *noun* (i) process of building; (ii) something built; **construction certificate** = official document allowing a person to build a property on empty land; **construction industry** = business *or* trade of building houses, offices, etc.; **construction workers** = people who work in the building trade, such as bricklayers, carpenters, scaffolders, etc.

consume [kən'sjuːm] *verb* to use up *or* to burn (fuel) *or* to eat (foodstuffs); ***the population consumes ten tonnes of foodstuffs per week; the new pump consumes only half the fuel which a normal pump would use***

◇ **consumable** [kən'sjuːməbl] *adjective* able to be consumed; **consumable goods** = goods which are bought by members of the public

◇ **consumables** [kən'sjuːməblz] *plural noun* = CONSUMABLE GOODS

◇ **consumer** [kən'sjuːmə] *noun* **(a)** person *or* company which buys and uses goods and services; ***gas consumers are protesting at the increase in prices; the factory is a heavy consumer of water;*** **consumer council** = group representing the interests of consumers; **consumer durables** = items such as washing machines *or* refrigerators *or* cookers which are bought and used by the public; **consumer goods** = goods bought by consumers *or* by members of the public; **consumer goods industry** = business of manufacturing and supplying consumer goods (NOTE: the opposite of **consumer goods** is **capital goods)** **consumer panel** = group of consumers who report on products they have used so that the manufacturers can improve them or use what the panel says about them in advertising; *US* **consumer price index** = index showing how prices of consumer goods have risen over a period of time, used as a way of measuring inflation and the cost of living; **consumer protection** = protecting consumers against unfair *or* illegal traders; **consumer research** = research into why consumers buy goods and what goods they really want to buy; **consumer society** = type of society where consumers are encouraged to buy goods **(b)** organism (such as an animal) which eats other organisms;

primary consumer = animal (such as a herbivore) which eats plants (which are producers in the food chain); **secondary consumer** = animal (such as a carnivore) which eats other consumers in the food chain; **tertiary consumer** = carnivore which only eats other carnivores

◊ **consumerism** [kən'sju:mərɪzm] *noun* movement for the protection of the rights of consumers; **green consumerism** = movement to encourage people to buy food and other products which are environmentally good (such as organic foods, lead-free petrol, etc.)

> QUOTE: the buildings sector, rather than transportation, is the single largest consumer of energy in the U.S. economy
> **Scientific American**

consumption [kən'sʌmpʃn] *noun* using a fuel *or* using goods or services; taking food *or* liquid into the body; ***a car with low petrol consumption; the factory has a heavy consumption of coal; the country's consumption of wood has fallen by a quarter; 3% of all food samples were found to be unfit for human consumption through contamination by lead;*** **home consumption** *or* **domestic consumption** = use of something in the home; **consumption residues** = waste matter left after manufactured goods are used

contact ['kɒntækt] **1** *noun* touching someone *or* something; **to have contact with someone** *or* **something** = actually to touch someone *or* something; **to be in contact with someone** = to touch someone *or* be in contact with someone; **contact herbicide** *or* **contact weedkiller** = substance (such as paraquat) which kills a plant whose leaves it touches but which may not kill a plant with large roots; **contact insecticide** = substance (such as DDT) which kills insects which touch it, used against insects such as mosquitoes **2** *verb* to get in touch with someone (as by post *or* telephone *or* visiting); ***contact your local office to find out details of membership of the group***

contagion [kən'teɪdʒn] *noun* spreading of a disease by touching an infected person *or* objects which an infected person has touched

◊ **contagious** [kən'teɪdʒəs] *adjective* (disease) which can be transmitted by touching an infected person *or* objects which an infected person has touched

container [kən'teɪnə] *noun* **(a)** box *or* bottle which holds something else; **disposable container** = box *or* bottle which is thrown away after use; **returnable container** = box *or* bottle which can be taken back to the place where it was bought and which can then be recycled **(b)** large case that can be transported by truck and then easily loaded on a ship; **container port** = port which has all the equipment for loading containers from trucks and trains onto ships

containment [kən'teɪnmənt] *noun* preventing the nuclei and electrons generated in nuclear fusion from reaching the walls of the reaction chamber

contaminate [kən'tæmɪneɪt] *verb* to make something impure by touching it *or* by adding something to it; ***supplies of drinking water were contaminated by refuse from the factories; a whole group of tourists fell ill after eating contaminated food; 7% of the soil is contaminated with heavy metals***

◊ **contaminant** [kən'tæmɪnənt] *noun* substance which contaminates

◊ **contamination** [kəntæmɪ'neɪʃn] *noun* **(a)** action of contaminating; ***the contamination resulted from drinking polluted water;*** **atmospheric contamination** = pollution of the air with harmful substances **(b)** state of something (such as water *or* food) which has been contaminated and so is harmful to living organisms

content ['kɒntent] *noun* proportion of a substance in something; ***these foods have a high starch content; dried fruit has a higher sugar content than fresh fruit; lead content should be reduced to 0.1 g/l by 1995;*** **sulphur content** = proportion of sulphur in something

continent ['kɒntɪnənt] *noun* one of the seven large landmasses on the earth's surface (Asia, Africa, North America, South America, Australia, Europe and Antarctica)

◊ **continental** [kɒntɪ'nentl] *adjective* referring to a continent; **continental climate** = type of climate found in the centre of a large continent away from the sea, with long dry summers, very cold winters and not much precipitation; **continental crust** = part of the earth's crust under the continents and continental shelves, lying above the oceanic crust; **continental drift** = geological theory that the present continents were once part of a single landmass and have gradually drifted away from each other over a period of millions of years; **continental margin** = area of sea and seabed around a continent, extending to a depth of about 2,000m; **continental shelf** = seabed surrounding a continent and covered with shallow water: usually taken to be the area between the shore and the 183m deep line; **continental slope** = area of the seabed which slopes down sharply from the edge of the continental shelf into deeper water

contour ['kɒntʊə] *noun* line drawn on a map to show ground of the same height above sea level; **contour farming** = method of cultivating sloping land, where the land is ploughed along a level rather than down the slope, so reducing soil erosion; **contour interval** = the space between two contour lines on a map; **contour line** = line drawn on a map to show ground of the same height above sea level; **contour map** = map showing the contours of a geographical area; **contour ploughing** *or* **contour ridging** *or* **contour tillage** = ploughing across the side of a hill so as to create ridges along the contours of the soil which will hold water and prevent erosion; **contour strip cropping** = planting different crops in bands along the contours of sloping land so as to prevent soil erosion

COMMENT: in contour farming, the ridges of earth act as barriers to prevent soil being washed away and the furrows retain the rainwater

contrail ['kɒntreɪl] *noun* = CONDENSATION TRAIL

control [kən'trəʊl] **1** *noun* **(a)** restraining *or* keeping in order; **biological control** = control of pests by using predators to eat them; **environmental control** = means of protecting an environment; **pollution control** = means of limiting pollution; **the specialists brought the epidemic under control** = they stopped it from spreading; **under state control** = (industry, farm, etc.) which is run by the government; **the epidemic rapidly got out of control** = it spread quickly; **(nuclear) control rod** = tube inserted into a nuclear reactor in order to alter the speed of the reaction **(b)** *(in experiments)* sample used as a comparison with the one being tested; **control group** = group of organisms *or* substances which are not being tested, but whose test data is used as a comparison **2** *verb* to keep in order; ***the medical authorities are trying to control the epidemic; they were unable to control the spread of the pest;*** **controlled drug** *or* **dangerous drug** = drug which is on the official list of drugs which are harmful and are not available to the public; **controlled tipping** = disposal of waste in special landfill sites, as opposed to throwing it away anywhere (NOTE: the US English is **sanitary landfill**)

conurbation [kɒnɜː'beɪʃn] *noun* large area covered with buildings (houses *or* factories *or* public buildings, etc.)

convection [kən'vekʃn] *noun* movement of air in a vertical direction (as opposed to advection, where the air moves horizontally)

convenience foods [kən'viːniəns 'fuːdz] *plural noun* foods which have been prepared so that they are ready to be served after simply reheating

Convention on International Trade in Endangered Species of Wild Fauna and Flora (CITES) *noun* agreement made in 1973 to protect wild animals and plants listed as endangered, by controlling international trading of these species and products derived from them

conventional [kən'venʃənl] *adjective* normal *or* traditional; not using new technologies; not using nuclear power; **conventional fuel** = traditional means of providing energy such as coal, wood, gas, etc., as opposed to alternative energy sources such as solar power, tidal power, wind power, etc.; **conventional medicine** = medical practice as taught in hospitals and medical schools (as opposed to alternative medicine); **conventional power station** = non-nuclear power station

convergence [kən'vɜːdʒəns] *noun* phenomenon which occurs whenever there is a net inflow of air into a region of the atmosphere, resulting in the accumulation of air and an increase in density; *compare* DIVERGENCE

convert [kən'vɜːt] *verb* to change to something different; ***photochemical reactions convert oxygen to ozone; he has converted his car to take unleaded petrol; sulphur dioxide and nitrogen oxides emitted into the atmosphere are chemically converted into sulphuric and nitric acid***

◇ **conversion** [kən'vɜːʃn] *noun* act of converting; ***the conversion of oxygen into ozone***

◇ **converter** [kən'vɜːtə] *noun* device *or* process which converts; **Bessemer converter** = type of furnace in which air is blown through molten metal; **catalytic converter** = device attached to the exhaust pipe of a motor vehicle which reduces the emission of carbon monoxide; **methane converter** = process which takes the gas produced by rotting waste in a landfill site and processes it into a usable form

COMMENT: a catalytic converter is a box filled with a catalyst, such as platinum. Converters can only be used on motor vehicles burning unleaded petrol, as the lead compounds in leaded petrol rapidly coat the catalyst in the converter and prevent it functioning

cool [kuːl] **1** *adjective* not very hot; ***cool weather is forecast for the rest of the week*** **2** *verb* to make less hot; to become less hot; **cooling pond** = (i) part of a nuclear reactor where irradiated elements are cooled; (ii) part of an industrial process where cooling water is allowed to cool in the open air; **cooling tower** = tall tower used for cooling the water used in industrial processes, such as at power stations; **cooling water** = water used to make something less hot, such as the irradiated elements from a nuclear reactor or the engine of a machine

◇ **coolant** ['kuːlənt] *noun* **(a)** substance used to cool something, such as the engine of a machine **(b)** substance used to take the heat generated from a nuclear power station to the boilers; ***ordinary water is used as coolant in certain types of nuclear reactor***

cooperation [kəʊɒpə'reɪʃn] *noun* working together in harmony to achieve a common goal

copper ['kɒpə] *noun* metallic trace element, essential to biological life and used in making alloys and in electric wiring; **copper ore** = mineral containing copper; **copper sulphate** ($CuSO_4$) = chemical occurring naturally, used in electroplating, dyeing and as a plant spray (NOTE: chemical symbol is **Cu;** atomic number is **29)**

coppice ['kɒpɪs] **1** *noun* area of trees which have been cut down to near the ground to allow shoots to grow which are then harvested; **coppice forest** *or* **coppice wood** = forest *or* wood containing trees which have grown up from trees which were once cut down to near the ground **2** *verb* to cut trees down to near the ground to produce strong straight shoots which are then harvested; ***coppicing is a traditional method of wood management; coppiced wood can be dried for use in wood-burning stoves*** *compare* POLLARD

COMMENT: the best trees for coppicing are those which naturally send up several tall straight stems from a bole, such as willow, alder or poplar. In coppice management, the normal cycle is about five to ten years of growth, after which the stems are cut back. Thick stems are dried and used as fuel, or for making charcoal. Thin stems are used for fencing

copra ['kɒprə] *noun* dried pulp of a coconut, from which oil is extracted by pressing

coral ['kɒrəl] *noun* rock-like substance composed of the skeletons of dead polyps; **coral reef** = reef formed of coral; **coral sand** = fine white particles which form tropical beaches, which is not sand at all but tiny pieces of dead coral

cordwood ['kɔ:dwʊd] *noun* pieces of cut tree trunks, all of the same length, ready for transporting

core [kɔ:] *noun* **(a)** central part of the earth; ***the earth's core is believed to be formed of nickel and iron*** **(b)** central part of a nuclear reactor, where the fuel rods are sited

Coriolis force [kɒrɪ'əʊlɪs 'fɔ:s] *noun* hypothetical force used to explain the phenomenon by which particles tend to move to the side as they travel forwards (bodies turn clockwise in the Northern Hemisphere and anticlockwise in the Southern)

◇ **Coriolis effect** [kɒrɪ'əʊlɪs ɪ'fekt] *noun* sideways movement of particles as explained by the Coriolis force

cork [kɔ:k] *noun* layer of dead cells under the bark of a tree, especially the thick layer of this in the cork oak; **cork oak** *or* **cork tree** = evergreen tree from which cork is obtained

corrie ['kɒri] *noun see* CIRQUE

corm [kɔ:m] *noun* bottom part of the stem of a plant which, like a bulb, can be preserved and from which the plant sprouts again in the spring

COMMENT: crocuses, gladioli and cyclamens have corms, not bulbs

corn [kɔ:n] *noun* **(a)** any cereal crop, but especially wheat **(b)** *(mainly US)* maize; **corn cob** = seed head of maize; **corn-on-the-cob** = head of maize, cooked and eaten as a vegetable; **corn oil** = vegetable oil obtained from maize grains, used for cooking and as a salad oil

◇ **cornflour** *or* US **corn starch** ['kɔ:nflaʊə *or* 'kɔ:n 'stɑ:tʃ] *noun* flour extracted from maize grain, used in cooking (it contains a high proportion of starch, and is used for thickening sauces)

corolla [kə'rəʊlə] *noun* ring of brightly coloured petals on a flower

corona [kə'rəʊnə] *noun* rings of colour round the sun or moon when seen through thin mist

corrasion [kə'reɪʒn] *noun* wearing away of rock by material carried by ice, water or wind

correction [kə'rekʃn] *noun* changing something to make it correct; **barometric corrections** = corrections made to the reading on a mercury thermometer to allow for altitude and outside temperature

corrode [kə'rəʊd] *verb* to change the composition of metals by exposure to water *or* air; ***iron corrodes and forms rust***

◇ **corrosion** [kə'rəʊʒn] *noun* (i) changing the composition of metals by exposure to water *or* air; (ii) erosion of rocks by the action of chemicals

◇ **corrosive** [kə'rəʊzɪv] *adjective* (substance) which corrodes; ***corrosive dry deposition can affect building materials***

cosmos ['kɒzmɒs] *noun* the whole universe

◇ **cosmic** ['kɒzmɪk] *adjective* referring to the cosmos; **cosmic rays** *or* **cosmic radiation** = radiation coming from outer space which strikes the earth

cost [kɒst] *noun* amount of money which has to be paid for something; **cost of living** = level of prices of the basic necessities of life, i.e. food, clothing, shelter and fuel; ***the cost of living has risen by 2% over the last year;*** **labour cost** = amount of money paid for work done

◇ **cost-benefit analysis** [kɒst'benɪfɪt ə'nælɪsɪs] *noun* examination in economic terms of the advantages and disadvantages of a certain course of action; **cost-benefit ratio** = comparison of costs to benefits

◇ **cost-effective** ['kɒstɪ'fektɪv] *adjective* which gives value, especially in comparison with something else, when the benefit is greater in value compared with the cost

◇ **cost-effectiveness** ['kɒstɪ'fektɪvnəs] *noun* measurement of how cost-effective something is

◇ **costly** ['kɒstli] *adjective* expensive; ***pollution control measures are costly to the community***

cottonwood ['kɒtənwʊd] *noun* North American hardwood tree

cotyledon [kɒtɪ'li:dən] *noun* first leaf of a plant as the seed sprouts

COMMENT: cotyledons are thicker than normal leaves, and contain food for the growing plant. Plants are divided into two groups: those which produce a single cotyledon (monocotyledons) and those which produce two cotyledons (dicotyledons)

Council for the Protection of Rural England (CPRE) organization which campaigns for the conservation of the English countryside

counter ['kaʊntə] *noun* instrument which measures something and shows the result on a dial or display; **Geiger counter** = instrument for the detection and measurement of radiation

countershading ['kaʊntəʃeɪdɪŋ] *noun* type of animal coloration where the back is darker than the belly, making it more difficult to see the animal's shape clearly

country ['kʌntri] *noun* area of land which is not a town or city; ***we live in the country; country areas are less polluted than urban areas;*** **country park** = area in the countryside set aside for the public to visit and enjoy

◇ **Country Landowners Association** organization representing the interests of landowners

◇ **countryside** ['kʌntrisaɪd] *noun* the land excluding towns and cities; **Countryside Commission** = official organization established in the UK in 1968, which supervises the planning of the countryside and how it can be used for the enjoyment of the population. It is particularly concerned with National Parks and Areas of Outstanding Natural Beauty, together with rights of way (such as footpaths and bridle paths) in the country

◇ **Countryside Council for Wales** *noun* branch of the Nature Conservancy Council responsible for the conservation of fauna and flora in Wales

coupe [ku:p] *noun* area of a forest in which trees have been cut

cover ['kʌvə] *noun* (i) something which is put over something to close it *or* protect it; (ii) amount of soil surface covered with plants; (iii) plants grown to cover the surface of the soil; ***grass cover will provide some protection against soil erosion;*** **ground cover** = plants which cover the surface of the soil; **plant cover** *or* **tree cover** = number of plants *or* trees growing on a certain area of land

cover crop ['kʌvə'krɒp] *noun* **(a)** crop which is sown to cover the soil and prevent it from drying out and being eroded; when the cover crop has served its purpose, it is usually ploughed in. Hence, leguminous plants which are able to enrich the soil are often used as cover crops **(b)** crop which is grown to give protection to another crop which is sown with it; in the tropics, bananas can be used as a cover crop for cocoa

CPRE = COUNCIL FOR THE PROTECTION OF RURAL ENGLAND

Cr *symbol for* chromium

cracking ['krækɪŋ] *noun* process by which crude oil is broken down into light oil

crater ['kreɪtə] *noun* round depression at the top of a volcano; **crater lake** = round lake which forms in a crater

creep [kri:p] *noun* slow downhill movement of soil, caused by gravity

creosote ['kri:əsəʊt] *noun* liquid derived from coal tar, used as a pesticide or to paint timber to preserve it from rot *or* to prevent attacks by insects

crepuscular rays [krɪ'pʌskʊlə 'reɪz] *plural noun* separate vertical rays of sunlight which pass through gaps in clouds

crest [krest] *noun* highest point along the length of a mountain ridge

crevasse [krə'væs] *noun* large crack in a glacier

crevice ['krevɪs] *noun* crack or little hole in rock

crisis ['kraɪsɪs] *noun* **(a)** important point *or* time when things are in a very bad state; ***a crisis caused by drought;*** **crisis centre** = building where people can go in time of crisis for food, shelter, advice, etc. **(b)** turning point in a disease after which the patient may start to become better or very much worse (NOTE: plural is **crises**)

criterion [kraɪ'tɪəriən] *noun* standard by which something can be judged (NOTE: plural is **criteria**)

critical ['krɪtɪkl] *adjective* **(a)** referring to crisis; **critical factor** = something in the environment, such as the introduction of a pollutant or a drop in temperature, etc. which causes a sudden change to occur; **critical link** = organism in a food chain which is responsible for taking up and storing nutrients which are then passed on down the chain; **critical load** = highest level of pollution which will not cause permanent harm to the environment; **critical mass** = minimum amount of fissile matter which can produce a chain reaction; **critical organ** = part of the body which is particularly sensitive to radiation; **critical pH** = level of acidity at which a sudden change will occur; **critical point** *or* **critical state** = moment at which a substance undergoes a change in temperature, volume or pressure; **critical temperature** = temperature below which a gas will normally become liquid **(b)** which criticizes; ***the report was critical of the steps taken to cut down pollution***

◊ **critically** ['krɪtɪkli] *adverb* in a way which criticizes; **critically ill** = very seriously ill when it is not known if the patient will get better

criticize ['krɪtɪsaɪz] *verb* to say what is wrong with something; ***the report criticized the government's attitude to building on green belt areas***

crocoite ($PbCrO_4$) ['krəʊkəʊaɪt] *noun* orange mineral containing lead chromate

crop [krɒp] **1** *noun* **(a)** plant grown for food; **cash crop** = crop which is grown to be sold rather than eaten by the person who grows it; **crop breeder** = person who specializes in developing new varieties of crops; ***crop breeders depend on wild plants to develop new and stronger strains;*** **crop dusting** *or* **crop spraying** = putting insecticide, herbicide, fungicide, etc. onto crops in the form of a fine dust *or* spray; **crop sprayer** = machine *or* aircraft which sprays insecticide, herbicide, fungicide, etc. onto crops **(b)** produce from plants; ***the tree has produced a heavy crop of apples; the first crop was a failure; the rice crop has failed*** **(c)** bag-shaped part of a bird's throat where food is stored before digestion **2** *verb (of plants)* to produce fruit; ***a strain of rice which crops heavily***

◊ **cropland** ['krɒplænd] *noun* agricultural land which is used for growing crops

◊ **cropper** ['krɒpə] *noun* **heavy cropper** = tree *or* plant which produces a large crop of fruit

◊ **crop rotation** *or* **rotation of crops** ['krɒp rəʊ'teɪʃn] *noun* system of cultivation where three different crops needing different nutrients are planted in three consecutive growing seasons to prevent the nutrients in the soil being totally used up. The land is then allowed to lie fallow for the fourth year

> COMMENT: the advantages of rotating crops are, firstly, that pests particular to one crop are discouraged from spreading and, secondly, that some crops actually benefit the soil. Legumes (peas and beans) increase the nitrogen content of the soil if their roots are left in the soil after harvesting

> QUOTE: most modern crops are not high-yielding, despite what their advocates may tell us. They are high response varieties. That is, given large inputs of artificial fertilisers to feed on, adequate water supplies, and protection from the epidemics they are so susceptible to, they can provide massive yields
> **Permaculture Magazine**

cross [krɒs] **1** *noun* **(a)** shape made with an upright line with another going across it, used as a sign of the Christian church; **the Red Cross** = international organization which provides emergency medical help **(b)** breeding of plants *or* animals from two different breeds or varieties; plant *or* animal bred from two different breeds or varieties; ***a cross between two strains of cattle*** **2** *verb* to breed a plant *or* animal from two different breeds or varieties; ***they crossed two strains of rice to produce a new strain which is highly resistant to disease***

◊ **crossbred** ['krɒsbred] *adjective* (animal) which is the result of breeding from two breeds; ***a herd of crossbred sheep***

◊ **crossbreed** ['krɒsbri:d] **1** *noun* animal which is bred from two different pure breeds **2** *verb* to breed new breeds of animals, by mating animals of different pure breeds

◊ **crossbreeding** ['krɒsbri:dɪŋ] *noun* mating *or* artificial insemination of animals of different breeds in order to combine the best characteristics of the two breeds

◊ **cross-fertilization** [krɒsfɜ:təlaɪ'zeɪʃn] *noun* fertilizing one individual plant from another of the same species

◊ **cross-pollination** [krɒspɒlɪ'neɪʃn] *noun* pollinating a plant with pollen from another plant of the same species. (The pollen goes from the anther of one plant to the stigma of another); *compare* SELF-POLLINATION

> COMMENT: all of these processes avoid inbreeding, which may weaken the species. Some plants are self-fertile (i.e. they are able to fertilize themselves) and do not need pollinators, but most benefit from cross-fertilization and cross-pollination

crown [kraʊn] *noun* top part of a tree where the main growing point is; ***the disease first affects the lower branches, leaving the crowns still growing***

◊ **crown-of-thorns starfish** ['kraʊnəv'θɔ:nz 'stɑ:fɪʃ] *noun* type of large starfish which lives on coral and destroys reefs

Cruciferae [kru:'sɪfəri:] *noun* family of common plants (such as the cabbage) whose flowers have four petals

crude oil *or* **crude petroleum** [kru:d] *noun* petroleum before it is refined and processed into petrol and other products

crusher ['krʌʃə] *noun* machine for breaking down rock, ore, seed, etc. into smaller pieces

crust [krʌst] *noun* hard top layer, especially the top layer of the surface of the earth, which is formed of rock and lies above the asthenosphere; **continental crust** = part of the earth's crust under the continents and continental shelves, lying above the oceanic crust

Crustacea [krʌs'teɪʃə] *noun* class of animals (such as crabs *or* lobsters) which have hard shells which are shed periodically as the animals grow

◊ **crustacean** [krʌs'teɪʃn] *noun* crab *or* lobster *or* shrimp, etc.

◊ **crustaceous** [krʌs'teɪʃəs] *adjective* referring to *or* belonging to the Crustacea

cryophilous [kraɪ'ɒfɪləs] *adjective* (plant) which needs a spell of cold weather to grow properly

COMMENT: cryophilous crops need a certain amount of cold weather in order to provide flowers later in the growing period. If such crops do not undergo this cold period, their growth remains vegetative, or they only form abortive flowers with no seeds. Cereal crops, such as wheat, barley and oats; peas; root crops like sugar beet and potatoes, are all cryophilous

cryophyte ['kraɪəfaɪt] *noun* plant which lives in cold conditions (as for example in snow)

cryptic coloration ['krɪptɪk kʌlə'reɪʃn] *noun* pattern of colouring which makes an animal less easy to see (such as the stripes on a zebra)

◊ **crypto-** ['krɪptəʊ] *prefix* meaning hidden

◊ **cryptozoic** [krɪptəʊ'zəʊɪk] *adjective* (animal) which lives hidden in holes in rocks or trees

crystal ['krɪstəl] *noun* regular shape formed by an element *or* compound when it becomes solid

◊ **crystalline** ['krɪstəlaɪn] *adjective* (rock) which is formed of crystals

Cs *chemical symbol for* caesium

CSF = CEREBROSPINAL FLUID

Cu *chemical symbol for* copper

cubic centimetre *or* **foot** *or* **inch** *or* **metre** *or* **yard** ['kju:bɪk] *noun* volume of a cube whose edge measures one centimetre *or* foot *or* inch *or* metre *or* yard

cuesta ['kwestə] *noun* ridge which has both scarp and dip slopes

cull [kʌl] **1** *noun* killing a certain number of animals in order to keep the population under control; **deer cull** *or* **seal cull** = killing a certain number of deer *or* seals **2** *verb* to kill a certain number of animals in order to keep the population under control

COMMENT: in the management of large wild animals without predators, such as herds of deer in Europe, it is usual to kill some mature animals each year to prevent a large population forming and overgrazing the pasture. Without culling the population would seriously damage their environment and eventually die back from starvation

QUOTE: introducing superior quality genotypes into the population can only improve the entire breed if inferior ones are culled
Appropriate Technology

QUOTE: the director of the fishermen's federation continues to call for a grey seal cull (he wants up to 15,000 killed each year)
BBC Wildlife

cullet ['kʌlɪt] *noun* broken glass (collected for recycling)

culm [kʌlm] *noun* **(a)** stem of a grass which produces flowers **(b)** type of waste from an anthracite processing plant

cultch [kʌltʃ] *noun* waste material placed in the sea to act as a breeding ground for oysters

cultivate ['kʌltɪveɪt] *verb* (i) to grow crops; (ii) to dig and manure the soil ready for growing crops; ***the fields are cultivated in the spring, ready for sowing corn;*** **cultivated land** = land which has been dug or manured for growing crops

◊ **cultivar** ['kʌltɪvɑ:] *noun* variety of a plant which has been developed under cultivation and which does not occur naturally in the wild, but which is a distinct sub-species

◊ **cultivation** ['kʌltɪ'veɪʃn] *noun* action of cultivating land; **land under cultivation** = land which is being cultivated *or* which has crops growing on it; ***the area under rice cultivation has grown steadily in the past 40 - 50 years;*** **to take land out of cultivation** = to stop cultivating land *or* growing crops on it and allow it to lie fallow; **shifting cultivation** = form of cultivation practised in some tropical countries, where land is cultivated until it is exhausted and then left as the farmers move on to another area

◊ **cultivator** ['kʌltɪveɪtə] *noun* **(a)** person who cultivates land **(b)** instrument *or* small machine for cultivating small areas of land

culture ['kʌltʃə] **1** *noun* bacteria *or* tissue grown in a laboratory; **culture medium** = liquid *or* gel used to grow bacteria *or* tissue; **stock culture** = basic culture of bacteria from which other cultures can be taken **2** *verb* to grow bacteria in a culture medium

culvert ['kʌlvət] *noun* covered drain for waste water

cumulus ['kju:mjʊləs] *noun* big fluffy white clouds which form at a low altitude (below 2,000m) as single clouds in warm summer weather, becoming eventually unbroken sheets of grey cloud (also called altocumulus or stratocumulus depending on the altitude); *see also* ALTOCUMULUS, STRATOCUMULUS

◊ **cumulonimbus** [kju:mjʊləʊ'nɪmbəs] *noun* dark low cumulus cloud whose top rises very high and produces rain from its lower part

curie ['kju:ri] *noun* former unit of measurement of radioactivity, now replaced by the becquerel; ***discharges of more than 20 curies per year are not permitted*** (NOTE: with figures usually written **Ci: 25Ci)**

current ['kʌrənt] *noun* flow (of water *or* air *or* electricity); ***dangerous currents make fishing difficult near the coast; a warm westerly current of air is blowing across the country***

curvature ['kɜ:vətʃə] *noun* way in which something bends from a straight line; **curvature of the earth** = where the horizon curves visibly (seen most clearly at sea)

curve [kɜ:v] *noun* bending line, as on a graph

cuspate foreland ['kʌspeɪt 'fɔ:lənd] *noun* large triangular area of coastal deposition

cut [kʌt] **1** *noun* act of felling a tree; **partial cut** = method of foresting where only some trees are felled, leaving others standing to seed the area which has been left clear; *see also* CLEARCUT **2** *verb* to fell (a tree)

◊ **cutting** ['kʌtɪŋ] *noun* small piece of a plant used for propagation; **leaf cutting** *or* **root cutting** *or* **stem cutting** = piece of a leaf *or* root *or* stem cut from a living plant and put in soil where it will sprout

> COMMENT: taking cuttings is a frequently used method of propagation which ensures that the new plant is an exact clone of the one from which the cutting was taken

cuticle ['kju:tɪkl] *noun* (i) thin waxy protective layer on plants; (ii) the epidermis, the outer layer of skin in animals

cwm [kʊm] *noun see* CIRQUE

cyan ['saɪən] *noun* primary blue colour

cyanide ['saɪənaɪd] *noun* any salt of hydrocyanic acid

> COMMENT: hydrogen cyanide *or* hydrocyanic acid (HCN) is a liquid *or* gas with a smell of almonds which kills very rapidly when drunk *or* inhaled. Cyanide is present in many effluents from industrial processes

cyano- ['saɪənəʊ] *prefix* meaning blue

◊ **cyanocobalamin** [saɪənəʊkəʊ'bæləmi:n] *noun* vitamin B_{12}

> COMMENT: Vitamin B_{12} is found in liver and kidney but not in vegetables. Lack of it can cause anaemia

Cyanophyta [saɪən'ɒfɪtə] *noun* blue-green algae, usually found in fresh water, which store starch in the form of glycogen

cyanosis [saɪə'nəʊsɪs] *noun* blue colour of the skin and mucous membranes, symptom of lack of oxygen in the blood (as in a blue baby)

cyclamate ['sɪkləmeɪt] *noun* sweetening substance (a salt of cyclamic acid) believed to be carcinogenic and banned in the USA as a food additive

cycle [saɪkl] *noun* series of events which recur regularly; ***industrial waste upsets the natural nutrient cycle;*** **carbon cycle** *or* **circulation of carbon** = carbon atoms from carbon dioxide are incorporated into organic compounds in plants during photosynthesis. They are then oxidized into carbon dioxide again during respiration by the plants or by herbivores which eat them and by carnivores which eat the herbivores, thus releasing carbon to go round the cycle again; **hydrologic(al) cycle** *or* **water cycle** = cycle of events when water in the sea evaporates in the heat of the sun, forms clouds which deposit rain as they pass over land, the rain then drains into rivers which return the water to the sea; **menstrual cycle** = period (usually about 28 days) during which a woman ovulates, then the walls of the uterus swell and bleeding takes place if the ovum has not been fertilized; **nitrogen cycle** = process by which nitrogen enters living organisms. The nitrogen is absorbed into green plants in the form of nitrates, the plants are then eaten by animals and the nitrates are returned to the ecosystem through the animal's excreta or when an animal or a plant dies

◊ **cyclical** ['sɪklɪkl] *adjective* referring to cycles

cyclo- ['sɪkləʊ] *prefix* meaning cyclical *or* referring to cycles; **cyclothymia** = mild form of manic depression where the patient suffers from alternating depression and excitement

cyclone ['saɪkləʊn] *noun* **(a)** any area of low pressure around which the air turns in the same direction as the earth; *see also* ANTICYCLONE **(b)** *(in the Indian Ocean)* **tropical cyclone** = tropical storm with masses of air turning rapidly round a low pressure area **(c)** device which removes solid particles from waste gases produced during industrial processes

◇ **cyclonic** [saɪ'klɒnɪk] *adjective* referring to a cyclone; **in a cyclonic direction** = flowing in the same direction as the rotation of the earth

◇ **cyclonically** [saɪ'klɒnɪkli] *adverb* in a cyclonic direction; ***waterspouts rotate cyclonically, that is anticlockwise in the Northern Hemisphere and clockwise in the Southern Hemisphere***

cylinder capacity ['sɪlɪndə kə'pæsɪti] *noun* total volume of a reciprocating engine's cylinders, expressed in litres, cubic centimetres *or* cubic inches

cyt- *or* **cyto-** [saɪt] *prefix* referring to cells

◇ **cytochemistry** [saɪtəʊ'kemɪstri] *noun* study of the chemical activity of living cells

◇ **cytogenetics** [saɪtəʊdʒə'netɪks] *noun* branch of genetics which studies the structure and function of cells, especially the chromosomes

◇ **cytokinesis** [saɪtəʊkaɪ'ni:sɪs] *noun* changes in the cytoplasm of a cell during cell division

◇ **cytology** [saɪ'tɒlədʒi] *noun* study of the structure and function of cells

◇ **cytolysis** [saɪ'tɒlɪsɪs] *noun* breaking down of cells

◇ **cytoplasm** ['saɪtəʊplæzm] *noun* jelly-like substance inside the cell membrane which surrounds the nucleus of a cell

◇ **cytoplasmic** [saɪtəʊ'plæzmɪk] *adjective* referring to the cytoplasm of a cell

◇ **cytosine** [saɪtəʊ'si:n] *noun* one of the nitrogenous bases of DNA

◇ **cytosome** [saɪtəʊ'səʊm] *noun* general term for a number of bodies in a cell

◇ **cytotoxin** [saɪtəʊ'tɒksɪn] *noun* substance which has a toxic effect on cells of certain organs

Dd

D *chemical symbol for* deuterium

Vitamin D ['vɪtəmɪn 'di:] *noun* vitamin which is soluble in fat and found in butter, eggs and fish (especially cod liver oil); it is also produced by the skin when exposed to sunlight

> COMMENT: Vitamin D helps in the formation of bones and lack of it causes rickets in children

dairy farming ['deəri 'fɑ:mɪŋ] *noun* keeping cows for milk production

dam [dæm] *noun* construction built to block a river

> COMMENT: dams are constructed either to channel the flow of water into hydroelectric power stations or to regulate the water supply to irrigation schemes. Dams can have serious environmental effects: the large heavy mass of water in the lake behind the dam may trigger earth movements if the rock beneath is unstable; in tropical areas, dams encourage the spread of bacteria, insects and parasites, leading to an increase in diseases such as bilharziasis; they may increase salinity in watercourses; in Arctic areas, the freezing of the large lake behind the dam may alter the whole climate of a region. A dam also retains silt which otherwise would be carried down the river and be deposited as fertile soil in the plain below. The silt also clogs the dam and in time may make it impossible to use the hydroelectric turbines

> QUOTE: construction of the dam, described by its developers as the world's biggest private power scheme, requires the clearance of more than 80,000 hectares, much of it tropical rainforest, and the resettlement of over 8,000 people
> **Financial Times**

damage ['dæmɪdʒ] **1** *noun* harm done to things, destruction; **damage threshold** = level of pollution above which harm is done; **desiccation damage** = harm done by the drying out (of the soil); **ecological damage** = harm done to an ecosystem; **environmental damage** = harm done to the environment, e.g. loss of wetlands, pollution of rivers, etc.; **genetic damage** = harm done to the genes of an organism, as caused by a pollutant, herbicide, etc. **2** *verb* to harm *or* spoil *or* destroy; ***the wind and rain damaged the barley crop***

damp [dæmp] **1** *adjective* slightly wet **2** *noun* slight wetness

◊ **dampen** [dæmpn] *verb* to make something slightly wet

◊ **damping** ['dæmpɪŋ] *noun* closing down of a blast furnace by cutting off the air supply

◊ **damp off** ['dæmp 'ɒf] *verb (of seedlings)* to die from a fungus which spreads in warm damp conditions and attacks the roots and lower stems

> COMMENT: damping off is a common cause of loss of seedlings in greenhouses

dangerous drugs ['deɪndʒərəs 'drʌgz] *noun* drugs (such as morphine and heroin) which may be harmful to people who take them and so can be prohibited from import or general sale

◊ **dangerous substance** ['deɪndʒərəs 'sʌbstəns] *noun* substance which is particularly hazardous because of its toxicity, bioaccumulation potential and persistence

Darwinism *or* **Darwinian theory** ['dɑ:wɪnɪzm *or* dɑ:'wɪniən 'θi:əri] *noun* theory of evolution, formulated by Charles Darwin (1809-82), which states that species of organisms arose by natural selection; *see also* NEO-DARWINISM; **Darwinian fitness** = measure of reproductive success *or* the success of organisms at passing on their genes to subsequent generations

data ['deɪtə] *noun* information, statistics; ***the data shows that a chemical change takes place***

daughter cell ['dɔ:tə 'sel] *noun* one of the cells that develop by mitosis from a single parent cell

daylight ['deɪlaɪt] *noun* light of the sun during the day; **daylight-saving time (DST)** *or US* **daylight time** = system of putting clocks forward one hour in the summer to provide extra daylight in the evening

dB = DECIBEL; **dBA scale** = DECIBEL A SCALE

DDT [di:di:'ti:] = DICHLORODIPHENYLTRICHLOROETHANE; **DDT-resistant insect** = insect which has built up a resistance to DDT

> COMMENT: DDT is a highly toxic insecticide which remains for a long time as a deposit in animal organisms. It gradually builds up in the food chain as smaller animals are eaten by larger ones, with the result that raptors (such as eagles and hawks) can be killed by the DDT which builds up in their systems from the bodies of smaller animals which they have eaten

deactivate [di'æktɪveɪt] *verb* to stop something being active

death [deθ] *noun* act of dying; **death rate** = number of deaths per year, shown per thousand of the population; ***a death rate of 15 per thousand; an increase in the death rate***

debacle [deɪ'bɑ:kl] *noun* breaking up of the ice on a large river as it melts in spring

debris ['debri:] *noun* rubbish *or* waste matter; ***volcanic debris from a volcano in eruption; mounds of debris left by open-cast mining;*** **debris flow** = landslide *or* mudslide of waste matter (as on a spoil heap)

decay [dɪ'keɪ] **1** *noun* **(a)** rotting *or* disintegration of dead organic matter; ***wood is treated to resist decay;*** **bacterial decay** = decay caused by the action of bacteria; **urban decay** = condition where part of a town becomes old *or* dirty *or* ruined because businesses and wealthy families have moved away from it **(b)** gradual disintegration of the nucleus of radioactive matter; **decay heat** = heat produced in a nuclear reactor as radioactive matter decays **2** *verb* **(a)** *(of organic matter)* to rot *or* to disintegrate **(b)** *(of radioactive matter)* to disintegrate

decibel (dB) ['desɪbel] *noun* degree of measurement of noise; **decibel A scale** *or* **dBA scale** = international scale of noise level

> COMMENT: the range of audible sounds is measured in decibels, which are calculated by measuring the sound pressure level. Decibels are calculated on a logarithmic scale, each step in the scale is an increase of ten times the previous step. The scale is further refined according to an A, B, C system, the dBA being the most used because it is the hearing level of humans

deciduous [de'sɪdjuəs] *adjective* **(a)** (tree) which loses its leaves *or* needles at some point during the year; ***deciduous woodlands grew up after the last Ice Age;*** **deciduous forest** = forest containing only deciduous trees **(b)** **deciduous dentition** *or* **deciduous teeth** = the first teeth of an animal, in humans called milk teeth, which are gradually replaced by permanent teeth

decimate ['desɪmeɪt] *verb* to reduce severely; ***overfishing has decimated the herring population in the North Sea***

decline [dɪ'klaɪn] **1** *noun* gradual reduction; ***the decline in the number of cases of pollution is due to better policing of factory emissions; the population of the birds now seems to be on the decline; ecologists are working to diagnose forest decline in its early stages;*** **decline in population** = reduction in the number of organisms living in one place; **on the decline** = gradually getting less **2** *verb* to become less; ***the fish population declined sharply as the water became more acid;* to**

decline in importance = to become less important

> QUOTE: much of the decline in energy consumption is due to the more efficient use of energy in homes and offices
> **Scientific American**

> QUOTE: the butterfly's alarming decline is probably the result of habitat reduction
> **Natural History**

declination [deklɪ'neɪʃn] *noun* **magnetic declination** = angle of difference between the direction of the North Pole and the North Magnetic Pole

decommission [di:kə'mɪʃn] *verb* to shut down (a nuclear reactor *or* a nuclear power station)

◇ **decommissioning** [di:kə'mɪʃənɪŋ] *noun* shutting down (a nuclear reactor *or* a nuclear power station); **decommissioning waste** = radioactive material no longer needed as a result of the shutting down of a nuclear reactor *or* a nuclear power station

decompose [di:kəm'pəʊz] *verb (of organic material)* to break down into simple chemical compounds

◇ **decomposable** [di:kəm'pəʊzəbl] *adjective* (material) which can be broken down into simple chemical compounds; **biologically decomposable** = (material) which can be broken down by the action of bacteria, sunlight, rain, the sea, etc.

◇ **decomposer** [di:kəm'pəʊzə] *noun* organism (such as the earthworm, a fungus *or* bacteria) which feeds on dead organic matter and breaks it down into simple chemicals

◇ **decomposition** [di:kɒmpə'zɪʃn] *noun* breaking down into simple chemical compounds

decontaminate [dikən'tæmɪneɪt] *verb* to remove a harmful substance, such as poison *or* radioactive material, from a building, a watercourse, a person's clothes, etc.

◇ **decontamination** [di:kəntæmɪ'neɪʃn] *noun* removal of a harmful substance, such as poison *or* radioactive material, from a building, a watercourse, a person's clothes, etc.

decrease 1 *noun* ['di:kri:s] reduction; ***there has been a decrease of 20% in the number of planning applications; the population of the lake now seems to be on the decrease;*** **decrease in the birth rate** = reduction in the number of births per annum; **decrease in the population** *or* **population decrease** = reduction in the number of organisms living in one place; **on the decrease** = reducing **2** *verb* [di:'kri:s] to reduce

DED = DUTCH ELM DISEASE

deep [di:p] *adjective* (water, well, mine, etc.) which goes down a long way; **deep ecology** = extreme form of ecological thinking whereby humans are considered as only one species among many in the environment and where large numbers of humans are seen as harmful to the environment in which they live

◇ **deep-freezing** [di:p'fri:zɪŋ] *noun* long-term storage at temperatures below freezing point; many crops, such as peas and beans, are grown specifically for the freezer market

◇ **deep-litter** ['di:p'lɪtə] *noun* system of using straw, wood shavings, sawdust or peat moss for bedding poultry or cattle; **deep-sea fisherman** = man who catches fish in the deepest parts of the sea; **deep-sea fishing** = catching fish in the deepest parts of the sea; **deep-sea plain** = ABYSSAL PLAIN; **deep-sea trench** = OCEANIC TRENCH; **deep-sea zone** = ABYSSAL ZONE

deficiency [dɪ'fɪʃənsi] *noun* lack of something essential; **iron deficiency** = lack of iron in the diet; **deficiency disease** = disease of plants *or* animals caused by the lack of an essential nutrient

◇ **deficient** [dɪ'fɪʃənt] *adjective* lacking something essential; ***the soil is deficient in important nutrients; scrub plants are well adapted to this moisture-deficient habitat; he has a calcium-deficient diet***

deficit ['defɪsɪt] *noun* situation in which the amount going out is larger than the amount coming in

definitive host [de'fɪnɪtɪv 'həʊst] *noun* host on which a parasite settles permanently

deflocculant [di:'flɒkjʊlənt] *noun* substance added to break up lumps which have formed in a liquid

◇ **deflocculation** [di:flɒkjʊ'leɪʃn] *noun* breaking up lumps which have formed in a liquid

defoliation [di:fəʊli'eɪʃn] *noun* removing leaves from plants

◇ **defoliant** [di:'fəʊliənt] *noun* type of herbicide which makes the leaves fall off plants

◇ **defoliate** [di:'fəʊlieɪt] *verb* to make the leaves fall off (a plant), especially by using a herbicide

deforest [di:'fɒrɪst] *verb* to cut down forest trees for commercial purposes *or* to make arable land; ***timber companies have helped to deforest the tropical regions; about 40,000 square miles are deforested each year***

◇ **deforestation** [di:fɒrɪs'teɪʃn] *noun* cutting down forest trees for commercial purposes *or* to make arable land

> QUOTE: the main reason for the drought and its severity is deforestation like that which has taken place in Ethiopia and Brazil. The destruction of the trees destroys the land's watersheds. The lack of trees also allows the fertile topsoil to blow away
> **Appropriate Technology**

degas [diː'gæs] *verb* to remove gas from something (as removing gas from a borehole)

◇ **degassing** [diː'gæsɪŋ] *noun* removing gas from something

degrade [dɪ'greɪd] *verb* (i) to reduce the quality of something; (ii) to make a chemical compound decompose into its elements; ***the land has been degraded through overgrazing; ozone may worsen nutrient leaching by degrading the water-resistant coating on pine needles***

◇ **degradable** [dɪ'greɪdəbl] *adjective* (substance) which can be degraded; *compare* BIODEGRADABLE

◇ **degradation** [degrə'deɪʃn] *noun* (i) reduction in the quality of something; making worse; becoming worse; (ii) decomposition of a chemical compound into its elements; ***chemical degradation of the land can be caused by overuse of fertilizers and by pollutants from industrial processes;*** **degradation of air** = pollution of clean air; **environmental degradation** = reduction in the quality of the environment

> QUOTE: it is possible to degrade the actinides to fission products with much shorter half-lives
> **New Scientist**

> QUOTE: around 900 million people in a hundred countries are threatened by the degradation of fragile land on the edges of the world's deserts
> **Appropriate Technology**

degree [dɪ'griː] *noun* **(a)** *(in science)* unit of measurement; ***a circle has 360°; the temperature is only 20° Celsius; water freezes at 0°C (zero degrees Celsius) or 32°F (thirty-two degrees Fahrenheit);*** **degree Celsius** = unit of measurement on a scale of temperature where the freezing and boiling points of water are 0° and 100°; *see also* CELSIUS; **degree Fahrenheit** = unit of measurement on a scale of temperature where the freezing and boiling points of water are 32° and 212°; *see also* FAHRENHEIT (NOTE: the word **degree** is written ° after figures: **40°C:** say: 'forty degrees Celsius') **(b)** level of how important *or* serious something is; **to a certain degree** *or* **to some degree** = to a certain extent; **to a minor degree** = in a small way

dehisce [diː'hɪs] *verb (of ripe seed pod or fruit)* to burst open to allow seeds to scatter

◇ **dehiscence** [diː'hɪsns] *noun* sudden bursting of a seed pod when it is ripe allowing the seeds to scatter

◇ **dehiscent** [diː'hɪsnt] *adjective* (seed pod) which bursts open to allow the seeds to scatter

dehumidifier [diːhjuː'mɪdɪfaɪə] *noun* machine which removes the moisture from air, usually part of an air-conditioning system

dehydrate [diːhaɪ'dreɪt] *verb* **(a)** to lose water; to make something lose water **(b)** to lose water from the body; ***after two days without food or drink, he became severely dehydrated***

◇ **dehydration** [diːhaɪ'dreɪʃn] *noun* **(a)** loss of water; making something lose water **(b)** loss of water from the body

> COMMENT: water is more essential than food for an animal's survival

deinking [diː'ɪŋkɪŋ] *noun* removing the ink from waste newspaper to allow it to be recycled; **deinking unit** = factory *or* machine which removes the ink from waste newspaper

deintensified farming [diːɪn'tensɪfaɪd 'fɑːmɪŋ] *noun* farming which was formerly intensive, using chemical fertilizers to increase production, but has now become extensive

delta ['deltə] *noun* **(a)** triangular piece of land at the mouth of a large river formed of silt carried by the river; ***the Nile Delta; the Mississippi Delta*** **(b)** fourth letter of the Greek alphabet; **delta ray** = electron forced from an atom by the effect of ionizing radiation

◇ **deltaic deposit** [del'teɪɪk diː'pɒzɪt] *noun* deposit of silt in a river delta

demand [dɪ'mɑːnd] *noun* need for something; ***they are building more power stations to satisfy the increasing demand for electricity; they calculate that demand will exceed supply within ten years;*** **chemical oxygen demand (COD)** = amount of oxygen taken up by organic matter in water, used as a measurement of the amount of organic matter in sewage; **supply and demand** = balance between things produced and required in a society, especially as affecting prices

deme [diːm] *noun* population of organisms in a small area

◇ **-deme** [diːm] *suffix* meaning section of a plant population which has distinct characteristics

demersal [dɪ'mɜːsəl] *adjective* (fish) which lives on *or* near the seabed; *compare* PELAGIC

demineralize [diː'mɪnərəlaɪz] *verb* to remove salts which are dissolved in water

◇ **demineralization** [diːmɪnərəlaɪ'zeɪʃn] *noun* removing salts which are dissolved in water

demography [dɪ'mɒgrəfi] *noun* study of human populations and their development

◇ **demographic** [demə'græfɪk] *adjective* referring to demography; **demographic transition** = pattern of change of population growth from high birth and death rates to low birth and death rates

denature [diː'neɪtʃə] *verb* (i) to make something change its nature; (ii) to convert a protein into an amino acid; **denatured alcohol** = alcohol which has unpleasant additives which are intended to make it unfit for human consumption

◇ **denaturation** [diːneɪtjə'reɪʃn] *noun* (i) making something change its nature; (ii) converting a protein into an amino acid

dendritic ['dendrɪtɪk] *adjective* with branches; **dendritic drainage** = system of drainage where smaller channels branch out from larger ones

◇ **dendrochronology** [dendrəʊkrɒ'nɒlədʒi] *noun* scientific method of finding the age of wood by the study of tree rings

◇ **dendroclimatology** [dendrəʊklaɪmə'tɒlədʒi] *noun* study of climate over many centuries, as shown in tree rings

> COMMENT: because tree rings vary in width depending on the weather during a certain year, it has been possible to show a pattern of yearly growth which applies to all wood from a certain region. This allows old structures to be dated even more accurately than with carbon-dating systems. It is also possible to chart past changes in climate in the same way, as rings vary in thickness according to rainfall, etc., allowing scientists to compare the climatic changes in various parts of the world over a very long period of time

dengue ['deŋgeɪ] *noun* breakbone fever, a tropical disease caused by an arbovirus, transmitted by mosquitoes, where the patient suffers a high fever, pains in the joints, headache and a rash

denitrification [diːnaɪtrɪfɪ'keɪʃn] *noun* releasing nitrogen from nitrates in the soil by the action of bacteria

dense [dens] *adjective* thick *or* close together; ***the animals live safely in dense tropical rainforest***

◇ **densely** ['densli] *adverb* **densely populated** = having many organisms inhabiting the same area

◇ **density** ['densɪti] *noun* (i) mass of the volume of a substance; (ii) number of species in a certain area; **density dependence** = situation where a population of a species is regulated by its density; **density-dependent** = (population) which is controlled by its own size; **density-independent** = (population) which is not controlled by its own size

denude [dɪ'njuːd] *verb* to make land *or* rock bare by cutting down trees *or* by erosion; ***the timber companies have denuded the mountains***

◇ **denudation** [dɪnjuː'deɪʃn] *noun* making land *or* rock bare by cutting down trees *or* by erosion

deoxygenate [diː'ɒksɪdʒəneɪt] *verb* to remove oxygen from (water, air, etc.)

◇ **deoxygenation** [diːɒksɪdʒə'neɪʃn] *noun* removing oxygen

deoxyribonucleic acid (DNA) [diːɒksɪraɪbəʊnjuː'kliːɪk 'æsɪd] *noun* one of the nucleic acids, the basic genetic material present in the nucleus of each cell; *see also* RNA

department [dɪ'pɑːtmənt] *noun* section of a government *or* government agency; *GB* **Department of the Environment (DoE)** = British government department concerned with the conditions in which people live, and also responsible for contacts between central government and certain aspects of local government; *US* **Department of Energy (DOE)** = US government department responsible for controlling energy (nuclear power, gas, electricity, etc.)

dependent [dɪ'pendənt] *adjective* which relies on something; ***the country's whole economy is dependent on its natural resources***

deplete [dɪ'pliːt] *verb* to remove (a resource); ***runoff from hillsides depletes the soil of nutrients***

◇ **depletion** [dɪ'pliːʃn] *noun* removing a resource from an area; ***a study of atmospheric ozone depletion; production of ice crystals from methane can deplete ozone still further;*** **ozone depletion** = removing ozone from the atmosphere; **soil depletion** = reduction of the soil layer by erosion

deposit [dɪ'pɒzɪt] **1** *noun* substance which has been laid down *or* has accumulated on something; ***deposits of coal have been found in the north of the country;*** **alluvial deposits** = deposits of silt on the bed of a river *or* lake; **coal deposits** = layers of coal in the rocks beneath the earth's surface; **lake deposits** = deposits of silt on the bed of a lake **2** *verb* to place (something) on; ***the volcanic eruption deposited a thin layer of ash over a wide area;*** **deposited matter** = fragments of rock, sand, shells, mud, etc. left by the action of rivers, the sea, a glacier *or* the wind

◇ **deposition** [depə'zɪʃn] *noun* action of placing something; ***the deposition of sediment at the bottom of a lake;*** **dry deposition** = falling of dry particles from polluted air (in the same way as acid rain falls) which form a harmful deposit on surfaces such as buildings or the leaves of trees; ***some of the sulphate particles settle to the ground in the process called dry deposition***

◊ **depository** [dɪ'pɒzɪtri] *noun* place where something is stored, such as refuse *or* nuclear waste; ***they carried out tests to establish the suitability of the rock formation as a waste depository;*** **deep depository** = place where nuclear waste is stored in very deep holes in the ground; *compare* REPOSITORY

depot ['depəʊ] *noun* place *or* building where something is stored; ***there has been a fire at the oil depot***

depression [dɪ'preʃn] *noun* **(a)** cyclone *or* area of low atmospheric pressure usually accompanied by rain **(b)** area of low land surrounded by higher ground

depth [depθ] *noun* how deep something is; distance downwards; **the water is two metres in depth** = is two metres deep; **depth of rainfall** = measurement of the amount of rain which has fallen

derelict ['derəlɪkt] *adjective* (land) which has been damaged and made ugly by mining *or* industrial processes *or* which has been neglected and is not used for anything; (building) which is neglected and in ruins; ***a plan to reconstruct derelict inner city sites***

◊ **dereliction** [derɪ'lɪkʃn] *noun* (i) being damaged, ugly, neglected, in ruins; (ii) not doing one's duty

derivative [də'rɪvətɪv] *noun* substance *or* product which is formed from something else; **petroleum derivative** = substance *or* product made from petroleum

derris ['derɪs] *noun* powdered insecticide extracted from the root of a tropical plant, used against fleas, lice and aphids

desalinate [diː'sælɪneɪt] *verb* to remove salt from (water *or* soil)

◊ **desalination** [diːsælɪ'neɪʃn] *noun* removing salt from a substance such as sea water *or* soil; **desalination plant** = factory which removes the salt from sea water to produce fresh drinking water

> COMMENT: desalination can refer simply to the removal of salts from soil by the action of rain. It is most commonly used to mean the process of removing salt from sea water to make it drinkable. Desalination plants work by distillation, dialysis or by freeze drying. The process is very costly, and is only cost-effective in desert countries where the supply of fresh water is minimal

desert ['dezət] *noun* area of land with very little rainfall, arid soil and little or no vegetation; **biological desert** = area where there is no life (not necessarily on land, as heavy pollution of a lake can turn the bottom of the lake into a biological desert); **cold desert** = area with little vegetation because of the cold temperatures; **hot desert** *or* **tropical desert** = desert situated in the tropics, such as the Sahara Desert *or* the Arabian Desert; **mid-latitude desert** *or* **warm desert** = desert situated between the tropics, such as the Gobi Desert *or* the Turkestan Desert; **rock desert** = desert where underlying rock has been exposed by the wind blowing away topsoil; **desert formation** = DESERTIFICATION; **desert soil** = the soil in a desert (it is normally sandy with little organic matter)

◊ **desertification** [dɪzɜːtɪfɪ'keɪʃn] *noun* process by which an area of land becomes a desert because of a change of climate *or* because of the action of man, e.g. through intensive farming; ***changes in the amount of sunlight reflected by different vegetation may contribute to desertification; increased tilling of the soil, together with long periods of drought have brought about the desertification of the area***

◊ **desertify** [dɪ'zɜːtɪfaɪ] *verb* to make land into a desert; ***it is predicted that half the country will be desertified by the end of the century***

> COMMENT: a desert will be formed in areas where rainfall is very low (less than 25 centimetres per annum) whether the region is hot or not. About 30% of all the land surface of the earth is desert or in the process of becoming desert. The spread of desert-type conditions in arid and semi-arid regions is caused not only by climatic conditions, but also by human pressures. So overgrazing of pasture and the clearing of forest for fuel and for cultivation, both lead to the loss of organic material, a reduction in rainfall by evaporation and to soil erosion. Reclaiming desert is a very expensive operation and usually not very successful

> QUOTE: desertification, broadly defined, is one of the principal barriers to sustainable food security and sustainable livelihoods in our world today
> **Environmental Conservation**

desiccate ['desɪkeɪt] *verb* to dry out (soil); ***cattle wander across the desiccated pastures in search of fodder***

◊ **desiccant** ['desɪkənt] *noun* (i) substance which dries; (ii) type of herbicide which makes leaves wither and die

◊ **desiccation** [desɪ'keɪʃn] *noun* act of drying out (the soil); ***the greenhouse effect may lead to climatic changes such as the vast desiccation of Africa;*** **desiccation damage** = harm done by the drying out (of the soil)

design [dɪ'zaɪn] **1** *noun* plan of how to lay something out *or* how to make something; **landscape design** = plan of how to lay out a landscape, where to plant trees and shrubs, etc.; **urban design** = plan of how to lay out a town, where to build certain buildings, etc. **2** *verb* to plan how to lay something out *or* how to make something

designate ['dezɪgneɪt] *verb* to name *or* appoint officially; ***an area of woodland which has been designated a Site of Special Scientific Interest; the city has been designated a nuclear-free zone;*** **designated development area** = area which has been officially appointed to be given special help from a government to encourage business and factories to be set up there

destroy [dɪ'strɔɪ] *verb* to kill *or* to remove completely; ***at this rate, all virgin rainforests will have been destroyed by the year 2020; the building of the motorway will destroy several areas of wild fen***

◊ **destruction** [dɪs'trʌkʃn] *noun* act of killing *or* removing completely; ***the destruction of the habitat has led to the almost complete extinction of the species***

desulphurization [diːsʌlfəraɪ'zeɪʃn] *noun* process of removing sulphur from a substance such as oil *or* iron ore *or* coal

detect [dɪ'tekt] *verb* to sense *or* to notice (usually something which is very small or difficult to see); ***an instrument to detect microscopic changes in cell structure; sensors detected an increase in radiation levels***

◊ **detectable** [dɪ'tektəbl] *adjective* which can be detected; ***some toxic substances are invisible and only detectable by analysis in a laboratory***

◊ **detection** [dɪ'tekʃn] *noun* action of sensing *or* noticing something; ***the detection of sounds by nerves in the ears; the detection of minute levels of lead in the atmosphere***

> QUOTE: the increase in the amount of carbon dioxide, together with trace gases such as methane and nitrogen oxide, is likely to cause a detectable global warming
>
> **Scientific American**

detergent [dɪ'tɜːdʒnt] *noun* cleaning substance which removes grease and bacteria from the surface of something; **detergent foam** *or* **detergent swans** = large masses of froth on the surface of rivers, canals and sewers, caused by hard detergent in effluent; **hard detergent** = detergent which is not broken down in water; **soft detergent** = detergent which is broken down in water but which can cause eutrophication and algal bloom

> COMMENT: most detergents do not trigger allergies but some biological detergents, which contain enzymes to remove protein stains, can cause skin rashes. The first detergents contained alkyl benzene sulphonate which does not degrade on contact with bacteria and so passed into sewage, creating large amounts of foam in sewers and rivers; some types of detergent are 'biological', that is, they are biodegradable and so can be degraded by bacteria

deteriorate [dɪ'tɪəriəreɪt] *verb* to become worse; ***the quality of the water in the river has deteriorated since the construction of factories on its banks***

◊ **deterioration** [dɪtɪəriə'reɪʃn] *noun* becoming worse; **deterioration of the environment** = reduction in the quality of the environment

> QUOTE: evidence for worldwide climatic deterioration is convincing and well-documented
>
> **New Scientist**

> QUOTE: according to recent estimates by some of the world's leading soil scientists, an area of about 1.2 thousand million hectares - about the size of China and India combined - has experienced moderate to extreme soil deterioration since World War II as a result of human activities
>
> **Environmental Conservation**

determine [dɪ'tɜːmɪn] *verb* to find out something correctly; ***inspectors are trying to determine the source of the pollution***

detinning [diː'tɪnɪŋ] *noun* removal of a coating of tin from something

detoxication *or* **detoxification** [diːtɒksɪ'keɪʃn *or* diːtɒksɪfɪ'keɪʃn] *noun* removal of harmful *or* poisonous substances

◊ **detoxify** [diː'tɒksɪfaɪ] *verb* to remove harmful *or* poisonous substances

detrimental [detrɪ'mentəl] *adjective* which can harm; ***conservation groups have criticized the introduction of red deer, a species which is highly detrimental to local flora***

detritus [dɪ'traɪtəs] *noun* waste matter (either organic *or* mineral)

◊ **detrital** [dɪ'traɪtəl] *adjective* formed from detritus; (crystal) which has been uncovered from weathered rock; **detrital food chain** = link between green plants and the decomposer organisms which feed on them

◊ **detritivorous** [dɪ'trɪtɪvɔːrəs] *adjective* (organism) which feeds on dead organic matter and breaks it down into simple chemicals

◊ **detrivore** [dɪ'trɪtɪvɔː] *noun* decomposer *or* organism (such as the earthworm, a fungus *or* bacteria) which feeds on dead organic matter and breaks it down into simple chemicals

deuterium [djuː'tɪəriəm] *noun* chemical element which is an isotope of hydrogen;

deuterium oxide (D_2O) = heavy water *or* water containing deuterium instead of the hydrogen atom, used as a coolant or moderator in certain types of nuclear reactor such as the CANDU (NOTE: chemical symbol is **D**; atomic number is **1**)

develop [dɪ'veləp] *verb* **(a)** to grow; to make grow *or* to create; ***the embryo developed quite normally in spite of the mother's illness; the company is trying to develop a new pesticide to deal with the problem*** **(b)** to plan and produce; **developed countries** = countries which have a high state of industrialization; **developing countries** = countries which are not fully industrialized **(c)** to plan and build on (an area of land); ***they are planning to develop the site as an industrial estate;*** **developed area** = area which has buildings on it

◊ **developer** [dɪ'veləpə] *noun* person *or* company that plans and builds structures such as roads, airports, houses, factories or office buildings; ***the land has been acquired by developers for an industrial park***

◊ **development** [dɪ'veləpmənt] *noun* **(a)** action of growing *or* creating; ***the development of the embryo takes place in the uterus; the development of new pesticides will take some time;*** **economic development** = process where a country changes from an economy based on agriculture to one based on industry; **industrial development** = planning and building of new industries in special areas **(b)** planning and building on an area of land; ***they are planning large-scale development of the docklands area;*** **development area** *or* **development zone** = area which has been given special help from a government to encourage business and factories to be set up there; **development plan** = plan drawn up by a government *or* local council showing how an area will be developed over a long period **(c)** area which has buildings on it; **housing development** = area which has houses built on it

device [dɪ'vaɪs] *noun* machine *or* tool; ***they have installed devices for measuring radiation on the site; the company has brought in a device for taking rock samples***

dew [dju:] *noun* water which condenses on surfaces in the open air as humid air cools at night; **dew point** = temperature at which dew forms on grass *or* leaves, etc.

◊ **dew pond** ['dju: 'pɒnd] *noun* small pond which forms on high chalky soil: it is formed from rainwater not dew

DHW = DOMESTIC HOT WATER

diagram ['daɪəgræm] *noun* chart *or* drawing which records information as lines or points; ***the book gives a diagram of the movement of air currents; the diagram shows the occurrence of cancer in the region of the nuclear power station***

dialysis [daɪ'æləsɪs] *noun* using a membrane as a filter to separate soluble waste substances from a liquid

◊ **dialyzing membrane** ['daɪəlaɪzɪŋ 'membreɪn] *noun* membrane used in dialysis

> COMMENT: in dialysis, larger dispersed particles are not allowed through the membrane but small dissolved particles will pass through

diameter [daɪ'æmɪtə] *noun* distance across a circle (such as a tube *or* pipe, etc.); **each droplet is less than 0.5mm in diameter** = measures less than 0.5mm across its widest part

diatom ['daɪətəm] *noun* type of tiny algae, with a skeleton formed of silica, which makes up some basic organisms such as plankton

◊ **diatomaceous** [daɪətə'meɪʃəs] *adjective* (earth, such as kieselguhr, *or* rock) formed from the fossil remains of diatoms

◊ **diatomist** ['daɪətɒmɪst] *noun* scientist who studies diatoms; ***diatomists have found that the lake is ten times more acid than it was ten years ago***

◊ **diatomite** [daɪ'ætəmaɪt] *noun* rock formed from the skeletons of diatoms

> COMMENT: most of the producers in sea water food chains are diatoms. Bodies of diatoms have contributed to the formation of oil reserves

dibromochlorpropane [daɪbrəʊməʊklɔ:'prəʊpeɪn] *noun* powerful insecticide

dichlorodiphenyltrichloroethane (DDT) [daɪklɒrəʊdaɪfenɪltraɪklɒrəʊ'i:θeɪn] *noun* highly toxic insecticide, no longer recommended for use

> COMMENT: DDT is an organochlorine insecticide which remains for a long time as a deposit in animal organisms. It gradually builds up in the food chain as smaller animals are eaten by larger ones, with the result that raptors (such as eagles and hawks) can be killed by the DDT which builds up in their systems from the bodies of smaller animals which they have eaten

dicotyledon [daɪkɒtɪ'li:dən] *noun* plant of which the cotyledon has two leaves: one of the two classifications of plants; ***dicotyledons form the largest group of plants*** *compare* COTYLEDON, MONOCOTYLEDON

◊ **dicotyledenous** [daɪkɒtɪ'li:dənəs] *adjective* referring to dicotyledons

die [daɪ] *verb* to stop living; ***the fish in the lake died poisoned by chemical discharge from the factory; scientists are trying to find out what makes the trees die***

◇ **die back** ['daɪ 'bæk] *verb (of a branch or shoot of a plant)* to die; ***silver fir are known to die back from time to time***

◇ **dieback** ['daɪbæk] *noun* **(a)** fungus disease of certain trees which kills shoots or branches **(b)** gradual dying of trees caused by acid rain (starting at the ends of branches); ***half the trees in the forest are showing signs of dieback*** *see also* WALDSTERBEN

COMMENT: there are many theories to try to explain the causes of dieback: sulphur dioxide, nitrogen oxides, ozone, are all possible causes; also acidification of the soil or acid rain on leaves

◇ **die down** ['daɪ 'daʊn] *verb* to become less strong; ***the strong winds have died down***

dieldrin [daɪ'eldrɪn] *noun* organochlorine insecticide which kills on contact. It is very persistent and can kill fish, birds and small mammals when it enters the food chain

diesel (oil *or* **fuel)** ['di:zəl] *noun* special type of fuel used in certain engines; ***diesel-powered vehicles are one of the main sources of air pollution;*** **diesel engine** = engine which uses diesel oil as a fuel; **diesel-engined vehicle** *or* **diesel-powered vehicle** = motor vehicle which uses diesel oil as a fuel

COMMENT: in a diesel engine, a quantity of oil is pumped to each cylinder in turn. Diesel-engined cars are usually more economical to run, but cause more pollution than petrol-engined cars

diet ['daɪət] **1** *noun* **(a)** amount and type of food eaten; ***it lives on a diet of insects and roots; the normal western diet is too full of carbohydrates;*** **balanced diet** = diet which contains the right quantities of basic nutrients; **low-calorie diet** = diet with few calories (which can help a person to lose weight); **salt-free diet** = diet which does not contain salt **(b)** measured amount of food eaten, usually to try to lose weight; **to be on a diet** = to reduce the quantity of food eaten in order to lose weight **2** *verb* to reduce the quantity of food eaten in order to lose weight *or* to change the type of food eaten in order to become healthier

◇ **dietary** ['daɪətri] *adjective* referring to a diet; **dietary fibre** *or* **roughage** = fibrous matter in food, which cannot be digested; **Dietary Reference Values (DRV)** = list published by the British government of nutrients that are essential for health

COMMENT: dietary fibre is found in cereals, nuts, fruit and some green vegetables. It is believed to be necessary to help digestion and avoid developing constipation, obesity and appendicitis

◇ **dietetic** [daɪə'tetɪk] *adjective* referring to diet

◇ **dietetics** [daɪə'tetɪks] *noun* study of food, nutrition and health, especially when applied to the food intake

◇ **dietitian** [daɪə'tɪʃn] *noun* specialist in dietetics

difference ['dɪfrəns] *noun* way in which two things are not the same; ***can you tell the difference between butter and margarine?***

◇ **different** ['dɪfrənt] *adjective* not the same; ***living in the country is very different from living in town; the landscape looks quite different since the mine was opened***

◇ **differential** [dɪfə'renʃl] *adjective* showing a difference; depending on a difference; **differential heating** = differences in the amount of heat received when different surfaces are heated by the sun: white surfaces, such as snow, heat less than dark ones, such as soil

diffuse 1 *verb* [dɪ'fju:z] to spread through tissue; ***some substances easily diffuse through the walls of capillaries*** **2** *adjective* [dɪ'fju:s] (disease) which is widespread in the body *or* which affects many organs *or* cells

◇ **diffusion** [dɪ'fju:ʒn] *noun* (i) spreading (of gas *or* light); (ii) mixing a liquid with another liquid *or* mixing a gas with another gas; (iii) passing of a liquid *or* gas through a membrane

digest [daɪ'dʒest] *verb* **(a)** to break down food in the alimentary tract and convert it into elements which can be absorbed by the body **(b)** to process waste, especially organic waste such as manure, to produce biogas; ***55% of UK sewerage sludge is digested; wastes from food processing plants can be anaerobically digested***

◇ **digester** [daɪ'dʒestə] *noun* machine which takes refuse and produces gas such as methane from it; **aerobic digester** = digester which operates in the presence of oxygen; **anaerobic digester** = digester which operates without oxygen; ***anaerobic digesters can be used to convert cattle manure into gas;*** **digester gas** = gas, such as methane, produced by a digester

◇ **digestible** [daɪ'dʒestəbl] *adjective* which can be digested; ***glucose is an easily digestible form of sugar;*** **digestible organic matter (DOM)** = substance, such as manure, which can be processed to produce biogas

◇ **digestion** [daɪ'dʒestʃn] *noun* (i) process by which food is broken down in the alimentary tract and converted into elements which can be absorbed by the body; (ii) conversion of organic matter into simpler chemical compounds, such as the production of biogas from manure

◇ **digestive** [daɪ'dʒestɪv] *adjective* referring to digestion; **digestive enzyme** = enzyme which speeds up the process of digestion; **digestive system** = all the organs in the body (such as the liver and pancreas) which are associated with the digestion of food

> QUOTE: the digester was developed by a group of scientists concerned with technological efficiency
> **Appropriate Technology**

> QUOTE: research has shown that up to 60% of the contaminants present in the sewage stream may be absorbed into the sludge removed in primary screening operations and then treated by digestion process
> **London Environmental Bulletin**

dilute [daɪ'lu:t] **1** *adjective* weak (solution) **2** *verb* to add water to a solution to make it weaker; ***the disinfectant must be diluted in four parts of water before it can be used on the skin;*** **dilute and disperse** = method of using unlined landfill sites where the waste is allowed to leak gradually into the surrounding soil (NOTE: opposite is **concentrate**)

◇ **diluent** ['dɪljuənt] *noun* substance (such as water) which is used to dilute a liquid

◇ **dilution** [daɪ'lu:ʃn] *noun* (i) action of diluting; (ii) liquid which has been diluted

> COMMENT: dilution is one of the processes of sewage treatment whereby effluent is passed into a large quantity of water

dimethyl sulphide [daɪ'meθɪl 'sʌlfaɪd] *noun* gas given off by water which is rich in sewage pollution

dinitro-ortho-cresol (DNOC) [daɪnaɪtrəʊɔ:θəʊ'kri:sɒl] *noun* chemical used in a compound with petroleum oil to destroy the eggs of fruit tree pests; DNOC is a poisonous yellow dye, and can also be used as a weedkiller

dioxide [daɪ'ɒksaɪd] *see* CARBON

dioxin [daɪ'ɒksɪn] *noun* extremely poisonous gas formed as a by-product of the manufacture of the herbicide 2,4,5-T. (This is the gas that escaped in the disaster at Seveso in 1976)

disaster [dɪ'zɑ:stə] *noun* terrible event which kills and causes massive destruction; **ecological disaster** = disaster which seriously disturbs the balance of the environment; **natural disaster** = phenomenon which occurs in nature (storm, flood, earthquake, etc.), which destroys property and kills people and livestock

discharge 1 *noun* ['dɪstʃɑ:dʒ] **(a)** action of passing waste material into the environment; **dust discharge** = release of dust into the atmosphere, especially from an industrial process; **heat discharge** *or* **thermal discharge** = release of waste heat into the atmosphere, especially from an industrial process; **discharge pipe** = pipe which carries waste material from an industrial process or in the form of sewage and deposits it somewhere else, e.g. in a tank or in the sea; ***discharge pipes take the liquid waste into the sea*** **(b)** waste material, as from an industrial process or in the form of sewage, which is passed into the environment; **radioactive discharges** = radiation **(c)** rate of flow of a liquid in a channel **2** *verb* [dɪs'tʃɑ:dʒ] **(a)** *(of a river)* to flow into a lake *or* the sea; ***the river Rhine discharges into the North Sea*** **(b)** to pass (waste material) into the environment; ***the factory discharges ten tonnes of toxic effluent per day into the river***

> QUOTE: the mill was poorly ventilated and such dust as was extracted from the air was discharged to an atmosphere in close proximity to houses and a local school
> **Environment Now**

disclimax [dɪs'klaɪmæks] *noun* stable ecological state which has been caused by human intervention (such as a desert caused by deforestation); **disclimax community** = stable plant community which is caused by human action (such as felling a rainforest for timber); *compare* CLIMAX

discolour *or US* **discolor** [dɪs'kʌlə] *verb* to change the colour of something, usually by making it paler

◇ **discoloration** *or* **discolouration** [dɪskʌlə'reɪʃn] *noun* change of colour

◇ **discoloured** *or US* **discolored** [dɪs'kʌləd] *adjective* with a changed colour

discontinuity [dɪskɒntɪ'nju:ɪti] *noun* band in the interior of the earth which separates two layers and through which seismic shocks do not pass; **Gutenberg discontinuity** = boundary separating the mantle from the core; **Mohorovičić discontinuity** = boundary layer in the interior of the earth between the crust and the mantle, below which seismic shocks move more rapidly

disease [dɪ'zi:z] *noun* illness (of people *or* animals *or* plants, etc.) where the body functions abnormally; ***he caught a disease in the tropics; the rice crop has been affected by a virus disease; he is a specialist in occupational diseases*** *or* ***in diseases which affect workers***

◇ **diseased** [dɪ'zi:zd] *adjective* (plant *or* person *or* part of the body) affected by an illness, not functioning normally, not whole or normal; ***to treat dieback, diseased branches should be cut back to healthy wood***

disinfect [dɪsɪn'fekt] *verb* to make (something *or* a place) free from germs *or* bacteria; ***all utensils must be thoroughly disinfected***

◇ **disinfectant** [dɪsɪn'fektənt] *noun* substance used to kill germs *or* bacteria

◇ **disinfection** [dɪsɪn'fekʃn] *noun* making something *or* a place free from germs *or* bacteria (NOTE: the words **disinfect** and **disinfectant** are used for substances which destroy germs on instruments, objects or the skin; substances used to kill germs inside infected people are **antibiotics, drugs, etc.**)

disinfest [dɪsɪn'fest] *verb* to get rid of vermin from a place

disintegrate [dɪs'ɪntɪgreɪt] *verb* to come to pieces; ***the wood had been eaten by termites and simply disintegrated***

◇ **disintegration** [dɪsɪntɪ'greɪʃn] *noun* act of coming to pieces

dismantle [dɪs'mæntl] *verb* to remove all parts of a nuclear reactor which have been contaminated by radiation and clean them as part of the decommissioning process

◇ **dismantlement** *or* **dismantling** [dɪs'mæntəlmənt *or* 'dɪsmæntlɪŋ] *noun* removing all parts of a nuclear reactor which have been contaminated by radiation and clean them as part of the decommissioning process

disorder [dɪs'ɔːdə] *noun* (i) disruption of the normal system *or* balance; (ii) illness; **environmental disorder** = disruption of the normal balance of the environment; **respiratory disorder** = illness which affects the patient's breathing

dispenser [dɪs'pensə] *noun* device *or* machine which gives out something; **aerosol dispenser** = container *or* device from which liquid *or* powder can be sprayed in tiny particles

disperse [dɪs'pɜːs] *verb* **(a)** *(of organisms)* to spread over a wide area **(b)** to send out over a wide area; ***power stations have tall chimneys to disperse the emissions of pollutants*** **(c)** **dispersed particles** = particles which are not dissolved in a liquid but remain in suspension

◇ **dispersal** [dɪs'pɜːsəl] *noun* spreading of individual plants *or* animals into or from an area; ***aphids breed in large numbers and spread by dispersal in wind currents;*** **water dispersal** *or* **wind dispersal** = spreading of plants by seed carried away by water *or* blown by the wind

◇ **dispersing agent** [dɪs'pɜːsɪŋ 'eɪdʒənt] *noun* chemical substance sprayed onto an oil slick to try and break up the oil into smaller particles

◇ **dispersion** [dɪs'pɜːʃn] *noun* pattern in which animals *or* plants are found over a wide area; **dispersion aerosol** = droplets of moisture which are blown into the air (as spray)

dispose of [dɪs'pəʊz ʌv] *verb* to get rid of something; ***the problem with nuclear reactors is how to dispose of the radioactive waste***

◇ **disposable** [dɪs'pəʊzəbl] *adjective* (cup, napkin, etc.) which is thrown away after use

◇ **disposal** [dɪs'pəʊzl] *noun* getting rid of something; ***the disposal of raw sewage into the sea is contaminating shellfish;*** **land disposal** = depositing waste in a hole in the ground; **marine disposal** = depositing waste at sea; **refuse disposal** *or* **waste disposal** = getting rid of refuse *or* waste; **sewage disposal** = getting rid of sewage

disrupt [dɪs'rʌpt] *verb* to upset the normal system *or* balance

◇ **disruption** [dɪs'rʌpʃn] *noun* upsetting the normal system *or* balance; **ecological disruption** = upsetting the balance of an ecosystem

dissemination [dɪsemɪ'neɪʃn] *noun* teaching about something *or* giving out information about a new process; ***the dissemination of new technology in third world countries***

dissolve [dɪ'zɒlv] *verb* to melt *or* to make something disappear in liquid; ***nitric acid readily dissolves in cloud droplets;*** **dissolved oxygen** = oxygen molecules dissolved in water, needed by organisms that live in water

◇ **dissolvable** [dɪ'zɒlvəbl] *adjective* (mineral) which can be dissolved

distil [dɪs'tɪl] *verb* **(a)** to produce a pure liquid by heating a liquid and condensing the vapour, as in the production of alcohol *or* essential oils; **distilled water** = pure water, used in certain industrial processes and in electric batteries **(b)** to produce by-products from coal

◇ **distillate** ['dɪstɪlət] *noun* substance produced by distillation

◇ **distillation** [dɪstɪ'leɪʃn] *noun* process of producing a pure liquid by heating a liquid and condensing the vapour, as in the production of alcohol *or* essential oils; **fractional distillation** = distillation process where different fractions are collected at different points during the process

distributary [dɪs'trɪbjuːtəri] *noun* stream *or* river flowing out from a larger river, as in a delta

distribute [dɪs'trɪbjuːt] *verb* to spread out over an area; ***the species is widely distributed throughout Northern Europe***

◇ **distribution** [dɪstrɪ'bjuːʃn] *noun* pattern in which a species is found in various areas, depending on climate, altitude, etc.; ***the***

distribution of crops in various regions of the world is a result of thousands of years of breeding and testing; **distribution area** = number of places in which a species is found

district ['dɪstrɪkt] *noun* part of an area, especially for administrative purposes; **district authority** = official body which controls the administration of the local area; **district heating** = system of heating all houses in a district from a central source (as from hot springs in Iceland or by cooling water from a power station)

disturb [dɪs'tɜːb] *verb* to change the condition of something; ***the building of the road has disturbed the balance of the ecosystem***

◇ **disturbance** [dɪs'tɜːbəns] *noun* change in an ecosystem caused by some alteration of the environmental conditions, such as drought, pollution, clearfelling of woodland, etc.; **disturbance threshold** = point at which an alteration of the environmental conditions causes change in an ecosystem

ditch [dɪtʃ] *noun* channel to take away rainwater

diuresis [daɪju'riːsɪs] *noun* increase in the production of urine

◇ **diuretic** [daɪju'retɪk] *adjective & noun* (substance) which makes the kidneys produce more urine

diurnal [daɪ'ɜːnəl] *adjective* happening in the daytime *or* happening every day; **diurnal cycle** = cycle (as of air pollution) during the daytime; **diurnal rhythm** = regularly recurring activities which take place every day (such as feeding, sleeping, hunting, etc.)

divergence [daɪ'vɜːdʒəns] *noun* phenomenon which occurs whenever there is a net outflow of air from a region of the atmosphere, resulting in the depletion of air and a reduction in density; *compare* CONVERGENCE

diversify [daɪ'vɜːsɪfaɪ] *verb* to do a number of different things; ***farmers are encouraged to diversify land use by using it for woodlands or for recreational facilities***

◇ **diversification** [daɪvɜːsɪfɪ'keɪʃn] *noun* changing to a different way of working

> COMMENT: the main alternative enterprises undertaken by farmers are: farm holidays and bed-and-breakfast; farm shops, selling produce from the farm; camping and caravan sites; country sports, such as horse riding, pony-trekking and fishing

> QUOTE: the scheme to help farmers diversify out of agriculture and so help farm income and rural development
>
> **Environment Now**

> QUOTE: all over the UK, farmers are seeking to boost tightening farm incomes by diversifying into other businesses
>
> **Farmers Weekly**

diversity [daɪ'vɜːsɪti] *noun* richness of the number of species; **alpha diversity** = number of species in a small area; **beta diversity** = number of species in a wide region; **genetic diversity** = richness of the variety and range of genes; **species diversity** = richness of the number of species

divide [dɪ'vaɪd] *verb* to separate into parts; ***the area was divided into quadrats for sampling purposes***

◇ **division** [dɪ'vɪʒn] *noun* **(a)** separating into parts; **cell division** = way in which a cell reproduces itself by mitosis **(b)** major category in the classification of plants, below kingdom

dizygotic twins [daɪzaɪ'gɒtɪk] *plural noun* fraternal twins, two offspring who are not identical (and not always of the same sex) because they come from two different ova fertilized at the same time; *compare* IDENTICAL, MONOZYGOTIC

DNA = DEOXYRIBONUCLEIC ACID

DNOC = DINITRO-ORTHO-CRESOL

DoE *GB* = DEPARTMENT OF THE ENVIRONMENT

DOE *US* = DEPARTMENT OF ENERGY

doldrums ['dɒldrʌmz] *plural noun* area of low pressure over the ocean near the equator where there is little wind

doline *or* **dolina** [dəʊ'liːnə] *noun* round *or* oval depression in the ground, found in limestone regions

dolomite ['dɒləmaɪt] *noun* alkaline carbonate of magnesium or calcium rock

dolphin ['dɒlfɪn] *noun* marine mammal belonging to the Delphinidae, which also includes killer whales and pilot whales

◇ **dolphinarium** [dɒlfɪ'neəriəm] *noun* display pool where dolphins are shown to the public

DOM = DIGESTIBLE ORGANIC MATTER, DRY ORGANIC MATTER

domestic [də'mestɪk] *adjective* referring to the home; used in the home; **domestic animal** = animal (such as a dog *or* cat) which lives with human beings; animal (such as a pig *or* goat) which is kept by human beings; **domestic heating oil** = petroleum oil used as fuel in a central heating boiler; **domestic hot water (DHW)** = hot water used for washing in a house; **domestic livestock** = animals (such as pigs *or* goats *or* sheep) which are kept by human beings; **domestic refuse** = waste material from houses; **domestic sewage** *or* **domestic waste** = sewage *or* waste from houses

◇ **domesticated** [də'mestɪkeɪtɪd] *adjective* (i) (wild animal) which has been trained to live near a house and not be frightened of human beings; (ii) species which was formerly wild, now selectively bred to fill human needs

dominance ['dɒmɪnəns] *noun* **(a)** state where one group in society rules other groups **(b)** situation where energy tends to flow through one channel rather than another

◇ **dominant** ['dɒmɪnənt] **1** *noun* plant *or* species which has most influence on the composition and distribution of other species; *see also* CODOMINANT, SUBDOMINANT **2** *adjective* (gene *or* genetic trait) which is more powerful; **dominant species** = species which has most influence on the composition and distribution of other species

> COMMENT: since each physical characteristic is governed by two genes, if one gene is dominant and the other recessive, the resulting trait will be that of the dominant gene. Traits governed by recessive genes will appear if genes from both parents are recessive

dormancy ['dɔːmənsi] *noun* state of an animal *or* plant, where metabolism is slowed during a certain period of the year

◇ **dormant** ['dɔːmənt] *adjective* not actively growing; **dormant plant** = plant which is not actively growing (e.g. during the winter); **dormant volcano** = volcano which is not erupting; *compare* EXTINCT

dormitory suburb *or* **dormitory town** ['dɔːmətri] *noun* suburb *or* town from which the residents travel to work somewhere else

dorsal ['dɔːsəl] *adjective* referring to the back (of an animal *or* leaf) (NOTE: the opposite is **ventral**)

dose [dəʊs] *noun* amount of a drug *or* of ionizing radiation received by an organism; ***the patients received more than the allowed dose of radiation;*** **exposure dose** *or* **irradiation dose** *or* **radiation dose** = amount of radiation to which an organism is exposed; **maximum permissible dose** = highest amount of radiation to which a person may safely be exposed during a certain period

◇ **dosimeter** *or* **dosemeter** [dəʊ'sɪmɪtə] *noun* instrument for measuring the amount of radiation

double ['dʌbl] *verb* to multiply by two; **doubling time** = time a population takes to double in size

Douglas fir ['dʌgləs 'fɜː] *noun* important North American softwood tree *(Pseudotsuga menziesii)* widely planted throughout the world, and yielding vast amounts of strong timber

downs [daʊnz] *plural noun* grass-covered chalky hills characterized by low bushes and few trees

◇ **downland** ['daʊnlænd] *noun* area of downs; **downland farm** = farm situated on the downs

down draught *or US* **down draft** ['daʊn drɑːft] *noun* (i) cool air which flows downwards as a rainstorm approaches; (ii) air which flows rapidly down the lee side of a building, mountain, etc.; (iii) air which blows down a chimney

◇ **downpipe** ['daʊnpaɪp] *noun* pipe carrying rainwater from a roof into a drain or soakaway

◇ **downstream** ['daʊnstriːm] *adverb & adjective* towards the mouth of a river; ***the silt is carried downstream and deposited in the delta; pollution is spreading downstream from the factory; downstream communities have not yet been affected***

◇ **downwash** ['daʊnwɒʃ] *noun* action which brings smoke from a chimney down to the ground as it is caught in a down draught

drain [dreɪn] **1** *noun* (i) underground pipe which takes waste water from buildings or from farmland; (ii) open channel for taking away waste water; **mole drain** = underground drain formed by a mole plough as it is pulled across a field; mole drains are normally drawn 3 to 4 metres apart, and are used in fields with a clay subsoil; **pipe drain** = underground drain made of lengths of clay tiles linked together; they may also be made of concrete or plastic pipes.; **storm drain** = specially wide channel for taking away large amounts of rainwater which fall during tropical storms **2** *verb* **(a)** to remove liquid **(b)** *(of liquid)* to flow into; ***the stream drains into the main river***

◇ **drainage** ['dreɪnɪdʒ] *noun* **(a)** removing of liquid; **drainage area** *or* **drainage basin** = catchment area *or* area of land which collects and drains the rainwater which falls on it (such as the area around a lake or river); **drainage ditch** = channel to take away rainwater **(b)** removing liquid waste and sewage from a building; **drainage system** = arrangement of pipes to carry liquid waste and sewage away from a building

◇ **drainpipe** ['dreɪnpaɪp] *noun* pipe carrying rainwater, sewage, etc. into a drain or soakaway

> QUOTE: dairy farms are responsible for most water pollution incidents caused by inadequate facilities for dirty water disposal; effective systems can be designed to handle yard drainage
> **Farmers Weekly**

draught animal ['drɑːft 'ænɪməl] *noun* animal used to pull vehicles or carry heavy loads

COMMENT: considerable use is made of draught animals in many areas of the world. Oxen, buffaloes, yaks, camels, elephants, donkeys, horses are all used as draught animals. The advantages of using animals are many: they produce young so do not always have to be bought; they are cheaper to buy than machines; they do not use expensive fuel, even though they eat large quantities of food; they may be slower than machines but they can work in difficult terrains. Their most important advantage is that they are appropriate to the local conditions

dredge [dredʒ] *verb* to remove silt and alluvial deposits from a river bed *or* harbour, etc.

drift [drɪft] *noun* slow movement; **drift current** = current in the sea caused by the wind; **drift ice** = large pieces of ice floating in the sea; **littoral drift** = movement of sand as it is carried by the sea along the coastline; **North Atlantic Drift** = warm current which travels northwards along the east coast of North America and then crosses the Atlantic to hit the British Isles; **spray drift** = blowing away of herbicide, insecticide, fertilizer, etc. (which is being applied in windy conditions) onto surrounding plants; ***wildlife in hedges can be harmed by spray drift from fields*** *see* CONTINENTAL, GLACIAL

drill [drɪl] *verb* to bore (a hole *or* an oil well); ***they are hoping to start drilling for oil by the end of the year***

◇ **drilling** ['drɪlɪŋ] *noun* boring a hole *or* an oil well; **drilling rig** *or* **drilling platform** = large metal construction with machinery for boring an oil well in the sea; **drilling ship** = ship which can bore an oil well

drink [drɪŋk] *verb* to swallow liquid

◇ **drinkable** ['drɪŋkəbl] *adjective* (water) which is safe to drink

◇ **drinking water** ['drɪŋkɪŋ wɔːtə] *noun* water which is safe to drink

drizzle ['drɪzl] *noun* light persistent rain with drops of less than 0.5mm in diameter

◇ **drizzly** ['drɪzli] *adjective* (weather) with a lot of drizzle

drop [drɒp] **1** *noun* **(a)** small quantity of liquid; ***a drop of acid burned through the material; drops of rain soon began to fall*** **(b)** reduction *or* fall in quantity of something; **drop in pressure** = sudden reduction in pressure; ***a drop in temperature of over 50° in one morning is uncommon*** **(c)** fall (of fruit); **June drop** = fall of small fruit in early summer, which allows the remaining fruit to grow larger **2** *verb* to fall *or* to let something fall; ***pressure began to drop as the storm approached***

◇ **droplet** ['drɒplət] *noun* very small drop of liquid

◇ **droppings** ['drɒpɪŋz] *plural noun* excreta from animals; ***the grass was covered with rabbit and sheep droppings***

drought [draʊt] *noun* long period without rain at a time when rain normally falls; ***relief agencies are bringing food to drought-stricken areas***

drowned valley ['draʊnd 'væli] *noun* valley which has been submerged by the advance of the sea *or* a lake

drumlin ['drʌmlɪn] *noun* small oval hill formed by the action of ice, with one end blunt and the other tapered, similar to half of a hard-boiled egg, cut lengthways

drupe [druːp] *noun* fruit with a single seed and a fleshy body (such as a peach)

DRV = DIETARY REFERENCE VALUES

dry [draɪ] **1** *adjective* not wet *or* with the smallest possible amount of moisture; **dry adiabatic lapse rate** = rate of temperature change in rising dry air; *see also* LAPSE; **dry deposition** = falling of dry particles from polluted air (in the same way as acid rain falls) which form a harmful deposit on surfaces such as buildings or the leaves of trees; **dry farming** = system of extensive agriculture, producing crops in areas of limited rainfall, without using irrigation; **dry ice** = solid carbon dioxide, used as a refrigerant; **dry organic matter (DOM)** = organic matter such as sewage sludge *or* manure which has been dried out and may be used as a fertilizer; **dry rot** = fungus disease causing rot in wood, potatoes, fruit, etc.; **dry season** = period in some countries when very little rain falls (as opposed to the rainy season); **dry steam** = steam that does not contain droplets of water **2** *verb* to remove moisture from something

◇ **drying** ['draɪɪŋ] *noun* removing moisture from something; **drying bed** = area where sewage sludge is spread out to dry; **the drying out of wetlands** = process whereby water is drained away from wetlands

◇ **dryness** ['draɪnəs] *noun* state of not being wet *or* having the smallest possible amount of moisture; ***the dryness of the atmosphere on very high mountains***

◇ **dry-stone wall** ['draɪstəʊn 'wɔːl] *noun* wall made of stones carefully placed one on top of the other without using any mortar

◇ **dry up** ['draɪ 'ʌp] *verb (of a river or lake)* to become dry; ***the river bed dries up completely in summer***

DST *UK* = DAYLIGHT-SAVING TIME

duct [dʌkt] *noun* tube *or* pipe which carries liquid *or* air; ***air-conditioning ducts need regular cleaning***

dump [dʌmp] **1** *noun* place where waste is thrown away; ***the mine is surrounded with dumps of excavated waste;*** **municipal dump** = place where a town's refuse is disposed of after it has been collected; **open dump** = place where waste is left on the ground and not buried in a hole **2** *verb* to throw away waste; ***the UK dumps its industrial waste into the North Sea***

◇ **dumping** ['dʌmpɪŋ] *noun* **(a)** throwing away waste; ***the dumping of nuclear waste into the sea should be banned;*** **controlled dumping** = throwing away waste on special sites as opposed to throwing it away anywhere; **no dumping** = sign telling people that they are not allowed to throw away waste in that particular place **(b)** selling of agricultural products at a price below the true cost, to get rid of excess produce cheaply, usually in an overseas market

◇ **dumpsite** ['dʌmpsaɪt] *noun* place where waste is dumped; ***sludge bacteria can survive in seawater for long periods and are widely dispersed from dumpsites***

dunes [dju:nz] *plural noun* area of sand blown by the wind into small hills and ridges which has very little soil or vegetation; ***the village was threatened by encroaching sand dunes***

dung [dʌŋ] *noun* solid waste excreta from animals (especially cattle), often used as fertilizer; ***they collected the dried dung in the fields and used it as fuel***

COMMENT: in some areas of the world, in particular in India, dried dung is used as a cooking fuel; this has the effect of preventing the dung being returned to the soil and so leads to depletion of soil nutrients

duramen [dju:'rɑ:mən] *noun* heartwood, the hard dead wood in the centre of a tree's trunk which helps support the tree

durum ['dju:rəm] *noun* hard type of wheat *(Triticum durum)* grown in southern Europe and used in making semolina for processing into pasta; also grown in the USA

dust [dʌst] *noun* fine powder made of particles of dry dirt *or* sand; **dust bowl** = large area of land where the strong winds blow away the dry topsoil (used to describe areas of Kansas, Oklahoma and Texas which lost topsoil through wind erosion in the 1930s); **dust burden** = amount of dust which is found suspended in a gas; **dust cloud** = mass of particles of dry dirt *or* sand suspended in the air; **dust devil** = rapidly turning column of air which picks up sand over a desert *or* beach and dust, leaves, litter, etc. elsewhere; **dust extractor** = machine which removes dust from a place

◇ **dustbin** ['dʌstbɪn] *noun* container into which household rubbish is placed to be collected by municipal refuse collectors (NOTE: US English is **garbage can** *or* **trash can**)

◇ **dustman** ['dʌstmən] *noun* refuse collector (NOTE: US English is **garbage man**)

◇ **dust storm** ['dʌst stɔ:m] *noun* storm of wind which blows dust and sand with it, common in North Africa

◇ **dust veil** ['dʌst veɪl] *noun* mass of dust in the atmosphere (created by volcanic eruptions, storms, burning fossil fuels), which cuts off solar radiation and so reduces the temperature of the earth's surface; *compare* GREENHOUSE EFFECT

Dutch elm disease [dʌtʃ 'elm dɪ'zi:z] *noun* fungus disease caused by *Ceratocystis ulmi.* It kills elm trees and is spread by a bark beetle

duty of care ['dju:ti ʌv 'keə] *noun* duty which every citizen and organization has not to act negligently; especially, the system for the safe handling of waste, introduced by the UK Environmental Protection Act 1990

dwelling ['dwelɪŋ] *noun* place where a person lives such as a house, apartment *or* flat

dye [daɪ] *noun* colouring substance

dyke [daɪk] **1** *noun* **(a)** long wall of earth built to keep water out **(b)** ditch for drainage **2** *verb* to build walls of earth to help prevent water from flooding land

dynamometer [daɪnə'mɒmɪtə] *noun* machine which measures exhaust fumes from cars

dysentery ['dɪsəntri] *noun* infection and inflammation of the colon causing bleeding and diarrhoea

◇ **dysenteric** [dɪsən'terɪk] *adjective* referring to dysentery

COMMENT: dysentery occurs mainly in tropical countries. The symptoms include diarrhoea, discharge of blood and pain in the intestines. There are two main types of dysentery: bacillary dysentery, caused by the bacterium *Shigella* in contaminated food; and amoebic dysentery or amoebiasis, caused by a parasitic amoeba *Entamoeba histolytica* spread through contaminated drinking water

dysphotic [dɪs'fɒtɪk] *adjective* referring to the area of water in a lake *or* the sea between the aphotic zone at the bottom of the water and the euphotic zone which sunlight can reach

dystrophic lake [dɪs'trɒfɪk 'leɪk] *noun* lake (as in a peat bog) where the water is acid, brown and peaty, and the dead vegetation does not decompose but settles at the bottom to form peat

dystrophy ['dɪstrəfi] *noun* wasting of an organ *or* muscle *or* tissue due to lack of nutrients in that part of the body

Ee

E number ['i: nʌmbə] classification of additives to food according to the European Union

COMMENT: additives are classified as follows: colouring substances: E100 - E180; preservatives: E200 - E297; antioxidants: E300 - E321; emulsifiers and stabilizers: E322 - E495; acids and bases: E500 - E529; anti-caking additives: E530 - E578; flavour enhancers and sweeteners: E620 - E637

Vitamin E ['vɪtəmɪn 'i:] *noun* vitamin found in vegetables, vegetable oils, eggs and wholemeal bread

earth [ɜ:θ] *noun* **(a) (the planet) Earth** = the planet on which we live; **earth science** = any science such as geochemistry, geodesy, geography, geology, geomorphology, geophysics or meteorology, which is concerned with the physical aspects of the earth **(b)** soil, soft substance formed of mineral particles, decayed organic matter, chemicals and water in which plants can grow; **earth closet** = toilet in which the excreta are covered with earth instead of being flushed away with water; **earth dam** = dam made of piled earth; **earth flow** *or* **slide** = sliding of wet earth and rocks down the side of a slope or mountain, often after heavy rainfall and caused, in some cases, by excavation

COMMENT: the planet earth can be divided into various zones: the lithosphere (solid rock and molten interior), the hydrosphere (the water covering the earth's surface), the atmosphere (the gaseous zone rising above the earth's surface) and the biosphere (those parts of the other zones in which living organisms exist). The interior of the earth is formed by a central core made of nickel and iron, part of which is solid. Above the core is the mantle, a layer about 2,700km thick of molten minerals. On top of the mantle is the crust, formed of solid rock between six and seventy kilometres thick

earthquake ['ɜ:θkweɪk] *noun* phenomenon where the earth's crust or the mantle beneath it shakes, and the surface of the ground moves (it is caused by movement inside the earth's crust along fault lines, and often causes damage to buildings); *see also* EPICENTRE, FOCUS, MODIFIED MERCALLI SCALE, RICHTER SCALE, SEISMIC

◊ **earth sciences** ['ɜ:θ 'saɪənsɪz] *noun* sciences dealing with the planet earth, including geology and climatology

◊ **earthworks** ['ɜ:θwɜ:ks] *noun* constructions such as walls made from soil

Earth Summit ['ɜ:θ 'sʌmɪt] *noun* popular name for the United Nations Conference on Environment and Development (UNCED) held in Rio de Janeiro, Brazil, in June 1992

COMMENT: The principal outcomes of the 'Earth Summit' were a declaration, conventions on controlling climate change and preserving biological diversity, and a lengthy agenda outlining the extent of global environmental problems and the measures needed to tackle them in order to achieve the agreed goal of 'sustainable development'. Opinion was divided as to how far the conference could be termed a success. Many environmentalists stressed its failure to set binding targets for resolving environmental problems, and to address adequately the links between third world poverty and environmental degradation. Others, such as many of the world leaders who attended the conference, stressed that it should be seen as a first step towards sustainable development, and an important acknowledgement of the seriousness of the problem and of the fact that global co-operation is needed to overcome it

earthworm ['ɜ:θwɜ:m] *noun* kind of invertebrate animal with a long thin body living in large numbers in the soil

COMMENT: earthworms provide a useful service by aerating the soil as they tunnel. They also eat organic matter and help increase the soil's fertility. They help stabilize the soil structure by compressing material and mixing it with organic matter and calcium. It is believed that they also secrete a hormone which encourages rooting by plants

east [i:st] *adjective, adverb & noun* one of the directions on the earth's surface, the direction facing towards the rising sun; ***the wind is blowing from the east; the river flows east into the ocean;*** **east wind** = wind which blows from the east

◇ **easterly** ['i:stəli] **1** *adjective* to *or* from the east; ***the hurricane was moving in an easterly direction*** **2** *noun* wind which blows from the east

◇ **eastern** ['i:stən] *adjective* in the east; towards the east; ***the main rain forests lie in the eastern half of the country***

ebb tide ['eb 'taɪd] *noun* tide which is going down *or* which is at its lowest; ***the plan is to build a barrage with 7gW of installed capacity, generating electricity on the ebb tide*** *compare* FLOOD TIDE

ebony ['ebəni] *noun* black tropical hardwood, now becoming scarce

EC = EUROPEAN COMMUNITY

echo ['ekəʊ] *noun* sound which is reflected back towards the person listening

◇ **echolocation** [ekəʊlə'keɪʃn] *noun* finding a location by sending out a sound signal and listening to the reflection of the sound. Bats and whales find their way by using echolocation

◇ **echosounder** ['ekəʊsaʊndə] *noun* device used to find the depth of water by sending a sound signal down to the bottom and calculating the distance from the time taken for the reflected sound to reach the surface again

◇ **echosounding** ['ekəʊsaʊndɪŋ] *noun* finding the depth of water using an echosounder

eclipse [ɪ'klɪps] *noun* situation when the moon passes between the sun and the earth (solar eclipse) *or* when the earth passes between the sun and the moon (lunar eclipse), in both cases cutting off the light visible from earth; **partial eclipse** = eclipse where only part of the sun *or* moon is hidden; **total eclipse** = eclipse where the whole of the sun *or* moon is hidden

eco- ['i:kəʊ] *prefix* concerning ecology

ecoclimate ['i:kəʊklaɪmət] *noun* climate seen as an ecological factor

ecocline ['i:kəʊklaɪn] *noun* changes which take place in a species as individuals live in different habitats

ecology [ɪ'kɒlədʒi] *noun* study of the relationships among organisms and the relationship between them and their physical environment; **animal ecology** = study of the relationship between animals and their environment; **deep ecology** = extreme form of ecological thinking where humans are considered as only one species among many in the environment and where large numbers of humans are seen as harmful to the environment in which they live; **human ecology** = study of man and communities of people, the place which they occupy in the natural world, and the ways in which they adapt to or change the environment; **marine ecology** = study of the relationship between organisms that live in the sea and their environment; **plant ecology** = study of the relationship between plants and their environment

◇ **ecological** [i:kə'lɒdʒɪkl] *adjective* referring to ecology; **ecological balance** *or* **balance of nature** = situation where relative numbers of organisms remain more or less constant. By polluting the environment, destroying habitats and increasing their own numbers, people can cause major changes in populations which may be irreversible; **ecological damage** = damage done to an ecosystem; **ecological disaster** = disaster which seriously disturbs the balance of the environment; **ecological disruption** = upsetting the balance of an ecosystem; **ecological efficiency** = measurement of how much energy is used at different stages in the food chain *or* at different trophic levels; **ecological factors** = factors which influence the distribution of a plant species in a certain habitat; **ecological indicator** = species that has particular requirements (acid soil, low temperature, high rainfall, etc.) and whose presence in an area indicates that those requirements are also satisfied in that area; **ecological niche** = all the characters (chemical, physical, biological) that determine the position of an organism *or* species in an ecosystem (commonly called the 'role' *or* 'profession' of an organism, e.g. an aquatic predator, a terrestrial herbivore); **ecological pyramid** = chart showing the structure of an ecosystem in terms of who eats what: the base is composed of producer organisms (usually plants), then herbivores, then carnivores. It may be measured in terms of number, biomass or energy; **ecological recovery** = return of an ecosystem to its former harmonious balance; **ecological succession** = series of stages, one after the other, by which a group of organisms living in a community reaches its final stable state (*or* climax)

◇ **ecologist** [i:'kɒlədʒɪst] *noun* **(a)** scientist who studies ecology **(b)** person who is in favour of maintaining a balance between living things and the environment in which they live in an attempt to improve the life of all organisms improve the life of all organisms

◇ **ecomovement** ['i:kəʊmu:vmənt] *noun* grouping of individuals and organizations dedicated to the protection of the environment

economic [iːkə'nɒmɪk] *adjective* **(a)** which provides enough money; ***the farm is let at an economic rent; it is hardly economic for the company to continue in business if pollution controls are so expensive*** **(b)** referring to finance *or* economics; **economic conservation** = management of nature to maintain a regular yield of natural resources; **economic crisis** *or* **economic depression** = state where a country is in financial collapse; **economic development** = expansion of the commercial and financial situation of a country, especially from a primary economy to an industrial economy; ***the economic development of the region has totally changed since they discovered oil there; economic development results in increased pollution of the environment;*** **economic efficiency** = making the best use of scarce natural resources; **economic ends** = aims which economic activity hopes to achieve; **economic geology** = study of rock *or* soil formation for the purpose of commercial mineral extraction; **economic growth** = increase in the national income expressed as the production of goods and services shown as a percentage of increase per member of the population; ***the country enjoyed a period of economic growth in the 1960s***

◇ **economical** [iːkə'nɒmɪkl] *adjective* which saves money or materials *or* which is cheap; **economical car** = car which does not use much petrol; **economical use of resources** = using resources as carefully as possible

◇ **economics** [iːkə'nɒmɪks] *plural noun* **(a)** study of production, distribution, selling and use of goods and services **(b)** study of financial structures to show how a product or service is costed and what returns it produces; ***the economics of town planning; I do not understand the economics of the coal industry***

◇ **economist** [ɪ'kɒnəmɪst] *noun* person who specializes in the study of economics; **agricultural economist** = person who studies the economics of the agricultural industry

◇ **economize** [ɪ'kɒnəmaɪz] *verb* **to economize on petrol** = to save petrol

◇ **economizer** [ɪ'kɒnəmaɪzə] *noun* device which saves waste heat in a boiler, by transferring heat from waste gases to the water being heated *or* where water is pre-heated before passing to the main boiler

◇ **economy** [ɪ'kɒnəmi] *noun* **(a)** being careful not to waste money or materials; **economy measure** = action to save money or materials; **to introduce economies** *or* **economy measures into the system** = to start using methods to save money or materials; **economy car** = car which does not use much petrol **(b)** financial state of a country *or* way in which a country makes and uses its money; **black economy** = work which is paid for in cash or goods, but not declared to the tax authorities; **capitalist economy** = system where each person has the right to invest money, to work in business, to buy and sell with no restrictions by the state; **controlled economy** = system where business activity is controlled by orders from the government; **free market economy** = system where the government does not interfere in business activity in any way; **mixed economy** = system which contains both nationalized industries and private enterprise; **planned economy** = system where the government plans all business activity

ecoparasite [iːkəʊ'pærəsaɪt] *noun* parasite which is adapted to a specific host; *compare* ECTOPARASITE, ENDOPARASITE

ecophysiology [iːkəʊfɪzɪ'ɒlədʒi] *noun* study of organisms and their functions and how they exist in their environment

◇ **ecospecies** [iːkəʊ'spiːʃiːz] *noun* subspecies of a plant

◇ **ecosphere** ['iːkəʊsfɪə] *noun* biosphere, part of the earth and its atmosphere where living organisms exist (including parts of the lithosphere, the hydrosphere and the atmosphere)

ecosystem ['iːkəʊsɪstəm] *noun* system which includes all the organisms of an area and the environment in which they live; ***European wetlands are classic examples of ecosystems that have been shaped by humans. For thousands of years they have provided pasture for livestock, reeds for thatching and fuel, silage and bedding for livestock***

> COMMENT: an ecosystem can be any size, from a pinhead to the whole biosphere. The term was first used in the 1930s to describe the interdependence of organisms among themselves and with the living and non-living environment

ecotone ['iːkəʊtəʊn] *noun* area between two different types of vegetation (such as the border between forest and moorland), which may share the characteristics of both

◇ **ecotype** ['iːkəʊtaɪp] *noun* species that has special characteristics which allow it to live in a certain habitat

ecto- ['ektəʊ] *prefix meaning* outside

ectoparasite [ektəʊ'pærəsaɪt] *noun* parasite which lives on the skin or outer surface of the host but feeds on the inside by piercing the skin; *compare* ENDOPARASITE

ectoplasm ['ektəʊplæzm] *noun (in cells)* the outer and densest layer of the cytoplasm

edaphic [ɪ'dæfɪk] *adjective* referring to soil; **edaphic climax** = climax caused by the type of soil in a certain area; **edaphic factors** = different soil conditions which affect the organisms living in a certain area

◇ **edaphon** [ɪ'dæfən] *noun* organisms living in soil

eddy ['edi] *noun* whirlpool of air *or* of water in a current

edible ['edəbl] *adjective* which can be eaten; **edible fungi** = fungi which can be eaten and are not poisonous

EEC = EUROPEAN ECONOMIC COMMUNITY

EER = ENERGY EFFICIENCY RATIO

EFA = ESSENTIAL FATTY ACID

effect [ɪ'fekt] *noun* result of an action; ***the effect of acid rain on the environment; radiation had a noticeable effect on animal life in the area;*** **greenhouse effect** = effect produced by the accumulation of carbon dioxide crystals and water vapour in the upper atmosphere, which insulates the earth and raises the atmospheric temperature by preventing heat loss. Chlorofluorocarbons also help create the greenhouse effect

◇ **effective** [ɪ'fektɪv] *adjective* which has an effect; ***is there a safe and effective way of stopping pollution from factory chimneys?;*** **effective height** = height above the ground when the plume of smoke from a factory chimney becomes horizontal

◇ **effectiveness** [ɪ'fektɪvnəs] *noun* being effective

efficient [ɪ'fɪʃənt] *adjective* which works well *or* which functions correctly; ***the new system of smoke control is extremely efficient;*** **energy-efficient machine** = machine which is efficient in its use of fuel

◇ **efficiency** [ɪ'fɪʃənsi] *noun* being efficient *or* making good use of fuel or nutrients; **ecological efficiency** = measurement of how much energy is used at different stages in the food chain *or* at different trophic levels; **economic efficiency** = making the best use of scarce natural resources; **end-use efficiency** = efficient way of using a form of energy by the end user; **fuel efficiency** = percentage of the heat from burning a fuel which is actually converted into energy

◇ **efficiently** [ɪ'fɪʃəntli] *adverb* in an efficient way

> QUOTE: suppose an efficient refrigerator conserves 1,000 kilowatt-hours per year, but costs $100 more than a less efficient model
> **Scientific American**

> QUOTE: much of the decline in energy consumption is due to the more efficient use of energy in homes and offices
> **Scientific American**

> QUOTE: most of the changes in demand are met by varying the output of the larger coal-fired power stations and by running the less efficient stations during the daytime only
> **Ecologist**

effluent ['efluənt] *noun* sewage, especially liquid *or* solid waste from industrial processes or liquid, solid or gas waste material such as slurry or silage effluent from a farm; **effluent charge** = fee paid by a company to be allowed to discharge waste into the sea *or* a river; **effluent monitor** = device which monitors the radioactivity in liquid waste from nuclear power stations; **effluent purification process** = method of purifying sewage; **effluent standard** = amount of sewage which is allowed to be discharged into a river *or* the sea

EHO = ENVIRONMENTAL HEALTH OFFICER

EIA = ENVIRONMENTAL IMPACT ASSESSMENT

EIS = ENVIRONMENTAL IMPACT STATEMENT

ejecta [ɪ'dʒektə] *plural noun* ash and lava thrown up by an erupting volcano

electricity [ɪlek'trɪsɪti] *noun* electron energy which can be converted to light *or* heat *or* power; ***the pump is run by electricity; electricity is used to pump water into the irrigation channels;*** **dynamic electricity** = electricity which is flowing in a current; **static electricity** = electricity which is in a static state as opposed to electricity which is flowing in a current; *see also* ELECTROSTATIC; **Electricity Board** = organization which administers the supply of electricity in a country *or* region

◇ **electric** [ɪ'lektrɪk] *adjective* worked by electricity; used for carrying electricity; ***the drying system is heated by an electric heater; the electric cables are buried underground to avoid spoiling the landscape;*** **electric fence** = thin wires supported by posts, the wires being able to carry an electric current; such a fence is easily moved around the farm, and makes strip grazing on limited areas possible; **electric storm** = thunderstorm *or* storm with thunder and lightning

◇ **electrical** [ɪ'lektrɪkl] *adjective* concerning electricity; worked by electricity; **electrical power generation** = production of electricity; **electrical appliance** = device *or* machine worked by electricity such as washing machine, vacuum cleaner, iron, toaster, etc.

> COMMENT: electricity, though it is clean and relatively cheap to use, is an inefficient form of energy since more than 50% of the heat needed to make it is wasted. On the other hand, electricity is used to drive

machines which could not be operated by other means

QUOTE: using electricity for heating is an extravagant use of our energy resources because, apart from the very small hydroelectric contribution, we use steam turbines to turn the generators that produce our electricity. In steam turbines, steam is raised by heating water by means of coal, oil or nuclear energy. The steam driving the turbines is then condensed and returned to the boilers to be used again. This reuse increases the heat efficiency, but over 95% of our electricity comes from about one-third of the heat available in the primary fuel. The rest is wasted, except in those countries where the heat is used in Combined Heat and Power stations. Britain has no CHP stations

Ecologist

electrodialysis [ɪletrəʊdaɪ'ælɪsɪs] *noun* process by which ions dissolved in sea water are removed, making the water fit to drink

electromagnet [ɪlektrəʊ'mægnɪt] *noun* magnet made from a coil of wire through which an electric current is passed

◇ **electromagnetic** [ɪlektrəʊmæg'netɪk] *adjective* having magnetic properties caused by a flow of electricity; containing *or* worked by an electromagnet

COMMENT: in a practical electromagnet the magnetic field produced by the coil carrying the electric current is concentrated by the insertion of a ferrous core into the coil

electromotive force [ɪlektrəʊ'məʊtɪv 'fɔːs] *noun* force of electricity which flows as a current

◇ **electron** [ɪ'lektrɒn] *noun* negative particle in an atom, balanced by a proton, but not found in the atom's nucleus; **electron microscope (EM)** = microscope which uses a beam of electrons instead of light

◇ **electronic** [elek'trɒnɪk] *adjective* referring to electrons; operated by devices using the motions of electrons to perform useful functions, as in a computer

◇ **electronics** [elek'trɒnɪks] *noun* science of applying the study of electrons and their properties to manufactured products such as calculators, computers, telephones, etc.

◇ **electrostatic** [ɪlektrəʊ'stætɪk] *adjective* referring to electricity which is not flowing; **electrostatic precipitator** = device for collecting minute particles of dust suspended in gas by charging the particles as they pass through an electrostatic field

element ['elɪmənt] *noun* chemical substance which cannot be broken down to a simpler substance; **essential element** = chemical element (such as hydrogen, carbon, nitrogen, oxygen and many others) which is necessary to an organism's growth and function; **trace element** = substance which is essential to organic growth, but only in very small quantities

◇ **elemental** [elɪ'mentl] *adjective* in pure form *or* existing as a pure element; ***the snail stores elemental sulphur in its shell***

COMMENT: there are 105 elements. The smallest form in which an element can exist is the atom

eliminate [ɪ'lɪmɪneɪt] *verb* to get rid of (pests *or* waste matter); ***they have eliminated smallpox in the country; mosquitoes were eliminated by a programme of spraying breeding grounds with oil***

◇ **elimination** [ɪlɪmɪ'neɪʃn] *noun* removal of pests *or* waste matter

elm [elm] *noun* large hardwood tree *(Ulmus* spp.) which grows in temperate areas; all species reproduce by root suckers, except the wych elm which is propagated by seed; *see also* DUTCH ELM DISEASE

El Niño [el 'niːnjəʊ] *see* NIÑO

eluvium [ɪ'luːviəm] *noun* gravel formed by rocks as they are broken down into fragments where they are lying; *compare* ALLUVIUM

◇ **eluviation** [ɪluːvi'eɪʃn] *noun* action of leaching out particles and chemicals from the topsoil down into the subsoil

EM = ELECTRON MICROSCOPE

embankment [ɪm'bæŋkmənt] *noun* wall made along a river bank to prevent the river from overflowing

embryo ['embriəʊ] *noun* living organism which develops from a fertilized egg or seed (such as an animal in the first weeks of gestation or a seedling plant with cotyledons and a root); **in embryo** = not yet developed

◇ **embryonic** [embri'ɒnɪk] *adjective* referring to an embryo; (something) in the first stages of its development; **embryonic abortion** = termination of pregnancy while the foetus is still an embryo. (A human offspring is considered to be an embryo in the first eight weeks from conception)

emergence [ɪ'mɜːdʒns] *noun* gradual upward movement of a land mass

emergency [ɪ'mɜːdʒənsi] *noun* dangerous situation where immediate action has to be taken; ***a state of emergency exists;*** **emergency shutdown** = stopping a nuclear reactor working when it seems that something dangerous may happen; **emergency ward** = hospital ward which deals with urgent cases (such as accident victims)

EMF = ELECTROMOTIVE FORCE

emigration [emɪ'greɪʃn] *noun* movement of a species out of an area

emit [ɪ'mɪt] *verb* to send out (smoke *or* radiation *or* waste, etc.); ***the chimney was found to be emitting twice as much pollution as was permitted***

◇ **emission** [ɪ'mɪʃn] *noun* act of emitting something; substance that has been emitted; ***gas emissions can cause acid rain;*** **emission charges** = fee paid by a company to be allowed to discharge waste into the environment; **emission standard** = amount of an effluent *or* pollutant which is permitted to be released into the environment, such as the amount of sewage which can be discharged into a river *or* the sea; **emission standards for vehicles** = amount of pollutants (hydrocarbons, carbon monoxide and nitrogen oxides) which can be released into the atmosphere by petrol and diesel engines

> QUOTE: the EEC Commission has proposed the establishment of emission limit values for large combustion plants as part of a set of measures designed to achieve a substantial reduction in overall emissions of sulphur dioxide and nitrogen oxides
>
> **Environment**

> QUOTE: 40% of gasoline vapour emissions occur at the gas pump. When gas is pumped into the car's fuel tank, built-up vapours are forced into the air
>
> **Environmental Action**

emulsify [ɪ'mʌlsɪfaɪ] *verb* to mix (two liquids) so thoroughly that they will not separate

◇ **emulsifier** *or* **emulsifying agent** [ɪ'mʌlsɪfaɪə *or* ɪ'mʌlsɪfaɪɪŋ 'eɪdʒənt] *noun* substance added to mixtures of food (such as water and oil) to hold them together (used in sauces, etc.), and also added to meat to increase the water content so that the meat is heavier. (In the EU, emulsifiers and stabilizers have E numbers E322 to E495); *see also* STABILIZER

◇ **emulsifying agent** *noun* = EMULSIFIER

encroach on [ɪ'krəʊtʃ 'ɒn] *verb* to come close to and gradually cover (something); ***the dunes are encroaching on the pasture land; the town is spreading beyond the by-pass, encroaching on good quality farming land; trees are spreading down the mountain and encroaching on the lower more fertile land in the valleys***

end [end] *noun* extreme limit; finish; purpose; **end product** = item *or* state produced by a manufacturing process *or* by radioactive decay, etc.; **end user** = consumer *or* person who uses the manufactured product; **end-use efficiency** = efficient way of using a form of energy by the end user

endanger [ɪn'deɪndʒə] *verb* to put (something) in danger; ***pollution from the factory is endangering the aquatic life in the lakes;*** **endangered species** = any species at risk of extinction

> COMMENT: The categories of the degree of risk of extinction as drawn up by the IUCN are: 1. Endangered: in danger of extinction and unlikely to survive if the causal factors continue to operate. 2. Vulnerable: will move into the endangered category in the near future if the causal factors continue to operate. 3. Rare: not endangered or vulnerable, but at risk of becoming so

endemic [ɪn'demɪk] *adjective* **(a)** (plant *or* animal) that exists in a certain area; ***the isolation of the islands has led to the evolution of endemic forms; the northern part of the island is inhabited by many endemic mammals and birds;*** **endemic population** = group of organisms existing in a certain area **(b)** (pest *or* disease) which is very common in a certain area; ***this disease is endemic to Mediterranean countries*** *see also* EPIDEMIC, PANDEMIC

endo- ['endəʊ] *prefix* meaning inside

◇ **endocrine gland** ['endəʊkriːn 'glænd] *noun* gland without a duct (such as the pituitary gland) which produces hormones which are introduced directly into the bloodstream

◇ **endoparasite** [endəʊ'pærəsaɪt] *noun* parasite which lives inside the host; *compare* ECTOPARASITE

◇ **endothelium** [endəʊ'θiːliəm] *noun* membrane of special cells which lines the internal passages and organs in the body, such as the heart, the blood vessels, etc.; *compare* EPITHELIUM

◇ **endothermic reaction** [endəʊ'θɜːmɪk rɪ'ækʃn] *noun* chemical reaction in which heat is removed from the surroundings; *compare* EXOTHERMIC

◇ **endotoxin** [endəʊ'tɒksɪn] *noun* poison from bacteria which pass into the body when contaminated food is eaten

endrin ['endrɪn] *noun* type of broad-based organochlorine insecticide which is extremely toxic and persistent

energy ['enədʒi] *noun* (i) force *or* strength to carry out activities; (ii) specifically, electricity; ***you need to eat certain types of food to give you energy;*** **primary energy** = energy required to produce other forms of energy, such as heat *or* electricity; **energy analysis** *or* **energy audit** = check of how much energy is used within a certain period in a factory *or* school, etc.; **energy balance** = measurements showing the movement of energy between organisms and their environment; **energy budget** = the levels of energy at different points in an ecosystem *or* an industrial process; **energy conservation** = avoiding wasting energy; **energy consumption** = amount of energy consumed by a person *or* an apparatus, shown as a unit; **energy crops** = crops, such as fast-growing trees, which are grown to be used to provide energy; **energy efficiency ratio (EER)** = measure of the efficiency of a heating *or* cooling system (such as a heat pump *or* an air-conditioning system) shown as the ratio of the output in Btu per hour to the input in watts; **energy-efficient** = using energy carefully and with minimum waste; **energy farm** = area of land *or* water where plants such as cassava, sugar cane, algae, etc. are cultivated to produce biofuels such as ethanol and methane; **energy flow** = flow of energy from one trophic level to another in a food chain; **energy output** = amount of energy produced; **energy requirements** = amount of energy needed; **energy reserves** = amount of energy stored; often refers to the stocks of non- renewable fuel, such as oil, which a nation, for example, possesses; **energy resources** = potential supplies of energy which have not yet been used (such as coal lying in the ground, solar heat, wind power, geothermal power, etc.); **energy value** *or* **calorific value** = heat value of a substance *or* number of Calories which a certain amount of a substance (such as a certain food) contains; ***a tin of beans has an energy value of 250 calories***

◇ **energize** ['enədʒaɪz] *verb* to supply energy e.g. electricity to a machine or system to make it work

COMMENT: energy is the capacity to do useful work and includes setting a process in motion, the production of heat and light, the emission of heat and light, the making of electricity, etc. It is measured in joules or calories. One joule is the amount of energy used to move one kilogram the distance of one metre, one calorie being the amount of heat needed to raise the temperature of one gram of water by one degree Celsius. The kilocalorie or Calorie is also used as a measurement of the energy content of food and to show the amount of energy needed by an average person. Energy conservation is widely practised to reduce excessive and costly consumption of energy. Reduction of heating levels in houses and offices, insulating buildings against loss of heat, using solar power instead of fossil fuels, increasing the efficiency of car engines, etc., are all examples of energy conservation

engineering [endʒɪ'ni:ərɪŋ] *noun* applying the principles of science to the design, construction and use of machines, buildings, etc.; **agricultural engineering** = applying the principles of science to farming; **civil engineering** = applying the principles of science to the design, construction and use of roads, bridges, dams, etc.; **genetic engineering** = techniques used to change the genetic composition of an organism so that certain characteristics can be created artificially; **engineering industry** = manufacture of machinery and machine parts

English Nature ['ɪŋglɪʃ 'neɪtʃə] *noun* alternative title given in April 1991 to the Nature Conservancy Council for England, official body which is responsible for the conservation of fauna and flora

enhance [ɪn'hɑ:ns] *verb* to make (something) better *or* stronger

◇ **enhancer** [ɪn'hɑ:nsə] *noun* artificial substance which increases the flavour of food, or even the flavour of artificial flavouring that has been added to food (in the EU, flavour enhancers added to food have the E numbers E620 to 637)

enrich [ɪn'rɪtʃ] *verb* (i) to make richer *or* stronger; (ii) to increase the amount of uranium-235 in the fuel of a nuclear reactor; ***fuel is enriched to 15% with fissile material***

◇ **enrichment** [ɪn'rɪtʃmənt] *noun* action of enriching the proportion of uranium-235 in nuclear fuel

ensilage *or* **ensiling** [ɪn'saɪlɪdʒ *or* ɪn'saɪlɪŋ] *noun* process of making silage for cattle by cutting grass and other green plants and storing it in silos

enter- *or* **entero-** ['entərəʊ] *prefix* referring to the intestine

enteric [ɪn'terɪk] *adjective* referring to the intestine; **enteric fermentation** = breaking down of food in the gut of ruminant animals, especially cattle, producing methane which is eliminated from the animal's body

◇ **enteritis** [entə'raɪtɪs] *noun* inflammation of the mucous membrane of the intestine; **infective enteritis** = enteritis caused by bacteria

◇ **Enterobacteria** [entərəʊbæk'tɪəriə] *noun* important family of bacteria, including Salmonella and Escherichia

◇ **Enterobius** [entə'rəʊbiəs] *noun* threadworm, small thin nematode which infests the intestine

◇ **enterotoxin** [entərəʊ'tɒksɪn] *noun* bacterial exotoxin which particularly affects the intestine

entomology [entə'mɒlədʒi] *noun* study of insects

◇ **entomological** [entəmə'lɒdʒɪkl] *adjective* referring to insects

◇ **entomologist** [entə'mɒlədʒɪst] *noun* scientist who specializes in the study of insects; **medical entomologist** = scientist who studies insects which may carry diseases

entrap [ɪn'træp] *verb* to catch and retain; ***90% of sulphur emissions can be entrapped; coal-fired power stations should be equipped with the means of entrapping sulphur***

environment [ɪn'vaɪrənmənt] *noun* the surroundings of any organism, including the physical world and other organisms; **the built environment** = built-up areas seen as the environment in which humans live; **living environment** = part of the environment made up of living organisms; **residential environment** = area characterized principally by the presence of houses and apartment blocks; **rural environment** = the countryside; **working environment** = surroundings, general ambience in which a person works; **environment protection** = act of protecting the environment by regulating the discharge of waste, the emission of pollutants, and other human activities

COMMENT: the environment is anything outside an organism in which the organism lives. It can be a geographical region, a certain climatic condition, the pollutants or the noise which surround an organism. Man's environment will include the country or region or town or house or room in which he lives; a parasite's environment will include the body of the host; a plant's environment will include a type of soil at a certain altitude

environmental [ɪnvaɪrən'mentl] *adjective* referring to the environment; **environmental annoyance** = nuisance caused by such environmental factors as traffic noise; **environmental audit** = assessment made by a company or organization of the financial benefits and disadvantages to be derived from adopting a more environmentally sound policy; **environmental biology** = study of living organisms in relationship to their environment; **environmental control** = means of protecting an environment; **environmental damage** = harm done to the environment e.g. loss of wetlands, pollution of rivers, etc.; **environmental forecasting** = forecasting the effects on the surrounding environment of new construction programmes; **environmental geology** = using the study of geology to solve problems concerned with the environment; **Environmental Health Officer (EHO)** = official of a local authority who examines the environment and tests for air pollution *or* bad sanitation *or* noise pollution, etc.; **environmental hygiene** = study of health and how it is affected by the environment; **environmental impact** = effect upon the environment (of large construction programmes, draining of marshes, etc.); **environmental impact assessment (EIA)** = evaluation of the effect upon the environment of a large construction programme; **environmental impact statement (EIS)** = statement required under US law for any major federal project; **normal environmental lapse rate** *see* LAPSE; **environmental policy** = document defining the way in which an organization will operate in relation to the environment; **environmental protection** = ENVIRONMENT PROTECTION; **environmental protection association** = organization dedicated to protecting the environment from damage, pollution, etc.; **environmental quality standards** = amount of an effluent *or* pollutant which is accepted in a certain environment, such as the amount of trace elements in drinking water or the amount of additives in food; **environmental resistance** = ability to withstand pressures like predation, competition, weather, food availability, which inhibit the potential growth of a population; **environmental science** = study of the relationship between man and the environment, the problems caused by pollution, loss of habitats, etc. and the proposed solutions. This relatively new science includes other traditional subjects such as geography, geology and economics; **environmental set-aside** = set-aside schemes which may improve the environment; for example, paying farmers to sow grass seed mixtures for a green cover and cutting a 15m-wide strip along the side of footpaths at the edge of cereal fields

◇ **Environmental Protection Act 1990** *noun* UK Act to allow the introduction of integrated pollution control, a duty of care for waste and other provisions

◇ **Environmental Protection Agency (EPA)** *noun US* administrative body in the USA which deals with pollution

QUOTE: Environmental audits can alert an organization to areas in which it does not meet current or proposed legislative requirements. This may avoid unnecessary expenditure on litigation costs, fines and damages. Such audits can also encourage more efficient techniques and waste reduction measures

Natural Resources Forum

◇ **environmentalism** [ɪnvaɪrən'mentəlɪzm] *noun* concern for the environment and its protection

◇ **environmentalist** [ɪnvaɪrən'mentəlɪst] *noun* person who is concerned with protecting the environment to keep it healthy; **environmentalist group** = association *or* society dedicated to the protection of the environment and to the increasing of people's awareness of environmental issues

◇ **environment-friendly** *or* **environmentally friendly** [ɪn'vaɪrənmənt 'frendli] *adjective* not harmful to the environment

QUOTE: the manufacturer claimed its aerosol products were environment-friendly
Green Magazine

Environmentally Sensitive Area (ESA) *noun* area designated by the Ministry of Agriculture on the basis of recommendations made by the Countryside Commission and the Nature Conservancy Council

COMMENT: ESAs were introduced under the 1986 Agriculture Act; they are selected for their landscape and wildlife value, and where traditional farming methods would help to maintain this value. Payments may be made to those farmers within these areas who agree to farm according to the practices set out in guidelines by the Ministry of Agriculture. ESAs are a means of slowing down and preventing agricultural landscape change, but also act as a mechanism for more positive change through the conversion of arable land to grass

enzyme ['enzaɪm] *noun* protein substance produced by living cells which catalyzes a biochemical reaction in living organisms (NOTE: the names of enzymes mostly end with the suffix **-ase**)

◇ **enzymatic** [enzaɪ'mætɪk] *adjective* concerning *or* referring to an enzyme

COMMENT: many different enzymes exist in organisms, working in the digestive system, in the metabolic processes and helping the synthesis of certain compounds. Some pesticides and herbicides work by interfering with enzyme systems or by destroying them altogether

eolian [iː'əʊliən] *US* = AEOLIAN

EPA = ENVIRONMENTAL PROTECTION AGENCY

ephemeral [ɪ'fiːmərəl] *adjective* (plant *or* insect) which does not have a long life; **ephemeral stream** = stream which flows only after rain or snowmelt

epibiosis [epɪbaɪ'əʊsɪs] *noun* state where an organism lives on the surface of another, but is not a parasite

◇ **epibiont** [epɪ'baɪɒnt] *noun* organism which lives on the surface of another

epicentre ['epɪsentə] *noun* point on the surface of the earth above the focus of an earthquake or in the centre of a nuclear explosion

epidemic [epɪ'demɪk] *adjective & noun* (infectious disease) which spreads quickly through a large part of the population; ***the disease rapidly reached epidemic proportions; the health authorities are taking steps to prevent an epidemic of cholera*** *or* ***a cholera epidemic*** *see also* ENDEMIC, PANDEMIC

◇ **epidemiology** [epɪdiːmɪ'ɒlədʒi] *noun* study of diseases in a population, in particular how they spread and how they can be controlled

◇ **epidemiological** [epɪdiːmiə'lɒdʒɪkl] *adjective* referring to epidemiology

epigeal [e'pɪdʒiəl] *adjective* (insect) which lives above ground; (plant) which appears as a seedling with cotyledons above the surface of the soil

◇ **epigenous** [e'pɪdʒənəs] *adjective* developing *or* growing on a surface

epilimnion [epɪ'lɪmniən] *noun* top layer of water in a lake: this layer contains more oxygen and is warmer than the water below

epilithic [epɪ'lɪθɪk] *adjective* growing on or attached to the surface of rocks or stones

epiphyte ['epɪfaɪt] *noun* plant which lives on another plant for physical support, but is not a parasite of it; **nest epiphyte** = plant whose aerial roots and stems collect rotting organic matter from which the plant takes nutrients

episode ['epɪsəʊd] *noun* occurrence *or* time when a certain phenomenon takes place; ***there have been three serious acid rain episodes in the last four months; high sulphur dioxide episodes have killed several hundred birds at one time***

QUOTE: pollution comes in periodic episodes, followed by intervals when the atmosphere remains fairly pure
Environment Now

QUOTE: although the normal amounts of pollutant do not damage vegetation, a number of episodes occurred when concentrations rose to more than 60 parts per billion parts of water
New Scientist

epithelium [epɪ'θiːliəm] *noun* layer(s) of cells covering an organ, including the skin and the lining of hollow cavities; *see also* ENDOTHELIUM, MESOTHELIUM

◇ **epithelial** [epɪ'θiːliəl] *adjective* referring to the epithelium; **epithelial layer** = the epithelium

epixylous [epɪ'ksaɪləs] *adjective* growing on wood

epizoite [epɪ'zəʊaɪt] *noun* animal which lives on the surface of another, without being a parasite

epoch ['iːpɒk] *noun* interval of geological time, the subdivision of a period

equator [ɪ'kweɪtə] *noun* **(terrestrial) equator** = imaginary line running round the surface of the earth, at an equal distance from the North and South Poles; ***the area of rainforests lies around the Equator;*** **celestial equator** = imaginary line in the sky above the terrestrial equator

◇ **equatorial** [ekwə'tɔːriəl] *adjective* referring to the equator; **equatorial current** = westward-moving current in the Atlantic Ocean; **equatorial region** = land area near the equator, mostly with a very hot and humid climate, except for land at high altitudes as in South America; **equatorial trough** = shallow low-pressure zone around the equator

equilibrium [ekwɪ'lɪbriəm] *noun* state of balance; **population equilibrium** = state where the population stays at the same level, because the number of deaths is the same as the number of births

equinox ['ekwɪnɒks] *noun* either of the two occasions in the year (spring and autumn) when the sun crosses the celestial equator and night and day are each twelve hours long

> COMMENT: the two equinoxes are the spring or vernal equinox, which occurs about March 21st and the autumn equinox which occurs about September 22nd

equip [ɪ'kwɪp] *verb* to provide the necessary apparatus; ***the operating theatre is equipped with the latest scanning devices***

◇ **equipment** [ɪ'kwɪpmənt] *noun* apparatus *or* tools which are required to do something; ***the centre urgently needs seismological equipment*** (NOTE: no plural: for one item say **'a piece of equipment'**)

equivalent [ɪ'kwɪvələnt] **1** *adjective* equal in size *or* importance; having the same effect **2** *noun* an equal thing *or* amount; **tonnes of oil equivalent (toe)** = unit of measurement of the energy content of a fuel, calculated by comparing its heat energy with that of oil

era ['ɪərə] *noun* major interval of geological time, divided into periods

eradicate [ɪ'rædɪkeɪt] *verb* to remove completely; ***international action to eradicate glaucoma***

◇ **eradication** [ɪrædɪ'keɪʃn] *noun* (i) removing completely; (ii) total extinction of a species

ergot ['ɜːgət] *noun* fungus which grows on cereals, especially rye

◇ **ergotamine** [ɜː'gɒtəmiːn] *noun* the poison which causes ergotism

◇ **ergotism** ['ɜːgətɪzm] *noun* poisoning by eating cereals *or* bread which have been contaminated by ergot

erode [ɪ'rəʊd] *verb* to wear away; ***the cliffs have been eroded by the sea***

◇ **erosion** [ɪ'rəʊʒn] *noun* wearing away of earth *or* rock by the effect of rain, wind, sea *or* rivers or by the action of toxic substances; ***grass cover provides some protection against soil erosion;*** **land erosion control** = method of preventing the soil from being worn away by irrigation, planting, mulching, etc.

> COMMENT: accelerated erosion is that caused by human activity which is in addition to the natural rate of erosion. Cleared land in drought-stricken areas can produce dry soil which can blow away; felling trees removes the roots which bind the soil particles together and so exposes the soil to erosion by rainwater; ploughing up and down slopes (as opposed to contour ploughing) can lead to the formation of rills and serious soil erosion

erupt [ɪ'rʌpt] *verb (of a volcano)* to become active, i.e. to produce lava, smoke and hot ash, etc.

◇ **eruption** [ɪ'rʌpʃn] *noun* sudden violent ejection of lava, smoke and ash by a volcano; ***several villages were destroyed in the volcanic eruption of 1978***

◇ **eruptive** [ɪ'rʌptɪv] *adjective* (i) (volcano) producing lava, smoke and ash; (ii) (rock) formed by the solidification of magma; (iii) (disease) causing boils, spots, etc. on the skin

ESA = ENVIRONMENTALLY SENSITIVE AREA

escape [ɪ'skeɪp] **1** *noun* **(a)** action of allowing toxic substances to leave a container; ***the area around the reprocessing plant was evacuated because of an escape of radioactive coolant*** **(b)** plant which was formerly cultivated but which has escaped by reproducing in the wild; animal which was formerly domesticated but has become wild; *compare* FERAL **2** *verb* to leave a container; to leave a cultivated *or* domestic area; ***feral cats are cats which have escaped from households and become wild***

escarpment [ɪ'skɑːpmənt] *noun* steep slope of a cuesta

Escherichia [eʃə'rɪkiə] *noun* one of the Enterobacteria commonly found in faeces; **Escherichia coli** = Gram-negative bacillus associated with acute gastroenteritis in infants

esker ['eskə] *noun* long winding ridge formed of gravel

essence ['esəns] *noun* concentrated oil from a plant, used in cosmetics and sometimes as an analgesic or antiseptic

◇ **essential** [ɪ'senʃəl] *adjective* extremely important *or* necessary; **essential amino acid** = amino acid which is necessary for growth but which cannot be synthesized by the body and has to be obtained from the food supply; **essential element** = chemical element (such as hydrogen, carbon, nitrogen, oxygen and many others) which is necessary to an organism's growth and function; **essential fatty acid (EFA)** = unsaturated fatty acid which is necessary for growth and health but which cannot be synthesized by the body and has to be obtained from the food supply; **essential oil** *or* **volatile oil** = concentrated oil from a scented plant used in cosmetics and as an antiseptic

> COMMENT: the essential amino acids are: isoleucine, leucine, lysine, methionine, phenylalanine, threonine, tryptophan and valine. The essential fatty acids are linoleic acid, linolenic acid and arachidonic acid

establish [ɪ'stæblɪʃ] *verb* **(a)** to settle permanently; ***the starling has become established in all parts of the USA; even established trees have been attacked by the disease*** **(b)** to decide what is correct *or* what is fact; ***the inspectors have tried to establish the cause of the disaster***

estate [ɪ'steɪt] *noun* (i) property consisting of a large area of land and a big house; (ii) area specially designed and constructed for residential or industrial use; **housing estate** = area of land with buildings specially designed and constructed for residential use; **industrial estate** = area of land with buildings specially designed and constructed for light industries; **real estate** = land, buildings

estimate 1 *noun* ['estɪmət] (i) approximate calculation of size, weight, extent, etc.; (ii) document giving details of how much a job will cost **2** *verb* ['estɪmeɪt] to make an approximate calculation of size, weight, extent, etc.

estivation [estɪ'veɪʃn] *US* = AESTIVATION

estuary ['estjʊri] *noun* part of a river where it meets the sea (partly composed of salt water)

◇ **estuarine** ['estjuərɪn] *adjective* referring to estuaries; **estuarine plants** = plants which live in estuaries where the water is alternately fresh and salty as the tide comes in and goes out

ethanol (C_2H_5OH) ['eθənɒl] *noun* alcohol *or* colourless inflammable liquid, produced by the fermentation of sugars and used as an ingredient of organic chemicals, intoxicating drinks and medicines

ethical ['eθɪkl] *adjective* concerning ethics; **ethical committee** = group of specialists who monitor experiments involving humans *or* who regulate the way in which members of the medical profession conduct themselves

◇ **ethics** ['eθɪks] *noun* code of working which shows how a professional group, such as doctors and nurses, should work, and in particular what type of relationship they should have with their patients

Ethiopian Region [i:θɪ'əʊpiən 'ri:dʒən] *noun* biogeographical region (part of Arctogea) comprising Africa south of the Sahara

ethno- ['eθnəʊ] *prefix* meaning man; **ethnobotany** = study of the way plants are used by man

ethology [i:'θɒlədʒi] *noun* study of the behaviour of living organisms (a branch of biology)

ethyl alcohol (C_2H_5OH) ['eθɪl 'ælkəhɒl] *noun* = ETHANOL

ethylene ['eθɪli:n] *noun* hydrocarbon used in the production of polythene, as an anaesthetic and occurring in natural gas

etiolation [i:tiə'leɪʃn] *noun* process by which a green plant grown in insufficient light becomes yellow and grows long shoots

etiology, etiological agent [i:ti'ɒlədʒi *or* i:tiə'lɒdʒɪkl] *US* = AETIOLOGY, AETIOLOGICAL AGENT

EU ['i:'ju:] = EUROPEAN UNION

eucalyptus [ju:kə'lɪptəs] *noun* Australian hardwood tree *(Eucalyptus* spp.), with strong-smelling resin; the trees are quick-growing and often used for afforestation. They are susceptible to fire

Euglenophyta [ju:glenə'fɪtə] *noun* type of algae similar to Protozoa

euphotic zone [ju:'fɒtɪk 'zəʊn] *noun* photic zone, the top layer of water in the sea *or* a lake, which sunlight can penetrate and in which photosynthesis takes place; *compare* APHOTIC

European Union (EU) [jʊərə'pi:ən 'ju:niən] *noun* organization established in 1957 by the Treaty of Rome. The original six members were Belgium, Netherlands, France, Italy, Luxembourg and West Germany. Denmark, Ireland and the United Kingdom joined in 1973; Greece joined in 1981; Spain and Portugal joined in 1986; Austria, Finland and Sweden joined in 1995. The organization is responsible for agricultural policy among its member states; *see also* COMMON AGRICULTURAL POLICY (NOTE: formerly called the **European Community** *or* **European Economic Community**)

euryhaline [ju:rɪ'hælaɪn] *adjective* (i) (organism) which can survive a wide range of salt levels in its environment; (ii) (organism) which can survive wide variations in osmotic pressure of soil water; *compare* STENOHALINE

eurythermous [juːrɪ'θɜːməs] *adjective* (organism) which can survive a wide range of temperatures in its environment; *compare* STENOTHERMOUS

eustatic changes [juː'stætɪk 'tʃeɪndʒɪz] *noun* changes relating to worldwide variations in sea level, as distinct from regional changes caused by earth movements in a particular area

eutrophy ['juːtrəfi] **1** *noun* = EUTROPHICATION **2** *verb* to fill up with nutrients; ***the sea is becoming eutrophied with nutrients***

◇ **eutrophic** [ju'trɒfɪk] *adjective* (water) which is rich in dissolved organic and mineral nutrients; **eutrophic lake** = lake which has a high decay rate in the epilimnion (the top layer of water), and so contains little oxygen at the lowest levels: it has few fish but is rich in algae; *compare* OLIGOTROPHIC

◇ **eutrophication** [jutrɒfɪ'keɪʃn] *noun* process by which water becomes full of phosphates and other nutrients which encourage the growth of algae and kill other organisms

> QUOTE: eutrophication takes two main forms in the Mediterranean. The first is 'red tides' caused by the build-up of dinoflagellates. These can release chemicals that are toxic to marine animals such as fish and anything eating fish. The second is the formation of mucus-like foam secreted by diatoms. The foam fouls beaches and removes oxygen from the water, killing many creatures on the seabed
>
> **New Scientist**

> QUOTE: in the North Sea, there have been problems of eutrophication in the German Bight, the angle between Denmark, Germany and the Netherlands. A major cause is probably the increased inputs of nutrients. There are two main sources: sewage discharges and leaching from agriculture, both arable and stock
>
> **Environment Times**

evaporate [ɪ'væpəreɪt] *verb* (i) to change liquid into vapour by heating; (ii); *(of liquid)* to change into vapour; ***in tropical areas up to 75% of rainfall will evaporate; surface water quickly evaporates in the sun***

◇ **evaporation** [ɪvæpə'reɪʃn] *noun* changing liquid into vapour; ***the evaporation of water from the surface of a lake;*** **surface evaporation** = evaporation of water from the surface of a lake, river, etc.

◇ **evaporative** [ɪ'væpərətɪv] *adjective* which can evaporate

◇ **evapotranspire** [ɪvæpəʊ'trænspaɪə] *verb* to lose water into the atmosphere by evaporation and transpiration

◇ **evapotranspiration** [ɪvæpəʊtrænspɪ'reɪʃn] *noun* loss of water in an area through evaporation from the soil and transpiration from plants

> QUOTE: deforestation is already disrupting the hydrological cycles that control rainfall. At least half of the rainwater that falls on most tropical forests is returned back to the atmosphere through evapotranspiration, hence the perpetual cloud that hangs over the world's great rainforests. The evaporated moisture is then carried by the wind to fall as rain in areas often many thousands of miles away
>
> **Ecologist**

event [ɪ'vent] *noun* time when a certain phenomenon takes place; ***three exceptionally warm years - 1980, 1983 and 1987 - years in which El Niño events were at their height***

evergreen ['evəgriːn] **1** *adjective* (plant) which does not lose its leaves in winter **2** *noun* tree *or* shrub which does not lose its leaves in winter; *compare* DECIDUOUS

evolution [iːvə'luːʃn] *noun* heritable changes in organisms which take place over a long period involving many generations

◇ **evolutionary** [iːvə'luːʃənəri] *adjective* referring to evolution; ***evolutionary changes which have taken place over millions of years***

examine [ɪg'zæmɪn] *verb* to look at *or* to investigate (someone *or* something) carefully; ***the water samples were examined in the laboratory***

◇ **examination** [ɪgzæmɪ'neɪʃn] *noun* looking at someone *or* something carefully; ***they carried out an examination of the problem; examination of the house revealed traces of radon***

exceed [ɪk'siːd] *verb* to be more than (a certain limit); ***the concentration of radioactive material in the waste exceeded the government limits;*** **it is dangerous to exceed the stated dose** = do not take more than the stated dose

excess [ɪk'ses] *noun & adjective* too much of a substance; ***excess phosphates run off into the river system;*** **in excess of** = more than; ***exposure to radiation levels in excess of the permitted maximum***

◇ **excessive** [ɪk'sesɪv] *adjective* more than normal; ***excessive ultraviolet radiation can cause skin cancer***

exchange [ɪks'tʃeɪndʒ] *noun* act of taking one thing and putting another in its place; **cation exchange** = exchange which takes place when the ions of calcium, magnesium and other metals found in soil replace the hydrogen atoms in acid; **gas exchange** = transfer of gases between an organism and its environment; **ion exchange** = exchanging of ions between a solid and a solution; **ion-exchange filter** *or* **water softener** =

device attached to the water supply to remove nitrates or calcium from the water

◇ **exchanger** [ɪks'tʃeɪndʒə] *noun* device which exchanges; **heat exchanger** = device which exchanges heat from one source and gives it to another (as in a heating system where a hot pipe goes through a water tank to heat it)

excrement ['ekskrɪmənt] *noun* faeces

excrete [ɪk'skriːt] *verb* to pass waste matter out of the body, especially to discharge faeces; ***the urinary system separates waste liquids from the blood and excretes them as urine***

◇ **excreta** [ɪk'skriːtə] *plural noun* waste material from the body (such as faeces)

◇ **excretion** [ɪk'skriːʃn] *noun* passing waste matter (faeces *or* urine *or* sweat *or* carbon dioxide) out of the system; *compare* SECRETE, SECRETION

exhale [eks'heɪl] *verb* to breathe out

◇ **exhalation** [ekshə'leɪʃn] *noun* breathing out

exhaust [ɪg'zɔːst] **1** *noun* (i) expulsion of gases (e.g. sending out waste gases from an engine); (ii) part of an engine through which waste gases pass; ***fumes from vehicle exhausts contribute a large percentage of air pollution in towns;*** **exhaust fumes** *or* **exhaust gases** = gases, including carbon dioxide and carbon monoxide, produced by the engine of a car *or* truck as it burns petrol *or* fuel; **exhaust gas filter system** = means of removing some or all of the harmful emissions from the exhaust gases of an engine by filtration or catalytic action; **exhaust pipe** *or* *US* **tailpipe** = tube at the back of a car *or* truck out of which gases produced by burning petrol are sent out into the atmosphere; **exhaust purification device** = = EXHAUST GAS FILTER SYSTEM; **exhaust system** = part of an engine through which waste gases pass **2** *verb* to use up completely; **exhausted fallow** = fallow land which is no longer fertile

◇ **exhaustion** [ɪg'zɔːstʃn] *noun* **(a)** using up completely; ***the exhaustion of the earth's natural resources*** **(b)** extreme tiredness *or* fatigue; **heat exhaustion** = collapse due to overexertion in hot conditions

exosphere ['eksəʊsfɪə] *noun* highest layers of the earth's atmosphere, more than 650km above the earth's surface, composed almost entirely of hydrogen

exothermic reaction [eksəʊ'θɜːmɪk rɪ'ækʃn] *noun* chemical reaction in which heat is given out to the surroundings; *compare* ENDOTHERMIC

exotoxin [eksəʊ'tɒksɪn] *noun* poison produced by bacteria, which affects parts of the body away from the place of infection (such as the toxins which cause botulism)

expanded [ɪk'spændɪd] *adjective* (plastic *or* polystyrene) made into hard lightweight foam by blowing air *or* gas into it

> COMMENT: expanded polystyrene and other plastics are extensively used for packaging. CFCs are sometimes used in their manufacture

expectancy [ɪk'spektənsi] *noun* something expected; **life expectancy** = number of years a person *or* animal, etc. is likely to live

expenditure [ɪk'spendɪtʃə] *noun* spending of money; amount of money spent; ***the expenditure of public funds for this project;*** **public expenditure** = amount of money spent by a government

experiment [ɪk'sperɪment] **1** *noun* test *or* trial to demonstrate or discover something; ***to carry out an experiment in order to analyse the component parts of a sample of air*** **2** *verb* to carry out a test *or* trial in order to demonstrate or discover something; ***they are experimenting with a new treatment for asthma***

◇ **experimental** [ɪksperɪ'mentəl] *adjective* using experiments; used in experiments; still being tested, still on trial; ***this process is still at the experimental stage***

◇ **experimentally** [ɪksperɪ'mentəli] *adverb* by carrying out experiments

explode [ɪk'spləʊd] *verb* to burst *or* make something burst violently, e.g. a bomb; *see also* EXPLOSION, EXPLOSIVE

exploit [ɪk'splɔɪt] *verb* to take advantage of something; ***large companies ruined the environment by exploiting the natural wealth of the forest; the birds have exploited the sudden increase in the numbers of insects***

◇ **exploitation** [ɪksplɔɪ'teɪʃn] *noun* taking advantage of something to make money; ***further exploitation of the coal deposits is not economic***

explosion [ɪk'spləʊʒn] *noun* sudden and violent expansion; **population explosion** = sudden increase in the number of organisms, especially of people, in a country

◇ **explosive** [ɪk'spləʊsɪv] **1** *adjective* **(a)** which may cause a rapid expansion (like a bomb); ***continuing deprivation is causing a potentially explosive situation in urban areas;*** **explosive hazard** = risk that a certain substance may blow up **(b)** (virus) which increases very rapidly **2** *noun* substance (such as gunpowder) which can explode

> QUOTE: the disaster started with heavy autumn rains in the Western Sahara, causing a population explosion among locusts
> **New Scientist**

exponential [ekspə'nenʃl] *adjective* increasing more and more rapidly; **exponential growth** = type of growth rate of a population which varies in proportion to the number of individuals present: at first the growth rate is slow, then it rapidly increases, theoretically to infinity, but in practice it falls off as the population exhausts its food, accumulates poisonous waste, etc.

◇ **exponentially** [ekspə'nenʃəli] *adverb* more and more rapidly; ***we can expect pollution to increase exponentially over the next years***

expose [ɪk'spəʊz] *verb* **(a)** to show something which was hidden; ***the report exposed a lack of supervision in maintenance of the reactor*** **(b) to expose something** *or* **someone to** = to place something *or* someone under the influence of something; ***she was exposed to a lethal dose of radiation***

◇ **exposure** [[ɪk'spəʊʒə] *noun* **(a)** being exposed; ***his exposure to radiation;*** **exposure dose** = amount of radiation to which someone has been exposed **(b)** being damp, cold and with no protection from the weather; ***the survivors of the crash were all suffering from exposure after spending a night in the snow***

> QUOTE: exposure to the general background level of airborne asbestos gives rise to a risk which is low in comparison with other risks of life
> **London Environmental Bulletin**

extended family [ɪk'stendɪd 'fæmɪli] *noun* family group which includes not only the nuclear family (the parents and their offspring) but also other blood relatives

extensive farming [ɪk'stensɪv 'fɑːmɪŋ] *noun* way of farming which is characterized by a low level of inputs per unit of land; yields per unit area are very low. Extensive farms are usually large in size; **extensive systems** = farming systems which use a large amount of land per unit of stock or output; ***an extensive system of pig farming***

◇ **extensification** [ɪkstensɪfɪ'keɪʃn] *noun* using less intensive ways of farming

> COMMENT: less intensive use of farming involves using fewer chemical fertilizers, leaving uncultivated areas at the edges of fields, reducing sizes of herds of cattle, etc. This allows lower yields from the same area of farmland, which is necessary if production levels are too high (as they are in the EU)

external [ɪk'stɜːnəl] *adjective* which is outside; **external effects** *or* **externalities** = costs to society of industrial processes which are not reflected in the price of the product sold (as in the case of the environmental effects of a power station)

exteroceptor ['ekstərəʊseptə] *noun* sensory nerve such as the eye *or* ear, which is affected by stimuli from outside the body; *see also* CHEMORECEPTOR, INTEROCEPTOR, RECEPTOR

extinct [ɪk'stɪŋkt] *adjective* **(a)** (species) which has died out; ***one species of animal or plant becomes extinct every day; several native species have become extinct since sailors in the nineteenth century introduced dogs to the island*** **(b)** (volcano) which is no longer able to erupt; *compare* DORMANT

◇ **extinction** [ɪk'stɪŋkʃn] *noun* dying out of a species; ***the last remaining examples of oryx were taken to zoos for breeding purposes, so as to save the species from extinction; the introduction of European species caused the extinction of the native species***

extract 1 *noun* ['ekstrækt] preparation made by removing water *or* alcohol from a substance, leaving only the essence **2** *verb* [ɪk'strækt] (i) to take out of the ground *or* to mine (minerals); (ii) to remove the essence from (a liquid)

◇ **extraction** [ɪk'strækʃn] *noun* action of extracting; ***the extraction of coal from the mine is becoming too costly;*** **gas extraction** = pumping gas out of a landfill site to use as fuel; **extraction fan** = EXTRACTOR FAN

◇ **extractor** [ɪk'stræktə] *noun* machine which extracts, such as one which removes fumes *or* gas; **extractor fan** = small fan which removes fumes *or* smoke, etc. from a kitchen *or* factory

extrusive [ɪk'struːsɪv] *adjective* (rock) which has formed from molten lava which has pushed up through the earth's crust

exurbia [eks'ɜːbiə] *noun* area lying outside a city *or* town

eye [aɪ] *noun* **(a)** part of the body with which an animal sees; **eye bank** = place where parts of eyes given by donors can be kept for use in grafts **(b)** central point of a tropical storm

> COMMENT: the eye of a typhoon is the point of lowest pressure, where the wind is light. As the storm approaches, winds grow in intensity until they are strongest just before the eye passes over the spot. The eye can be several kilometres wide and may take some minutes to pass. When it has passed, the winds begin again, but from the opposite direction

eyesore ['aɪsɔː] *noun* something which is unpleasant to look at; ***the lake full of old bottles and cans is an eyesore and should be cleaned up***

Ff

F 1 *abbreviation for* Fahrenheit **2** *chemical symbol for* fluorine

◇ **F_1** ['ef 'wɒn] *noun (in breeding experiments)* the first generation of offspring; **F_1 hybrid** = plant produced by breeding two parent plants, which is stronger than the parents, but which will not itself breed true

COMMENT: F_1 hybrids can be crossbred to produce F_2 hybrids and the process can be continued for many generations

facies ['feɪʃi:z] *noun* appearance of something *or* 'face' of an animal; **neritic facies** = appearance of sedimentary rocks laid down in shallow water, where the ripple marks made by waves are clearly visible

facilities [fə'sɪlɪtɪz] *plural noun* equipment *or* resources which can be used to do something; ***there are no sports facilities in the town;*** **public facilities** = facilities which are open to the general public; **recreational facilities** = facilities which can be used for leisure activities

factor ['fæktə] *noun* **(a)** something which has an influence, which makes something else take place; **ecological factor** = factor which influences the distribution of a plant species in a habitat; **growth factor** = chemical substance produced in one part of the body which encourages the growth of a type of cell (such as red blood cells); **hereditary factor** = characteristic which is passed from parent to offspring in genes **(b)** the number that multiplies in a multiplication; **by a factor of ten** = ten times *or* multiplied by ten

QUOTE: nitrates and sulphates have increased by factors of 2 and 3.5 respectively in the Arctic in the 200 years since the Industrial Revolution
New Scientist

QUOTE: the dilution of the acid lowers the concentration of sulphur and nitrogen compounds in precipitation by a factor of between 3 and 30
Scientific American

factory ['fæktəri] *noun* building where products are manufactured; ***smoke from factory chimneys blew across the town; the planning department has refused permission to site the new chemical factory in the middle of a residential district; illegal discharge of slurry from factory farms can contaminate the water system;*** **factory farming** = highly intensive rearing of animals characterized by keeping large numbers of animals indoors in confined spaces and feeding them artificial foods, with frequent use of drugs to control diseases which are a constant threat under these conditions; **factory inspector** *or* **inspector of factories** = government official who inspects factories to see if they obey government regulations; **factory ship** = large ship which processes fish caught by a fishing fleet

faeces ['fi:si:z] *plural noun* excreta, solid waste matter passed from the bowels after food has been eaten and digested

◇ **faecal** ['fi:kəl] *adjective* referring to faeces; **faecal matter** = solid waste matter from the bowels (NOTE: also spelt **feces, fecal** especially in the USA)

Fahrenheit ['færənhaɪt] *noun* scale of temperatures where the freezing and boiling points of water are 32° and 212°; *compare* CELSIUS, CENTIGRADE (NOTE: used in the USA, but less common in the UK. Normally written as an **F** after the degree sign: **32°F** (say: 'thirty-two degrees Fahrenheit')

COMMENT: to convert Fahrenheit temperatures to Celsius, subtract 32, multiply by 5 and divide by 9. So 68°F is equal to 20°C. As a quick rough estimate, subtract 30 and divide by two

fall [fɔ:l] **1** *noun* **(a)** reduction in quantity of something; ***after installing filters, there was a fall in the emission of particles of ash*** **(b)** something which comes down; **fall of snow** *or* **snowfall** = quantity of snow which comes down at any one time; ***there was a heavy fall of snow during the night*** *see also* RAINFALL **(c)** *US* = AUTUMN **2** *verb* **(a)** to become less; ***barometric pressure is falling, and a storm is likely to pass over the area during the next twenty-four hours; radiation levels have fallen*** **(b)** to drop down; ***snow fell all night; no rain has fallen in the area for over six months***

◇ **fallout** ['fɔ:laʊt] *noun* polluting matter which falls from the atmosphere as particles, either in rain or as dust; **nuclear** *or* **radioactive fallout** = radioactive material which falls from the atmosphere after a nuclear explosion; **fallout shelter** = building where people are protected from radioactive material which falls from the atmosphere after a nuclear explosion

◇ **falls** [fɔ:lz] *plural noun* waterfall (NOTE: often used in names: **Niagara Falls, Victoria Falls**)

◇ **fallspeed** ['fɔ:lspi:d] *noun* speed at which raindrops *or* dry particles fall through the air

◇ **fallstreak** ['fɔ:lstri:k] *noun* column of ice particles which are falling through a cloud

> QUOTE: the overall level of radioactivity in the vegetation eaten by sheep has fallen by one-fifth over the last year
>
> **New Scientist**

fallow ['fæləʊ] **1** *adjective* (land) which is not being used for growing crops for a period so that the nutrients can build up again in the soil; **to let land lie fallow** = to allow land to stand without being cultivated **2** *noun* period when land is not used for cultivation; ***shifting cultivation is characterized by short cropping periods and long fallows;*** **fallow crop** = crop grown in widely spaced rows, so that it is possible to hoe and cultivate between the rows; **fallow cultivation** = type of shifting cultivation: the period under crops is increased and the length of the fallow is reduced; **bush fallow** = subsistence type of agriculture in which land is cultivated for a few years until its natural fertility is exhausted, then allowed to rest for a considerable period during which the natural vegetation regenerates itself, after which the land is cleared and cultivated again

family ['fæməli] *noun* **(a)** group composed of parents and offspring; **extended family** = family group which includes not only the nuclear family but also other blood relatives; **nuclear family** = main family group composed of two parents and their offspring **(b)** group of genera which have certain characteristics in common (several families form an order); ***tigers and leopards are members of the cat family*** (NOTE: names of families of animals end in **-idae** and of families of plants in **-ae**)

famine ['fæmɪn] *noun* period of severe shortage of food; ***when the monsoon failed for the second year, the threat of famine became more likely;*** **famine relief** = sending supplies of basic food to help people who are starving

fan out ['fæn 'aʊt] *verb* to spread out from a central point

◇ **fanning** ['fænɪŋ] *noun* spreading out of a horizontal layer of smoke and other pollutants from a chimney

FAO = FOOD AND AGRICULTURE ORGANIZATION

far [fɑː] *adjective* distant, situated a long way away; **the Far East** = countries of eastern Asia, including China, Japan, Korea, Indochina, etc.; **the Far North** = Arctic and subarctic regions

farm [fɑːm] **1** *noun* area of land used for growing crops and keeping animals to provide food; also the buildings on it; **family farm** = a farm unit which ideally supports one family; **farm buildings** = buildings on a farm: such buildings are needed to shelter animals, get stock off the ground to improve grass production, provide an artificial environment for stock or crops, improve working conditions for staff, house machinery, and protect stored materials from weather and vermin; **farm produce** = fruit, vegetables, meat, milk, butter, etc. which is produced on a farm **2** *verb* to run a farm; to keep animals for sale *or* to grow crops for sale

◇ **farmed** [fɑːmd] *adjective* grown *or* produced by a farmer, not in the wild; ***farmed venison***

◇ **farmer** ['fɑːmə] *noun* person who owns *or* runs a farm; **dairy farmer** = farmer who keeps cattle for milk; **farmer's lung** = form of asthma caused by an allergy to rotting hay

◇ **farmhouse** ['fɑːmhaʊs] *noun* house where a farmer and his family live

◇ **farming** ['fɑːmɪŋ] *noun* running a farm by keeping animals for sale *or* growing crops for sale and producing cereals, vegetables, meat, dairy products, etc.; **arable farming** = growing crops for sale; **contour farming** = method of cultivating sloping land, where the land is ploughed along a level rather than down the slope, so reducing soil erosion; **dairy farming** = keeping cows for milk production; **fruit farming** = growing fruit for sale; **mixed farming** = farming involving arable and dairy farming; **organic farming** = method of farming which does not involve using chemical fertilizers *or* pesticides; **sea farming** = systematic production of food from the sea; **vegetable farming** = growing vegetables for sale; **farming community** = group of families living in a certain area where the main source of income is from farming; *see also* FACTORY FARMING, FISH FARMING, LEY FARMING

◇ **farmland** ['fɑːmlænd] *noun* cultivated land *or* land which is used for growing crops or rearing animals for food

◇ **farmstead** ['fɑːmsted] *noun* farmhouse and the farm buildings around it

◇ **farmworker** ['fɑːmwɜːkə] *noun* person who works on a farm

◇ **farmyard** ['fɑːmjɑːd] *noun* area around the farm buildings; **farmyard manure (FYM)** = manure formed of cattle excreta mixed with straw, used as a fertilizer

> QUOTE: in the last 10 years, the number of farmers in the U.K. has dropped by 12 % to 189,000, while there has been a reduction in the farming workforce to 130,000
>
> **Guardian**

fast [fɑːst] *adjective* (process) which takes place in a short period of time; (neutron) which has a lot of energy; **fast breeder reactor (FBR)** *or* **fast (neutron) reactor** = nuclear reactor which produces more fissile material than it consumes, using fast-moving neutrons and making plutonium-239 from uranium-238, thereby increasing the reactor's efficiency; **fast (neutron) fission** = nuclear fission of the uranium-238 isotope, which is much faster than that of uranium-235

COMMENT: uranium-238 is a natural uranium isotope and is fertile, i.e. it can be used to produce the fissile plutonium- 239. In a breeder reactor, uranium-238 is used as a blanket round the plutonium fuel and when the plutonium is fissioned high-speed neutrons are produced which change on contact with the uranium- 238 and eventually produce a slightly greater quantity of plutonium-239 than that originally used as fuel. The excess plutonium can be used as a fuel in another breeder reactor or in an ordinary burner reactor

fat [fæt] **1** *adjective* having a lot of (or too much) flesh on the body; (animal) which has been reared for meat production and which has reached the correct standard for sale in a market; ***the price of fat cows has increased;*** **fat lamb** = lamb in condition satisfactory for slaughter **2** *noun* **(a)** white *or* oily substance in the body which stores energy and protects the body against cold; **body fat** *or* **adipose tissue** = tissue where the cells contain fat which replaces the normal fibrous tissue when too much food is eaten; **brown fat** = animal fat which can easily be converted to energy and is believed to offset the effects of ordinary white fat; **saturated fat** = fat which has the largest amount of hydrogen possible; **unsaturated fat** = fat which does not have a large amount of hydrogen, and so can be broken down more easily **(b)** type of food which supplies protein and Vitamins A and D, especially that part of meat which is white *or* solid substances (like lard *or* butter) produced from animals and used for cooking *or* liquid substances like oil

◇ **fat-soluble** [fæt'sɒljʊbl] *adjective* which can dissolve in fat; ***Vitamin D is fat-soluble; PCBs are fat-soluble and so collect in the blubber of seals***

◇ **fatstock** ['fætstɒk] *noun* livestock which has been fattened for meat production

◇ **fatten** ['fætn] *verb* to give animals more food so as to prepare them for slaughter; ***he buys lambs for fattening and then sells them for meat***

◇ **fatty** ['fæti] *adjective* containing fat; **fatty acid** = acid (such as stearic acid) which is an important substance in the body; **essential fatty acid (EFA)** = unsaturated fatty acid which is necessary for growth and health but which cannot be synthesized by the body and has to be obtained from the food supply

COMMENT: fat is a necessary part of diet because of the vitamins and energy-giving calories which is contains. Fat in the diet comes from either animal fats or vegetable fats. Animal fats such as butter, fat meat or cream, are saturated fatty acids. It is believed that the intake of unsaturated and polyunsaturated fats (mainly vegetable fats and oils and fish oil) in the diet, rather than animal fats, helps keep down the level of cholesterol in the blood and so lessens the risk of atherosclerosis. A low-fat diet does not always help to reduce weight

QUOTE: modern strains of broiler are not only heavier but also contain a greater proportion of fat at killing age compared with broilers of 20 years ago. Actual levels of fatness can vary considerably between flocks, but body fat contents of 12-25% are not uncommon

Poultry Science Symposium 18

fatal ['feɪtəl] *adjective* causing death; ***a fatal accident; the bite of that snake is fatal***

fault [fɔːlt] *noun* line of a crack in the earth's crust along which movements can take place leading to major earthquakes; **reversed fault** *or* **thrust fault** = fault in which the upper layers of rock have been pushed forward over the lower layers; **San Andreas Fault** = crack in the earth's crust in California, running parallel to the coast and passing close to San Francisco

◇ **fault plane** ['fɔːlt 'pleɪn] *noun* face of the rock at a fault where one mass of rock has slipped against another

fauna ['fɔːnə] *noun* (i) wild animals and birds which live naturally in a certain area; (ii) book *or* list describing the animals and birds of a certain area; *see also* FLORA; **deep-sea fauna** = fauna which lives in the deepest part of the sea; **soil fauna** = fauna which lives in soil (NOTE: plural is **fauna** or **faunas**)

FBC = FLUIDIZED-BED COMBUSTION

FBR = FAST BREEDER REACTOR

FC = FIBRE-CONCRETE

FCO = FOREIGN AND COMMONWEALTH OFFICE

FDA = FOOD AND DRUG ADMINISTRATION

Fe *chemical symbol for* iron

feces, fecal *see* FAECES, FAECAL

feed [fiːd] **1** *noun* food given to animals and birds; ***traces of pesticide were found in the cattle feed;*** **feed additives** = supplements added to the feed of farm livestock, particularly pigs and poultry, to promote growth; **feed concentrates** = animal feedingstuffs which have a high food value relative to volume (oats, wheat, maize, oilseed meal, etc., are used to make feedingstuffs); **feed grains** = cereal crops, such as corn, which are fed to animals and birds; **feed intake** = amount of food eaten by an animal **2** *verb* **(a)** to eat food; to give an animal *or* a person food to eat; **feeding grounds** = areas where animals come to feed;

estuaries are winter feeding grounds for thousands of migratory birds **(b)** to lead into and so make larger; ***several small streams feed into the river***

◇ **feeder** ['fi:də] *noun* **(a)** animal which feeds; **bottom-feeder** = fish which eats food found at the bottom of the water; **filter-feeder** = fish which eats particles of detritus floating in water; **plant-feeder** = animal which eats grass and other plants **(b)** device which feeds; **feeder reservoir** = small reservoir from which water flows into another reservoir; **feeder stream** = small stream which leads into a river

feedback ['fi:dbæk] *noun* (i) information; (ii) mechanism by which a process can regulate itself by the product it produces; ***we are getting some feedback from workers in the area about the distribution of relief supplies;*** **positive feedback** = situation where the result of a process stimulates the process which caused it (as when a change at one trophic level affects all levels above it)

> QUOTE: the changing carbon dioxide content of the atmosphere amplifies the change in a feedback
> **New Scientist**

feedingstuffs ['fi:dɪŋstʌfs] *noun* various types of food available for farm animals

feedlot ['fi:dlɒt] *noun* field with small pens in which cattle are fattened

> COMMENT: a new type of feedlot is an area of land surrounded by an earth embankment, which protects the cattle from cold winds while they are being fed intensively. See also CHILLSHELTER

feedstuff ['fi:dstʌf] *see* FEEDINGSTUFFS

feldspar *or* **feldspath** ['feldspɑ: *or* 'feldspæθ] *noun* common type of crystal rock formed of silicates

fell [fel] **1** *noun* high moor and mountain in the North of England **2** *verb* to cut down (a tree); *see also* CLEARFELL

female ['fi:meɪl] *adjective & noun* (animal) which produces ova and bears young; (flower) which has carpels but not stamens

fen [fen] *noun* area of flat marshy land, with reeds and mosses growing in alkaline water

◇ **fenland** ['fenlænd] *noun* area of land covered by fens

fence [fens] **1** *noun* barrier put round a field, either to mark the boundary or to prevent animals entering or leaving; **fence post** = wooden post which supports the wire of a fence **2** *verb* to put a fence round an area of land

> COMMENT: various methods are used to fence field boundaries, most commonly woven wire and wooden fence posts. Movable electric fences are an efficient way of limiting areas of a field for grazing purposes

feral ['ferəl] *adjective* (animal) which was formerly domesticated and has since reverted to living wild; ***the native population of rabbits was exterminated by feral cats***

fermentation [fɜ:mən'teɪʃn] *noun* process whereby carbohydrates are broken down by enzymes from yeast and produce heat and alcohol

fern [fɜ:n] *noun* type of green plant which propagates itself by spores from its leaves, and not seeds, therefore it has no flowers

ferro- ['ferəʊ] *prefix* referring to *or* containing iron

◇ **ferric oxide** (**Fe_2O_3**) ['ferɪk 'ɒksaɪd] *noun* red insoluble oxide of iron, such as rust

◇ **ferrous** ['ferəs] *adjective* referring to *or* containing iron

◇ **ferruginous** [fe'ru:dʒɪnəs] *adjective* (rock *or* water) which contains iron

fertile ['fɜ:taɪl] *adjective* (i) (plant) which is able to produce fruit; (ii) (animal) which is able to produce young; (iii) (soil) which is able to produce good crops

◇ **fertility** [fɜ:'tɪlɪti] *noun* (i) being fertile; (ii) proportion of eggs which develop into young; **fertility rate** = number of births per year, shown, in humans, per thousand females aged between 15 and 44 (NOTE: the opposite is **sterile, sterility**)

◇ **fertilization** [fɜ:təlaɪ'zeɪʃn] *noun* joining of an ovum and a sperm to form a zygote and so start the development of an embryo

◇ **fertilize** ['fɜ:təlaɪz] *verb* **(a)** to put fertilizer on (land) **(b)** *(of a sperm)* to join with (an ovum) **(c)** *(of a male)* to make a female pregnant

◇ **fertilizer** ['fɜ:təlaɪzə] *noun* chemical *or* natural substance spread and mixed with soil to make it richer and stimulate plant growth; **artificial fertilizer** *or* **chemical fertilizer** = fertilizer manufactured from chemicals; **nitrogen fertilizer** = fertilizer containing mainly nitrogen; **compound fertilizer** *or US* **mixed fertilizer** = fertilizer that supplies two or more nutrients (fertilizers that supply only one nutrient are called 'straights')

> COMMENT: organic materials used as fertilizers include manure, slurry, rotted vegetable waste, bonemeal, fishmeal, seaweed. Inorganic fertilizers are also used, such as powdered lime or sulphur. In commercial agriculture, artificially prepared fertilizers (manufactured compounds containing nitrogen, potassium and other chemicals) are most often used, but excessive use of them can cause

pollution, when all the chemicals are not taken up by plants and the excess is leached out of the soil into rivers and may cause algal bloom

QUOTE: excessive amounts of potassium lead to poor fertility in dairy cows grazing intensively fertilized pastures. High levels of nitrate in grass are suspected of causing liver damage in cows
Ecologist

fetus ['fi:təs] *US* = FOETUS

FGD = FLUE GAS DESULPHURIZATION

fibre *or US* **fiber** ['faɪbə] *noun* **(a)** organic structure shaped like a thread (in plants forming the structure of the stems, and in animals forming connective tissue) **(b) dietary fibre** = fibrous matter in food, which cannot be digested; **high-fibre diet** = diet which contains a large amount of cereals, nuts, fruit and vegetables

◇ **fibre-concrete (FC)** ['faɪbə 'kɒŋkri:t] *noun* construction material made of sand, cement and fibre, used for making roofing tiles, light walls, etc.

◇ **fibreglass** *or US* **fiberglass** ['faɪbəglɑ:s] *noun* material made of fine fibres of glass, used as a heat and sound insulator and for building bodies of cars, boats, etc.

◇ **fibrous** ['faɪbrəs] *adjective* made of a mass of fibres; **fibrous-rooted plant** = plant with roots which are masses of tiny threads, with no major roots like taproots

COMMENT: dietary fibre is found in cereals, nuts, fruit and some green vegetables. There are two types of fibre in food: insoluble fibre (in bread and cereals) which is not digested and soluble fibre (in vegetables and pulses). Foods with the highest proportion of fibre are bread, beans and dried apricots. Fibre is thought to be necessary to help digestion and avoid developing constipation, obesity and appendicitis

fidelity [fɪ'delɪti] *noun* degree to which an organism stays in one type of environment

field [fi:ld] *noun* **(a)** area of cultivated land, usually surrounded by a fence *or* hedge, used for growing crops *or* for pasture; **field botanist** = botanist who examines plants in their growing habitat; **field botany** = scientific study of plants in their growing habitat; **field capacity** = maximum possible amount of water remaining in the soil after excess water has drained away; **field crop** = crop grown over a wide area, such as most agricultural crops and some market-garden crops; **field drainage** = building drains in or under fields to remove surplus water; **field-grown** = (crop, fruit, vegetable, etc.) which is grown in a field as opposed to in a greenhouse, etc.; **field observation** *or* **observation in the field** = examination made in the open air, looking at organisms in their natural habitat, as opposed to in a laboratory; **field research** *or* **field study** = scientific study made in the open air as opposed to in a laboratory; **field test** = test carried out on a substance *or* on an organism in the open air as opposed to in a laboratory **(b)** (i) physical area; (ii) area of natural resources (such as an oilfield); **field of vision** = area which can be seen without moving the eye; **magnetic field** = area round a body which is under the influence of its magnetic effect.(The Earth's magnetic field is concentrated round the two magnetic poles) **(c)** area of interest; ***he specializes in the field of environmental health***

◇ **fieldwork** ['fi:ldwɜ:k] *noun* scientific study made in the open air as opposed to in a laboratory

filament ['fɪləmənt] *noun* stalk of a stamen

filariasis [fɪlə'raɪəsɪs] *noun* tropical disease caused by parasitic threadworms in the lymph system, transmitted by mosquito bites

film [fɪlm] *noun* **(a)** roll of material which is put into a camera for taking photographs **(b)** very thin layer of a substance, especially on the surface of a liquid; ***a film of oil on the surface of water***

filter ['fɪltə] **1** *noun* piece of paper *or* cloth through which a liquid is passed to remove solid substances; **filter basin** = large tank through which drinking water is passed to be filtered; **filter bed** = layer of charcoal *or* clinker, etc. through which liquid sewage is passed to clean it; **filter-feeder** = fish which eats particles of detritus floating in water **2** *verb* to pass a liquid through a piece of paper *or* cloth to remove solid substances; ***impurities are filtered from the liquid by passing it through layers of charcoal***

◇ **filtrate** ['fɪltreɪt] *noun* liquid which has passed through a filter

◇ **filtration** [fɪl'treɪʃn] *noun* passing a liquid through a filter to remove solid substances

fine [faɪn] *adjective* **(a)** (particle *or* grain of sand) which is very small; ***particles of fine ash are carried into the upper atmosphere*** **(b)** (weather) which is warm, sunny, with few clouds and no rain or fog

finite ['faɪnaɪt] *adjective* which has an end; ***coal supplies are finite and are forecast to run out in 1995;*** **finite resources** = resources which will in the end be used up (such as reserves of coal *or* oil *or* gas) (NOTE: opposite is **renewable**)

fiord [fjɔ:d] = FJORD

fir (tree) [fɜ:] *noun* common evergreen softwood tree

◇ **fir cone** ['fɜː 'kəʊn] *noun* hard case containing the seeds of a fir tree

fire ['faɪə] *noun* substance in the state of burning, which can be seen in the form of flames *or* a glow; ***forest fires create huge amounts of atmospheric pollution;*** **fire climax** = condition by which an ecosystem is maintained by fire

◇ **firebreak** ['faɪəbreɪk] *noun (in a forest)* area where no trees are planted, so that a forest fire cannot pass across and spread to other parts of the forest

◇ **-fired** [faɪəd] *suffix* meaning burning as a fuel; ***coal-fired power station; gas-fired central heating***

◇ **firedamp** ['faɪədæmp] *noun* methane, colourless inflammable gas found in coal mines

◇ **fireproof** ['faɪəpruːf] *adjective* (material) which does not burn easily

◇ **fire-retardant** ['faɪə rɪ'tɑːdənt] *adjective* (substance) which slows down the rate at which a material burns

◇ **firewood** ['faɪəwʊd] *noun* wood which can be burnt to provide heat

COMMENT: in shifting cultivation, the practice of clearing vegetation by burning is widespread. One of the simplest forms involves burning off thick and dry secondary vegetation. Immediately after burning, a crop like maize is planted and matures before the secondary vegetation has recovered. Where fire clearance methods are used, the ash acts as a fertilizer

QUOTE: ecologists accept that forest fires are, over the long term, essential for vigorous tree growth. Burning clears out undergrowth that might choke saplings, removes old dead wood and recycles nutrients into the soil
Nature

firn [fɜːn] *noun* névé, spring snow on high mountains which becomes harder during the summer

fish [fɪʃ] **1** *noun* cold-blooded aquatic vertebrate, some species of which are eaten for food. (Fish are high in protein, phosphorus, iodine and vitamins A and D; white fish have very little fat); **fish farm** = place where edible fish are bred in special pools for sale as food; **fish farming** = breeding edible fish in special pools for sale as food; **fish kill** = instance of fish being killed (as by pollution); ***aluminium is a critical factor in fish kills;*** **fish ladder** = series of pools, each one higher than the other, specially built to allow fish (such as salmon) to swim up or down the river; **fish pond** = pool in which fish are kept; **fish population** = number of fish in the sea or in a lake, etc.; **fish stock** = quantity of fish held for future use **2** *verb* to try to catch fish (for food)

◇ **fisherman** ['fɪʃəmən] *noun* man who catches fish as an occupation

◇ **fishery** ['fɪʃri] *noun* **(a)** catching fish (as a business) **(b)** area of sea where fish are caught; ***the boats go each year to the rich fisheries off the north coast;*** **coastal fishery** = area of sea next to the shoreline where fish are caught; **fishery protection zone** = area of sea round a country where fishing is regulated; ***a 200-mile fishery protection zone has been established round the islands***

◇ **fishing** ['fɪʃɪŋ] *noun* catching fish; **fishing area** = area of water where fish are caught; **fishing grounds** = area of sea where fish are caught; **fishing rights** = permission to catch fish in a certain area; **fishing zone** = area of sea which can be fished

◇ **fishmeal** ['fɪʃmiːl] *noun* dried fish reduced to a powder, used as an animal feed or as a fertilizer

◇ **fish pass** *or* **fishway** ['fɪʃ 'pɑːs *or* 'fɪʃweɪ] *noun* channel near a dam, specially built to allow fish (such as salmon) to by-pass the dam and swim up or down the river

QUOTE: since the early 1950s extraction rates from marine fisheries have more than quadrupled: from 20 million to nearly 90 million tonnes
Appropriate Technology

fissile ['fɪsaɪl] *adjective* **(a)** (rock) which can split *or* can be split **(b)** (isotope) which can split on impact with a neutron

◇ **fission** ['fɪʃən] **1** *noun* **(a)** splitting (as of the cells of bacteria) **(b) atomic fission** *or* **nuclear fission** = splitting of the nucleus of an atom, such as uranium-235, into several small nuclei which then releases energy and neutrons; **fast (neutron) fission** = nuclear fission of the uranium-238 isotope, which is much faster than that of uranium-235 **2** *verb* to split (in nuclear fission); ***when the plutonium is fissioned, fast neutrons are produced***

◇ **fissionable** ['fɪʃənəbl] *adjective* fissile; (isotope) which can split on impact with a neutron

COMMENT: uranium-238 is a natural uranium isotope and is fertile, i.e. it can be used to produce the fissile plutonium- 239. In a breeder reactor, uranium-238 is used as a blanket round the plutonium fuel and when the plutonium is fissioned high-speed neutrons are produced which change on contact with the uranium- 238 and eventually produce a slightly greater quantity of plutonium-239 than that originally used as fuel. The excess plutonium can be used as a fuel in another breeder reactor or in an ordinary burner reactor

fissure ['fɪʃə] *noun* crack *or* groove in rock

fitness ['fɪtnəs] *noun* measure of evolutionary success, applied to genes, traits, organisms and populations; **Darwinian fitness** = the measure of reproductive success *or* the success of organisms at passing on their genes to subsequent generations; **inclusive fitness** = the sum of an organism's Darwinian fitness with the fitness of its relatives

fix [fɪks] *verb* to attach something permanently; **nitrogen-fixing plants** = plants, such as lucerne or beans, which form an association with bacteria which convert nitrogen from the air into nitrogen compounds which pass into the soil; **fixed costs** = costs (such as rent) which do not increase with the quantity of a product produced

fixation [fɪk'seɪʃn] *noun* act of fixing something; **nitrogen fixation** = process by which nitrogen in the air is converted by bacteria in certain plants into nitrogen compounds: when the plants die the nitrogen is released into the soil and acts as a fertilizer

fjord [fjɔːd] *noun* long inlet of the sea among mountains (in temperate or arctic regions)

flake [fleɪk] *noun* a unit of snow, small piece of snow which falls from the sky

flammability [flæmə'bɪlɪti] *noun* ability of a material to catch fire easily

◇ **flammable** ['flæməbl] *adjective* (material) which catches fire easily

flash [flæʃ] *noun* something which happens very rapidly; ***a flash of lightning;*** **flash flood** = sudden rush of water in a small stream, often causing damage; **flash point** = lowest temperature at which a vapour over a liquid can catch fire in air

flatfish ['flætfɪʃ] *noun* type of fish (such as a sole) which lives on the bed of the sea *or* of a lake and has both eyes on the top of the body

◇ **flatworm** ['flætwɜːm] *noun* any of several types of parasitic worm with a flat body (such as a tapeworm)

flavouring agent ['fleɪvərɪŋ 'eɪdʒənt] *noun* substance added to give flavour

flea [fliː] *noun* small insect which lives as a parasite on animals, sucking their blood and causing disease, such as bubonic plague

fleet [fliːt] *noun* number of ships *or* aircraft *or* buses, etc.; **fishing fleet** = number of boats used for catching fish; ***the country's fishing fleet is half its former size***

flightless bird ['flaɪtləs 'bɜːd] *noun* bird (such as the ostrich) which has small wings and cannot fly

flocculation [flɒkjʊ'leɪʃn] *noun* **(a)** gathering together in lumps (used of impurities in water which form lumps before settling to the bottom of a tank) **(b)** grouping of small particles of soil together to form larger ones; ***the flocculation of particles is very important in making clay soils easy to work***

◇ **flocculant** ['flɒkjʊlənt] *noun* substance added to water as it is treated to encourage impurities to settle

flock [flɒk] *noun* large group of birds *or* herbivorous animals; ***a flock of geese; a flock of sheep*** (NOTE: the word 'flock' is used for sheep, goats, and domesticated birds such as chickens or geese. The word used for cattle is 'herd')

floe [fləʊ] *noun* **ice floe** = sheet of ice floating in the sea

flood [flʌd] **1** *noun* large amount of water covering land which is normally dry (caused by melting snow, heavy rain, high tides, storms, etc.); ***after the rainstorm there were floods in the valleys; the spring floods have washed away most of the topsoil;*** **river in flood** = river which contains an unusually large quantity of water; **flood alleviation** *or* **flood control measures** = helping to reduce the possibility of flooding by controlling the flow of water in rivers; **flood damage** = damage caused by floodwater; **flood warning** = alert that there is likely to be a flood **2** *verb* to cover dry land with a large amount of water; ***the river bursts its banks and floods the whole valley twice a year in the rainy season***

◇ **flooding** ['flʌdɪŋ] *noun* action of covering land with a large amount of water; ***severe flooding has been reported in the north of the country***

◇ **flood plain** ['flʌd 'pleɪn] *noun* wide flat part of the bottom of a valley which is usually covered with water when the river floods

◇ **flood tide** ['flʌd 'taɪd] *noun* tide which is rising *or* which is at its highest; *compare* EBB TIDE

◇ **floodwater** ['flʌdwɔːtə] *noun* water which floods land; ***after the floodwater receded the centre of the town was left buried in mud;*** **floodwater level** = highest point that the water reaches in a flood

floor [flɔː] *noun* ground at the bottom of something; ***fish which live on the floor of the ocean; the forest floor is covered with decaying vegetation***

flora ['flɔːrə] *noun* (i) wild plants which grow naturally in a certain area; (ii) book *or* list describing the plants of a certain area; *compare* FAUNA

◇ **floral** ['flɔːrəl] *adjective* referring to plants *or* flowers

◇ **floret** ['flɒrɪt] *noun* little flower which forms parts of a larger composite flower head

flourish ['flʌrɪʃ] *verb* to live well and spread; ***the colony of rabbits flourished in the absence of any predators; the island has a flourishing plant community***

flow [fləʊ] **1** *noun* **(a)** movement of liquid *or* gas *or* heat, etc.; **energy flow** = movement of energy from one trophic level to another in a food chain **(b)** amount of liquid *or* gas *or* heat, etc. which is moving; ***the meter measures the flow of water through the pipe;*** **lava flow** = lava moving down the sides of a volcano **2** *verb (of liquid)* to move past; ***the water flows through the pipe and turns the turbine***

◇ **flowmeter** ['fləʊmi:tə] *noun* meter attached to a pipe to measure the speed at which a liquid *or* gas moves in the pipe

flower ['flaʊə] *noun* usually brightly-coloured reproductive part of a plant, with an external calyx, petals, stamens and pistils which bear pollen

fluctuate ['flʌktjueɪt] *verb* to vary *or* to rise and fall; ***the average annual temperature of the earth fluctuates very little; most populations fluctuate within fairly narrow limits***

◇ **fluctuation** [flʌktʃu'eɪʃn] *noun* variation *or* change; ***small fluctuations in the earth's average temperature can have important climatic results***

flue [flu:] *noun* tube through which gas *or* smoke is released from a furnace, stove, etc.; ***flue gases are passed directly into the atmosphere;*** **flue gas desulphurization (FGD) plant** = device which traps sulphur in emissions from coal-burning furnaces and prevents it reaching the atmosphere; **flue gas scrubber** = device for cleaning flue gas of particles of pollutant

> QUOTE: in flue gas desulphurization wet limestone is sprayed into the plant's hot exhaust gases, where it scavenges as much as 90% of the sulphur dioxide.
> **Scientific American**

fluid ['flu:ɪd] *noun* liquid *or* substance which can flow

◇ **fluidized-bed combustion (FBC)** ['flu:ɪdaɪzd bed kəm'bʌstʃən] *noun* method of burning low-grade fuel while keeping the emission of pollutant gases to a minimum; *see also* PRESSURIZED FBC

fluke [flu:k] *noun* parasitic flatworm which settles inside the liver (liver fluke) *or* in the bloodstream (Schistosoma) *or* in other parts of the body

fluoride ['flʊəraɪd] *noun* any chemical compound of fluorine (usually found with sodium *or* potassium *or* tin); **fluoride toothpaste** = toothpaste containing fluoride as a means of helping to prevent tooth decay

◇ **fluoridate** ['flu:ərɪdeɪt] *verb* to add sodium fluoride to drinking water (to help prevent tooth decay); **water fluoridated to 1ppm** = water which has had 1ppm of fluoride added to it

◇ **fluoridation** [flu:ərɪ'deɪʃn] *noun* adding sodium fluoride to drinking water (to help prevent tooth decay)

◇ **fluorine** ['flʊəri:n] *noun* chemical element (a yellowish gas); **fluorine compound** = chemical substance containing fluorine (NOTE: chemical symbol is **F**; atomic number is **9**)

◇ **fluorocarbon** [flʊərə'kɑ:bən] *noun* compound of fluorine and carbon; *see also* CHLOROFLUOROCARBON

◇ **fluorosis** [flu:ə'rəʊsɪs] *noun* condition caused by excessive fluoride in drinking water or in eaten vegetable matter (it causes discoloration of the teeth and affects the milk yields of cattle)

> COMMENT: fluorides such as hydrogen fluoride are emitted as pollutants from certain industrial processes and can affect plants (especially citrus fruit) by reducing chlorophyll. They also affect cattle by reducing milk yields. On the other hand, sodium fluoride will reduce decay in teeth and is often added to drinking water or to toothpaste. In some areas, the water contains fluoride naturally and here fluoridation is not carried out. Some people object to fluoridation, although tests have proved that instances of dental decay are fewer in areas where fluoride is present in drinking water

flush [flʌʃ] *verb* to clear (a sewer) by sending water through it; **flush toilet** = toilet where the excreta are washed into the sewage system by sending water through it; *compare* CHEMICAL TOILET, EARTH CLOSET

fluvial *or* **fluviatile** ['flu:viəl *or* 'flu:viətaɪl] *adjective* referring to rivers; **fluvial deposits** *or* **fluviatile deposits** = sediments deposited by rivers

flux [flʌks] *noun* rate at which heat *or* energy *or* radiation flows; **heat flux** = flow of heat in heat exchange

fly [flaɪ] *noun* **1 (a)** general term for a small insect with two wings, of the order *Diptera* ***flies can walk on the ceiling; flies can carry infection onto food*** **(b) fly ash** = fine ash which is carried in smoke and fumes from burning processes (which can be collected and used to make bricks); ***the cloud contained particles of fly ash*** **2** *verb* to move through the air

◇ **fly-tipper** ['flaɪ 'tɪpə] *noun* person who dumps rubbish anywhere

◇ **fly-tipping** ['flaɪtɪpɪŋ] *noun* dumping of rubbish anywhere (and not at official rubbish dumps)

> QUOTE: fly-tipping is the worst side-effect of the construction boom, due to the total absence of government regulations
> **London Environment Alert**

FO = FOREIGN OFFICE

foam [fəʊm] *noun* liquid filled with bubbles of air; ***detergent foam drifted along the river;*** **foam plastic** *or* **plastic foam** = plastic with bubbles blown into it, making a light material used for packing. (CFCs are used in the manufacture of plastic foam)

focus ['fəʊkəs] *noun* point in the interior of the earth which is the centre of an earthquake; *compare* EPICENTRE

fodder ['fɒdə] *noun* plants such as grass, clover, dry hay, which are grown and given to animals as food; **fodder crop** = a crop, such as kale, lucerne or hay, grown for use as animal feed

FoE = FRIENDS OF THE EARTH

foehn [fɜːn] = FÖHN

foetus *or US* **fetus** ['fiːtəs] *noun* unborn animal in the womb

fog [fɒg] *noun* thick mist made up of millions of tiny drops of water; **advection fog** = fog which forms when a warmer, moist air mass moves over a colder surface (land *or* sea); **coastal fog** *or* **sea fog** = type of advection fog which forms along the coast *or* over sea; **freezing fog** = fog which forms from supercooled vapour droplets (i.e. droplets which remain liquid in the air, even though the temperature is below 0°C) which turn to ice when they touch a surface; **radiation fog** = fog which forms when the air just above ground level is cooled as the land surface immediately beneath it cools at night due to radiation

◇ **foggy** ['fɒgi] *adjective* (weather) when there is fog; ***foggy days are common in November***

> COMMENT: fog is caused by a fall in the temperature of damp air, making the moisture in the air condense into droplets. This can happen when a warmer, moist air mass moves over a colder surface (advection fog), or when the air just above ground level is cooled as the land surface immediately beneath it cools at night due to radiation (radiation fog). Technically speaking, a fog occurs when visibility falls to below 1,000m. Above this, moisture in the atmosphere is called 'mist'. Different levels of fog thickness may be classified as: thick fog (with visibility down to 200m, where road traffic is affected) and dense fog (visibility down to 50m, where it is not safe for any vehicles to move about)

föhn [fɜːn] *noun* warm dry wind which blows down the lee side of a mountain, caused when moist air rises up the mountain on the windward side, loses its moisture as precipitation and then goes down the other side as a dry wind; *similar to* CHINOOK

foliage ['fəʊliɪdʒ] *noun* leaves (on plants); ***in a forest, much of the rainfall is lost through evaporation from foliage surfaces***

◇ **foliar** ['fəʊliə] *adjective* referring to leaves; **foliar feed** *or* **foliar spray** = liquid nutrient used by gardeners to spray onto the leaves of plants which then absorb it

folic acid ['fɒlɪk 'æsɪd] *noun* vitamin of the B complex

food [fuːd] *noun* something which can be eaten; **health food** = food with no additives *or* food consisting of natural cereals, dried fruit and nuts; **food additive** = chemical substance added to food, especially one which is added to food to improve its appearance or to prevent it going bad; **food aid** = help in the form of food given to a developing country; **Food and Agriculture Organization (FAO)** = international organization based in Rome; an agency of the United Nations, it was established with the purpose of encouraging higher standards of nutrition and eradicating malnutrition and hunger; **food allergy** = reaction caused by sensitivity to certain foods (some of the commonest being strawberries, chocolate, milk, eggs, oranges); **Food and Drug Administration (FDA)** = *US* government department which protects the public against unsafe foods, drugs and cosmetics; **food balance** = the balance between food supplies and the demand for food from the population; **food chain** = series of organisms which pass energy from one to another as each provides food for the next. The first organism in the food chain is the producer and the rest are consumers; **food crop** = plant grown for food; **food dye** = substance used to colour food; **food grains** = cereal crops, such as corn, barley, rye, etc.; **food material** = substance which can be absorbed by an organism to produce energy; **food poisoning** = illness caused by eating food which is contaminated with bacteria; ***the hospital had to deal with six cases of food poisoning; all the people at the party went down with food poisoning;*** **food processing industry** = treating raw materials to produce various foodstuffs; **food shortage** = lack of food; **food supply** = (i) production of food and the way in which it gets to the consumer; (ii) stock of food; **food value** = amount of energy produced by a certain amount of a certain food; **food web** = series of food chains which are linked together in an ecosystem

◇ **foodstuff** ['fu:dstʌf] *noun* type of food

> COMMENT: two basic kinds of food chain exist: the grazing food chain and the detrital food chain in which plant-eaters and detritus-eaters participate. In practice, food chains are interconnected, making up food webs

footpath ['fʊtpɑ:θ] *noun* way along which people can walk on foot; ***long-distance footpaths have been created through the mountain regions***

forage ['fɒrɪdʒ] **1** *noun* crops grown for consumption by livestock **2** *verb* to look for food; ***the woodpecker forages in the forest canopy while searching for insects***

forecast ['fɔ:kɑ:st] **1** *noun* description of what will happen in the future; **weather forecast** = description of what the weather will be for a period in the future; **medium-range weather forecast** = forecast covering two to five days; **long-range weather forecast** = forecast covering a period more than five days ahead **2** *verb* to describe what will happen in the future

◇ **forecasting** ['fɔ:kɑ:stɪŋ] *noun* describing what will happen in the future; **weather forecasting** = scientific study of weather conditions and patterns, which allows you to describe what the weather will be for a period in the future

foreign ['fɒrɪn] *adjective* belonging to *or* coming from another country; **foreign aid** = help in the form of people, food, medicines, equipment, etc. given to *or* received from another country; *UK* **Foreign and Commonwealth Office (FCO)** *or* **Foreign Office (FO)** = government department which deals with overseas countries

foreshock ['fɔ:ʃɒk] *noun* small shock which comes before a main earthquake shock; *see also* AFTERSHOCK

foreshore ['fɔ:ʃɔ:] *noun* area of sand *or* pebbles which is only covered by the sea when there are very high tides

forest ['fɒrɪst] **1** *noun* **(a)** natural assembly of plants and animals, of which the dominant organisms are trees (a large area is a forest, a small area is a wood); ***the forests of Germany are affected with Waldsterben; the whole river basin is covered with tropical forest; a virgin tropical forest is a very different ecosystem from the managed forests of Europe; forest fires are widespread in the dry season and can sometimes be started by lightning;*** **high forest** = forest made up of tall trees which block the light to the forest floor; **mixed forest** = forest containing more than one species of tree; **rainforest** = thick tropical forest which grows in regions where the rainfall is very high; **forest floor** = ground at the base of the trees in a forest; **forest ranger** = person in charge of the management and protection of a forest; *US* **Forest Service** = government agency responsible for the management of national forests; **forest tree** = large tree suitable for growing in a forest **(b)** another name for what is really a plantation (trees and shrubs planted in rows, usually for commercial exploitation) **2** *verb* to manage a forest, by cutting wood as necessary, and planting new trees

◇ **forester** ['fɒrɪstə] *noun* person who manages woods and plantations of trees

◇ **forestry** ['fɒrɪstri] *noun* management of forests, woodlands and plantations of trees; *UK* **Forestry Commission** = government agency responsible for the management of state-owned forests

> QUOTE: one of the most vital functions fulfilled by forests is the control of water runoff to rivers. In a well-forested watershed, 95% of the annual rainfall is trapped in the network of roots that underlies the forest floor. This water is then released slowly over the year, keeping streams and rivers flowing during the dry season. When the forest is removed, the rains rush down the denuded slopes, straight into the local streams and rivers, only 5% of the rainwater being absorbed into the soil
>
> **Ecologist**

> QUOTE: forests are home to an estimated 60% of the world's species
>
> **Nature**

formation [fɔ:'meɪʃn] *noun* **(a)** act of taking a form *or* shape; ***millions of years of warm damp climate led to the formation of coal seams*** **(b)** way in which something is formed *or* shaped; **cloud formation** = shapes of clouds as seen from the ground

fossil ['fɒsəl] *noun* remains of an ancient animal *or* plant found preserved in rock; **fossil fuel** = fuel (such as coal *or* oil *or* gas) formed from the remains of plants in rock; **fossil-fuelled power station** = power station which burns fossil fuel to generate electricity

◇ **fossilized** ['fɒsɪlaɪzd] *adjective* (animal *or* plant) which has become a fossil

foul [faʊl] *adjective* dirty; **foul air** = air which has been circulated in a building *or* mine without being changed; **foul water** = water containing waste or sewage

fowl [faʊl] *noun* bird (especially a hen) raised on a farm for food; **fowl pest** = virus disease of chickens

Fr *chemical symbol for* francium

fraction ['frækʃn] *noun* component of a mixture separated out by a fractional process; ***the fraction of petroleum which is a gas is natural gas***

◇ **fractional distillation** ['frækʃənəl dɪstɪ'leɪʃn] *noun* distillation process where different fractions are collected at different points during the process

◇ **fractional process** ['frækʃənəl 'prəʊses] *noun* process where components of a mixture are separated out

fracture ['fræktʃə] *noun* breaking of rock (as at a fault line)

fragile ['frædʒaɪl] *adjective* easily broken *or* easily destroyed

Framework Convention on Climate Change ['freɪmwɜːk kən'venʃn] *noun* one of the two binding treaties agreed at the Earth Summit requiring states to take steps to limit the emission of greenhouse gases, especially carbon dioxide, believed to be responsible for global warming

francium ['frænsiəm] *noun* natural radioactive element (NOTE: chemical symbol is **Fr**; atomic number is **8**)

fraternal twins [frə'tɜːnəl 'twɪnz] *plural noun* two offspring who are not identical (and not always of the same sex) because they come from two different ova fertilized at the same time

free [friː] *adjective* **(a)** not attached *or* not controlled; **free acceleration test** = test to analyse the exhaust fumes produced by a motor vehicle, where the engine is accelerated while the vehicle is stationary, and samples of the exhaust fumes are collected; **free heat** = heat which is present in a building, coming not from heating but from appliances, people, heat kept in walls or floors, etc.; **free-living animal** = animal which exists in its environment without being a parasite on another; **free-range eggs** = eggs from hens that are allowed to run about in the open and eat more natural food (as opposed to battery hens); **free temperature rise** = difference in temperature between the temperature outside a building and the free heat inside it (if the building is well insulated, it can be as much as 10°C) **(b)** chemically uncombined; **free oxygen** = oxygen not combined with any other element

freeze [friːz] *verb* **(a)** (i) to make something very cold; (ii) to become very cold **(b)** *(of liquid)* to become solid due to a drop in temperature; **freezing fog** = fog formed from supercooled vapour droplets (i.e. droplets which remain liquid in the air, even though the temperature is below 0°C) which turn to ice when they touch a surface; **freezing level** = altitude at which the atmospheric temperature has fallen to 0°C; **freezing point** = temperature at which a liquid becomes solid; ***the freezing point of water is 0°C or 32°F*** **(c)** to be so cold that water turns to ice; ***it's freezing outside; they say it will freeze tomorrow*** *compare* MELT

◇ **freeze-drying** [friːz 'draɪɪŋ] *noun* method of preserving food by freezing rapidly and drying in a vacuum

freon ['friːɒn] *noun* trade name for chlorofluorocarbon *or* compound of chlorine, carbon and fluorine used as a propellant in aerosol cans, as a coolant in refrigerators, for cleaning electronic devices and for making expanded plastic foam

frequency ['friːkwənsi] *noun* density *or* number of species in a certain area

fresh [freʃ] *adjective* **(a)** not used *or* not dirty; **fresh air** = open air; ***they came out of the mine into the fresh air*** **(b)** not tinned *or* frozen; ***fresh fish is less fatty than tinned fish; fresh vegetables are expensive in winter***

◇ **fresh water** [freʃ 'wɔːtə] *noun* water in rivers and lakes which contains almost no salt (as opposed to salt water in the sea)

◇ **freshwater** ['freʃwɔːtə] *adjective* (lake) containing fresh water; (animal) living in fresh water; ***some freshwater fish such as pike can withstand certain levels of acidity***

friable ['fraɪəbl] *adjective* (soil) which is light and crumbles easily into fragments

Friends of the Earth (FoE) *noun* pressure group formed to influence local and central governments on environmental matters

front [frʌnt] *noun* line marking the point where two masses of air meet; **cold front** = edge of an advancing mass of cold air, associated with an area of low pressure, bringing clouds and rain as it meets the warmer air which it displaces; **occluded front** = front where warm and cold air masses meet and mix together, with the warm air rising away from the surface of the ground; **warm front** = movement of a mass of warm air which displaces a mass of cold air and gives rain

◇ **frontal** ['frʌntəl] *adjective* referring to a front; **frontal system** = series of cold *or* warm fronts linked together; **frontal wave** = movement of air at the edge of a warm front

frost [frɒst] *noun* freezing weather when the temperature is below the freezing point of water. (It may lead to a deposit of crystals of ice on surfaces); ***there was a frost last night;*** **air frost** = condition where the air temperature above ground level is below 0°C; **ground frost** = condition where the air temperature at ground level is below 0°C; **hard frost** = weather when the temperature falls well below 0°C; **hoar frost** = frozen dew which forms on outside surfaces when the temperature falls below 0°C; **frost hollow** *or* **frost pocket** = low-lying area where cold air collects and frosts are frequent; **frost-free region** = region where there are no frosts; **frost point** = temperature at which moisture in saturated air turns to ice

◇ **frosty** ['frɒsti] *adjective* (weather) when the air temperature is below 0°C

frostbite ['frɒstbaɪt] *noun* injury caused by very severe cold which freezes tissue

◇ **frostbitten** ['frɒstbɪtən] *adjective* suffering from frostbite

COMMENT: in very cold conditions, the outside tissue of the fingers, toes, ears and nose can freeze, becoming white and numb. Thawing of frostbitten tissue can be very painful and must be done very slowly

fruit [fru:t] **1** *noun* (i) ripe ovary of a plant and its contents of seeds; (ii) in general usage, the fleshy material round the fruit which is eaten as food; ***a diet of fresh fruit and vegetables;*** **fruit fly** = a fly which attacks fruit; **fruit tree** = tree which produces edible fruit (NOTE: no plural when referring to the food: **you should eat a lot of fruit**) **2** *verb (of a tree)* to have fruit; ***some varieties of apple fruit very early; the drought has damaged fruiting trees;*** **fruiting season** = time of year when a particular tree has fruit

◇ **fruitwood** ['fru:twʊd] *noun* wood from a fruit tree (such as apple *or* cherry) which may be used to make furniture

COMMENT: fruit contains fructose which is a good source of vitamin C and some dietary fibre. Dried fruit has a higher sugar content but less vitamin C than fresh fruit

fuel ['fju:əl] **1** *noun* substance (such as wood *or* coal *or* gas *or* oil) which can be burnt to provide heat *or* power; **fossil fuel** = fuel (such as coal *or* oil) formed from the remains of plants in rock; **nuclear fuel** = substance which is fissile (such as uranium-238) and can be used to create a controlled reaction in a nuclear reactor; **solid fuel** = coal *or* coke *or* wood; **solid fuel heating** = heating which is provided by burning solid fuels; **fuel additive** = substance (such as tetraethyl lead) which is added to petrol to prevent knocking; **fuel efficiency** = percentage of the heat from burning a fuel which is actually converted into energy; **fuel element** = piece of nuclear fuel in a reactor; **fuel injection** = spraying pressurized liquid fuel into the combustion chambers of an internal combustion engine to increase the engine's performance; **fuel oil** = petroleum oil used as fuel in a domestic heating boiler, industrial furnace, etc.; **fuel rod** = piece of nuclear fuel, in the form of a rod, placed in the core of a nuclear reactor; **fuel tank** = container for holding liquid fuel, such as in a motor vehicle **2** *verb* to used as a fuel; ***the boilers are fuelled by natural gas***

◇ **fuel-efficient** [fju:lɪ'fɪʃənt] *adjective* which uses fuel efficiently

◇ **fuel-saving** ['fju:lseɪvɪŋ] *adjective* (device *or* measure, etc.) which uses less fuel

◇ **fuelwood** ['fju:lwʊd] *noun* wood which is grown to be used as fuel

QUOTE: over 200,000 woodstoves have been sold in the UK, and the fuelwood market is estimated at 250,000 tonnes per annum
Environment Now

QUOTE: fuelwood and charcoal remain the principal sources of energy in most developing countries
Forestry Chronicle

fumarole ['fju:mərəʊl] *noun* small hole in the earth's crust near a volcano from which gases *or* smoke *or* steam are released

fumes [fju:mz] *plural noun* **(a)** gas *or* vapour; **toxic fumes** = poisonous gases or smoke given off by a substance **(b)** solid particles produced by a chemical reaction which pass into the air as smoke

fumigate ['fju:mɪgeɪt] *verb* to kill germs *or* insects by using fumes

◇ **fumigation** [fju:mɪ'geɪʃn] *noun* **(a)** killing germs *or* insects by applying pesticides in the form of fumes **(b)** high amount of air pollution near the ground, caused when the morning sun heats the air and forces polluted air from higher levels down to the ground

◇ **fumigant** ['fju:mɪgənt] *noun* chemical compound which becomes volatile when heated and is used to kill insects

functional food ['fʌŋkʃənl 'fu:d] *noun* food designed to be medically beneficial, helping to protect against serious diseases such as diabetes, cancer, heart disease, etc.

fungus ['fʌŋgəs] *noun* simple plant organism with thread-like cells (such as yeast, mushrooms, mould) and without green chlorophyll; **edible fungus** = fungus which is not poisonous to humans; **fungus disease** = disease caused by a fungus; **fungus poisoning** = poisoning by eating a poisonous fungus (NOTE: plural is **fungi** ['fʌŋgi:]
For other terms referring to fungi, see words beginning with **myc-**)

◇ **fungal** ['fʌŋgəl] *adjective* referring to fungi; ***he had a fungal skin infection***

◇ **fungicidal** [fʌŋgɪ'saɪdl] *adjective* (substance) which kills fungi

◇ **fungicide** ['fʌŋgɪsaɪd] *adjective & noun* (substance) which kills fungi

◇ **fungoid** ['fʌŋgɔɪd] *adjective* like a fungus

COMMENT: some fungi can become parasites of animals and cause diseases such as thrush. Other fungi, such as yeast, react with sugar to form alcohol. Some antibiotics (such as penicillin) are derived from fungi

fur [fɜː] *noun* **(a)** coat of hair, covering an animal; ***the rabbit has a thick coat of winter fur*** **(b)** skin and hair removed from an animal, used to make clothes; ***she wore a fur coat and fur gloves***

COMMENT: trapping of wild animals for their fur is considered cruel by many people. Artificial fur can be used as an alternative. Some native peoples, such as Eskimos, need to kill animals both for food and for their skins, so a total ban on killing is not practicable

furnace ['fɜːnəs] *noun* container for burning fuel and ore; **blast furnace** = heating device for producing iron or copper from ore, in which the ore, coke and limestone are heated together, air is blown through the mixture and the molten metal is drawn off into moulds. The waste matter from this process is known as slag

furrow ['fʌrəʊ] *noun* long trench and ridge cut in the soil by a plough

fuse [fjuːz] *verb* to join together to form a whole

◊ **(nuclear) fusion** [fjuːʒn] *noun* joining together of several nuclei to form a single large nucleus, creating energy (as in a hydrogen bomb); **fusion reactor** = nuclear reactor producing energy from the fusion of two atoms, such as deuterium and tritium

COMMENT: fusion is the opposite process to fission, fission being currently used in all nuclear power stations. Fusion is used to create bombs such as the hydrogen bomb and research is being carried out into ways of using it in power stations

FYM = FARMYARD MANURE

Gg

G *abbreviation for* giga-

g *abbreviation for* GRAM; **g/l** = grams per litre

Gaia theory ['gaɪə 'θiːəri] *noun* theory that the biosphere is like a single organism where the living fauna and flora of the earth, its climate and geology, all function together and are interrelated, influencing the development of the whole environment

gain [geɪn] *noun* increase in quantity *or* size, etc.; **energy gain** *or* **heat gain** = increase in the amount of energy *or* heat

gale [geɪl] *noun* very strong wind (force 8 on the Beaufort scale), usually blowing from a single direction; ***gales are forecast for the areas; two oaks were blown down in the October gales;*** **gale warning** = alert that there is going to be a gale

galena (PbS) [gə'liːnə] *noun* ore (lead sulphide) from which lead is produced

gall [gɔːl] *noun* hard growth on a plant caused by a parasitic insect

gallon ['gælən] *noun* **(a)** a measure of capacity, equivalent to eight pints or 4.55 litres; used both for liquids and for measuring dry goods, such as grain (NOTE: this is also called the 'imperial gallon') **(b)** *US* measure of capacity, equal to 3.78 litres; used only for liquids (NOTE: written **gal(l)** with figures: **80 gal(l)**)

galvanized iron ['gælvənaɪzd 'aɪən] *noun* iron which has been coated with zinc to prevent it rusting; galvanized iron sheeting is widely used for roofs

game [geɪm] *noun* animals which are hunted and killed for sport (and food); **big game** = large wild animals (such as elephants, tigers, etc.) which are hunted and killed for sport; **game bird** = any bird which is hunted and killed for sport (and food); **game reserve** = area of land where wild animals are kept to be hunted and killed for sport

◊ **the Game Conservancy** [geɪm kən'sɜːvənsi] an organization concerned with the conservation of game species; it advises on shoots and woodland management

◊ **gamekeeper** ['geɪmkiːpə] *noun* person working on a private estate who protects wild birds and animals bred to be hunted

◊ **game warden** ['geɪmwɔːdən] *noun* person who protects big game for photographers or for hunters

gamete ['gæmiːt] *noun* sex cell, in animals a spermatozoon and an ovum, in plants the pollen and an ovule

◊ **gametocide** [gə'miːtəsaɪd] *noun* drug which kills gametocytes

◇ **gametocyte** [gə'mi:təsaɪt] *noun* cell which develops into a gamete

gamma ['gæmə] *noun* third letter of the Greek alphabet

◇ **gamma radiation** ['gæmə reɪdɪ'eɪʃn] *noun* radiation from gamma rays

◇ **gamma rays** ['gæmə 'reɪz] *plural noun* rays which are shorter than X-rays and are given off by radioactive substances

COMMENT: gamma rays form part of the high-energy radiation which the earth receives from the sun. Gamma rays are also given off by radioactive substances and can penetrate very thick metal

garbage ['gɑ:bɪdʒ] *noun US* rubbish *or* household waste

◇ **garbage can** ['gɑ:bɪdʒ 'kæn] *noun US* container into which household waste is placed to be collected by municipal refuse collectors (NOTE: British English is **dustbin**)

◇ **garbage grinder** ['gɑ:bɪdʒ 'graɪndə] *noun US* device which fits into the plughole of a kitchen sink and which grinds up household waste so that it can be flushed away (NOTE: British English is **waste disposal unit**)

◇ **garbage man** ['gɑ:bɪdʒ mæn] *noun US* refuse collector (NOTE: British English is **dustman**)

garden ['gɑ:dən] *noun* land cultivated as a hobby *or* for pleasure, rather than to produce an income; **botanical garden** = place where plants are grown for showing to the public and for study; **flower garden** = garden where only flowers are grown; **kitchen garden** = garden with herbs and small vegetables, ready for use in the kitchen; **market garden** = place for the commercial growing of plants, usually vegetables, soft fruit, salad crops and flowers, found near a large urban centre which provides a steady outlet for the sale of its produce; **garden city** *or* **garden suburb** = town planned in the early 20th century on farmland near a large town, with gardens (both public parks and private gardens round each house), the aim being to mix the urban and rural environments

◇ **gardener** [gɑ:dnə] *noun* person who looks after a garden

◇ **gardening** ['gɑ:dnɪŋ] *noun* horticulture *or* looking after a garden; **market gardening** = growing fresh vegetables and salad crops for sale in a nearby town

gas [gæs] *noun* **(a)** substance intermediate between a liquid and a solid, which will completely fill a container it occupies; ***heating turned the liquid into a gas; if a gas is cooled it will become liquid;*** **air gas** *or* **producer gas** = mixture of carbon monoxide and nitrogen made by passing air over hot coke and used as a fuel; **gas carrier** = inert gas used to transport the sample in gas chromatography; **gas chromatography** = scientific method for analysing a mixture of volatile substances; **gas cleaning** = removing pollutants from gas, especially from emissions from factories and power stations; **gas-cooled reactor** = nuclear reactor in which carbon dioxide or helium is used as the coolant and is passed into water tanks to create the steam which will drive the turbines; **gas exchange** = transfer of gases between an organism and its environment; **gas gangrene** = complication of severe wounds in which the bacterium *Clostridium welchii* breeds in the wound and then spreads to healthy tissue which is rapidly decomposed with the formation of gas; **gas mask** = protective covering for the face, where air and poisonous gas is filtered through a charcoal filter which extracts the poison; **gas pipe** = pipe which carries gas; **gas poisoning** = poisoning by breathing in carbon monoxide or other toxic gas; **gas turbine** = internal combustion engine where expanding gases from combustion chambers drive a turbine; a rotary compressor driven by the turbine sucks in the air used for combustion **(b)** substance often produced from coal or found underground and used to cook or heat; ***a gas cooker; we heat our house by gas;*** **coal gas** = gas produced by processing coal, giving coke as a residue; **natural gas** = gas (usually methane and its compounds) found underground and not manufactured, brought to towns for domestic use; **town gas** = COAL GAS; **gas coal** = coal used for making coal gas; **gas coke** = coke resulting from the processing of coal for gas; **gas engine** = type of internal combustion engine using a flammable gas as a fuel; **gas-fired power station** = power station which burns gas, as opposed to a coal-fired station or nuclear power station; **gas supply line** = pipe which carries gas from its source to the consumer; **gasworks** = place where gas, especially coal gas, is made **(c)** *US* = GASOLINE (NOTE: British English is **petrol**)

◇ **gaseous** ['gæsiəs] *adjective* formed of gas; **water in the gaseous state** = steam; **gaseous pollutant** = pollutant in the form of gas

◇ **gasholder** ['gæshəʊldə] *noun* large tank for storing coal gas *or* natural gas

◇ **gasification** [gæsɪfɪ'keɪʃn] *noun* process of converting coal into gas to be used as fuel in gas-fired power stations

◇ **gasifier** ['gæsɪfaɪə] *noun* factory which can convert coal into gas to be used as fuel

◇ **gasohol** ['gæsəhɒl] *noun* mixture of petrol and ethyl alcohol, used as a fuel in internal combustion engines

◇ **gasoline** ['gæsəli:n] *noun US* liquid made from petroleum, used as a fuel in internal combustion engines; **gasoline-powered** =

(engine) which works on gasoline (NOTE: British English is **petrol**)

◇ **gasometer** [gə'sɒmɪtə] = GASHOLDER

gastr- [gæstr] *prefix* referring to the stomach

◇ **gastric** ['gæstrɪk] *adjective* referring to the stomach; **gastric acid** = hydrochloric acid secreted into the stomach by acid-forming cells; **gastric juices** = mixture of hydrochloric acid, pepsin, intrinsic factor and mucus secreted by the cells of the lining membrane of the stomach to help the digestion of food

◇ **gastroenteritis** [gæstrəʊentə'raɪtɪs] *noun* inflammation of the membrane lining the intestines and the stomach, caused by a viral infection and resulting in diarrhoea and vomiting

gatherers ['gæðərəz] *noun* lowest order of economic activity in which people collect food and materials (practised by a few primitive groups only)

GCV = GROSS CALORIFIC VALUE

GDP = GROSS DOMESTIC PRODUCT

Geiger counter ['gaɪgə kaʊntə] *noun* instrument for detection and measurement of radiation

COMMENT: a Geiger counter is made of a tube (forming a negative electrode) filled with argon, with a wire running through the centre which forms the positive electrode. The presence of radiation causes a discharge of electricity between the electrodes which creates an audible pulse which is used as a measure of the radiation: the greater the frequency of the pulses, the higher the level of radiation

gene [dʒiːn] *noun* unit of DNA on a chromosome which governs the synthesis of one protein, usually an enzyme, and determines a particular characteristic of an organism; **dominant gene** = gene which is more powerful; **recessive gene** = gene which is weaker than and hidden by a dominant gene; **gene bank** = collection of seeds from potentially useful wild plants, which may be used in the future for breeding new varieties; **gene mutation** = change in the DNA which changes the gene; **gene pool** = genetic information carried by the sex cells of all the individual organisms in a population; *see* GENETIC

COMMENT: genes are either dominant, when the characteristic is always passed on to the offspring, or recessive, when the characteristic only appears if both parents have contributed a recessive gene

genera ['dʒenərə] *see* GENUS

generalist ['dʒenərəlɪst] *noun* **(a)** species which can live in many different environments **(b)** person who studies many different subjects, rather than specializing in one

generate ['dʒenəreɪt] *verb* to make something exist; ***carbon monoxide is generated by car engines; the nuclear reaction generates a huge amount of heat;*** **generating plant** = factory which produces something, such as electricity *or* a chemical substance

◇ **generator** ['dʒenəreɪtə] *noun* device which produces electricity or other forms of power; **electric generator** = device which produces electricity

COMMENT: in a power station, various forms of fuel are used to create steam to turn turbines. The turbines turn a central shaft (the rotor) surrounded by a casing (the stator). Both rotor and stator are covered with windings of wire and the electricity is generated by the movement of the shaft inside the casing

QUOTE: in a biogas generator, organic material such as weeds or manure are fermented by bacteria in a tank. This produces a sludge that can be used for fertilizer and biogas that can be used for heating and lighting

New Scientist

QUOTE: nearly 80% of Britain's electricity is generated in coal-fired power stations. Of the rest, some 17% is generated by nuclear power stations, at those times when they are not out of commission, and a mere 2% by hydroelectricity

Ecologist

generation [dʒenə'reɪʃn] *noun* group of individual organisms which have usually derived from the same parents, leading back to a common ancestor

generic [dʒə'nerɪk] *adjective* referring to a genus; **generic name** = the name of a genus

COMMENT: organisms are usually identified by using their generic and specific names, e.g. *Homo sapiens* (man) and *Felix catus* (domestic cat). The generic name is written or printed with a capital letter. Both names are usually given in italics or are underlined if written or typed

genet ['dʒenɪt] *noun* (i) individual organism which is genetically different from others; (ii) clone from such an individual organism

genetic [dʒə'netɪk] *adjective* referring to the genes; ***breeders of new crop plants are dependent on genetic materials from wild forms of maize and wheat;*** **genetic code** *or* **genetic information** = information which determines the synthesis of a cell, is held in the DNA of a cell and is passed on when the cell divides; **genetic damage** = harm done to the genes of an organism, as caused by a pollutant, herbicide, etc.; **genetic engineering** = techniques used to change the genetic composition of an organism so that certain characteristics can be created artificially

◊ **genetics** [dʒə'netɪks] *noun* study of the way the characteristics of an organism are inherited through the genes; **population genetics** = genetic features of all the organisms in a community

QUOTE: the emergent field of genetic engineering by which science devises new variations of life forms, does not render wild genes useless. This new science must be based on existing genetic material and makes such material more valuable
Brundtland Report

QUOTE: Genetic engineering offers the agrochemical industry prospects not seen since the pesticides boom of the 1950s. Scientists are on the brink of rapid growth in the development and use of designer insects and pest-resistant plants
Report of the Royal Commission on Environmental Pollution (1989)

QUOTE: given the factors of instability and uncertainty of genetic engineering, the 'safety' of genetically engineered organisms cannot be taken as an *a priori* assumption
Third World Resurgence

QUOTE: human error has been suggested as the most likely cause of the contamination problems which led to the abandonment of experiments involving the release of a new, genetically engineered pesticide, strengthened with scorpion poison, into a field of caterpillar-infested cabbages
Pesticides News

-genic ['dʒenɪk] *suffix* produced by *or* which produces; **photogenic** = produced by light *or* which produces light

genome ['dʒi:nəʊm] *noun* (i) all the genes in an individual; (ii) set of genes which are inherited from one parent

genotype ['dʒenətaɪp] *noun* genetic composition of an organism; *compare* PHENOTYPE

genus ['dʒi:nəs] *noun* group of closely-related species, the members of which are more closely related to each other than they are to members of other genera. (Several genera form a family) (NOTE: plural is **genera**)

geo- ['dʒi:əʊ] *prefix* referring to the earth

geochemistry [dʒi:əʊ'kemɪstri] *noun* scientific study of the chemical composition of the earth

◊ **geochemical** [dʒi:əʊ'kemɪkl] *adjective* referring to geochemistry

◊ **geochemist** [dʒi:əʊ'kemɪst] *noun* scientist who specializes in the study of geochemistry

geocline ['dʒi:əʊklaɪn] *noun* changes that take place in a species across different geographical environments

◊ **geodesy** [dʒi:'ɒdɪsi] *noun* science of the measurement of the earth or of very large sections of it

◊ **geodetic** *or* **geodesic** [dʒi:əʊ'detɪk *or* dʒi:əʊ'di:sɪk] *adjective* referring to geodesy; ***geodetic surveys employ precise observations of angles and distances to determine the exact location of points on the earth's surface***

geography [dʒi'ɒgrəfi] *noun* scientific study of the earth's surface, its climate, its physical features, etc.; **human geography** = scientific study of the distribution of human populations with reference to their geographical environment; **physical geography** = GEOMORPHOLOGY

◊ **geographer** [dʒi:'ɒgrəfə] *noun* person who specializes in the study of the geography

◊ **geographical** *or* **geographic** [dʒi:əʊ'græfɪkl] *adjective* referring to geography; **geographical pole** = one of two points (the North and South Poles) where longitudinal lines meet and which are the most northerly *or* southerly points on the earth. They are near to, but not identical with, the geomagnetic *or* magnetic poles

geology [dʒi'ɒlədʒi] *noun* scientific study of the composition of the earth's surface and its underlying strata; **economic geology** = study of rock *or* soil formation for the purpose of commercial mineral extraction

◊ **geological** [dʒi:əʊ'lɒdʒɪkl] *adjective* referring to geology; **geological era** *or* **geological period** = span of millions of years during which the earth's surface and its underlying strata underwent certain changes; *US* **Geological Survey** = government department which studies water and mineral resources and carries out land surveys

◊ **geologist** [dʒi:'ɒlədʒɪst] *noun* scientist who studies the earth's surface and its underlying strata

geomagnetic [dʒi:əʊmæg'netɪk] *adjective* referring to the earth's magnetic field; **geomagnetic pole** = one of the two poles of the earth (near to, but not identical with, the geographical poles) which are the centres of the earth's magnetic field and to which a compass points

◊ **geomagnetism** [dʒi:əʊ'mægnɪtɪzm] *noun* study of the earth's magnetic field

geomorphology [dʒi:əʊmɔ:'fɒlədʒi] *noun* study of the physical features of the earth's surface, their development and how they are related to the core beneath

geophone ['dʒi:əʊfəʊn] *noun* sensitive device which records sounds of seismic movements below the earth's surface

geophysics [dʒi:əʊ'fɪzɪks] *noun* scientific study of the physical properties of the earth

◇ **geophysicist** [dʒi:əʊ'fɪzɪsɪst] *noun* scientist who studies the physical properties of the earth

geophyte ['dʒi:əʊfaɪt] *noun* perennial plant propagated by underground buds

geopolitics [dʒi:əʊ'pɒlɪtɪks] *noun* (study of) the influence of geographical factors on the politics of a country

geoscience [dʒi:əʊ'saɪəns] *noun* any science such as geochemistry, geodesy, geography, geology, geomorphology, geophysics or meteorology, which is concerned with the physical aspects of the earth

◇ **geoscientist** [dʒi:əʊ'saɪəntɪst] *noun* person who specializes in one *or* several of the geosciences

geosphere ['dʒi:əʊsfi:ə] *noun* central part of the earth, which contains no living organisms (as opposed to the atmosphere, the biosphere and the hydrosphere)

geostrophic wind [dʒi:əʊ'strɒfɪk 'wɪnd] *noun* wind which blows horizontally along the isobars, across the surface of the earth

geosyncline [dʒi:əʊ'sɪnklaɪn] *noun* long fold in the earth's crust, forming a basin filled with a sediment of volcanic rocks

geothermal [dʒi:əʊ'θɜ:məl] *adjective* referring to heat from the interior of the earth; **geothermal deposits** = heat-producing matter inside the earth; **geothermal energy** *or* **geothermal power** = energy *or* electricity generated from the heat inside the earth (such as the energy in hot springs); **geothermal expert** = person with specialized knowledge about the heat inside the earth; **geothermal field** = area of the earth where there is heat beneath the surface, such as near hot springs or a volcano; **geothermal gradient** = increasing temperature with increasing depth inside the earth; **geothermal installation** = equipment *or* establishment where hot water and steam are extracted from inside the earth and used to heat buildings or to generate electricity; **geothermal power plant** = power station generating electricity from the heat inside the earth (such as the energy in hot springs)

◇ **geothermally** [dʒi:əʊ'θɜ:məli] *adverb* from geothermal sources; ***geothermally heated water can be used for domestic heating***

COMMENT: apart from channelling water from hot springs, geothermal energy can also be created by pumping cold water into deep holes in the ground at points where hot rocks lie relatively close to the surface. The water is heated and becomes steam which returns to the surface and is used for domestic heating. See also HOT ROCKS

germ [dʒɜ:m] *noun* **(a)** part of an organism which develops into a new organism; **germ cell** = gamete *or* cell which is capable of developing into a spermatozoon or ovum **(b)** microbe (such as a virus *or* bacterium) which causes a disease; ***germs are not visible to the naked eye;*** **germ carrier** = person who carries bacteria of a disease in his body and who can transmit the disease to others without showing any signs of it himself; **germ warfare** = bacteriological warfare *or* war where one side tries to kill *or* affect the people of the enemy side by infecting them with bacteria (NOTE: in this sense **germ** is not a medical term)

◇ **germicide** ['dʒɜ:mɪsaɪd] *noun* substance which kills germs *or* other microbes

germinate ['dʒɜ:mɪneɪt] *verb (of a plant seed or spore)* to start to grow

◇ **germination** [dʒɜ:mɪ'neɪʃn] *noun* starting to grow

gestation (period) [dʒes'teɪʃn] *noun* period from conception to birth, when a female animal has living young in her uterus: usually 266 days in humans

geyser ['gi:zə] *noun* phenomenon which happens when hot water and steam are sent out of a hole in the ground at regular intervals; **geyser field** = area of land where there are geysers

COMMENT: a geyser (such as Old Faithful in Yellowstone National Park in the USA) is caused when water deep below the earth's surface is heated to steam and rises rapidly up the pipe leading to the surface, pushing the water already in the channel up with it

giga- ['gɪgə] *prefix* meaning one thousand million; **gigawatt (GW)** = one thousand million watts; **gigawatt-hour (GWh)** = one thousand million watts of electricity used for one hour; ***air-conditioning accounts for one-third of the 500GW peak demand in the USA***

gills [gɪlz] *plural noun* **(a)** breathing apparatus of fish and other animals living in water, consisting of a series of layers of tissue which extract oxygen from water as it passes over them **(b)** series of thin leaf-like structures on the underside of the cap of a fungus, carrying the spores

glacier ['glæsiə] *noun* (i) mass of ice moving slowly across land, like a frozen river; (ii) large amount of stationary ice covering land in the Arctic regions

◇ **glacial** ['gleɪʃl] *adjective* referring to a glacier; ***the rocks are marked by glacial action;*** **glacial climatologist** = scientist who studies climate as recorded in the ice of glaciers; **glacial deposits** *or* **glacial drift** = sand *or* soil *or* gravel deposited by a glacier; **glacial striation** = scratches made on rocks by a moving glacier

◇ **glaciation** [gleɪsi'eɪʃn] *noun* (i) formation of glaciers; (ii) formation of ice crystals at the top of a rain cloud

◇ **glaciologist** [gleɪsɪ'ɒlədʒɪst] *noun* scientist who specializes in the study of glaciers

◇ **glaciology** [gleɪsɪ'ɒlədʒi] *noun* study of glaciers

COMMENT: during the Ice Ages, glaciers covered large parts of the Northern Hemisphere, depositing sands in the form of glacial moraines and boulder clay. Glaciers are still found in the highest mountain areas and in the Arctic and Antarctic regions

glass [glɑːs] *noun* substance made from sand and soda *or* lime, etc., usually transparent and used for making bottles, windowpanes, etc.; **glass collection** = picking up, gathering together and taking away of used glass for recycling, especially empty bottles and jars; **glass fibre** = FIBREGLASS

glaucoma [glɔː'kəʊmə] *noun* condition of the eyes, caused by abnormally high pressure of fluid inside the eyeball, resulting in disturbances of vision and blindness

glauconite ['glɔːkənaɪt] *noun* green mineral composed of iron, potassium, aluminium and magnesium

glaze [gleɪz] *noun US* = BLACK ICE

glen [glen] *noun (in Scotland)* long narrow mountain valley with a stream running along it

gley [gleɪ] *noun* thick rich soil found in waterlogged ground

◇ **gleyed soil** ['gleɪd 'sɔɪl] *noun* soil which is waterlogged

◇ **gleying** ['gleɪɪŋ] *noun* series of properties of soil which indicate poor drainage and lack of oxygen (anaerobism); the signs are a blue-grey colour, rusty patches and standing surface water

globe [gləʊb] *noun* the earth, seen as a large ball

◇ **global** ['gləʊbəl] *adjective* referring to the whole earth; ***global temperatures will rise over the next fifty years;*** **global solar radiation (GSR)** = rays emitted by the sun which fall on the earth; **global warming** = gradual rise in temperature over the whole of the earth's surface, caused by the greenhouse effect

QUOTE: fossil fuel power plants and internal combustion engines throw millions of tons of carbon, sulphur dioxide and nitrogen oxides into the air each year, contributing to air pollution, acid rain and global warming

Greenpeace

QUOTE: climatologists now predict that the combined effect of deforestation and the burning of fossil fuels will cause levels of carbon dioxide in the atmosphere to double, bringing about a 2 or 3 degree rise in global temperatures. Rising global temperatures could completely alter the face of the earth.

Ecologist

globule ['glɒbjuːl] *noun* round drop (of oil *or* fat)

gloom [gluːm] *noun* dark and miserable weather; **anticyclonic gloom** = darkness during the daytime, when low stratocumulus clouds form at the approach of an anticyclone

glow-worm ['gləʊwɜːm] *noun* beetle of which the female and larva produce a greenish light

gluten ['gluːtən] *noun* the protein in a seed, such as wheat or maize; the nitrogenous part of flour which remains when the starch is extracted; **corn gluten** *or* **maize gluten** = type of animal feed made from the residues after starch has been extracted from grain

COMMENT: the gluten is what makes dough elastic and bread soft; millet and rice do not contain gluten and so cannot be used for making bread

glycogen ['glaɪkədʒen] *noun* type of starch, converted from glucose by the action of insulin and stored in the liver as a source of energy

GMT = GREENWICH MEAN TIME

gneiss [naɪs] *noun* rough rock with layers of different minerals

GNP = GROSS NATIONAL PRODUCT

gob [gɒb] *noun* waste matter from coal processing plants which use bituminous coal; *see also* CULM

gold [gəʊld] *noun* heavy yellow metal, used in jewellery (NOTE: chemical symbol is **Au;** atomic number is **79)**

goods [gʊdz] *plural noun* articles, products, merchandise; **consumer goods** = goods bought by consumers *or* by members of the public

gorge [gɔːdʒ] *noun* narrow valley with steep sides

gr = GRAIN (c)

graben ['grɑːbən] *noun* form of rift valley, made where land between fault lines has sunk

gradient ['greɪdiənt] *noun* **(a)** angle of a slope; ***plant roots cannot retain the soil on very steep gradients*** **(b)** rate of increase *or* decrease of a measurement; **geothermal gradient** = increasing temperature with increasing depth inside the earth; **pressure gradient** = change in atmospheric pressure from one place to another on the ground (as shown on a map by isobars); **temperature gradient** = gradual increase in temperature as you travel from the North or South Pole towards the equator

graft [grɑ:ft] **1** *noun* piece of plant *or* animal tissue transplanted onto another plant *or* animal and growing there **2** *verb* to transplant a piece of tissue from one plant *or* animal to another

> COMMENT: many cultivated plants are grafted. The piece of tissue from the original plant (the scion) is placed on a cut made in the outer bark of the host plant (the stock) so that a bond takes place. The aim is usually to ensure that the hardy qualities of the stock are able to benefit the weaker cultivated scion

grain [greɪn] *noun* **(a)** cereal crop, such as corn, of which the seeds are dried and eaten; **grain reserves** = amount of grain held in a store by a country which is estimated to be above the country's requirements for one year; **grain storage** = keeping grain until it is sold or used. Most grain is stored on the farm until it is sold, and is kept in bins or in bulk on the floor of the granary. The system of storage depends on whether the grain is to be used for feedingstuff on the farm, or is to be sold **(b)** size of crystals in a rock; size of particles of sand **(c)** measure of weight equal to .0648 grams (NOTE: when used with numbers, **grain** is usually written **gr)**

gram [græm] *noun* **(a)** measure of weight; ***a thousand grams make one kilogram; I need 5g of morphine*** (NOTE: when used with numbers, **gram** is usually written **g: 50g)** **(b)** *(in India and Pakistan)* the chick pea, the most important pulse grown in the Indian subcontinent

-gram [græm] *suffix* meaning a record in the form of a picture; **cardiogram** = X-ray picture of the heart

Gramineae *or* **Graminales** [græ'mɪnii: *or* græmɪ'nɑ:li:z] *noun* the grasses, a very large family of plants including cereals such as wheat, maize, etc.

Gram's stain *or* **Gram's method** [græmz] *noun* method of staining bacteria so that they can be identified; **Gram-positive bacterium** = bacterium which retains the first dye and appears blue-black when viewed under the microscope; **Gram-negative bacterium** = bacterium which takes up the red counterstain, after the alcohol has washed out the first violet dye

> COMMENT: the tissue sample is first stained with a violet dye, treated with alcohol, and then counterstained with a red dye

granite ['grænɪt] *noun* hard grey rock with pieces of quartz, feldspar and other minerals in it

grant [grɑ:nt] *noun* financial aid

granule ['grænju:l] *noun* (i) small particle *or* grain of a mineral; (ii) small artificially made particle; fertilizers are produced in granule form, which is easier to handle and distribute than powder

◇ **granular** ['grænjʊlə] *adjective* in the form of granules

graph [grɑ:f] *noun* diagram which shows the relationship between different quantities; **temperature graph** = graph showing how the temperature rises and falls

-graph [grɑ:f] *suffix* meaning a machine which records by drawing

◇ **-grapher** ['græfə] *suffix* meaning a person skilled in a subject *or* a technician who operates a machine which records

◇ **-graphy** ['græfi] *suffix* meaning the process of drawing

graphite ['græfaɪt] *noun* mineral form of carbon occurring naturally as crystals or as a soft black deposit; used as a moderator in certain types of nuclear reactor and mixed with clay to make lead pencils

grass [grɑ:s] *noun* flowering monocotyledon of which there are a great many genera, including wheat, barley, rice, oats; an important food for herbivores and humans; *see also* GRAMINALES

◇ **grassland** ['grɑ:slænd] *noun* land covered by grasses, especially wide open spaces such as the prairies of North America or the pampas of South America

> QUOTE: the first problem for the pasture agronomist is to find a grass that can produce dense ground cover quickly. Legumes must not compete too much with the grass because cows prefer to eat grasses
>
> **ew Scientist**

gravel ['grævəl] *noun* sand and small pebbles occurring as deposits. (On the Wentworth-Udden scale, gravel has a diameter of between 2 and 4 millimetres)

graveyard ['greɪvjɑ:d] *noun* place where nuclear waste is buried

gravity ['grævɪti] *noun* force of nature by which bodies are attracted to each other; **the**

earth's gravity = the natural force which attracts bodies to fall to the ground

◇ **gravitational** [grævɪ'teɪʃnəl] *adjective* referring to gravity; **gravitational field** = area of space in which gravity is felt; ***tides are caused by the moon's gravitational pull***

gray [greɪ] *noun* SI unit of measurement of absorbed radiation equal to 100 rads (NOTE: **gray** is written **Gy** with figures. See also RAD)

graze [greɪz] *verb (of animals)* to feed on grass

◇ **grazier** ['greɪziə] *noun* farmer who looks after grazing animals

◇ **grazing** ['greɪzɪŋ] *noun* **(a)** action of feeding animals on growing grass; **grazing food chain** = food chain in which the energy in vegetation is eaten by animals, digested, then passed into the soil as dung and so taken up again by plants which are eaten by animals, etc.; *see also* OVERGRAZING **(b)** land covered with grass, suitable for animals to feed on; ***there is good grazing on the mountain pastures***

green [gri:n] *adjective & noun* **(a)** of a colour like the colour of leaves; ***the green colour in plants is provided by chlorophyll;*** **green algae** = algae which contain chlorophyll; **green manure** = rapid-growing green vegetation (such as mustard plants or rape) which is grown and ploughed into the soil to rot and act as manure; **green space** = area of land which has not been built on, containing grass, plants and trees **(b)** immature; **green wood** = new shoots on a tree, which have not ripened fully **(c)** referring to an interest in ecological and environmental problems; **green consumerism** = movement to encourage people to buy food and other products which are environmentally good (such as organic foods, lead-free petrol, etc.); **green energy** = power produced by alternative technology (such as tidal power, wind power, etc.); **Green Party** *or* **the Greens** = political party which is mainly concerned with environmental issues; **green petrol** = petrol containing fewer pollutants than ordinary petrol; **green politics** = political proposals put forward by environmentalists **(d) green currencies** *or* **green rates** = fixed exchange rates for currencies used for agricultural payments in the EU; **green pound** = the fixed sterling exchange rate as used for agricultural payments in sterling between the UK and other members of the EU

> QUOTE: a study of products being promoted in the Uk as 'green' has concluded that the public are being misled by many of the claims being made
> **Independent**

Green Belt ['gri:n belt] *noun* area of agricultural land *or* woodland *or* parkland which surrounds an urban area

> COMMENT: Green Belt land is protected and building is restricted and often prohibited completely. The aim of setting up a Green Belt is to prevent urban sprawl and reduce city pollution

greenfield site ['gri:nfi:ld 'saɪt] *noun* fields which are chosen as the site for a new housing development *or* factory; ***urban fringe sites are less attractive to developers than greenfield sites***

greenhouse effect ['gri:nhaʊs ɪ'fekt] *noun* effect produced by the accumulation of carbon dioxide crystals and water vapour in the upper atmosphere, which insulates the earth and raises the atmospheric temperature by preventing heat loss. Chlorofluorocarbons also help create the greenhouse effect; **greenhouse gases** = gases (carbon dioxide, methane, CFCs and nitrogen oxides) which are produced by burning fossil fuels and which rise into the atmosphere, forming a barrier which prevents heat loss; ***the EU is planning to introduce a tax to inhibit greenhouse gas emissions***

> COMMENT: carbon dioxide particles allow solar radiation to pass through and reach the earth, but prevent heat from radiating back into the atmosphere. This results in a rise in the earth's atmospheric temperature, as if the atmosphere were a greenhouse protecting the earth. Even a small rise of less than one degree Celsius in the atmospheric temperature could have serious effects on the climate of the earth as a whole. The temperature over the poles would rise, melting the polar ice caps and causing sea levels to rise everywhere. This would cause permanent flooding in many parts of world. In addition, temperate areas in Asia and America would experience hotter and drier conditions, causing crop failures. Carbon dioxide is largely formed from burning fossil fuels. Other gases contribute to the greenhouse effect, for instance, methane is increasingly produced by rotting vegetation in swamps, from paddy fields, from termites' excreta and even from the stomachs of cows. There is an opposite effect to the greenhouse, which is the cooling of the earth's atmosphere caused by the dust veil, a mass of particles of dust which circulates in the upper atmosphere and prevents solar radiation from passing through to the earth's surface

> QUOTE: in the next 40 or 50 years, the greenhouse gases in the atmosphere will double and the world will get warmer by between 1 and 4.5°C
> **Guardian**

> QUOTE: in general, a temperature rise will decrease the time available for the specific phases of cereal crops from leaf growth through to reproduction. Yields will probably decrease as temperatures increase. But more CO_2 will put yields up and if temperature and CO_2 were the only factors, output from most crops would not change much
>
> **Farmers Weekly**

Greenpeace ['gri:npi:s] *noun* international pressure group, which takes physical action to prevent the pollution of the environment

Green Revolution ['gri:n revə'lu:ʃn] *noun* development of new forms of cereal plants such as wheat and rice and the use of more powerful fertilizers, which give much higher yields and increase the food production especially in tropical countries

> QUOTE: in Asia, the Green Revolution, which featured high-input, high-yield agricultural technology, heralded a remarkable increase in grain production to the point that India and Pakistan became self-sufficient having formerly been large importers
>
> **Appropriate Technology**

Greenwich Mean Time (GMT) ['grenɪtʃ 'mi:n taɪm] *noun* local time on the 0° meridian where it passes through Greenwich, England; used to calculate international time zones

grid [grɪd] *noun* **(a)** series of crossing lines which form regular squares, used on maps; **grid reference** = numbers which refer to a point on a map, used for accurate location of places **(b) electricity grid** *or* **national grid** = system for carrying electricity round a country, using power lines from various power stations. (The electricity is at a high voltage, which is reduced by transformers to low voltage by the time the electricity is brought into use)

grinder ['graɪndə] *noun* device *or* machine which reduces a substance to fine particles by crushing

◊ **garbage grinder** ['gɑ:bɪdʒ 'graɪndə] *noun US* device which fits into the plughole of a kitchen sink and which grinds up household waste so that it can be flushed away (NOTE: British English is **waste disposal unit**)

◊ **grinding** ['graɪndɪŋ] *noun* reducing a substance to fine particles by crushing; **coal grinding** = reducing coal to fine particles in a crushing machine

grit [grɪt] *noun* **(a)** sharp-grained sand **(b)** tiny solid particle in the air, larger than dust

groove [gru:v] *noun* long shallow depression in a surface

gross [grəʊs] *adjective* total *or* with no deductions; **gross calorific value (GCV)** = total number of Calories which a certain amount of a substance contains; **gross domestic product (GDP)** = annual value of goods sold and services paid for inside a country; **gross national product (GNP)** = annual value of goods and services in a country including income from other countries; **gross output** = total amount produced; **gross productivity** = rate at which energy is produced by plants through photosynthesis, before the plant uses any of the energy itself

ground [graʊnd] *noun* **(a)** surface layer of soil *or* earth; **ground clearance** = removal of trees, undergrowth, etc. in preparation for ploughing, building, etc.; **ground frost** = condition where the air temperature at ground level is below 0°C; **ground level** = height of the ground; **ground-level concentration** = amount of a pollutant measured at the height of the ground *or* just above it *or* just below it; **ground moraine** = deposit of gravel and sand left under a glacier; **ground plant** = plant which covers the soil; **ground pollution** = presence of abnormally high concentrations of harmful substances in the soil; **ground water** = water which stays in the top layers of soil or in porous rocks and can collect pollution (as opposed to surface water which drains away rapidly and leaches pollutants from the surface); **ground-water basin** = area of land where water stays in the top layers of soil or in porous rocks and can collect pollution; **ground-water level** = point below the surface of the earth where ground water lies; **ground-water runoff** = water which enters streams and rivers from below ground **(b)** area; **breeding ground** = area where birds *or* animals come each year to breed; **spawning ground** = area of water where fish come each year to produce their eggs; **wintering ground** = area where birds come each year to spend the winter **(c)** area of land; **dumping ground** = place where waste is thrown away; **recreation ground** = area of land open to the public to play games on; **sports ground** = area of land open to the public *or* to a private club, etc. to play games on; **waste ground** = area of land which is not used for any purpose

group [gru:p] **1** *noun* **(a)** several people *or* animals *or* things which are all close together; **environmentalist group** = association *or* society dedicated to the protection of the environment and to the increasing of people's awareness of environmental issues; **pressure group** = group of people with similar interests who try to influence politicians; **social group** = several people *or* animals living together in an organized society; group of people with a similar position in society **(b)** way of bringing similar things together; **age group** = all people of a certain age; **blood group** = one of the different types of blood by which people are identified; **taxonomic group** = several organisms classified together scientifically **2** *verb* to bring close together; **blood grouping**

= classifying people according to their blood groups

grove [grəʊv] *noun* small group of trees

grow [grəʊ] *verb* **(a)** *(of a plant)* to exist and flourish; ***bananas grow only in warm humid conditions; tomatoes grow well in greenhouses*** **(b)** *(of plants and animals)* to increase in size, to become taller *or* bigger; ***the tree grows a new ring each year; he grew three centimetres in one year;*** **growing point** = point on the stem of a plant where growth occurs (often at the tip of the stem *or* branch); **growing season** = time of year when a plant grows; ***alpine plants have a short growing season*** **(c)** *(of people)* to cultivate (plants); ***peasant farmers are encouraged to grow new strains of rice; farmers here grow two crops in a year; he grows peas for the local canning factory; growing early vegetables is difficult because of the cold winters*** **(d)** to become; ***it's growing colder at night now; she grew weak with hunger***

◇ **growth** [grəʊθ] *noun* increase in size; amount by which something increases in size; ***the disease stunts the conifers' growth; the growth in the population since 1960; the rings show the annual growth of the tree;*** **population growth** = increase in the size of a population; **urban growth** = increase in the size and number of towns; **growth hormone** = natural *or* artificial substance which makes an organism grow; **growth inhibitor** = substance which stops an organism growing; **growth promoter** = substance which makes an organism grow; **growth rate** = amount *or* speed of the increase in size; **growth regulator** = substance which controls the growth of an organism: mainly used for weed control; **growth retardant** = substance used to make an organism grow more slowly; **growth rings** = rings seen in the cross-section of the trunk of a felled tree, each one added during a single year; *see also* ANNUAL RINGS; **growth stimulant** = substance which makes an organism start to grow

groyne [grɔɪn] *noun* breakwater which is built from the shore into the sea in order to block the force of waves and so prevent erosion

grub [grʌb] **1** *noun* small caterpillar *or* larva **2** *verb* **to grub up** = to dig up a plant with its roots; ***miles of hedgerows have been grubbed up to make larger fields***

GSR = GLOBAL SOLAR RADIATION

guano ['gwɑːnəʊ] *noun* mass of accumulated bird droppings, found especially on small islands in the sea, gathered and used as fertilizer

guaranteed prices [gærən'tiːd 'praɪsɪs] *noun* a feature of national agricultural policy in which the producers of a certain commodity are guaranteed a certain minimum price for their produce

guide [gaɪd] **1** *noun* person *or* book which shows you how to do something *or* what to do; ***read this guide to services offered by the local authority*** **2** *verb* to show someone where to go *or* how to do something

◇ **guidelines** ['gaɪdlaɪnz] *plural noun* general rules laid down, telling people how something should be done; ***the minister has issued guidelines for the planning of inner cities***

guild [gɪld] *noun* group of plants *or* animals of the same species which live in the same type of environment

gulf [gʌlf] *noun* very large area of sea enclosed partly by land

◇ **Gulf Stream** ['gʌlf 'striːm] *noun* current of warm water in the Atlantic Ocean, which flows north along the east coast of the USA, then crosses the Atlantic Ocean to hit Northern Europe

◇ **gulfweed** ['gʌlfwiːd] *noun* floating seaweed which grows in the Sargasso Sea

gull [gʌl] *noun* general name for a number of species of aquatic bird, having a stout build, rather hooked bills and webbed feet

gully ['gʌli] *noun* **(a)** deep channel formed by soil erosion, and which cannot be removed by cultivation **(b)** small channel for water (such as an artificial channel dug at the edge of a field *or* a natural channel in rock); **gully cleaning** = clearing rubbish from drainage gullies at the edge of roads

gum [gʌm] *noun* substance existing in a liquid state in the trunks and branches of trees, which hardens on contact with air and is used in confectionery, pharmacy and stationery

gust [gʌst] **1** *noun* sudden violent wind *or* wind and rain **2** *verb (of wind)* to blow suddenly; ***the met office is forecasting winds gusting to 90 miles per hour***

◇ **gusty** ['gʌsti] *adjective* (wind) which is blowing in gusts

Gutenberg discontinuity ['guːtənbɜːg dɪskɒntɪ'njuːɪti] *noun* boundary between the mantle and the core inside the earth

GW *or* **gW** = GIGAWATT

◇ **GWh** = GIGAWATT-HOUR

Gy *abbreviation for* gray

gyre [dʒaɪə] *noun* circular or spiral motion of ocean water

Hh

H *chemical symbol for* hydrogen

ha = HECTARE

haar [hɑː] *noun* sea mist occurring during the summer in the north of the British Isles

habit ['hæbɪt] *noun* way in which a plant grows; ***a bush with an erect habit*** *or* ***with a creeping habit***

habitat ['hæbɪtæt] *noun* type of environment in which an organism lives; **habitat loss** *or* **habitat reduction =** permanent disappearance of *or* decrease in the amount of suitable environment available to an organism

QUOTE: because of its strange topography the area is richly endowed with contrasting habitats: bare rock and grassland, vine thickets and savanna
New Scientist

QUOTE: throughout India, elephant habitat is suffering encroachment by coffee and tea plantations
BBC Wildlife

haematite ['hiːmətaɪt] *noun* an iron oxide (Fe_2O_3), the most common form of iron ore

haemoglobin (Hb) [hiːmə'gləʊbɪn] *noun* red respiratory pigment containing iron in red blood cells which gives blood its red colour; *see also* CARBOXYHAEMOGLOBIN

COMMENT: haemoglobin absorbs oxygen in the lungs and carries it in the blood to the tissues. Haemoglobin is also attracted to carbon monoxide and readily absorbs it instead of oxygen, causing carbon monoxide poisoning

hail [heɪl] **1** *noun* water falling from clouds in the form of small round pieces of ice **2** *verb* to fall as small pieces of ice

◇ **hailstone** ['heɪlstəʊn] *noun* piece of ice which falls from clouds like rain

◇ **hailstorm** ['heɪlstɔːm] *noun* storm, where the precipitation is hail and not rain

COMMENT: hail occurs when rain forms in cumulonimbus clouds and is carried upwards into colder air where it freezes. It then falls back through the cloud, growing in size as it accumulates moisture

half-hardy [hɑːf'hɑːdi] *adjective* (plant) which can stand a certain amount of cold

half-life *or* **half-life period** *or* **half-value period** ['hɑːflaɪf *or* hɑːf'væljuː] *noun* time taken for half the atoms in a radioactive isotope to decay

COMMENT: radioactive substances decay in a constant way and each has a different half-life: strontium-90 has a half-life of 28 years, radium-226 one of 1,620 years and plutonium-239 has a half-life of 24,360 years

halo ['heɪləʊ] *noun* circle of light seen round the sun or moon, caused by ice crystals in the earth's atmosphere; *compare* CORONA

halo- ['hæləʊ] *prefix* meaning salt

◇ **halobiotic** [hæləʊbaɪ'ɒtɪk] *adjective* (organism) which lives in salt water

◇ **halocline** ['hæləʊklaɪn] *noun* line where two masses of water meet, as between a mass of fresh water and the sea

◇ **halogen** ['hælədʒən] *noun* one of a series of chemically-related elements (fluorine, chlorine, iodine, bromine and astatine)

◇ **halogenated** [hə'lɒdʒəneɪtɪd] *adjective* (chemical compound) which contains one of the halogens

◇ **halogenous** [hə'lɒdʒənəs] *adjective* referring to *or* containing a halogen

◇ **halomorphic soil** [hæləʊ'mɔːfɪk sɔɪl] *noun* soil which contains large amounts of salt

halon ['heɪlɒn] *noun* chemical compound which contains bromine and resembles a chlorofluorocarbon

halophile ['hæləfaɪl] *noun* freshwater species which can also live in salt water

◇ **halophyte** ['hæləfaɪt] *noun* plant which lives in salty soil

hanging valley ['hæŋɪŋ 'væli] *noun* valley formed when a glacier flows into a larger glacier

COMMENT: hanging valleys are the result of glaciation. The main valley has been cut deeper by the large glacier, leaving the smaller valley to join it at a cliff; this is one of the ways in which waterfalls are formed

hard [hɑːd] *adjective* **(a)** not soft; ***a diamond is one of the hardest minerals; granite is a very hard stone;*** **hard water =** tap water which contains a high percentage of calcium and magnesium, which makes it difficult for soap to lather and also causes deposits in pipes, boilers and kettles **(b)** difficult; ***it is hard for grazing animals to find enough to eat in drought conditions*** **(c) hard winter =** very cold winter; ***in a hard winter, many smaller birds may be killed***

◇ **harden** ['hɑːdən] *verb* to make hard *or* to become hard; **to harden off =** to make plants become gradually more used to cold; ***after***

seedlings have been grown in the greenhouse, they need to be hardened off before planting outside in the open ground; **cold hardening** = process by which a conifer prepares for the winter by reducing the amount of water contained in its needles

◇ **hardener** ['hɑːdnə] *noun* substance which causes another to become hard by chemical reaction

◇ **hardness** ['hɑːdnəs] *noun* **(a)** way of showing the percentage of calcium in water. (Water hardness is permanent if it remains after the water has been boiled (this is caused by calcium and magnesium). Temporary hardness is caused by carbonates of calcium and can be removed by boiling the water) **(b)** measurement of how hard a mineral is

> COMMENT: hardness of minerals is shown on a scale of 1 to 10: diamond is the hardest (i.e. it has hardness 10) and talc is the softest. A mineral of one grade is able to scratch or mark a mineral of the grade below

hardpan ['hɑːdpæn] *noun* hard soil surface, usually formed of dried clay

◇ **hardwood** ['hɑːdwʊd] *noun* (i) slow-growing broad-leaved tree (such as oak *or* teak) which produces a hard wood with a fine grain; (ii) wood from such a tree; **hardwoods** = (forests of) hardwood trees; *compare* SOFTWOOD

◇ **hardy** ['hɑːdi] *adjective* (plant *or* animal) which can withstand the cold; *see also* HALF-HARDY

harmattan [hɑː'mætən] *noun* hot dry winter wind which blows from the north-east and causes dust storms in the Sahara

harness ['hɑːnəs] *verb* to control a natural phenomenon and make it produce energy; ***a tidal power station harnesses the power of the tides***

harrow ['hærəʊ] **1** *noun* farm implement, made of a series of spikes which is dragged across the surface of ploughed soil to level it **2** *verb* to level the surface of ploughed soil, covering seeds which have been sown in furrows

harvest ['hɑːvɪst] **1** *noun* (i) time when a crop is gathered; (ii) crop which is gathered; ***we think this year's rice harvest will be a good one*** **2** *verb* to gather a crop which is ripe; ***they are harvesting the rice crop; clearcutting is one of the methods of harvesting timber***

◇ **harvester** ['hɑːvɪstə] *noun* machine which harvests a crop; most crops are now harvested by machines such as combine harvesters or sugar beet harvesters

hatch [hætʃ] *verb (of an animal in an egg)* to become mature and break out of the egg

◇ **hatchery** ['hætʃəri] *noun* place where eggs are kept warm artificially until the animal inside becomes mature and breaks out

hay [heɪ] *noun* grass mowed and dried before it has flowered

◇ **haymaking** ['heɪmeɪkɪŋ] *noun* cutting grass in fields to make hay

> COMMENT: hay is cut before the grass flowers; at this stage in its growth it is a nutritious fodder. If it is mowed after it has flowered it is called straw, and is of much less use as a food and so is used for bedding

hay fever ['heɪfiːvə] *noun* inflammation of the nose and eyes caused by an allergic reaction to pollen *or* fungus spores *or* dust in the atmosphere; ***when he has hay fever, he has to stay indoors; the hay fever season starts in May***

> COMMENT: tree pollen is most prevalent in spring, followed by the pollen of flowers and grasses during the summer months and fungal spores in the early autumn. The pollen is released by the stamens of a flower and floats in the air until it finds a female flower. Pollen in the air is a major cause of hay fever. It enters the nose and eyes and chemicals in it irritate the mucus and force histamines to be released by the sufferer, causing the symptoms of hay fever to appear

hazard ['hæzəd] *noun* risk *or* danger; ***stagnant water poses a health hazard;*** **fire hazard** = risk of something catching fire; **health hazard** = danger to the health of an organism; **at hazard** = at risk *or* in danger

◇ **hazardous** ['hæzədəs] *adjective* risky *or* dangerous; **hazardous to health** = dangerous to the health of an organism; **hazardous waste** = rubbish which can pose a risk to people's health

> QUOTE: some 2.3 million tonnes of hazardous waste were transported for disposal in another country in 1983
>
> **Guardian**

haze [heɪz] *noun* dust suspended in the atmosphere which reduces visibility; **heat haze** = reduction in visibility caused by warm air rising from the ground

Hb = HAEMOGLOBIN

HCl = HYDROCHLORIC ACID

HCN = HYDROCYANIC ACID

HD = HIGH-DENSITY

He *symbol for* helium

head [hed] *noun* point where a river starts to flow

◇ **headland** ['hedlænd] *noun* **(a)** high mass of land sticking into the sea **(b)** area of soil at the edge of a field (where the tractor turns when ploughing)

◇ **headstream** ['hedstri:m] *noun* stream which flows into the river near its head

◇ **headwaters** ['hedwɔ:təz] *plural noun* tributary streams feeding into a river at the point where it rises

health [helθ] *noun* being well *or* not being ill; state of being free from physical *or* mental disease; ***the council said that fumes from the factory were a danger to public health; all cigarette packets carry a government health warning;*** **Health and Safety Executive (HSE)** = British government organization responsible for checking the conditions of work of workers, including farmworkers; **Health and Safety at Work Act** = Act of Parliament which rules how the health of workers should be protected by the companies they work for; **health authority** = official body in an area *or* region dealing with all matters concerning health, including the administration of hospitals; **health education** = teaching people (school children and adults) to do things to improve their health, such as taking more exercise, stopping smoking, etc.; **Environmental Health Officer (EHO)** *or* **Public Health Inspector** = official of a local authority who examines the environment and tests for air pollution *or* bad sanitation *or* noise pollution, etc.

◇ **healthy** ['helθi] *adjective* (i) well *or* not ill; (ii) likely to make you well; ***being a farmer is a healthy job; people are healthier than they were fifty years ago; this town is the healthiest place in England; if you eat a healthy diet and take plenty of exercise there is no reason why you should fall ill***

heartwood ['hɑ:twʊd] *noun* hard dead wood in the centre of a tree's trunk which helps support the tree; *compare* SAPWOOD

heat [hi:t] **1** *noun* **(a)** being hot *or* energy which is moving from a source to another point; ***the heat of the sun made the surface of the road melt;*** **heat balance** = state in which the earth loses as much heat by radiation and reflection as it gains from the sun, making the earth's temperature constant from year to year; **heat capacity** = amount of heat required to raise the temperature of a substance by 1°; **heat conduction** = passing of heat from one part to another; **heat efficiency** = ability of a process to heat efficiently; **heat engine** = (i) machine *or* organism which consumes fuel and produces heat which can be converted into work; (ii) phenomenon which produces the earth's climatic pattern, caused by the difference in temperature between the hot equatorial zone and the cold polar regions, making warm water and air from the tropics move towards the poles; **heat exchanger** = device which transfers heat from one source to another (as in a heating system where a hot pipe goes through a water tank to heat it); **heat exhaustion** = collapse due to overexertion in hot conditions; **heat flow** = movement of heat (as through a metal *or* from the earth into space); **heat haze** = reduction in visibility caused by warm air rising from the ground; **heat island** = increase in temperature experienced in the centre of a large urban area, caused by the release of heat from buildings; **heat insulation** = preventing escape of heat; **heat loss** = amount of heat lost (as through inadequate insulation); **heat pump** = device which cools or heats by transferring heat from cold areas to warm ones (used in refrigerators or for heating large buildings); **heat recovery** = collecting heat from substances heated during a process and using it to heat further substances, so as to avoid heat loss; **heat-proof** = (material) which is not affected by heat *or* through which heat cannot pass; **heat-sealing** = method of closing plastic food containers; air is removed from a plastic bag with the food inside and the bag is then pressed by a hot plate which melts the plastic and seals the contents in the vacuum **(b)** the oestrus period, when a female will allow mating; **an animal on heat** = female animal in the oestrus period, when she will accept a male **2** *verb* to make hot; ***the solution should be heated to 25°C; in the sun, the soil surface can heat to 60°C***

◇ **heating** ['hi:tɪŋ] *noun* making hot; **central heating** = system for heating a building where hot water *or* hot air is circulated round from a single source of heat, usually a boiler; **district heating** = system of heating all houses in a district from a central source (as from hot springs in Iceland or by cooling water from a power station); **heating appliance** = device which supplies heat; **heating oil** = petroleum oil used as fuel in a central heating boiler; **heating power** *or* **heating value** = measurement of the amount of heat which a substance *or* process can supply

heath [hi:θ] *noun* area of dry sandy acid soil with low shrubs such as heather and gorse growing on it

◇ **heathland** ['hi:θlænd] *noun* wide area of heath

heavy ['hevi] *adjective* which weighs a lot; **heavy hydrogen** = deuterium; **heavy industry** = industry (such as steelmaking or engineering) which takes raw materials and makes them into finished products *or* which extracts raw materials (such as coal); **heavy metal** = metal such as lead, cadmium and zinc which has a high atomic number and which, if present in soil, has a stunting effect on plants; **heavy oil** = mixture of hydrocarbons which is distilled from coal tar and is heavier than water; **heavy water** = deuterium oxide (D_2O) *or* water containing deuterium instead of the hydrogen atom, used as a coolant or moderator in certain types of nuclear reactor such as the CANDU

> QUOTE: the government agreed to reduce by 50% its discharges into the North Sea of heavy metals such as cadmium, iron, lead zinc and arsenic
> **New Scientist**

hectare ['hekteə] *noun* area of land measuring 100 by 100 metres, i.e. 10,000 square metres or 2.47 acres (NOTE: usually written **ha** after figures: **2,500 ha)**

> QUOTE: the UK is well-placed to grow up to 200,000ha (0.5m acres) of linseed to replace imports and provide exports to other EU countries
> **Farmers Weekly**

hedge [hedʒ] *noun* row of bushes planted and kept trimmed to provide a barrier around a field *or* garden

◇ **hedgelaying** ['hedʒleɪɪŋ] *noun* traditional method of cultivating hedges, where tall saplings are cut through halfway and then bent over so that they lie horizontally and make a thick barrier

◇ **hedgerow** ['hedʒrəʊ] *noun* line of bushes forming a hedge

◇ **hedging** ['hedʒɪŋ] *noun* art of cultivating hedges

> COMMENT: it is said that you can judge the age of a hedge by counting the species in it and multiplying by 100; the more species there are, the older the hedge is

> QUOTE: a recent Friends of the Earth report states that 50% of the hedgerows in Britain have been grubbed out since 1945
> **Permaculture Magazine**

helio- ['hiːliəʊ] *prefix* meaning sun

◇ **heliophyte** ['hiːliəfaɪt] *noun* plant which prefers to grow in sunlight

◇ **heliotropic** [hiːliə'trɒpɪk] *adjective* (plant) which grows *or* turns towards the light

helium ['hiːliəm] *noun* light inert gas, used in balloons and as a coolant in some types of nuclear reactor (NOTE: chemical symbol is **He;** atomic number is **2)**

helophyte ['heləʊfaɪt] *noun* plant typical of marshy or lake-edge environments

hemisphere ['hemɪsfiə] *noun* half of a sphere; one of the two main sections of the brain; **Northern Hemisphere** *or* **Southern Hemisphere** = the two halves of the earth, north and south of the equator

hemlock ['hemlɒk] *noun* **(a)** North American softwood tree **(b)** poisonous plant *(Conium maculatum)*

hemp [hemp] *noun* plant *(Cannabis* sp) used to make rope; also producing an addictive drug

herb [hɜːb] *noun* **(a)** plant which has no perennial stem above the ground during the winter period **(b)** plant which can be used as a medicine *or* to give a certain taste to food *or* to give a certain scent

◇ **herb-** [hɜːb] *prefix* referring to plants *or* vegetation

◇ **herbaceous** [hɜː'beɪʃəs] *adjective* (plant) without perennial stems above the ground; **herbaceous border** = bed of herbaceous plants growing along the edge of a lawn

◇ **herbage** ['hɜːbɪdʒ] *noun* green foodstuffs eaten by grazing animals

◇ **herbal** ['hɜːbəl] *adjective* referring to herbs; **herbal remedy** = remedy made from plants, such as an infusion made from dried leaves or flowers in hot water

◇ **herbalism** ['hɜːbəlɪzm] *noun* treatment of illnesses *or* disorders by the use of herbs or by medicines extracted from herbs

◇ **herbalist** ['hɜːbəlɪst] *noun* person who treats illnesses *or* disorders by the use of herbs

◇ **herbarium** [hɜː'beəriəm] *noun* collection of plant specimens

◇ **herbicide** ['hɜːbɪsaɪd] *noun* chemical which kills plants, especially weeds; **contact herbicide** = herbicide (such as paraquat) which kills a plant by contact with the leaves, but which may not kill a plant with large roots; **residual herbicide** = herbicide applied to the surface of the soil which acts through the roots of the plants; not only growing weeds are killed but also new plants as they germinate; **selective herbicide** = herbicide which is supposed to kill only certain plants and not others; **systemic herbicide** = herbicide which is absorbed into a plant's sap system through its leaves

◇ **herbivore** *or* **herbivorous animal** ['hɜːbɪvɔː *or* hɜː'bɪvərəs] *noun* animal which eats plants; *compare* CARNIVORE, OMNIVORE

herd [hɜːd] **1** *noun* group of herbivorous animals which live together; ***a herd of wildebeest*** *or* ***of elephants*** *or* ***of cows*** (NOTE: the word 'herd' is usually used with cattle; for sheep, goats, and birds such as hens or geese, the word to use is 'flock') **2** *verb* to tend a herd of animals; to crowd animals together; ***livestock herding is the main source of revenue for the people of the area***

◇ **herdsman** ['hɜːdzmən] *noun* farm worker who looks after a herd of animals

hereditary [hə'redɪtəri] *adjective* which is transmitted from parents to offspring

◇ **heredity** [hə'redɪti] *noun* transmission of physical *or* mental characteristics from parents to offspring

heritage ['herɪtɪdʒ] *noun* the environment, including the countryside, historic buildings, etc. seen as something to be passed on in good condition to future generations

Her Majesty's Inspectorate of Pollution (HMIP) *noun* regulatory body attached to, but with autonomy from, the UK Department of the Environment

hetero- ['hetərəʊ] *prefix* meaning different

◇ **heterogeneous** [hetərəʊ'dʒiːniəs] *adjective* having different characteristics *or* qualities

◇ **heterogenous** [hetə'rɒdʒənəs] *adjective* coming from a different source

◇ **heterophyte** ['hetərəʊfaɪt] *noun* **(a)** plant that grows in a wide range of habitats **(b)** plant that lacks chlorophyll (i.e. it is parasitic)

◇ **heterotrophic organism** *or* **heterotroph** [hetərəʊ'trɒfɪk *or* 'hetərətrɒf] *noun* organism (animal *or* parasite *or* fungus, etc.) which obtains its energy from carbon, by breaking down organic matter in the ecosystem and eating other organisms

> QUOTE: ice crystals form in the polar vortex of air that swirls over Antarctica in the southern winter. This allows a series of heterogeneous chemical reactions to occur on the surface of the crystals, creating large amounts of chlorine monoxide which destroys ozone
>
> **New Scientist**

hexachlorocyclohexane ($C_6H_6Cl_6$) [heksəklɔːrəʊsaɪkləʊ'heksein] *noun* white *or* yellow powder containing lindane, used as an insecticide

Hg *chemical symbol for* mercury

hibernate ['haɪbəneɪt] *verb (of an animal)* to survive the cold winter months by a big reduction in metabolic rate and activity, and by using up stored body fat for food

◇ **hibernaculum** [haɪbə'nækjʊləm] *noun* place (such as a nest) where an animal hibernates; ***as the first frosts occur, small animals go into their winter hibernacula*** (NOTE: plural is **hibernacula**)

◇ **hibernation** [haɪbə'neɪʃn] *noun* big reduction in metabolic rate and activity, and the using up of stored body fat to survive the cold winter months

> COMMENT: during the cold weather, many small mammals (such as hedgehogs) and reptiles hibernate. Their blood temperature falls and their metabolism slows

> QUOTE: hibernating bats don't simply shut down for the winter. Their body temperatures and activity levels vary widely and from time to time they emerge from their hibernacula
>
> **BBC Wildlife**

hickory ['hikəri] *noun* North American hardwood tree

hide [haɪd] *noun* shelter where birdwatchers can stay hidden while watching birds

high [haɪ] **1** *adjective* **(a)** tall *or* reaching far from the ground level; ***altocumulus clouds form at higher levels than cumulus;*** **high-level inversion** = situation where warm air lies above cold air relatively high above the ground; **high water** = point when the level of the sea *or* of a river, etc. is at its highest **(b)** *(referring to numbers)* of greater than average intensity, etc.; ***the sample gave a high reading of radioactivity; the soil is red and high in aluminium and iron oxide; an area of high pressure is moving south across the country; the regions with highest rainfall are those of the western mountains; UV rays are part of the high-energy radiation from the sun;*** **high-alumina cement** = cement made of bauxite and limestone, used because it resists heat; **high-energy foods** = foods containing a large number of calories, such as fats or carbohydrates, which give a lot of energy when they are broken down by the digestive system **2** *noun* area of high pressure; **subtropical high** = area of high pressure normally found in the subtropics

◇ **high-grade** ['haɪ 'greɪd] *adjective* good *or* of very high quality; **high-grade ore** = ore which contains a large percentage of metal

highland ['haɪlənd] *noun* area of high land *or* mountains; ***vegetation in the highlands*** *or* ***highland vegetation is mainly grass, heather and herbs*** (NOTE: opposite is **lowland)**

high-density ['haɪ'densɪti] *adjective* (i) (substance) which has a large degree of mass per unit of volume; (ii) (housing, habitat, etc.) where a lot of people *or* organisms live closely together; **high-density polythene** *or* **HD polythene** = very thick, heavy-duty plastic

high-level radioactive waste ['haɪlevəl 'reɪdiəʊæktɪv 'weɪst] *noun* waste which is hot and emits strong radiation

> COMMENT: high-level radioactive waste is potentially dangerous and needs special disposal techniques. It is sealed in special containers, sometimes in glass, and is sometimes disposed by dumping at sea

> QUOTE: consideration should be given to abandoning the site and declaring it unsuitable for the permanent disposal of high-level nuclear waste
>
> **Environmental Action**

high-risk ['haɪrɪsk] *adjective* (person) who is very likely to catch a disease *or* develop a cancer *or* suffer an accident; ***high-risk categories of worker;*** **high-risk patient** = patient who is very likely to catch an infection

high-tension ['haɪ 'tenʃn] *adjective* (electricity cable) which is carrying a high voltage

high-yielding ['haɪ 'ji:ldɪŋ] *adjective* which yields a high crop; ***they have started to sow high-yielding varieties (HYV) of seed with very good results***

highway ['haɪweɪ] *noun* main public road; **Highways Department** = local government office dealing with all matters affecting the planning, construction and maintenance of roads in the area

hill [hɪl] *noun* ground higher than the surrounding areas but not as high as a mountain; **hill farms** = farms in mountainous country, with 95% or more of their land classified as rough grazing, mainly breeding ewe flocks

◇ **hillside** ['hɪlsaɪd] *noun* sloping side of a hill

hinterland ['hɪntəlænd] *noun* area of land lying behind the shore of the sea *or* of a river

histamine ['hɪstəmi:n] *noun* substance released from mast cells throughout the human body which stimulates tissues in various ways; ***the presence of substances to which someone is allergic releases large amounts of histamine into the blood*** *see also* ANTIHISTAMINE

◇ **histaminic** [hɪs'tæmɪnɪk] *adjective* referring to histamine

histo- ['hɪstəʊ] *prefix* referring to tissue

◇ **histochemistry** [hɪstəʊ'kemɪtstri] *noun* study of the chemical constituents of cells and tissues and also their function and distribution

◇ **histology** [hɪst'ɒlədʒi] *noun* study of anatomy of tissue cells and minute cellular structures, done using a microscope after the cells have been stained

history ['hɪstəri] *noun* study of what happened in the past; **case history** = details of what has happened to a patient undergoing treatment

HMIP = HER MAJESTY'S INSPECTORATE OF POLLUTION

hoar (frost) ['hɔ: frɒst] *noun* frozen dew which forms on outside surfaces when the temperature falls below freezing point

hoe [həʊ] **1** *noun* garden implement, with a small sharp blade, used to break up the surface of soil or cut off weeds **2** *verb* to cultivate land with a hoe

holarctic region [hɒl'ɑ:ktɪk] *noun* biogeographical region which includes the Nearctic region (i.e. North America) and the Palaearctic region (i.e. Europe, North Africa and North Asia)

holistic [hɒ'lɪstɪk] *adjective* (medical treatment) involving all the patient's mental and family circumstances rather than just dealing with the illness from which he is suffering; (policy, programme, etc.) dealing with a subject as a whole rather than looking at one aspect; ***the Trust has been calling for a more holistic transport policy***

hollow ['hɒləʊ] **1** *adjective* (space) which is empty *or* which has nothing inside; ***water collects in the hollow stems of plants*** **2** *noun* place which is lower than the rest of the surface; ***frost occurs in low-lying areas, called frost hollows***

holoplankton [hɒləʊ'plæŋktən] *noun* organisms such as algae and diatoms which remain as plankton throughout their entire life cycle

home [həʊm] *noun* place where a person *or* animal lives; environment, habitat; **second home** = house where a person only lives for part of the time, such as at weekends *or* during a holiday; **home consumption** = eating *or* drinking *or* using something at home; ***home consumption of alcohol has increased greatly over the last twenty years;*** **home range** = area occupied by an animal during the day-to-day course of activity (not to be confused with territory, which is a defended area, and may be part or all of the home range)

homeo- *or* **homoeo-** ['həʊmiəʊ] *prefix* meaning like *or* similar

◇ **homeostasis** [həʊmiəʊ'steɪsɪs] *noun* (i) tendency of a system to resist change and maintain itself in a state of equilibrium; (ii) process by which the functions and chemistry of a cell *or* organism are kept stable, even when external conditions vary greatly

◇ **homeotherm** [həʊmiəʊ'θɜ:m] *noun* = HOMOIOTHERM

homing ['həʊmɪŋ] *noun (of an animal)* returning to a particular site which is used for sleeping or breeding

homo- ['həʊməʊ] *prefix* meaning the same

◇ **homograft** ['həʊməʊgrɑ:ft] *noun* graft of tissue from one specimen to another of the same species

◇ **homosphere** ['həʊməʊsfiə] *noun* zone of the earth's atmosphere, including the troposphere, the stratosphere and the mesosphere, where the composition of the atmosphere remains relatively constant

homoiotherm [hɒ'mɔɪəθɜ:m] *noun* organism which has warm blood and a steady body temperature; *compare* POIKILOTHERM

hookworm ['hʊkwɜ:m] *noun* parasitic worm in the intestine which holds onto the wall of the intestine with its teeth and lives on the blood and protein of the carrier

horizon [hə'raɪzən] *noun* **(a)** the boundary where the earth and sky appear to meet; ***because of the curvature of the earth, the horizon looks curved when seen from a ship*** **(b)** layer of soil which is of a different colour *or* texture from the rest: the topsoil containing humus, the subsoil containing minerals leached from the topsoil, etc.

hormone ['hɔːməʊn] *noun* substance produced in animals and plants in one part of the body which has a particular effect in another part of the body; **juvenile hormone** = hormone in an insect larva which regulates its development into an adult; **plant hormones** = hormones (such as auxin) which particularly affect plant growth; **hormone rooting powder** = powder containing plant hormones (auxins) into which cuttings can be dipped to encourage the formation of roots; **hormone weedkiller** = substance which is absorbed by a plant and prevents its growth

◇ **hormonal** [hɔː'məʊnəl] *adjective* referring to hormones; **hormonal deficiency** = lack of necessary hormones

horticulture ['hɔːtɪkʌltʃə] *noun* gardening, growing of plants for food *or* decoration

◇ **horticultural** [hɔːtɪ'kʌltʃərəl] *adjective* referring to horticulture; ***allotment holders show their vegetables at the local horticultural show***

◇ **horticulturist** [hɔːtɪ'kʌltʃərɪst] *noun* gardener, person who specializes in horticulture

host [həʊst] *noun* plant *or* animal on which a parasite lives; **definitive host** = host on which a parasite settles permanently; **intermediate host** = host on which a parasite lives for a time before passing on to another host

> QUOTE: all mistletoes contain chlorophyll and conduct their own photosynthesis, but they vary in the extent to which they rely upon their host for organic material
> **Nature**

hot [hɒt] *adjective* very warm; with a high temperature; ***really hot weather only occurs in summer; the hottest part of the reactor is the core;*** **hot rocks** = rock of high temperature beneath the earth's surface. They can be used to create geothermal energy by pumping down cold water and making use of the rising hot water which the rocks have heated; **hot spot** = place where background radiation is particularly high; **hot spring** = hot water running out of the earth continuously

> QUOTE: Scottish scientists have started on a map of radiation hot spots in Britain following the Chernobyl disaster
> **Farmers Weekly**

> QUOTE: hot dry rock systems, where hot rocks alone at accessible depths are used to heat water pumped down from the surface
> **Environment**

house [haʊs] *noun* building where a person lives; **house moth** = small moth which sometimes lives in houses and whose larvae can destroy clothes and blankets, etc. kept in cupboards

◇ **housefly** ['haʊsflaɪ] *noun* common fly living in houses, which can spread disease by laying its eggs in decaying meat and vegetables

◇ **household** ['haʊshəʊld] *noun* all the people living together in the same house

◇ **housing** ['haʊzɪŋ] *noun* buildings where people live

HSE = HEALTH AND SAFETY EXECUTIVE

human ['hjuːmən] **1** *adjective* referring to a member of *Homo sapiens,* that is a man, woman or child; **a human being** = a person; **the human race** = all human beings; **human ecology** = study of man and communities of people, the place which they occupy in the natural world, and the ways in which they adapt to or change the environment **2** *noun* person; ***most animals are afraid of humans***

humate ['hjuːmeɪt] *noun* salt which is derived from humus

humid ['hjuːmɪd] *adjective* (air) which is damp *or* which contains moisture vapour; ***decomposition of organic matter is rapid in hot and humid conditions***

◇ **humidify** [hjuː'mɪdɪfaɪ] *noun* to make damp

◇ **humidifier** [hjuː'mɪdɪfaɪə] *noun* device for making dry air damp (as in air conditioning *or* central heating systems)

◇ **humidity** [hjuː'mɪdɪti] *noun* measurement of how much water vapour is contained in the air; **absolute humidity** = vapour concentration *or* mass of water vapour in a given quantity of air; **relative humidity** = ratio between the amount of water vapour in air and the amount which would be present if the air was saturated (shown as a percentage); **specific humidity** = ratio between the amount of water vapour in air and the total mass of the mixture of air and water vapour; **humidity control** = method of making the air humidity remain at a certain level, often by adding moisture to the air circulating in central heating systems

humify ['hjuːmɪfaɪ] *verb* to break down rotting organic waste to form humus

◇ **humification** [hjuːmɪfɪ'keɪʃn] *noun* breaking down of rotting organic waste to form humus

humus ['hju:məs] *noun* **(a)** fibrous organic matter in soil which makes the soil dark and binds it together **(b)** dark organic residue left after sewage has been treated in sewage works

hurricane ['hʌrɪkən] *noun* tropical storm in the Caribbean or Eastern Pacific Ocean with extremely strong winds; **hurricane force wind** = wind blowing at force 12 on the Beaufort scale; *compare* CYCLONE, TYPHOON

husbanding ['hʌzbəndɪŋ] *noun* using and keeping carefully; ***a policy of husbanding scarce natural resources***

◇ **husbandry** ['hʌzbəndri] *noun* farming *or* looking after farm animals and crops; ***a new system of intensive cattle husbandry;*** **animal husbandry** = breeding and looking after farm animals

> QUOTE: there is an added risk of the spread of infection under intensive farming methods, both by the stock being housed under more concentrated conditions and by the greater use of artificial feeding in this method of husbandry
>
> **Farmers Weekly**

hybrid ['haɪbrɪd] *adjective & noun* cross between two varieties *or* species of plant *or* animal; **hybrid vigour** = increase in size *or* rate of growth *or* fertility *or* resistance to disease found in offspring of a cross between two species

◇ **hybridization** [haɪbrɪdaɪ'zeɪʃn] *noun* production of hybrids

hydr- *or* **hydro-** ['haɪdrəʊ] *prefix* referring to water; **hydropower** *or* **hydroelectric power** = electricity produced by using a flow of water to drive the turbines

hydric ['haɪdrɪk] *adjective* wet (environment); *compare* XERIC

hydrocarbon [haɪdrəʊ'kɑ:bən] *noun* compound formed of hydrogen and carbon; **aliphatic hydrocarbons** = paraffins, acetylenes and olefins, i.e. compounds that do not contain benzene; **aromatic hydrocarbons** = benzenes; **liquid hydrocarbon** = organic compound in liquid form, such as fuel oil

> COMMENT: hydrocarbons are found in fossil fuels such as coal, oil, petroleum and natural gas. Hydrocarbons form a large part of exhaust fumes from cars and contribute to the formation of smog. When released into the air from burning coal or oil they react in the sunlight with nitrogen dioxide to form ozone. Hydrocarbons are divided into aliphatic hydrocarbons (paraffins, acetylenes and olefins) and aromatic hydrocarbons (benzenes).

hydrochloric acid (HCl) [haɪdrə'klɒrɪk 'æsɪd] *noun* aqueous solution of hydrogen chloride, used widely in the chemical, food and oil industries. It also forms in the stomach as part of the digestive process. It is extremely corrosive

hydrocyanic acid (HCN) [haɪdrəʊsaɪ'ænɪk 'æsɪd] = HYDROGEN CYANIDE

hydroelectricity [haɪdrəʊɪlek'trɪsɪti] *noun* electricity produced by water power

◇ **hydroelectric** [haɪdrəʊɪ'lektrɪk] *adjective* referring to hydroelectricity; ***the valley was flooded to construct the hydroelectric scheme;*** **hydroelectric energy** *or* **power** = electricity produced by using a flow of water to drive the turbines; **hydroelectric power station** = power station which produces electricity using a flow of water to drive the turbines

hydrofluoric acid [haɪdrəʊflu'ɒrɪk 'æsɪd] *noun* aqueous solution of hydrogen fluoride which is extremely corrosive and attacks glass and stone

hydrogen ['haɪdrədʒən] *noun* chemical element, a gas which combines with oxygen to form water, with other elements to form acids, and is present in all animal tissue; it is also used as a moderator in some nuclear reactors; **hydrogen chloride (HCl)** = (i) colourless strong-smelling gas; (ii) hydrochloric acid; **hydrogen cyanide (HCN)** = liquid *or* gas with a smell of almonds which kills very rapidly when drunk *or* inhaled; **hydrogen fluoride (HF)** = (i) colourless poisonous strong-smelling gas; (ii) hydrofluoric acid; **hydrogen fuel** = liquid hydrogen, proposed as an alternative fuel for use in cars and aircraft; **hydrogen sulphide (H_2S)** = colourless gas with an unpleasant smell, produced during industrial processes, which corrodes metal and is toxic to animals (NOTE: chemical symbol of hydrogen is **H**; atomic number is **1**)

hydrograph ['haɪdrəgrɑ:f] *noun* graph showing the level of water *or* flow of water in a river *or* lake

◇ **hydrography** [haɪ'drɒgrəfi] *noun* science of measuring and charting rivers, lakes and seas

◇ **hydrological** [haɪdrə'lɒdʒɪkl] *adjective* referring to hydrology; **hydrologic cycle** *or* **hydrological cycle** = cycle of events when water in the sea evaporates in the heat of the sun, forms clouds which deposit rain as they pass over land, the rain then drains away into rivers which return the water to the sea

◇ **hydrology** [haɪ'drɒlədʒi] *noun* study of water, its composition and properties

◇ **hydrometer** [haɪ'drɒmɪtə] *noun* instrument which measures the density of a liquid

◇ **hydromorphic soil** [haɪdrə'mɔ:fɪk 'sɔɪl] *noun* waterlogged soil found in bogs and marshes

◇ **hydrophyte** ['haɪdrəfaɪt] *noun* plant which lives in water *or* in marshy conditions

◇ **hydroponics** [haɪdrə'pɒnɪks] *noun* science of growing plants without soil but in sand *or* vermiculite *or* other granular material, using a liquid solution of nutrients to feed them

◇ **hydrosere** ['haɪdrəsiːə] *noun* series of plant communities growing in water *or* in wet conditions

◇ **hydrosphere** ['haɪdrəsfiːə] *noun* all the earth's water in the sea, the atmosphere and on land

◇ **hydrostatic** [haɪdrə'stætɪk] *adjective* referring to water which is not moving; **hydrostatic pressure** = pressure of water which is not moving

hydrothermal [haɪdrə'θɜːməl] *adjective* referring to water and heat under the earth's crust; **hydrothermal formation** = rock formation where water pockets or porous water-filled rock come into contact with magma, so creating steam; **hydrothermal vent** = place on the ocean floor where hot water and gas flow out of the earth's crust

> QUOTE: hydrothermal energy is used to produce electricity - more than 2,000 megawatts - at a geyser field north of San Francisco
> **Environment**

hydroxide [haɪ'drɒksaɪd] *noun* compound containing inorganic OH^- groups; these compounds are basic

hygiene ['haɪdʒiːn] *noun* (i) being clean and keeping healthy conditions; (ii) science of health; ***nurses have to maintain strict personal hygiene;*** **environmental hygiene** = study of health and how it is affected by the environment; **tropical hygiene** = hygiene relating to tropical regions

◇ **hygienic** [haɪ'dʒiːnɪk] *adjective* (i) clean; (ii) which produces healthy conditions; ***don't touch the food with dirty hands - it isn't hygienic***

hygro- ['haɪgrəʊ] *prefix* meaning wet

◇ **hygrometer** [haɪ'grɒmɪtə] *noun* scientific instrument for measuring humidity (in the atmosphere, in grain, etc.)

◇ **hygrometry** [haɪ'grɒmɪtri] *noun* scientific measurement of humidity

◇ **hygroscope** ['haɪgrəskəʊp] *noun* device *or* substance which gives an indication of humidity (in some cases by changing colour)

◇ **hygroscopic** [haɪgrə'skɒpɪk] *adjective* (substance) which absorbs moisture from the atmosphere

hyp- [haɪp] *prefix* meaning under *or* less *or* too little *or* too small; *opposite is* HYPER-

◇ **hypalgesia** [haɪpəl'dʒiːziə] *noun* low sensitivity to pain

hyper- ['haɪpə] *prefix* meaning over *or* above *or* higher *or* too much; *opposite is* HYP- *or* HYPO-

◇ **hyperactive** [haɪpə'æktɪv] *adjective* being very active

◇ **hyperparasite** [haɪpə'pærəsaɪt] *noun* parasite which is a parasite on other parasites

◇ **hypersensitive** [haɪpə'sensɪtɪv] *adjective* (person) who reacts more strongly than normal to an antigen

◇ **hypersensitivity** [haɪpəsensɪ'tɪvɪti] *noun* condition where an organism reacts unusually strongly to something (such as an allergic substance *or* food additive); ***hypersensitivity to dairy products is quite common in children***

hypo- ['haɪpəʊ] *prefix* meaning under *or* less *or* too little *or* too small; *opposite is* HYPER-; **hypodermic syringe** = instrument for injecting liquids under the skin

◇ **hypolimnion** [haɪpəʊ'lɪmniən] *noun* lowest layer of water in a lake, which is cold and stationary and contains less oxygen than the epilimnion

◇ **hypotension** [haɪpəʊ'tenʃn] *noun* condition where the pressure of the blood is abnormally low

◇ **hypothalamus** [haɪpəʊ'θæləməs] *noun* part of the brain above the pituitary gland, which controls the production of hormones by the pituitary gland and regulates important bodily functions such as hunger, thirst and sleep

◇ **hypothermia** [haɪpəʊ'θɜːmiə] *noun* reduction in body temperature below normal, for official purposes taken to be below 35°C

hypothesis [haɪ'pɒθɪsɪs] *noun* suggestion that something is true, though without proof (NOTE: plural is **hypotheses)**

HYV = HIGH-YIELDING VARIETY

Ii

I *chemical symbol for* iodine

IAEA = INTERNATIONAL ATOMIC ENERGY AGENCY

-iasis ['aɪəsɪs] *suffix* meaning disease caused by something; **amoebiasis** = infection caused by amoeba, which can result in amoebic dysentery in the large intestine (intestinal amoebiasis) and can sometimes infect the lungs (pulmonary amoebiasis)

ice [aɪs] *noun* **(a)** frozen water; **Ice Age** = long period of time when the earth's temperature was cool and large areas of the surface were covered with ice; **polar ice cap** = large area of thick ice covering the North or South polar regions; **ice crystal** = hard particle which forms as a liquid freezes; **ice floe** = sheet of ice floating in the sea; **ice sheet** = large area of thick ice covering the North or South polar regions; **ice shelf** = outer margin of an ice cap or ice sheet that extends into and over the sea **(b) dry ice** = solid carbon dioxide, used as a refrigerant

◇ **iceberg** ['aɪsbɜːg] *noun* very large block of ice floating in the sea (formed when ice breaks away from an arctic glacier or ice sheet)

COMMENT: ice is formed when water freezes at 0°C. Ice is less dense than water and so floats. Because the ice in the polar ice caps is very thick and has been formed over many thousands of years, scientists are able to discover information about the climate over a very long period of time by examining core samples obtained by drilling into the ice

QUOTE: scientists studying the massive new iceberg in Antarctica have given a warning that almost every ice sheet on its coast is shrinking, as warmer summer temperatures melt the continent's frozen wastes

Times

-icide [ɪ'saɪd] *suffix* meaning which kills; **pesticide** = substance which kills pests

ICRP = INTERNATIONAL COMMISSION ON RADIOLOGICAL PROTECTION

identical [aɪ'dentɪkl] *adjective* exactly the same; **identical twins** = monozygotic twins, two offspring born at the same time and from the same ovum, therefore exactly the same in appearance and sex

igneous ['ɪgniəs] *adjective* (rock such as basalt *or* granite) formed from molten lava; *compare* METAMORPHIC, SEDIMENTARY

COMMENT: igneous rocks are formed from lava which has either broken through the earth's crust (as in a volcanic eruption) or has entered the crust from below and formed a layer inside the crust

illegal [ɪ'liːgəl] *adjective* not done according to the law; ***the illegal killing of a protected species; it is illegal to take eggs from the nests of rare birds***

◇ **illegally** [ɪ'liːgəli] *adverb* in a way which is not according to the law; ***the company was accused of illegally felling protected forest***

illuviation [ɪluːvi'eɪʃn] *noun* deposition of particles and chemicals leached out from the topsoil into the subsoil

imbalance [ɪm'bæləns] *noun* lack of balance *or* situation where one substance *or* species is dominant; ***lack of vitamins A and E creates hormonal imbalances in farm animals***

immature [ɪmə'tjʊə] *adjective* not mature, still developing; **immature cell** = cell which is still developing

immigration [ɪmɪ'greɪʃn] *noun* movement of an organism into a new area

immune [ɪ'mjuːn] *adjective* protected against an infection *or* against an allergic disease; ***this strain is not immune to virus;*** **immune body** = substance which protects against an infection *or* against an allergic disease; **immune system** = arrangement of organs, cells and substances (including the spleen, lymph tissue and white blood cells) which protect the body against an infection *or* against an allergic disease

◇ **immunity** [ɪ'mjuːnɪti] *noun* **(a)** *(in animals)* ability to resist attacks of a disease because antibodies are produced; ***the vaccine gives immunity to tuberculosis;*** **acquired immunity** = immunity which a body acquires and which is not congenital; **natural immunity** = immunity from disease inherited by newborn offspring from birth, acquired in the womb *or* from the mother's milk **(b)** *(in plants)* resistance to disease through a protective covering on leaves *or* through the formation of protoplasts *or* through the development of inactive forms of viruses

◇ **immunization** [ɪmjuːnaɪ'zeɪʃn] *noun* making a person immune to an infection, either by injecting an antiserum (passive immunization) *or* by giving the body the disease in such a small dose that the body does not develop the disease, but produces antibodies to counteract it

◇ **immunize** ['ɪmju:naɪz] *verb* to make a person immune to an infection (NOTE: you immunize someone **against** a disease)

impact ['ɪmpækt] *noun* effect *or* impression made on something; **impact assessment** *or* **impact study** = evaluation of the effect upon the environment of a large construction programme

◇ **impacted area** [ɪm'pæktɪd 'eəriə] *noun* area of land affected *or* concerned, as by a large-scale building project, etc.

impair [ɪm'peə] *verb* to harm something so that it does not function properly

◇ **impairment** [ɪm'peəmənt] *noun* harming something so that it does not function properly

impermeable [ɪm'pɜ:miəbl] *adjective* (substance) which does not allow a liquid *or* gas to pass through; (membrane) which allows a liquid to pass through, but not solid particles suspended in the liquid; ***rocks which are impermeable to water***

impervious [ɪm'pɜ:viəs] *adjective* (surface of a substance) which does not allow water to pass through

import 1 *noun* ['ɪmpɔ:t] (i) bringing something into a country from abroad; (ii) something brought into a country from abroad; **import duty** = tax which has to be paid on goods brought into a country from abroad **2** *verb* [ɪm'pɔ:t] to bring something into a country from abroad

impoverish [ɪm'pɒvərɪʃ] *verb* to make (soil) less fertile; ***overfarming has impoverished the soil***

◇ **impoverished** [ɪm'pɒvərɪʃt] *adjective* (soil) which is less fertile; ***if impoverished soil is left fallow for some years, nutrients may build up in the soil again***

◇ **impoverishment** [ɪm'pɒvərɪʃmənt] *noun* making less fertile; ***overexploitation led to the impoverishment of the soil***

impregnate ['ɪmpregneɪt] *verb* **(a)** to fertilize a female, by introducing male spermatozoa into the female's body so that they link with the female's ova **(b)** to fill something with a substance by passing it inside through the outer surface; ***many fruits on sale are impregnated with pesticides even if they have been washed; wooden posts are impregnated with creosote***

imprint [ɪm'prɪnt] *verb (of young animals)* to learn from a source (usually the mother) by imitation at a very early age, usually occurring during a very brief period, such as the first few hours after a bird has hatched

improvement grant [ɪm'pru:vmənt 'grɑ:nt] *noun* grant available to improve the standard of a building or of a farm; grants are available for a variety of purposes, e.g. for putting in domestic drainage and bathrooms, for the eradication of bracken on pasture land, for draining, for providing a water supply to fields

I/M program *noun US* = INSPECTION AND MAINTENANCE PROGRAM

impulse turbine ['ɪmpʌls 'tɜ:baɪn] *noun* turbine where jets of water are directed at bucket-shaped blades which catch the water

impure [ɪm'pjʊə] *adjective* not pure

◇ **impurities** [ɪm'pjʊrɪtɪz] *plural noun* substances which are not pure *or* clean; ***a filter which removes impurities from drinking water***

in. = INCH

inactive [ɪn'æktɪv] *adjective* **(a)** (substance) which is not active **(b)** (volcano) which is not erupting or likely to erupt, though not necessarily extinct

◇ **inactivate** [ɪn'æktɪveɪt] *verb* to make something inactive; ***biopesticides are easily inactivated in sunlight***

inbreeding ['ɪnbri:dɪŋ] *noun* breeding between a closely related male and female, who have the same parents *or* grandparents, so increasing the possibility of congenital defects; **inbreeding depression** = reduction of characteristic strengths by inbreeding (NOTE: opposite is **outbreeding**)

◇ **inbred** [ɪn'bred] *adjective* resulting from inbreeding

incentive [ɪn'sentɪv] *noun* something which encourages; **investment incentive** = something which encourages people to invest money in a project; **tax incentive** = reduction in tax, offered as an encouragement to do something *or* to use something

inch [ɪnʃ] *noun* unit of measurement of length (= 0.0254 metre)

incidence ['ɪnsɪdəns] *noun* number of times something happens in a certain area *or* in a certain population over a period of time; ***the incidence of radiation leaks has increased over the past four years;*** **incidence rate** = number of new cases of a disease during a given period, per thousand of population

◇ **incident** ['ɪnsɪdənt] *noun* something which happens; ***last year six hundred incidents of oil pollution were reported;*** **the Three Mile Island incident** = the accident at the Three Mile Island power station

> QUOTE: water pollution incidents more than doubled in England and Wales during the Eighties
> **Green Magazine**

incinerate [ɪn'sɪnəreɪt] *verb* to burn (waste)

◊ **incineration** [ɪnsɪnə'reɪʃn] *noun* burning (of waste); ***uncontrolled incineration can contribute to atmospheric pollution; controlled incineration of waste is one of the most effective methods of disposal;*** **ocean incineration** = burning of toxic waste in special ships at sea (at present only permitted in the North Sea); **incineration ash** = powder left after a substance has been burnt; **incineration facility** *or* **incineration plant** = establishment where waste is burnt

◊ **incinerator** [ɪn'sɪnəreɪtə] *noun* device which burns waste

> QUOTE: sludge incinerators have been shown to emit up to 1.8 µg per cubic metre of dioxins in the stack gases. A local incineration facility sited to reduce transport costs could therefore pose a toxicological hazard to residents
> **London Environmental Bulletin**

incipient [ɪn'sɪpiənt] *adjective* which is just beginning *or* which is in its early stages; **incipient lethal level** = level of toxic substances at which 50% of the affected organisms will die

inclusive fitness [ɪn'klu:sɪv 'fɪtnəs] *noun* sum of an organism's Darwinian fitness with the fitness of its relatives

incompatible [ɪnkɒm'pætɪbl] *adjective* which does not go together with something else; (tissue) which is genetically different from other tissue, making it impossible to transplant into that tissue; (graft) which does not take with a certain stock

◊ **incompatibility** [ɪnkəmpætɪ'bɪlɪti] *noun* being incompatible

incubation [ɪŋkju:'beɪʃn] *noun* (i) hatching of eggs either by a sitting bird *or* by artificial means in an incubator; (ii) growing of bacteria; **incubation period** = time during which a bacterium *or* a virus develops in the body after infection *or* contamination, before the appearance of the symptoms of the disease

◊ **incubate** ['ɪŋkju:beɪt] *verb* (i) to hatch eggs either by a sitting bird *or* by artificial means in an incubator; (ii); *(of bacteria)* to grow

◊ **incubator** ['ɪŋkju:beɪtə] *noun* special unit providing artificial heat, used to hatch eggs *or* to grow bacteria

indicator ['ɪndɪkeɪtə] *noun* (i) substance which shows something, especially a substance secreted in body fluids which shows which blood group a person belongs to; (ii) substance which shows that another substance is present; ***lichens act as indicators for atmospheric pollution;*** **ecological indicator** = species that has particular requirements (acid soil, low temperature, high rainfall, etc.) and whose presence in an area shows that these requirements are satisfied in that area; **indicator species** = species which is very sensitive to changes in the environment, which can warn that environmental changes are taking place

indigenous [ɪn'dɪdʒənəs] *adjective* which is native to a place; ***there are six indigenous species of monkey on the island; oaks are indigenous to the British Isles; the government encourages the planting of indigenous hardwoods where possible*** (NOTE: something is indigenous **to** a place)

> QUOTE: the Gambia is believed to be the first country in Africa to have adopted regulations supporting the use of indigenous low-cost building materials
> **Appropriate Technology**

individual [ɪndɪ'vɪdʒuəl] *noun & adjective* single plant *or* animal of a species; ***it is a native of North America, but individuals have been found as far south as Brazil***

induce [ɪn'dju:s] *verb* to cause *or* bring about; **human-induced stress** = stress in animals caused by humans (such as ringing birds *or* driving cars in deer parks, etc.)

◊ **inducer** [ɪn'dju:sə] *noun* (i) substance which changes the way in which an enzyme acts; (ii) substance which helps certain genes to be passed on from parent to offspring

◊ **induction** [ɪn'dʌkʃn] *noun* changing the way in which an enzyme acts

> COMMENT: organochlorine type pesticides work by induction; they can affect the enzymes in animals in a food chain (birds, for instance) which they are not intended to kill

industry ['ɪndʌstri] *noun* all factories *or* companies *or* processes involved in the manufacturing of products; **building industry** = business *or* trade of constructing houses, offices, etc.; **heavy industry** = industry (such as steelmaking or engineering) which takes raw materials and makes them into finished products *or* which extracts raw materials (such as coal); **light industry** = industry which makes small *or* lightweight products

◊ **industrial** [ɪn'dʌstriəl] *adjective* referring to industries *or* factories; **industrial dereliction** = ugly and neglected state of the landscape and environment damaged by industrial processes; **industrial disease** = disease which is caused by the type of work done by a worker (such as by dust produced *or* chemicals used in the factory); **industrial estate** = area of land with buildings specially designed and constructed for light industries; **industrial melanism** = phenomenon where certain animals (such as butterflies and moths) become darker in colour in industrial areas, allowing them to match the trees and leaves which are covered with soot; **industrial park** = = INDUSTRIAL ESTATE

◇ **industrialist** [ɪn'dʌstriəlɪst] *noun* person who owns *or* manages an industrial business

◇ **industrialization** [ɪndʌstriəlaɪ'zeɪʃn] *noun* process of developing industry in a country

◇ **industrialize** [ɪn'dʌstriəlaɪz] *verb* to develop industry in (a country), to equip with factories; **industrialized country** = country equipped with factories whose economy depends on industry and not on agriculture; ***industrialized countries are trying to control soot emission***

◇ **industrial waste** *or* **industrial effluent** *or* **industrial sewage** *noun* liquid *or* solid waste from industrial processes

COMMENT: industrial wastes can be divided into various types: solid waste, such as dust particles or slag from coal; liquid wastes from various processes, including radioactive coolants from power stations; and gas wastes, largely produced by the chemical industry

inert [ɪ'nɜːt] *adjective* without the power of action; **inert gases** = noble gases *or* gases (helium, neon, argon, krypton, xenon and radon) which do not react chemically with other substances

infect [ɪn'fekt] *verb* to contaminate with disease-producing microorganisms *or* toxins; to transmit infection

◇ **infection** [ɪn'fekʃn] *noun* entry of microbes into the body, which then multiply in the body; ***as a carrier he was spreading infection to other people in the office***

◇ **infectious** [ɪn'fekʃəs] *adjective* (disease) which is caused by microbes and can be transmitted to other persons by direct means; ***this strain of flu is highly infectious***

◇ **infective** [ɪn'fektɪv] *adjective* (disease) caused by a microbe, which can be caught from another person but which cannot always be directly transmitted

◇ **infectivity** [ɪnfek'tɪvɪti] *noun* being infective

infertile [ɪn'fɜːtaɪl] *adjective* (i) not able to bear fruit; (ii); *(of animal)* not able to produce young; (iii); *(of soil)* barren *or* not able to produce good crops

◇ **infertility** [ɪnfə'tɪlɪti] *noun* not being fertile, not being able to reproduce *or* have offspring

COMMENT: an infertile soil is one which is deficient in plant nutrients. The fertility of a soil at any one time is partly due to its natural makeup, and partly to its condition, which is largely dependent on the management of the soil in recent times. Application of fertilizers can raise soil fertility; bad management can decrease it. Infertility in cattle can be a serious problem, and may be caused by disease such as contagious abortion, or by lack of or imbalance in minerals in the diet

infest [ɪn'fest] *verb (of parasites)* to be present in large numbers; ***pine forests have been infested with beetles; infested plants should be dug up and burnt; the child's hair was infested with lice***

◇ **infestation** [ɪnfes'teɪʃn] *noun* having large numbers of parasites; ***the crop showed serious infestation with greenfly; the condition is caused by infestation of the hair with lice***

infiltration [ɪnfɪl'treɪʃn] *noun* passing of water into the soil *or* into a drainage system; **infiltration basin** = depression in the ground where infiltration occurs; **infiltration capacity** = maximum rate at which water is absorbed by soil; **infiltration water** = water which passes into the soil *or* into a drainage system

infinite ['ɪnfɪnət] *adjective* which will never end *or* be used up

inflammable [ɪn'flæməbl] *adjective* (material) which catches fire easily

inflorescence [ɪnflə'resəns] *noun* flower *or* group of small flowers, together with the stem

inflow ['ɪnfləʊ] *noun* action of flowing in; ***an inflow of effluent into a river***

influent ['ɪnfluənt] *noun* **(a)** stream *or* river flowing into a larger river **(b)** organism which has an important effect on the balance of its community

infrared rays *or* **infrared radiation** [ɪnfrə'red] *noun* long invisible rays, below the visible red end of the colour spectrum, which form part of the warming radiation which the earth receives from the sun; **infrared photography** = taking photographs using an infrared camera, which shows up heat sources or which can be used to take pictures of wild animals at night

infrastructure ['ɪnfrəstrʌktʃə] *noun* (i) basic framework of a system *or* organization; (ii) basic facilities and systems of a country *or* city, such as roads, pipelines, electricity and telecommunications networks, schools, hospitals, etc.

ingest [ɪn'dʒest] *verb* to take (food *or* liquid) into the body

◇ **ingestion** [ɪn'dʒestʃn] *noun* taking food *or* liquid into the body

inhabit [ɪn'hæbɪt] *verb* to live in a place

◇ **inhabitant** [ɪn'hæbɪtənt] *noun* plant *or* animal which lives in a place

inhale [ɪn'heɪl] *verb* to breathe in

◇ **inhalation** [ɪnhə'leɪʃn] *noun* breathing in

inhibit [ɪn'hɪbɪt] *verb* to make it difficult for something to happen; ***lack of nutrients may inhibit plant growth; the government is planning to introduce a tax to inhibit greenhouse gas emissions***

◇ **inhibitor** [ɪn'hɪbɪtə] *noun* **growth inhibitor** = substance which stops an organism growing

inlet ['ɪnlət] *noun* passage *or* opening for sea water into the land

inner ['ɪnə] *adjective* further inside *or* further towards the centre; **inner city** = part of a city at *or* near the centre, especially a slum area where poor people live in bad housing

inorganic [ɪnɔː'gænɪk] *adjective* (substance) which does not come from an animal *or* a plant; (substance) which does not contain carbon; ***inorganic substances include acids, alkalis and metals;*** **inorganic chemistry** = branch of chemistry dealing with inorganic compounds; **inorganic fertilizer** = artificial synthesized fertilizer (as opposed to manure, compost and other organic fertilizers which are produced from bones, blood and other parts of formerly living matter); **inorganic pesticide** *or* **fungicide** *or* **herbicide** = pesticide *or* fungicide *or* herbicide made from inorganic substances such as sulphur

input ['ɪnpʊt] *noun* something which is needed for industrial production, such as a raw material *or* labour force, etc.; **inputs** = substances put into the soil (such as fertilizers which are applied by a farmer)

inquire [ɪŋ'kwaɪə] *verb* to ask questions about something; ***he inquired about the environmental effects of the proposed scheme***

◇ **inquiry** [ɪŋ'kwaɪri] *noun* official investigation; ***there has been a government inquiry into the radiation leaks***

insanitary [ɪn'sænɪtri] *adjective* not clean, unhygienic; ***cholera spread rapidly because of the insanitary conditions in the town***

insect ['ɪnsekt] *noun* small animal with six legs and a body in three parts; ***insects were flying round the lamp; he was stung by an insect;*** **insect bite** = sting caused by an insect which punctures the skin and in so doing introduces irritants

◇ **insect-borne** ['ɪnsekt 'bɔːn] *adjective* (disease) which is carried and transmitted by insects

◇ **insecticide** [ɪn'sektɪsaɪd] *noun* substance which kills insects; **natural insecticide** = insecticide produced from plant extracts; **persistent insecticide** = insecticide that remains toxic (either in the soil or in the body of an animal) and is passed from animal to animal through the food chain; **synthetic insecticide** = insecticide which is made artificially by chemical reaction

◇ **insectivore** *or* **insectivorous animal** [ɪn'sektɪvɔː *or* ɪnsek'tɪvərəs] *noun* animal (such as a hedgehog) which eats insects; **insectivorous plant** = plant (such as the sundew) which attracts insects, traps them and then digests them

COMMENT: Insects form the class Insecta. The body of an insect is divided into three distinct parts: the head, the thorax and the abdomen. The six legs are attached to the thorax. There are two antennae on the head. Most insect bites are simply irritating, but some people can be extremely sensitive to the bites of certain types of insect (such as bee stings). Other insect bites can be more serious, as insects can carry the bacteria which produce typhus, sleeping sickness, malaria, filariasis, etc. Natural insecticides (i.e. those produced from plant extracts) are less harmful to the environment than the synthetic insecticides which, though effective, are often persistent and kill not only insects but also other larger animals when they get into the food chain. In agriculture, most pesticides are chemically based: they are either chlorinated hydrocarbons, organophosphorous compounds or carbamate compounds. Insecticides are used in a number of ways, including spraying and dusting, or in granular forms as seed dressings. In the form of a gas, insecticides are used to fumigate greenhouses and granaries

QUOTE: the perennial problem of locust infestation in North Africa can be contained by comprehensive spraying of insecticides. The widespread banning of the use of dieldrin, a potent agent against the locust, is adding to costs, since other more benign chemicals require more frequent application

Middle East Agribusiness

inselberg ['ɪnselbɜːg] *noun* steep-sided isolated hill that stands above nearby hills

inshore ['ɪnʃɔː] *adjective* on the water, but near the coast; **inshore fishing** = catching fish from boats, but near the coast; **inshore waters** = water close to the coast

insolation [ɪnsə'leɪʃn] *noun* radiation from the sun

insoluble [ɪn'sɒljʊbl] *adjective* which cannot be dissolved; ***gold is insoluble in water***

inspect [ɪn'spekt] *verb* to examine *or* to look at something carefully; ***the kitchens are inspected by health inspectors***

◇ **inspection** [ɪn'spekʃn] *noun* act of examining something; ***the officials have***

carried out an inspection of the factory waste to see if it can be safely incinerated; **inspection chamber** = large hole built above a drain, allowing someone to look at the inside of the pipes to see if they are blocked; *US* **Inspection and Maintenance (I/M) program** = programme of testing cars to check if their emission levels are within government safety standards

◇ **inspector** [ɪn'spektə] *noun* person who examines something; **Animals Inspector** = official whose job is to inspect animals to see if they have notifiable diseases, are being kept in acceptable conditions, etc.; **Government Health Inspector** = government official who examines offices *or* factories to see if they are clean and healthy

installation [ɪnstə'leɪʃn] *noun* apparatus *or* machinery that has been installed in a certain place; ***a geothermal installation is being built on the geyser field; the oil installations were bombed***

insulate ['ɪnsjuleɪt] *verb* to prevent heat *or* cold *or* sound escaping *or* entering; ***well-insulated houses need less heating***

◇ **insulation** [ɪnsju'leɪʃn] *noun* preventing heat *or* cold *or* sound escaping *or* entering; **insulation material** = material used to insulate something, especially a building

COMMENT: insulation can cut the cost of fuel needed for heating, by preventing heat escaping through windows, roofs, etc.

intake ['ɪnteɪk] *noun* amount of a substance taken into an organism (either eaten *or* absorbed); ***a study of food intake among grassland animals; the bird's daily intake of insects is more than half its own weight***

integrated pest management (IPM) ['ɪntɪgreɪtɪd 'pest mænɪdʒmənt] *noun* appropriate combination of all methods of pest control, i.e. good cultivation practices (such as crop rotation), use of chemical pesticides, choosing resistant crop varieties and biological control

intelligence quotient (IQ) [ɪn'telɪdʒəns 'kwəʊʃənt] *noun* ratio of the result of an intelligence test shown as a relationship of the mental age to the actual age of the person tested (the average being 100)

intensity [ɪn'tensɪti] *noun* degree to which land is used; **low-intensity land** = land on which crops are not intensively cultivated

◇ **intensification** [ɪntensɪfɪ'keɪʃn] *noun* **(a)** doing something in an intensive way; ***intensification of farming has contributed to soil erosion*** **(b)** becoming deeper (of a depression)

◇ **intensify** [ɪn'tensɪfaɪ] *verb* **(a)** to start doing something intensively **(b)** *(of depression)* to become deeper

◇ **intensive** [ɪn'tensɪv] *adjective* **intensive animal breeding** *or* **intensive livestock production** = specialized system of breeding animals where the livestock are kept indoors and fed on concentrated foodstuffs, with frequent use of drugs to control diseases which are a constant threat under these conditions; **intensive cultivation** *or* **intensive farming** = farming in which as much use is made of the land as possible by growing crops close together *or* by growing several crops in a year *or* by using large amounts of fertilizer

◇ **intensively** [ɪn'tensɪvli] *adverb* in an intensive way (NOTE: the opposite is **extensive, extensification**)

QUOTE: the diet of animals kept in intensive units is usually quite different from what the animals would eat if foraging for themselves. Grazing animals have the ability to seek out certain plants which contribute to their health

Ecologist

inter- ['ɪntə] *prefix* meaning between

◇ **interaction** [ɪntə'ækʃn] *noun* relationship between two *or* more organisms *or* things; **host-parasite interaction** = relationship between the host and the parasite

◇ **interbreed** [ɪntə'bri:d] *verb* to breed together (individuals from the same species can interbreed; those from different species cannot)

◇ **interdependence** [ɪntədɪ'pendəns] *noun* state of two *or* more organisms *or* processes which depend on each other

◇ **interdependent** [ɪntədɪ'pendənt] *adjective* (two *or* more organisms *or* processes) which depend on each other

interest group ['ɪntrəst gru:p] *noun* group of people who are all concerned about the same issue and who try to influence the opinions of politicians, local officials and businessmen on this particular issue

interferon [ɪntə'fɪərɒn] *noun* protein produced by cells, usually in response to a virus and which then reduces the spread of viruses

interfluve [ɪntə'flu:v] *noun* land area between two rivers

interglacial (period) [ɪntə'gleɪsiəl 'pɪəriəd] *noun* period between two Ice Ages when the climate becomes warmer

interior [ɪn'tɪəriə] *adjective & noun* inside; ***reserves in the interior of the continent; the interior walls of the intestine*** *US* **Department of the Interior** = government department dealing with the conservation and development of natural resources

intermediate [ɪntə'mi:diət] *adjective* between two extremes; **intermediate host** = host on which a parasite lives for a time before passing on to another; **intermediate technology** = technology which is between the advanced electronic technology of industrialized countries and the primitive technology in developing countries; **intermediate waste** = waste from a nuclear reactor in the form of sludge, metal parts, etc., which are more radioactive than low-level waste and less radioactive than high-level waste

internal combustion engine [ɪn'tɜ:nəl kʌm'bʌstʃən 'endʒɪn] *noun* type of engine used in motor vehicles, where the fuel is a mixture of petrol and air burnt in a closed chamber (the combustion chamber) to give energy to the pistons

international [ɪntə'næʃnəl] *adjective* between countries; **International Atomic Energy Agency (IAEA)** = agency of the United Nations Organization dealing with all aspects of nuclear energy; **International Commission on Radiological Protection (ICRP)** = group of scientists who try to decide on worldwide safety standards for the nuclear industry by fixing a maximum allowable dose of radiation; **International Union for the Conservation of Nature (IUCN)** = international organization which coordinates the protection of species throughout the world; **International Whaling Convention** = agreement (signed in 1946) to control the commercial killing of whales

interoceptor ['ɪntərəʊseptə] *noun* nerve cell which reacts to a change taking place inside the body; *see also* CHEMORECEPTOR, EXTEROCEPTOR, RECEPTOR

interspecific [ɪntəspə'sɪfɪk] *adjective* between species

intertidal zone [ɪntə'taɪdəl 'zəʊn] *noun* area of sea water and shore between the high and low water marks

intertropical convergence zone (ITCZ) [ɪntə'trɒpɪkl kən'vɜ:dʒəns 'zəʊn] *noun* boundary between the trade winds and tropical air masses of the Northern and Southern hemispheres

intervention [ɪntə'venʃn] *noun* purchase by the EU of surplus agricultural produce when the price drops to a certain predetermined level; **intervention stocks** = agricultural produce held by the EU in store

intra- ['ɪntrə] *prefix* meaning inside

◇ **intraspecific** [ɪntrəspe'sɪfɪk] *adjective* among species

intrinsic factor [ɪn'trɪnsɪk 'fæktə] *noun* protein produced in the gastric glands which reacts with vitamin B_{12} and which, if lacking, causes pernicious anaemia

introduce [ɪntrə'dju:s] *verb* to bring something to a new place; ***several of the species of plant now common in Britain were introduced by the Romans; settlers introduced dogs to the island, with the result that all flightless birds became extinct; starlings were introduced to the USA in 1891***

◇ **introduction** [ɪntrə'dʌkʃn] *noun* **(a)** bringing something to a new place; ***before the introduction of grey squirrels, the red squirrel was widespread; the death rate from malaria was very high before the introduction of new anti-malarial techniques*** **(b)** plant *or* animal which has been brought to a new place; ***it is not an indigenous species but a 19th century introduction***

intrusion [ɪn'tru:ʒn] *noun* area of rock which has pushed into other rocks; **igneous intrusion** = igneous rock which solidified before it reached the surface of the earth and remains as a layer among other rocks

invasion [ɪn'veɪʒn] *noun* (i) arrival of large numbers of pests in an area; (ii) entry of bacteria into a body *or* first attack of a disease

inventory ['ɪnvəntri] *noun* list of organisms found in a particular area

inversion [ɪn'vɜ:ʃn] *noun* (i) turning something in the opposite direction; (ii) atmospheric phenomenon, where cold air is nearer the ground than warm air, making the temperature of the air rise as it gets further from the ground, trapping pollutants between the layers of air; **inversion layer** = upper limit of a layer of warm air trying to rise through a layer of cold air

COMMENT: air normally cools at a rate of 6.4°C per 1,000 metres of altitude. During the night, the ground cools as it loses heat by radiation and the air at ground level becomes cooler than the air above. This thermal inversion can cause smog, when the cooler ground-level air cannot move because there is no wind and remains trapped with its pollutants between the earth and the warm air above it

QUOTE: the government's decision to delay the start of morning classes was to protect schoolchildren from the hazards of atmospheric inversions, which are most common on winter mornings
Environment

invertebrate [ɪn'vɜ:tɪbrət] *adjective & noun* (animal) which has no backbone (NOTE: opposite is **vertebrate**)

iodine ['aɪədiːn] *noun* chemical element which is essential to the body, especially to the functioning of the thyroid gland, found in seaweed; **tincture of iodine** = weak solution of iodine in alcohol, used as an antiseptic (NOTE: chemical symbol is **I**; atomic number is **53**)

◇ **iodize** ['aɪədaɪz] *verb* to treat *or* impregnate with iodine

ion ['aɪən] *noun* atom which has an electric charge. (Ions with a positive charge are called cations and those with a negative charge are anions); **ion exchange** = exchanging of ions between a solid and a solution; **ion-exchange filter** *or* **water softener** = device attached to the water supply to remove nitrates *or* calcium from the water; *compare* NEUTRON

◇ **ionization** [aɪənaɪ'zeɪʃn] *noun* production of atoms with electric charges; **radiation ionization** = IONIZING RADIATION

◇ **ionize** ['aɪənaɪz] *verb* to give an atom an electric charge

◇ **ionizer** *or* **negative ion generator** ['aɪənaɪzə] *noun* machine that increases the amount of negative ions in the atmosphere of a room, so counteracting the effect of positive ions

◇ **ionizing radiation** ['aɪənaɪzɪŋ reɪdɪ'eɪʃn] radiation (such as alpha particles *or* X-rays) which produces atoms with electrical charges as it passes through a medium

◇ **ionosphere** [aɪ'ɒnəsfiə] *noun* layer of the atmosphere more than 90 kilometres above the surface of the earth, composed of nitrogen (70%), oxygen (15%) and helium (15%), in which atoms are ionized by solar radiation

> COMMENT: it is believed that living organisms (including human beings) react to the presence of ionized particles in the atmosphere. Hot dry winds contain a higher proportion of positive ions than normal and these winds (such as the föhn in the Alps) cause headaches and other illnesses. If negative ionized air is introduced into an air-conditioning system, the incidence of headaches and nausea among people working in the building may be reduced

IPM = INTEGRATED PEST MANAGEMENT

IQ = INTELLIGENCE QUOTIENT

iroko [ɪ'rəʊkəʊ] *noun* African hardwood, formerly widely used, but now becoming rarer

iron ['aɪən] *noun* metallic element essential to biological life and an essential part of human diet (found in liver, eggs, etc.); **cast iron** = very hard alloy of iron, silicon and 1.7% to 4.5% carbon, which is smelted and poured into moulds to shape it; **iron ore** = rock which contains iron compounds, and from which the metal can be extracted; **iron oxide (Fe_2O_3)** = FERRIC OXIDE; **iron pyrites** *or* **iron sulphide (FeS_2)** = PYRITE; **ironworks** = place where iron is smelted and worked (NOTE: chemical symbol of iron is **Fe**; atomic number is **26**)

> COMMENT: iron is an essential part of the red pigment in red blood cells. Lack of iron in haemoglobin results in iron-deficiency anaemia. The metal and alloys made from it are magnetic. Compass needles made from it point to magnetic north

irradiate [ɪ'reɪdieɪt] *verb* (i) to subject something to radiation; (ii) to treat food with radiation to prevent it going bad

◇ **irradiation** [ɪreɪdi'eɪʃn] *noun* **(a)** spread from a centre, as nerve impulses **(b)** use of rays to treat patients *or* to kill bacteria in food; ***the irradiation of materials used in the construction of the power station caused severe damage;*** **solar irradiation** = rays which are emitted by the sun; **total body irradiation** = treating the whole body with radiation; **irradiation dose** = amount of radiation to which an organism is exposed

> COMMENT: food is irradiated with gamma rays from isotopes which kill bacteria. It is not certain, however, that irradiated food is safe for humans to eat, as the effects of irradiation on food are not known. In some countries irradiation is only permitted as a treatment of certain foods

irreversible [ɪrɪ'vɜːsɪbl] *adjective* (process) which cannot be reversed

irrigate ['ɪrɪgeɪt] *verb* to supply water to land to allow plants to grow, usually through a system of man-made channels

◇ **irrigation** [ɪrɪ'geɪʃn] *noun* supplying water to land to allow plants to grow, usually through a system of man-made channels; ***new areas of land must be brought under irrigation to meet the rising demand for food;*** **sprinkler irrigation** = supplying water to land with a spray

> COMMENT: irrigation can be carried out using powered sprinklers or simply by channelling water along small irrigation canals from reservoirs or rivers. Irrigation is not necessarily an advantage, as it can cause salinization of the soil. This happens when the soil becomes waterlogged so that salts in the soil rise to the surface. At the surface, the irrigated water rapidly evaporates, leaving the salts behind in the form of a saline crust. Irrigation also has the further disadvantage of increasing the spread of disease. Water insects easily spread through irrigation canals and reservoirs

> QUOTE: in 1950 about 2 million hectares of the Aral Sea zone were irrigated. These days, irrigation covers some 7 million hectares
> **Guardian**

island ['aɪlənd] *noun* piece of land surrounded by water, in the sea *or* in a river *or* in a lake

iso- ['aɪsəʊ] *prefix* meaning equal

◇ **isobar** ['aɪsəbɑː] *noun* line on a map linking points which are of equal barometric pressure at a given time

◇ **isobaric chart** [aɪsə'bærɪk 'tʃɑːt] *noun* weather map showing the isobars at a given time

◇ **isohaline** [aɪsəʊ'hælaɪn] *noun* line on a map linking areas of equal salt content

◇ **isohyet** [aɪsəʊ'haɪət] *noun* line on a map linking points of equal rainfall

◇ **isoleucine** [aɪsəʊ'luːsiːn] *noun* essential amino acid

◇ **isotach** [aɪsəʊ'tæʃ] *noun* line on a map linking points where the wind is blowing at the same speed

◇ **isotherm** ['aɪsəʊθɜːm] *noun* line on a map linking points of equal temperature

isolate ['aɪsəleɪt] *verb* **(a)** to keep one patient apart from others (usually because he *or* she has a dangerous infectious disease) **(b)** to identify a single virus *or* bacteria among many; ***scientists have been able to isolate the virus which causes legionnaires' disease; candida is easily isolated from the mouths of healthy adults***

◇ **isolated** ['aɪsəleɪtɪd] *adjective* (area) which is a long way from any other place *or* from any human settlement; ***specimen plants have been found in isolated points on cliffs above the North Sea***

◇ **isolation** [aɪsə'leɪʃn] *noun* keeping one patient apart from others (usually because he *or* she has a dangerous infectious disease; **radioactive waste isolation** = keeping radioactive waste separate so that it does not contaminate other things; **isolation hospital** *or* **isolation ward** = special hospital *or* special ward in a hospital where patients suffering from dangerous infectious diseases can be isolated

isotope ['aɪsətəʊp] *noun* form of a chemical element which has the same chemical properties as other forms, but a different atomic mass; **radioactive isotope** = natural *or* artificial isotope which sends out radiation, used in radiotherapy (NOTE: in formulae, an isotope is usually written as the letters of the element, with the mass and atomic number before it. Uranium-235 is therefore $^{235}_{92}U$)

> COMMENT: uranium exists in several isotopes: uranium-238 is the commonest and is used in fast breeder reactors because of its property of releasing energy very slowly under normal conditions; uranium-235 is the isotope used in fission, because of its ability to release energy rapidly

isthmus ['ɪsməs] *noun* narrow piece of land linking two larger areas of land

itai-itai ['ɪtaɪ 'ɪtaɪ] *noun* painful bone disease, caused by cadmium poisoning, found first in Japan, 'itai' is the Japanese expression for pain

ITCZ = INTERTROPICAL CONVERGENCE ZONE

IUCN = INTERNATIONAL UNION FOR THE CONSERVATION OF NATURE

ivory ['aɪvəri] *noun* smooth whitish substance forming the tusks of the elephant, walrus, etc. and which is used to make piano keys and ornaments; **ivory nut** = seed of a South American palm tree, which is used as a substitute for real ivory

> COMMENT: the main source of ivory is the African elephant which is an endangered species. Opinion is divided on whether there should be a total ban or only a partial ban on the trade in ivory. There are several ivory substitutes including plastic, ceramic and the ivory nut

Jj

J = JOULE

jam [dʒæm] *noun* blockage *or* congestion caused by too many things in a certain space, as, for example, too many motor vehicles on a road

jet [dʒet] *noun* narrow spray of liquid *or* gas; **jet engine** = engine (used on aircraft) which produces forward motion by sending out a jet of hot gases backwards; **jet fuel** = kerosene *or* fuel used in jet engines, such as on aircraft; **jet-powered** *or* **jet-propelled** = driven by jet propulsion; **jet propulsion** = making something move forward by sending out backwards a jet of hot gases *or* air *or* water; **jet stream** = (i) wide belt of fast-moving air occurring at the top limit of the troposphere (at about 15 kilometres above the earth's surface); (ii) flow of gases from a jet engine

jigger ['dʒɪgə] *noun* sandflea, a tropical insect, whose larvae enter the skin between the toes and travel under the skin, causing intense irritation

joule [dʒu:l] *noun* SI unit of measurement of energy (NOTE: usually written **J** with figures: **25J**)

> COMMENT: one joule is the amount of energy used to move one kilogram the distance of one metre, using the force of one newton. 4.184 joules equals one calorie

jungle ['dʒʌŋgl] *noun (informal)* tropical rainforest

juniper ['dʒu:nɪpə] *noun* small coniferous tree of the Northern Hemisphere, whose cones are like berries and are used as a flavouring; **juniper berry** = purple cone of the juniper, used as a flavouring

juvenile ['dʒu:vənaɪl] *adjective & noun* young (animal *or* sometimes plant); **juvenile hormone** = hormone in an insect larva which regulates its development into an adult; **juvenile phase** = period of development of a plant before it flowers *or* of an animal before it becomes adult

Kk

K 1 = KELVIN **2** *chemical symbol for* potassium

k *symbol for* KILO-

Vitamin K [vɪtəmɪn 'keɪ] *noun* vitamin found in green vegetables like spinach and cabbage, which helps the clotting of blood and is needed to activate prothrombin

kala-azar [kæləə'zɑ:] *noun* severe infection, occurring in tropical countries

> COMMENT: kala-azar is a form of leishmaniasis, caused by the infection of the intestines and internal organs by a parasite *Leishmania* spread by flies. Symptoms are fever, anaemia, general wasting of the body and swelling of the spleen and liver

kaolin ['keɪəlɪn] *noun* china clay, fine white clay used for making china and also for coating shiny paper; ***spoil heaps from kaolin workings are bright white***

karst [kɑ:st] *noun* terrain typical of limestone country, with an uneven surface and holes and cracks due to weathering

karyotype ['kæriəʊtaɪp] *noun* the chromosome make-up of a cell, shown as a diagram *or* as a set of letters and numbers

katabatic wind [kætə'bætɪk 'wɪnd] *noun* cold wind which flows downhill as the ground surface cools at night (NOTE: opposite is **anabatic**)

katadromous [kə'tædrəməs] = CATADROMOUS

kelp [kelp] *noun* variety of large brown seaweed which is a source of iodine and potash

kelvin ['kelvɪn] *noun* base SI unit of measurement of thermodynamic temperature (NOTE: temperatures are shown in kelvin without a degree sign: **20K.** On the kelvin scale 0°C is equal to 273.15K)

kerosene ['kerəsi:n] *noun* fuel distilled from petroleum, used in jet engines

ketose [kɪ'təʊz] *noun* a simple sugar

kg = KILOGRAM

khamsin ['kæmsiːn] *noun* hot wind which brings dust storms in North Africa

kieselguhr ['kiːzəlguːr] *noun* diatomaceous type of earth formed from the skeletons of diatoms

kill [kɪl] **1** *noun* **(a)** act of making someone *or* something die; ***pollutants in water are one of the main causes of fish kills*** **(b)** prey which has been killed; ***the vultures surrounded the remains of the lion's kill*** **2** *verb* to make someone *or* something die; ***she was given the kidney of a person killed in a car crash; heart attacks kill more people every year; antibodies are formed to kill bacteria***

◇ **kill off** ['kɪl 'ɒf] *verb* to kill all the individual members of a species (usually one by one); ***dodos were killed off by eighteenth century sailors***

kiln [kɪln] *noun* furnace used for making pottery *or* bricks, etc.; ***the smoke from the brick kilns was taken away by the prevailing winds;*** **cement kiln** = furnace where lime is burnt to make cement

kilo ['kiːləʊ] *abbreviation for* KILOGRAM

kilo- ['kɪləʊ] *prefix* meaning one thousand

◇ **kilocalorie** [kɪləʊ'kæləri] *noun* SI unit of measurement of heat (= 1,000 calories) (NOTE: with figures usually written **Cal.** Note that it is now more usual to use the term **joule)**

◇ **kilogram** *or* **kilo** ['kɪləgræm] *noun* base SI unit of measurement of weight (= 1,000 grams); ***two kilos of sugar; he weighs 62 kilos (62kg)*** (NOTE: with figures usually written **kg)**

◇ **kilogray** ['kɪləgreɪ] *noun* SI unit of measurement of absorbed radiation (= 1,000 grays)

◇ **kilojoule** ['kɪlədʒuːl] *noun* SI measurement of energy *or* heat (= 1,000 joules) (NOTE: with figures usually written **kJ)**

◇ **kilometre** *or US* **kilometer** [[kɪ'lɒmɪtə] *noun* unit of measurement of length (= 1,000 metres *or* approximately 0.62 miles) (NOTE: with figures usually written **km)**

◇ **kilowatt** ['kɪləwɒt] *noun* unit of measurement of electricity (= 1,000 watts) (NOTE: with figures usually written **kW) kilowatt-hour** = one thousand watts of electricity used for one hour; ***the electricity consumption is 250kWh*** (NOTE: with figures usually written **kWh)**

kinetic [kaɪ'netɪk] *adjective* concerning motion; **kinetic energy** = energy of motion

◇ **kinetics** [kaɪ'netɪks] *noun* scientific study of bodies in motion

kingdom ['kɪŋdəm] *noun* largest category in the classification of organisms; ***the largest species in the animal kingdom is the whale;*** **animal** *or* **plant kingdom** = category of all organisms classed as animals *or* plants; **mineral kingdom** = category of all non-living substances, such as minerals and rocks

> COMMENT: the different kingdoms are: Kingdom Monera, Kingdom Protista, Kingdom Plantae, Kingdom Fungi and Kingdom Animalia. See the chart in the Supplement

kJ = KILOJOULE

km = KILOMETRE

knock [nɒk] *verb (of petrol engine)* to make a loud noise as the mixture of petrol and air explodes, caused when the mixture is not rich enough in petrol; **antiknock additive** *or* **antiknocking additive** = substance (such as tetraethyl lead) which is added to petrol to prevent knocking

> COMMENT: antiknock additives increase the power of petrol but create dangerous lead pollution through exhaust fumes

knot [nɒt] *noun* unit of measurement of speed

> COMMENT: one knot is equal to 1.85km/h. Knots are used for measuring the speed of water currents, winds and ships

Kr *chemical symbol for* krypton

Krebs cycle ['krebz 'saɪkl] *noun* citric acid cycle, important series of reactions in which the intermediate products of fats, carbohydrates and amino acid metabolism, are converted to carbon dioxide and water in the mitochondria

krill [krɪl] *noun* masses of tiny shrimps which live in cold seas and form the basic diet of many marine animals including whales

krotovina [krɒ'tɒvɪnə] *noun* animal burrow that has been filled with organic or mineral material from another soil horizon

krypton ['krɪptən] *noun* inert gas found in very small quantities in the atmosphere (NOTE: chemical symbol is **Kr;** atomic number is **36)**

kW = KILOWATT

kwashiorkor [kwɒʃi'ɔːkɔː] *noun* malnutrition of small children, mostly in tropical countries, causing anaemia, wasting of the body and swollen liver; it is caused by protein deficiency in the diet, especially where cassava is the staple foodstuff, since the protein content of cassava is almost nil

kWh = KILOWATT-HOUR

Ll

lab [læb] *noun informal* = LABORATORY ***we'll send the specimens away for a lab test; the lab report is negative; the samples have been returned by the lab***

label ['leɪbəl] **1** *noun* piece of paper attached to produce, showing the price and other details; **quality label =** label which shows that the produce is of good quality, or that it meets government requirements **2** *verb* to stick a label onto something

> COMMENT: government regulations cover the labelling of food; it should show not only the price and weight, but also where it comes from, the quality grade, and a sell-by date

laboratory [lə'bɒrətri] *noun* special room where scientists can do research *or* can test chemical substances *or* can grow tissues in cultures, etc.; ***the samples of water from the reservoir have been sent to the laboratory for testing;*** **laboratory officer =** qualified person in charge of a laboratory; **laboratory technician =** person who does practical work in a laboratory and has particular care of equipment; **laboratory techniques =** methods *or* skills needed to perform experiments in a laboratory

lacustrine [lə'ku:stri:n] *adjective* referring to a lake *or* pond

lag [læg] *verb* to cover (water pipes) with an insulating material to protect them against cold *or* to stop heat escaping; ***boilers and pipes should be carefully lagged to prevent heat loss***

◇ **lagging** ['lægɪŋ] *noun* material used to lag pipes

lagoon [lə'gu:n] *noun* **(a)** shallow part of the sea in the tropics, surrounded *or* almost surrounded by reefs **(b)** artificial lake used for purifying sewage *or* the runoff from silage; **maturation lagoon =** pond used in the final stages of sewage treatment; **stabilization lagoon =** pond used to purify sewage by allowing sunlight to fall on the mixture of sewage and water

◇ **lagooning** [lə'gu:nɪŋ] *noun* creation of artificial lakes for the purifying of sewage, etc.

lake [leɪk] *noun* **(a)** large area of fresh water surrounded by land; **salt lake =** low-lying lake (not connected with the sea) with water which has a very high salt content because of evaporation and insufficient inflow of fresh water; **lake bloom =** mass of algae which develops rapidly in a lake due to eutrophication; **lake deposits =** deposits of silt on the bed of a lake; *see also* DYSTROPHIC, EUTROPHIC, MESOTROPHIC, OLIGOTROPHIC **(b)** large quantity of liquid produce stored because of overproduction; ***a wine lake*** *or* ***a milk lake*** *see also* MOUNTAIN

land [lænd] *noun* solid part of the earth's surface (as opposed to the sea); **back to the land =** encouragement given to people who once lived in the country and moved to urban areas to find work, to return to the country; **land breeze =** light wind which blows from the land towards the sea, for example, during the day when the land is warm; **land burial =** depositing waste in a hole in the ground; **land clearance =** removal of trees, undergrowth, etc. in preparation for ploughing, building, etc.; **land consolidation =** joining small plots of land together to form larger farms *or* large fields; **land disposal =** getting rid of waste on land, as opposed to at sea; **land improvement =** making the soil more fertile; **land reclamation =** bringing back into productive use a piece of land which was not usable before (e.g. a waste site); **land reform =** government policy of splitting up agricultural land and dividing it up between those people who do not own any land; **land use =** way in which the land is used; ***they are carrying out a study of land use in northern areas***

landfill ['lændfɪl] *noun* way of disposing of rubbish by putting it into holes in the ground and covering it with earth; **lined landfill =** hole in the ground covered on the inside with nylon sheets to prevent leaks of dangerous liquids from waste deposited there; **landfill site =** area of land where domestic rubbish is put into holes in the ground and covered with earth; ***the council has decided to use the old gravel pits as a landfill site; landfill sites can leak pollutants into the ground water; more than 2,000 landfill sites have been closed due to tighter environmental laws; landfill sites, if properly constructed, can be used to provide gas for fuel*** (NOTE: in US English this is called **sanitary landfill**)

◇ **landfilling** ['lændfɪlɪŋ] *noun* disposing of rubbish by putting it into holes in the ground and covering it with earth

landlocked ['lændlɒkt] *adjective* (lake) which is entirely surrounded by land

◇ **landmass** ['lændmæs] *noun* large area of land (as opposed to the ocean); ***the continental landmass of the United States***

◇ **landrace** ['lændreɪs] *noun* native species of plant *or* animal which has not been cultivated

Landsat ['lændsæt] *noun* series of US satellites which scan the land surface of the Earth, particularly the vegetation cover

landscape ['lændskeɪp] *noun* scenery *or* general shape, structure and contents of the surface of an area of land; **urban landscape** = landscape of a town; **landscape design** *or* **landscape planning** *or* **landscaping** = plan of how to lay out a landscape, where to plant trees and shrubs, etc.

◇ **landslide** *or* **landslip** ['lændslaɪd *or* 'lændslɪp] *noun* sudden fall of large amounts of soil and rocks down the side of a mountain *or* of waste matter down the side of a spoil heap

lapse [læps] *noun* short period of time which separates two happenings

◇ **lapse rate** ['læps 'reɪt] *noun* rate at which temperature changes according to altitude; **adiabatic lapse rate** = rate of temperature change in rising air (10°C per thousand metres for dry air and 5.8°C per thousand metres for damp air); **normal environmental lapse rate** *or* **temperature lapse rate** = rate at which the temperature of the air falls as one rises above the earth (about 6.4°C per thousand metres, under conditions where there are no upward air currents or wind)

larch [lɑːtʃ] *noun* a coniferous European softwood tree *(Larix decidua)* ; a fast-growing tree used as a timber crop

larva ['lɑːvə] *noun* stage (caterpillar *or* grub) in the development of an insect *or* animal, after the egg has hatched but before the animal becomes adult (NOTE: plural is **larvae)**

◇ **larval** ['lɑːvəl] *adjective* referring to larvae; **larval stage** = early stage in the development of an insect after it has hatched from an egg

laser ['leɪzə] *noun* instrument which produces a highly concentrated beam of light which can be used to cut *or* attach materials; **laser beam** = highly concentrated beam of light produced by a laser

Lassa fever ['læsə 'fiːvə] *noun* highly infectious and often fatal virus disease found in Central and West Africa

latent ['leɪtənt] *adjective* present but not yet developed; hidden; **latent heat** = heat taken in *or* given out when a solid changes into a liquid *or* vapour *or* when a liquid changes into a vapour at a constant temperature and pressure

lateral ['lætərəl] *adjective* at the side; **lateral moraine** = deposit of sand and gravel left at the sides of a glacier as it moves forwards

laterite ['lætəraɪt] *noun* hard rock-like clay found in the tropics, formed when latosol dries out; ***in some countries houses are built of laterite***

◇ **lateritic** [lætə'rɪtɪk] *adjective* (soil) which contains laterite

◇ **laterization** [lætəraɪ'zeɪʃn] *noun* process of weathering tropical soil into hard laterite

◇ **laterize** ['lætəraɪz] *verb* to weather tropical soil into hard laterite

◇ **latosol** ['lætəsɒl] *noun* type of reddish soft soil, found in tropical areas, and characterized by deep weathering and hydrous oxide material

> COMMENT: when tropical rainforests are cleared, the soil beneath rapidly turns to laterite as nutrients are leached out by rain. Laterized land is incapable of cultivation and such areas turn to desert

latex ['leɪteks] *noun* (i) white fluid from a plant such as the poppy; (ii) thick white fluid from the rubber tree, which is treated and processed to make rubber

latitude ['lætɪtjuːd] *noun* position on the earth's surface measured in degrees north *or* south of the equator; ***the pine tree grows in temperate latitudes;*** **line of latitude** *or* **parallel of latitude** = imaginary line running round the earth, linking points at an equal distance from the equator; **high latitudes** = areas near the poles; **low latitudes** = areas near the equator; **mid-latitudes** = areas halfway between the poles and the equator; **at a latitude of 46°N** = at a position on the earth's surface which is 46 degrees north of the equator

> COMMENT: together with longitude, latitude is used to indicate an exact position on the earth's surface. Latitude is measured in degrees, minutes and seconds. The centre of London is latitude 51°30'N, longitude 0°5'W. The lines of latitude are numbered and some of them act as national boundaries: the 49th parallel marks most of the border between the USA and Canada

lava ['lɑːvə] *noun* molten rock and minerals which flow from an erupting volcano and solidify into various types of igneous rock; **lava flow** = lava moving down the sides of a volcano; ***lava flows from the volcano destroyed sugar plantations***

law [lɔː] *noun* (i) rule *or* set of rules by which a country is governed; (ii) basic principle of science *or* mathematics; **law of supply and demand** = basic principle of economics that there is a balance between things produced and required in a society, especially as affecting prices; **law of (universal) gravitation** = basic principle of physics that any two masses attract each other with a force equal to a constant multiplied by the product of the two masses and divided by the square of the distance between them; **energy conservation law** = rule which makes it illegal to waste energy; **by law** = legally; **under British** *or*

French, etc. law = according to British *or* French, etc. law

lawrencium [lə'rensiəm] *noun* one of the transuranic elements (NOTE: chemical symbol is **Lr;** atomic number is **103)**

layer ['leɪə] **1** *noun* **(a)** flat area of a substance under *or* over another area (NOTE: in geological formations, layers of rock are called **strata;** layers of soil are called **horizons) (b)** stem of a plant which has made roots where it touches the soil **2** *verb* to propagate a plant by bending a stem down until it touches the soil and letting it form roots there

LD50 = LETHAL DOSE 50%

leach [li:tʃ] *verb* to wash a substance out of the soil by passing water through it; ***excess chemical fertilizers on the surface of the soil leach into rivers and cause pollution; nitrates have leached into ground water and contaminated the water supply***

◇ **leachate** ['li:tʃeɪt] *noun* (i) substance which is washed out of the soil; (ii) liquid which forms at the bottom of a landfill site

◇ **leaching** ['li:tʃɪŋ] *noun* process by which a substance is washed out of the soil by water passing thought it; **leaching field** = area round a septic tank with pipes which allow the sewage to drain away underground

lead [led] *noun* very heavy soft metallic element which is poisonous in compounds and is used as a shield in nuclear reactors (NOTE: chemical symbol is **Pb;** atomic number is **82)** **airborne lead** = particles of lead carried in polluted air; **tetraethyl lead** *or* **lead tetraethyl** **($Pb(C_2H_5)_4$)** = additive added to petrol to prevent knocking

◇ **lead-based additive** ['ledbeɪst] *noun* substance, such as tetraethyl lead, added to petrol to prevent knocking

◇ **leaded petrol** ['ledɪd 'petrɒl] *noun* petrol to which a lead-based additive, such as tetraethyl lead, has been added to prevent knocking

◇ **lead-free** ['led'fri:] *adjective* (paint, etc.) which has no lead in it; **lead-free petrol** *or* **unleaded petrol** = petrol with a low octane rating, which has no lead additives in it and therefore creates less lead pollution in the atmosphere

◇ **lead poisoning** [led 'pɔɪzənɪŋ] *noun* poisoning caused by taking in lead salts

COMMENT: small children are particularly vulnerable to lead pollution, as lead affects brain development. Lead can enter the diet through drinking water which has been kept in lead pipes; it can also enter the body through paint (children's toys must be painted with lead-free paint). Lead is added to petrol to prevent knocking and give more power, but it causes lead fumes which are toxic and can be avoided by using lead-free petrol. Lead poisoning also occurs in birds, such as swans, which have eaten the lead pellets used by fishermen to weight their lines

leader stroke ['li:də 'strəʊk] *noun* first lightning flash, which makes a path for other flashes to follow

leaf [li:f] *noun* green, usually flat, part of a plant, growing from a stem, whose purpose is to activate photosynthesis, the means by which the plant gets energy from sunlight

leak [li:k] **1** *noun* escape of liquid *or* gas from a sealed container into the environment; amount of liquid *or* gas escaped; ***the government is worried about the leaks of radioactive gas into the atmosphere;*** **leak rate** = speed at which a liquid *or* gas escapes from a sealed container **2** *verb (of liquid or gas)* to escape from a sealed container into the environment; ***gallons of lethal acid leaked into the river from the chemical plant***

◇ **leakage** ['li:kɪdʒ] *noun* = LEAK 1 ***the monitoring team reported the leakage of oil from the tanker***

> QUOTE: children were warned to stay away from three watercourses yesterday after several hundred gallons of sulphuric acid leaked into a stream
> **Guardian**

lean [li:n] *adjective* (mixture of fuel and air) which has a relatively low ratio of fuel to air

◇ **lean-burn engine** ['li:n 'bɜ:n] *noun* type of internal combustion engine adapted to use less fuel than normal engines, and so release less carbon monoxide and nitrogen oxide into the atmosphere

lee [li:] *adjective & noun* (side) which is protected from the wind; ***the trees on the lee side of the house grow better than those on the windward side***

◇ **leeward** ['li:wəd] *adjective* protected from the wind *or* away from the wind

legionnaires' disease [li:dʒiə'neəz dɪ'zi:z] *noun* bacterial disease similar to pneumonia

COMMENT: the disease is thought to be transmitted in droplets of moisture in the air, and so the bacterium is found in central air-conditioning systems. It can be fatal to old or sick people and so is especially dangerous if present in a hospital

legislation [ledʒɪs'leɪʃn] *noun* laws by which a country is governed; **building legislation** *or* **planning legislation** = laws controlling all aspects of putting up new buildings *or* altering existing buildings

legume ['legju:m] *noun* any member of the plant family *Leguminosae* which produce seeds in pods (such as pea and bean plants); ***grass and legume associations are common in European pastureland***

◇ **Leguminosae** [legu:mɪn'əʊsɪ] *noun* family of plants (including pea and bean plants) which produce seeds in pods

◇ **leguminous** [lɪ'gju:mɪnəs] *adjective* (plant) which produces seeds in pods

COMMENT: there are many species of legume, and some are particularly valuable because they have root nodules that contain nitrogen-fixing bacteria. Such legumes have special value in maintaining soil fertility and are used in crop rotation. Peas, beans, peanuts, alfalfa, clover and vetch are all legumes

QUOTE: the cowpea is a leguminous crop grown in tropical or semi-tropical climates for fodder, pulse, vegetable and green manuring
Indian Farming

leishmaniasis [li:ʃmə'naɪəsɪs] *noun* any of several diseases (such as kala-azar) caused by the parasite *Leishmania,* one form giving disfiguring ulcers, another attacking the liver and bone marrow

leisure ['leʒə] *noun* free time which can be used for relaxation, sport, etc.; **leisure centre** = place which provides equipment *or* resources for activities such as swimming, playing games, canoeing, etc.; **leisure facilities** = equipment *or* resources (such as a sports centre, swimming pool, park, cinema, etc.) which can be used during free time

lentic ['lentɪk] *adjective* referring to stagnant water

Lepidoptera [lepɪ'dɒptərə] *noun* order of insects which includes butterflies and moths

-less [ləs] *suffix* meaning without; **a fishless lake** = a lake with no fish

lessen ['lesən] *verb* to reduce *or* to make less strong; ***the company is taking steps to lessen the release of pollutants into the atmosphere***

lethal ['li:θəl] *adjective* which can kill; ***these fumes are lethal if inhaled;*** **lethal dose 50% (LD50)** *or* **mean lethal dose** = dose of a substance which will kill half the organisms which absorb it; **lethal gene** = gene which can kill the person who inherits it; **lethal yellowing (LY)** = disease which attacks and kills coconut palms

leucine ['lu:si:n] *noun* essential amino acid

levee ['levi] *noun (in the USA)* dyke *or* embankment built up along the bank of a river to prevent flooding

level ['levəl] **1** *adjective* flat *or* horizontal; ***the mountains give way to miles of level grassy plain*** **2** *noun* amount; ***smoke levels in the city are only 10% of what they were twenty years ago; forest will gradually die if pollution levels remain constant; present levels of water pollution are not acceptable; the fish contained higher than permitted levels of mercury;*** **concentration level** = amount of a substance in a solution *or* in a certain volume; **noise level** *or* **sound level=** loudness of a noise *or* sound which can be measured; **trophic level** = one of the levels in a food chain **3** *verb* to make something become flat

QUOTE: atmospheric levels of carbon dioxide are increasing at about 0.35 per cent a year
New Scientist

ley [leɪ] *noun* field in which crops are grown in rotation with periods when the field is under pasture

COMMENT: ley farming is an essential part of organic farming: pasture land is fertilized by the animals which graze on it and then is ploughed for crop growing. When the land has been exhausted by the crops, it is put back to pasture to recover

LH2 [el'eɪtʃ'tu:] *noun* liquid hydrogen, proposed as an alternative fuel for use in cars and aircraft

Li *chemical symbol for* lithium

liana [lɪ'ɑ:nə] *noun* climbing plant found in tropical rainforest

lias ['laɪəs] *noun* type of rock formation consisting of shale and limestone

lichen ['laɪkən] *noun* primitive plant which grows on the surface of stones *or* trunks of trees and can survive in arctic climates

COMMENT: lichens are formed of two organisms: a fungus which provides the outer shell and algae which provide chlorophyll and give the plant its colour. Lichens are very sensitive to pollution, especially sulphur dioxide, and act as indicators for atmospheric pollution. They also provide food for many arctic animals

life [laɪf] *noun* **(a)** state of active metabolism *or* of being alive; **life cycle** = all the changes an organism goes through between a certain stage in its development and the same stage in the next generation; **life expectancy** = number of years a person *or* animal, etc. is likely to live; **life forms** = different types of organisms; **life history** = all the changes an organism goes through from fertilization to death; **life science** = science (such as biology *or* botany) which studies living organisms; **life style** = habits, behaviour, attitudes, etc. of a particular individual *or* group; **life system** = part of an ecosystem which is formed of a living organism and the parts of the environment which support it; **life table** = chart showing how long a person *or* animal *or* plant is likely to live; **life-threatening disease** = disease which may kill a person *or* animal; **life zone** = place *or* area in which the type and number of organisms differ slightly from neighbouring areas because of variations in the environmental conditions **(b)** plants, animals and all kinds of living organisms; ***deforestation has destroyed animal life in the area;*** **marine life** = animals and plants which live in the sea; *see also* WILDLIFE

light [laɪt] **1** *adjective* **(a)** not heavy; **light industry** = industry which makes small *or* lightweight products; **light water** = ordinary water (as opposed to heavy water *or* deuterium oxide) used as a coolant in some types of power station; **light water reactor (LWR)** = nuclear reactor which uses ordinary water as a coolant **(b)** bright so that one can see well; ***at six o'clock in the morning it was just getting light*** **2** *noun* radiation in the part of the electromagnetic spectrum which eyes are adapted to receive; ***the light of the sun makes plants green; there's not enough light in here to take a photo;*** **light adaptation** = changes in the eye to adapt to an abnormally bright or dim light *or* to adapt to normal light after being in darkness; **light reflex** = reaction of the pupil of the eye which changes size according to the amount of light going into the eye; **light waves** = waves travelling in all directions from a source of light which stimulate the retina and are visible

lightning ['laɪtnɪŋ] *noun* discharge of electricity between clouds and the earth, which gives a bright flash and makes the sound of thunder; **sheet lightning** = lightning where the flash cannot be seen, but where the clouds are lit by it; **lightning arrester** = device which prevents surges of the electrical current which are caused when lightning strikes a building and which can damage equipment; **lightning conductor** *or* **lightning rod** = length of metal running down the outside wall of a building to the ground, which acts as a channel for the electric current when lightning strikes the building

lignite ['lɪgnaɪt] *noun* brown coal *or* type of soft coal with a low carbon content

lime [laɪm] **1** *noun* **(a)** calcium compound (calcium oxide) made from burnt limestone, used to spread on soil to reduce acidity and add calcium, in the composition of cement and in many industrial processes; **lime slurry** = mixture of lime and water added to hard water to make it softer; **lime treatment** = spreading lime on soil to reduce acidity and add calcium **(b)** common European hardwood tree *(Tulia* sp) **(c)** *Citrus aurantifolia,* a citrus fruit tree, with green fruit similar to, but smaller than, a lemon **2** *verb* to treat acid soil by spreading lime on it

> QUOTE: pulses are generally susceptible to soils which are acidic in nature and need liming if the soil pH is below 6
>
> **Indian Farming**

limestone ['laɪmstəʊn] *noun* common sedimentary rock

◇ **limewater** ['laɪmwɔːtə] *noun* solution of calcium hydroxide in water

◇ **liming** ['laɪmɪŋ] *noun* spreading lime on soil to reduce acidity and add calcium

> COMMENT: limestone is formed of calcium minerals and often contains fossilized shells of sea creatures. It is used in agriculture and building. Limestone is porous in its natural state and may form large caves by being weathered by water

limit ['lɪmɪt] **1** *noun* furthest point *or* place beyond which you cannot go; ***they have set strict limits on the amount of fish which foreign fishing boats are allowed to catch;*** **freshwater limit** = place in the course of river at which the salinity of the water has decreased below a certain level and upstream of which the water can be considered fresh **2** *verb* to set a limit to something; ***the government has limited the number of barrels of oil to be extracted each day;*** **limiting factor** = factor which limits the growth of an organism

limn- [lɪmn] *prefix* referring to fresh water

◇ **limnetic** [lɪm'netɪk] *adjective* referring to deep fresh water; **limnetic zone** = area of deep water away from the edge of a lake, in which plants cannot live, but where phytoplankton can exist

◇ **limnic** ['lɪmnɪk] *adjective* (deposits) in fresh water

◇ **limnology** [lɪm'nɒlədʒi] *noun* study of fresh water (such as rivers and lakes)

lindane ['lɪndeɪn] *noun* organochlorine pesticide, which is harmful to some animals such as bees and fish. Lindane is used in Britain as a farm insecticide and as a chemical for treating wood. It is banned in the USA, and its use is being examined in the UK (NOTE: also called **HCH**)

line [laɪn] **1** *noun* row of things, one after another; **line squall** *or* **line storm** = belt of thunderstorms advancing in a line **2** *verb* to cover the inside of a container to prevent the contents escaping; ***landfill sites are lined with nylon to prevent leaks of dangerous liquids***

◇ **liner** ['laɪnə] *noun* material (such as nylon sheets) used to line something

link [lɪŋk] *noun* something which joins two things; **critical link** = organism in a food chain which is responsible for taking up and storing nutrients which are then passed on down the chain

Linnaean system [lɪ'ni:ən 'sɪstəm] *noun* scientific system of naming organisms devised by the Swedish scientist, Carolus Linnaeus (1707-1778)

> COMMENT: the Linnaean system (or binomial classification) gives each organism a name made up of two Latin words. The first is a generic name referring to the genus to which the organism belongs, and the second is a specific name referring to the particular species. Organisms are usually identified by using both their generic and specific names, e.g. *Homo sapiens* (man), *Felix catus* (domestic cat) and *Sequoia sempervirens* (redwood). A third name can be added to give a subspecies. The generic name is written or printed with a capital letter. Both names are usually given in italics or are underlined if written or typed

linoleic acid [lɪnə'li:ɪk 'æsɪd] *noun* one of the essential fatty acids which cannot be synthesized and has to be taken into the body from food (such as vegetable oil)

linolenic acid [lɪnə'lenɪk 'æsɪd] *noun* one of the essential fatty acids

lipid ['lɪpɪd] *noun* fat *or* fatlike substance which exists in human tissue and forms an important part of the diet; **lipid metabolism** = chemical changes where lipids are broken down into fatty acids

> COMMENT: lipids are not water-soluble. They float in the blood and can attach themselves to the walls of arteries causing atherosclerosis

liposoluble [lɪpəʊ'sɒlju;bl] *adjective* which can dissolve in fat

liquid ['lɪkwɪd] *adjective & noun* matter (like water) which is not solid and is not a gas and has no shape; **liquid hydrogen** = alternative fuel proposed for use in cars and aircraft; **liquid fertilizers** = solutions of normal solid fertilizers; liquid fertilizers are easier, quicker and cheaper to handle and apply than solid fertilizers; **liquid manure** = manure consisting of dung and urine in a liquid form (manure in semi-liquid form is slurry); **liquid paraffin** = oil distilled from petroleum and used as a laxative (NOTE: US English is **mineral oil)**

◇ **liquefaction** [lɪkwɪ'fækʃn] *noun* making a solid *or* gas into liquid

◇ **liquefy** ['lɪkwɪfaɪ] *verb* to make a gas into liquid; **liquefied natural gas (LNG)** = natural gas, extracted from the earth, cooled and transported in containers; **liquefied petroleum gas (LPG)** = gas (either propane *or* butane *or* a combination of both) produced by refining crude petroleum oil and used for domestic heating, camp stoves, etc.

listeria [lɪs'tɪəriə] *noun* genus of bacteria found on domestic animals and in prepared foods, which can cause meningitis

lith- [lɪθ] *prefix* meaning stone

◇ **lithosere** ['lɪθəʊsi:ə] *noun* succession of communities growing on rock

◇ **lithosol** ['lɪθəʊsɒl] *noun* soil which forms on the surface of rock, with no soil horizons

◇ **lithosphere** ['lɪθəʊsfiə] *noun* (i) the earth's solid surface, together with the molten interior; (ii) solid surface of the earth which lies above the asthenosphere

◇ **lithospheric** [lɪθə'sferɪk] *adjective* referring to the lithosphere; **lithospheric plates** = solid masses of the earth's surface, of the size of continents, which move slowly in the low velocity zone

lithium ['lɪθiəm] *noun* soft silvery metallic element, the lightest known metal, used in batteries; **lithium battery** = electric battery containing lithium (NOTE: chemical symbol is **Li**; atomic number is **3)**

litmus ['lɪtməs] *noun* organic matter used to indicate acidity: it becomes red when the pH factor falls below pH7, indicating acid, and becomes blue when pH is above pH7, indicating alkaline

litre *or US* **liter** ['li:tə] *noun* unit of measurement of liquids (= 1.76 pints) (NOTE: with figures usually written **l: 2.5l)**

litter ['lɪtə] **1** *noun* **(a)** (i) dead leaves lying on the floor of a forest; (ii) rubbish left by people; **litter basket** *or* **litter bin** = container in a public place in which rubbish is disposed of **(b)** group of young mammals born to one mother at the same time **(c)** bedding for livestock (straw is the best type of litter) **2** *verb* **(a)** to lie all over the place; ***the valley is littered with huge boulders*** **(b)** to give birth; ***the bear litters in early spring***

littoral ['lɪtərəl] **1** *adjective* referring to the coast; **littoral current** = current which moves along the shore; **littoral drift** = movement of sand as it is carried by the sea along the coastline; **littoral zone** = (i) area of fresh water at the edge of a lake where plants can exist; (ii) area of the sea and shore between the high and low water marks **2** *noun* coast *or* part of the land at the edge of the sea

live [laɪv] **1** *adjective* **(a)** active *or* carrying out metabolism; **live well** = well from which water *or* oil is being extracted **(b)** carrying electricity; ***he was killed when he touched a live wire*** **2** *verb* to be alive; ***animals which live partly in water and partly on land are amphibians;*** **living conditions** = surroundings *or* circumstances in which a person lives; **living environment** = part of the environment made up of living organisms

◇ **live off** ['lɪv 'ɒf] *verb* to exist by eating (something); ***these fish live off the debris which sinks to the bottom of the lake***

◇ **live on** ['lɪv 'ɒn] *verb* **(a)** to exist by eating (something); ***most apes live on berries and roots*** **(b)** to exist on the surface of (something); ***lice live on the skin of their host***

◇ **livestock** ['laɪvstɒk] *noun* cattle and other farm animals which are reared to produce meat *or* milk *or* other products; ***livestock production has increased by 5%; pastoralists move their livestock from place to place to find new grazing***

LNG = LIQUEFIED NATURAL GAS

load [ləʊd] *noun* **(a)** cargo carried by a ship *or* aircraft *or* truck; **load on top** = way of carrying crude oil, where water used to clean out tanks is left in a ballast tank and crude oil is carried on top of it **(b)** amount of something which is carried; solid material carried along by water *or* by wind *or* by a glacier; **critical load** = highest level of pollution which will not cause permanent harm to the environment

loam [ləʊm] *noun* good dark soil, with medium-sized grains of sand, which crumbles easily and is very fertile

◇ **loamy** ['ləʊmi] *adjective* (soil) which is dark, crumbly and fertile

lobby ['lɒbi] **1** *noun* group of people *or* pressure group trying to influence the opinions of politicians, local officials and businessmen on a particular issue; **car lobby** = group of people who try to persuade politicians that cars should be encouraged and not restricted; **environmentalist lobby** = group of people who try to persuade politicians that the environment must be protected, that pollution must be stopped, etc. **2** *verb* to ask someone (such as a politician *or* local official) to do something on your behalf; ***they went to lobby their MP about the plan to build houses on the Green Belt***

◇ **lobbyist** ['lɒbɪɪst] *noun* person who is paid by a pressure group to act on the group's behalf

local ['ləʊkəl] *adjective* concerning a particular area; **local authority** = *GB* official body which controls a particular area of a country; **local government** = (i) *GB* government of a particular area of a country; (ii) *US* = LOCAL AUTHORITY

location [lə'keɪʃn] *noun* (i) place, site; (ii) action of finding a place *or* site *or* of placing *or* siting; ***several locations are being considered for the new airport***

loch [lɒk] *noun (in Scotland)* lake; **Loch Ness monster** = large prehistoric aquatic animal believed by some people to live in Loch Ness, in Scotland

lockjaw ['lɒkdʒɑː] *noun see* TETANUS

locust ['ləʊkəst] *noun* flying insect which occurs in subtropical areas, flies in swarms and eats large amounts of vegetation

lode [ləʊd] *noun* deposit of metallic ore

loess ['ləʊɪs] *noun* fine fertile soil formed of tiny clay and silt particles deposited by the wind

log [lɒg] *noun* large piece of wood cut from the trunk *or* from a main branch of a tree

◇ **logger** [lɒgə] *noun* person who cuts down trees

◇ **logging** ['lɒgɪŋ] *noun* cutting down trees; **logging company** = company which cuts down trees and transports them for sale as timber *or* for paper manufacture; **logging residues** = material left on the ground after cut logs have been removed

> QUOTE: logging is central to deforestation in the tropics. Loggers open up the forest for local people to exploit
>
> **Environment Now**

longitude ['lɒnʒɪtjuːd] *noun* position on the earth's surface measured in degrees east *or* west; **line of longitude** = meridian *or* imaginary line on the surface of the earth running from the North Pole to the South Pole, at right angles to the equator

> COMMENT: longitude is measured from Greenwich (a town in England, just east of London) and, together with latitude, is used to indicate an exact position on the earth's surface. Longitude is measured in degrees, minutes and seconds. The centre of London is latitude 51°30'N, longitude 0°5'W

long-lived ['lɒŋ 'lɪvd] *adjective* which lives a long time

long-range ['lɒŋ 'reɪnʒ] *adjective* (weather forecast) which covers a period more than five days ahead

◇ **longshore** ['lɒŋʃɔː] *adjective* (current) which flows along the shore; **longshore drift** = movement of sand particles along the shore, caused by longshore currents

◇ **long-term** ['lɒŋ'tɜːm] *adjective* which lasts for a long time; ***the long-term effects of exposure to radiation from power lines are not yet known;*** **long-term consequences** = results which will not be known for a long time

looping ['luːpɪŋ] *noun* situation where a plume of smoke from a tall chimney is brought down to ground level by air currents and then rises again

loss [lɒs] *noun* not having something any more; **energy loss** = amount of energy lost; **heat loss** = amount of heat lost (as through inadequate insulation); **vegetation loss** = death of plants, as caused by pollution, clearfelling, etc.

lough [lɒk] *noun (in Ireland)* lake

low [ləʊ] **1** *adjective & adverb* near the bottom *or* towards the bottom; not high; ***the temperature is too low here for oranges to grow;*** **low blood pressure** *or* **hypotension** condition where the pressure of the blood is abnormally low; **low-calorie diet** = diet with few calories (which can help a person to lose weight); **low-fat diet** = diet with little animal fat (for example, to help skin conditions); **low-level waste** = waste which is only slightly radioactive and does not cause problems for disposal; **low-octane petrol** = petrol which does not contain very much octane (a hydrocarbon) and therefore produces less pollution; **area of low pressure** = cyclone *or* area in which the atmospheric pressure is low and around which the air turns in the same direction as the earth; **low-velocity zone** = area of the interior of the earth (below the crust) where earthquake shock waves travel particularly slowly; **low-waste technology** = technology which produces little waste; **low water** = point when the level of the sea *or* of a river, etc. is at its lowest **2** *noun* depression *or* area of low atmospheric pressure usually accompanied by rain; ***a series of lows are crossing the Atlantic towards Iceland***

◇ **lower** ['ləʊə] *verb* to make something go down; to reduce; ***they covered the patient with wet cloth to try to lower his body temperature***

◇ **low-grade** ['ləʊgreɪd] *adjective* poor *or* not of very high quality; **low-grade ore** = ore which contains a small percentage of metal; **low-grade petrol** = petrol with a low octane rating, which produces less pollution

◇ **lowland** ['ləʊlænd] *noun* area of low land (as opposed to hills and mountains *or* highlands); ***vegetation in the lowlands*** *or* ***in the lowland areas is sparse***

LPG = LIQUEFIED PETROLEUM GAS

Lr *chemical symbol for* lawrencium

lubricant ['lu:brɪkənt] *noun* oily *or* greasy substance applied to moving parts (as in an engine) to make them run smoothly

lubricating oil ['lu:brɪkeɪtɪŋ 'ɔɪl] *noun* oil applied to moving parts (as in an engine) to make them run smoothly

lucerne [lu'sɜ:n] *noun* plant *(Medicago sativa)* of the Leguminosae family, grown to use as fodder

> COMMENT: lucerne is the most important forage legume; it is called lucerne in Europe, Oceania and South Africa, and elsewhere it is called alfalfa. Lucerne is perennial, drought-resistant and rich in protein. It is mainly used for cutting, either for green feed or for hay or silage

lumber ['lʌmbə] *noun (in the USA and Canada)* timber, trees which have been cut down

◇ **lumberjack** ['lʌmbədʒæk] *noun (in the USA and Canada)* logger, person who cuts down trees

lunar ['lu:nə] *adjective* referring to the moon; **lunar eclipse** = situation when the earth passes between the sun and the moon and the shadow of the earth falls across the moon, so cutting off the moon's light; **lunar phases** = changes in the appearance of the moon as it moves from new to full and back again every 29 days. (The phases are: new moon, first quarter, full moon and last quarter)

lush [lʌʃ] *adjective* (vegetation) which is thick and rich; ***the cattle were put to graze on the lush grass by the river; the lush tropical vegetation rapidly covered the clearing***

LWR = LIGHT WATER REACTOR

LY = LETHAL YELLOWING

lymph [lɪmf] *noun* colourless liquid containing white blood cells, which circulates in the body, carrying waste matter away from tissues to the veins, and is an essential part of the body's defence against infection; **lymph node** *or* **lymph gland** = mass of tissue which produces white blood cells and filters waste matter (such as infection) from the lymph as it passes through

lyophilize [laɪ'ɒfɪlaɪz] *verb* to freeze-dry food, a method of preserving food by freezing it rapidly and drying in a vacuum

lysine ['laɪsi:n] *noun* essential amino acid in protein foodstuffs, essential for animal growth

Mm

m = METRE

MAC = MAXIMUM ALLOWABLE CONCENTRATION

mackerel sky ['mækrəl 'skaɪ] *noun* pattern of wavy cirrocumulus or altocumulus cloud with holes which looks like the body markings of mackerel fish

macro- ['mækrəʊ] *prefix* meaning large (NOTE: opposite is **micro-**)

macrobiotics [mækrəʊbaɪ'ɒtɪks] *noun* system of eating which promotes the use of food without artificial additives *or* preservatives, especially whole grains, fruit and vegetables; **macrobiotic food** = food which has been produced naturally without artificial additives *or* preservatives

COMMENT: macrobiotic diets are usually vegetarian and are prepared in a special way; they consist of beans, coarse flour, fruit and vegetables

◇ **macroclimate** ['mækrəʊklaɪmət] *noun* climate covering a large area

◇ **macroeconomic** [mækrəʊiːkə'nɒmɪk] *adjective* concerning macroeconomics

◇ **macroeconomics** [mækrəʊiːkə'nɒmɪks] *noun* branch of economics dealing with broad general aspects, such as national income *or* investment, etc.; *compare* MICROECONOMICS

◇ **macronutrient** [mækrəʊ'njuːtriənt] *noun* nutrient which an organism uses in very large quantities, such as oxygen, carbon, hydrogen, nitrogen, phosphorus, potassium, calcium, magnesium and iron

◇ **macrophyte** ['mækrəʊfaɪt] *noun* large water plant (as opposed to phytoplankton)

◇ **macroplankton** [mækrəʊ'plæŋktən] *noun* plankton of about 1mm in length

mad cow disease [mæd 'kaʊ dɪ'ziːz] *see* BOVINE SPONGIFORM ENCEPHALOPATHY

magma ['mægmə] *noun* molten substance in the earth's mantle, which escapes as lava during volcanic eruptions and solidifies to form igneous rocks

◇ **magmatic** [mæg'mætɪk] *adjective* referring to magma

COMMENT: magma is formed of silicate materials which include crystals and dissolved gases

magnesium [mæg'niːziəm] *noun* chemical element, a white metal which is used in making alloys and is also an essential element in biological life: in human diets it is found in green vegetables (NOTE: chemical symbol is **Mg**; atomic number is **12**)

◇ **magnesium oxide (MgO)** [mæg'niːziəm 'ɒksaɪd] *noun* white substance with no taste, used as a laxative

magnet ['mægnɪt] *noun* piece of iron *or* alloy containing iron, with the property of attracting other pieces of iron; *see also* ELECTROMAGNET

◇ **magnetic** [mæg'netɪk] *adjective* having the attraction of a magnet; **magnetic anomaly** = way in which the local magnetic field differs from the normal magnetic field in a certain area; **magnetic attraction** = power of a body to attract other substances to it; **magnetic declination** *or* **magnetic variation** = angle of difference between the direction of the North Pole and that of the North Magnetic Pole; **magnetic field** = area round a body which is under the influence of its magnetic effect. (The Earth's magnetic field is concentrated round the two magnetic poles); **magnetic north** = direction in which the needle of a compass points (magnetic North Pole), as opposed to true north (North Pole); **magnetic pole** = one of the two poles of the earth (near to, but not identical with, the geographical poles) which are the centres of the earth's magnetic field and to which a compass points; **magnetic storm** = sudden disturbance of the earth's magnetic field which affects compasses and radio and TV waves

◇ **magnetism** ['mægnətɪzm] *noun* (i) property of attraction possessed by a naturally magnetic substance *or* by a conductor carrying an electric current (electromagnet); (ii) study of magnetic properties

◇ **magnetohydrodynamics (MHD)** [mægniːtəʊhaɪdrəʊdaɪ'næmɪks] *noun* scientific study of charged gases and liquids in a magnetic field which may be used to generate electricity

magnification [mægnɪfɪ'keɪʃn] *noun* measure of increase, especially in the apparent size of an image seen through a microscope; **biological magnification** = way in which a pollutant increases in concentration at each level of a food chain

magnox ['mægnɒks] *noun* alloy of magnesium, aluminium and other metals, used to surround uranium fuel rods in a nuclear reactor; **magnox (power) station** = nuclear power station with a gas-cooled reactor in which the uranium fuel rods are surrounded with magnox; **magnox reactor** = type of gas-cooled nuclear reactor in which the uranium fuel rods are surrounded with magnox

COMMENT: the first magnox power station was built in the UK in 1956. The safety record has been very good, but they are now coming to the end of their commercial life and decommissioning began in 1990

mahogany [mə'hɒgəni] *noun* dark tropical hardwood, now becoming rare

main [meɪn] **1** *adjective* most important; **main drain** *or* **main sewer** = large sewer which collects sewage from an area; **main drainage system** = system of collection of sewage from houses by pipes which take the sewage to treatment plants (as opposed to each house being provided with a septic tank) **2** *noun* principal pipe *or* cable; **gas main** *or* **water main** *or* **electricity main** = pipe *or* cable connected to an extensive system which brings gas *or* water *or* electricity to a house

◊ **mains** [meɪnz] *noun* system of pipes which bring gas *or* water *or* electricity to a house; **mains gas** *or* **mains water** *or* **mains electricity** = gas *or* water *or* electricity brought to a house by a pipe *or* cable connected to an extensive system

maintain [meɪn'teɪn] *verb* to keep up

◊ **maintenance** ['meɪntənəns] *noun* (i) keeping at a certain level; (ii) keeping the population of a species at a certain level; **maintenance ration** = quantity of food needed to keep a farm animal in good condition

maize [meɪz] *noun* widely grown cereal crop *(Zea mays)* (NOTE: in US English called **corn**)

make up water ['meɪk ʌp 'wɔːtə] *noun* water introduced into an irrigation *or* sewage system to make up for water lost by leaking *or* evaporation

malaria [mə'leəriə] *noun* paludism, tropical disease caused by a parasite *Plasmodium* which enters the body after a bite from a mosquito

COMMENT: malaria is a recurrent disease which produces regular periods of shivering, vomiting, sweating and headaches as the parasites develop in the body; the patient also develops anaemia

◊ **malarial** [mə'leəriəl] *adjective* referring to malaria; **malarial parasite** = parasite transmitted to the human bloodstream by the bite of a mosquito

◊ **malarious** [mə'leəriəs] *adjective* (region) where malaria is endemic

malathion [mælə'θaɪən] *noun* organophosphorous insecticide used to kill small aphids and mites

male [meɪl] *noun & adjective* (animal) which produces sperm; (plant) that produces pollen

malnutrition [mælnjuː'trɪʃn] *noun* (i) bad nutrition, as a result of starvation *or* wrong diet *or* bad absorption of food; (ii) not having enough to eat

mammal ['mæməl] *noun* animal of the class Mammalia (such as the human being) which gives birth to live young, secretes milk to feed them, keeps a constant body temperature and is covered with hair

man [mæn] *noun* (i) adult male human being (as opposed to a woman); (ii) all human beings; ***man has existed for a very short time compared with fish***

◊ **man-caused** ['mæn 'kɔːzd] *adjective* (disaster, etc.) which has been brought about by human beings

◊ **manhole** ['mænhəʊl] *noun* hole in a roadway *or* pavement leading to a shaft down which workmen can go to inspect the sewers

◊ **mankind** [mæn'kaɪnd] *noun* all human beings

◊ **man-made** [mæn'meɪd] *adjective* (object, substance, disaster, etc.) which has been created by human beings

manage ['mænɪdʒ] *verb* **(a)** to control; to be in charge of; ***the department is in charge of managing land resources;*** **managed woodland** = woodland which is controlled by felling, coppicing, planting, etc. **(b)** to be able to do something; to succeed in doing something; ***protesters managed to prevent the new road being built***

◊ **management** ['mænɪdʒmənt] *noun* organized use of resources *or* materials; **pasture management** = the control of pasture by grazing, cutting, reseeding, etc.; **woodland management** = controlling an area of woodland so that it is productive (by regular felling, coppicing, planting, etc.)

◊ **manager** ['mænədʒə] *noun* person who is in charge of an organization, etc.; **farm manager** = person who runs a farm on behalf of the owner

manganese ['mæŋgəniːz] *noun* metallic trace element which is used in making steel and is essential for biological life (NOTE: chemical symbol is **Mn;** atomic number is **25**)

COMMENT: manganese deficiency is associated with high pH and soils which are rich in organic matter

mangrove ['mæŋgrəʊv] *noun* genus of tropical shrub *or* tree which grows in saltwater swamps in the estuaries of rivers in Asia and America; ***mangrove forests cover muddy tidal marshes, lagoons and estuaries;*** **mangrove swamp** = swamp covered with mangroves

COMMENT: some mangrove trees produce adventitious roots which take root near the parent tree and form new trees; others produce seeds which have roots even before they fall from the tree. The result is that mangrove swamps are very thick and spread quickly

mantle ['mæntl] *noun* layer of the interior of the earth, between the solid crust and the core, formed of magma

manufacture [mænju'fæktʃə] *verb* (i) to make a product for sale, using machines; (ii) to produce a chemical naturally; ***ozone is constantly being manufactured and constantly being destroyed by natural processes in the atmosphere***

◇ **manufacturer** [mænju'fæktʃərə] *noun* person *or* company which produces machine-made products for sale; ***the company is a large manufacturer of farm machinery***

◇ **manufacturing** [mænju'fæktʃərɪŋ] *noun* producing machine-made products for sale; **manufacturing industry** = industry which takes raw materials and makes them into finished products

manure [mə'njʊə] **1** *noun* animal dung used as fertilizer (in liquid form it is called 'slurry'); **artificial manure** = artificial fertilizer, manufactured chemical substance used to increase the nutrient level of the soil; **green manure** = rapid-growing green vegetation (such as mustard plants *or* rape) which is grown and ploughed into the soil to rot and act as manure **2** *verb* to spread animal dung on land as fertilizer

COMMENT: all farm manures and slurries are valuable, and should not be regarded as a problem for disposal, but rather as assets to be used in place of expensive fertilizers which would otherwise need to be bought. Manure and slurry has to be spread in a controlled way, or dangerous pollution can result from runoffs into streams after rainfall

QUOTE: Farm pollution of rivers rose last year (1988) to a record 4,141 reported cases in England and Wales. Animal slurry, the main form of farm pollution, is up to 100 times more damaging than untreated sewage, while the liquor from silage is 200 times more damaging. Mild and wet conditions last winter were partially blamed for forcing farmers to stockpile slurry until it could be spread
Guardian

maple ['meɪpl] *noun* hardwood tree of northern temperate regions, some varieties of which produce sweet sap which is used for making sugar and syrup

marble ['mɑ:bl] *noun* form of limestone which has been metamorphosed, used especially in building as it can be polished to give a flat shiny surface

margarine [mɑ:gə'ri:n] *noun* substance made from vegetable fat, which looks like butter and is used instead of butter

margin ['mɑ:dʒɪn] *noun (of a field or a piece of paper)* edge *or* border

◇ **marginal** ['mɑ:dʒɪnəl] *adjective* **(a)** (land) which is at the edge of cultivated land (such as edges of fields, banks beside roads, etc.; (plant) which grows at the edge of two types of habitat **(b)** land of poor quality which results from bad physical conditions, such as poor soil, high rainfall, steep slopes, and where farming is often hazardous; ***cultivating marginal areas can lead to erosion***

mariculture ['mærɪkʌltʃə] *noun* type of fish farming where sea fish *or* shellfish are grown in sea-water farms

marijuana [mærɪ'hwɑ:nə] *noun* addictive drug made from the leaves *or* flowers of the Indian hemp plant

marine [mə'ri:n] *adjective* referring to the sea; (animal, vegetation, etc.) which lives in the sea; ***seals and other marine mammals;*** **marine fauna** *or* **marine flora** = animals *or* plants which live in the sea; **marine park** = natural park created on the bottom of the sea (as on a tropical reef) where visitors go into observation chambers under the sea to look at the fish and plant life; *compare* OCEANARIUM; **marine science** = science which studies all aspects of the sea, including animals and plants which live in the sea, underwater rock formations, etc.

maritime climate ['mærɪtaɪm 'klaɪmət] *noun* climate that is much modified by oceanic influences

market garden ['mɑ:kɪt 'gɑ:dən] *noun* place for the commercial growing of plants, usually vegetables, soft fruit, salad crops and flowers, found near a large urban centre which provides a steady outlet for the sale of its produce; **market gardening** = growing vegetables, salad crops, fruit for sale; **market gardener** = person who runs a market garden

marl [mɑ:l] *noun* fine soil formed of a mixture of clay and lime, used for making bricks

marram grass ['mærəm 'grɑ:s] *noun* type of grass *(Ammopila arenaria)* which is planted on sand dunes to stabilize them and prevent them being extended by the wind

marsh [mɑ:ʃ] *noun* area of permanently wet land and the plants which grow on it; **marsh gas (CH_4)** = methane *or* colourless inflammable gas, produced naturally from rotting organic waste and also found in coal mines where it is called firedamp

◇ **marshland** ['mɑ:ʃlænd] *noun* area of land covered with marsh

◇ **marshy** ['mɑːʃi] *adjective* (land) which is permanently wet and supports a natural community of plants and animals

COMMENT: a marsh usually has a soil base, as opposed to a bog or fen which is composed of peat

marsupial [mɑː'suːpiəl] *noun* type of Australian mammal with a pouch in which the young are carried

COMMENT: marsupials give birth to young at a much earlier stage of development than other mammals so that the young need to be protected in the mother's pouch for some months until they become able to look after themselves

mass [mæs] *noun* **(a)** body of matter; **critical mass** = minimum amount of fissile material which can produce a chain reaction; **mass number** = all the protons and neutrons in the nucleus of an atom; **mass spectrometer** = device used in chemical analysis, which separates particles according to their masses **(b)** large quantity or large number; **mass flow** = slide of sediment down a slope; **mass production** = manufacturing large quantities of products; **mass radiography** = taking X-ray photographs of large numbers of people to check for tuberculosis; **mass screening** = testing large numbers of people for the presence of a disease; **mass wasting** = downhill movement of weathered rock (such as a landslide)

◇ **massive** ['mæsɪv] *adjective* very large; ***in the accident some of the workers received massive doses of radiation***

mast [mɑːst] *noun* **(a)** seeds of the beech tree **(b)** **mast cell** = large cell in connective tissue which carries histamine and reacts to allergens

material [mə'tiəriəl] *noun* **(a)** matter which can be used to make something; **raw material** = substance which is used to manufacture something, such as ore for making metals, wood for making furniture, etc. **(b)** cloth

matter ['mætə] *noun* substance *or* material; **organic matter** = substance which comes from an animal *or* a plant; substance which contains carbon; **inorganic matter** = substance which does not come from an animal *or* a plant; substance which does not contain carbon

maturation [mætjʃu'reɪʃn] *noun* becoming mature *or* fully developed; **maturation lagoon** = pond used in the final stages of sewage treatment

◇ **mature** [mə'tʃʊə] *adjective* fully developed

◇ **maturing** [mə'tʃʊrɪŋ] *adjective* developing

◇ **maturity** [mə'tʃʊrɪti] *noun* being fully developed

maximum ['mæksɪməm] **1** *adjective* largest possible **2** *noun* largest possible amount; **maximum allowable concentration (MAC)** = largest amount of a pollutant which workers are allowed to be in contact with in their work environment; **maximum-minimum thermometer** = thermometer which shows the highest and lowest temperatures reached since it was last checked, as well as the current temperature; **maximum permissible dose** = highest amount of radiation to which a person may safely be exposed during a certain period; **maximum permissible level** = highest level of radiation which is allowed to be present in a certain environment (NOTE: plural is **maximums** *or* **maxima**)

MCPA 2-methyl-4chloro-phenoxy-acetic acid; a translocated herbicide, introduced in 1942

COMMENT: MCPA kills all the worst and strongest broad-leaved weeds: nettle, buttercups, charlock, dock seedlings, plantains, thistles, etc.

MCPP *see* MECOPROP

meadow ['medəʊ] *noun* field of grass and other wild plants, grown for fodder; **water meadow** = grassy field near a river, which is subject to flooding

mean [miːn] *adjective & noun* average; ***the temperature has been above the mean for the time of year; mean crop yields are down compared with last year;*** **mean lethal dose** = dose of a substance which will kill half the organisms which absorb it; **mean temperatures** = average temperatures over a period of time

meander [mɪ'ændə] *noun* bend in the course of a river; **meander belt** = total width of the area covered by a river which meanders; *see also* OX-BOW LAKE

means [miːnz] *noun* way *or* method of doing something; ***is there no means of reducing emission levels any further?;*** **means of transport** = way of transporting someone *or* something

meat [miːt] *noun* animal flesh which is eaten as food; **meat-eating animal** = carnivore *or* animal (such as a member of the cat family) which eats almost exclusively meat

mecoprop (MCPP) ['mekəʊprɒp] *noun* herbicide used to control weeds such as chickweed and cleavers, as well as the weeds controlled by MCPA

medical ['medɪkl] *adjective* (i) referring to the study of disease; (ii) referring to the study and treatment of disease which does not involve surgery; **medical assistance** = help provided by a nurse *or* by ambulancemen *or* by a member of the Red Cross, etc.; **medical officer** = person who deals with all matters affecting health in an organization such as a local authority

◇ **medication** [medɪ'keɪʃn] *noun* (i) method of treatment by giving drugs to a patient; (ii) drug *or* preparation taken to treat a disease *or* condition

◇ **medicine** ['medsɪn] *noun* **(a)** drug *or* preparation taken to treat a disease *or* condition **(b)** (i) study of diseases and how to cure *or* prevent them; (ii) study and treatment of disease which does not involve surgery

◇ **medicinal** [me'dɪsɪnəl] *adjective* referring to medicine; (substance) which has healing properties; **medicinal herb** = herb which can be used to cure a disease; **medicinal spring** = water coming naturally out of the ground and which is thought to be beneficial in the treatment of disease

◇ **medicinally** [me'dɪsɪnəli] *adverb* used as a medicine; ***the herb can be used medicinally***

mega- ['megə] *prefix* meaning large (NOTE: the opposite is **micro-. The prefix mega-** is used in front of SI units to indicate one million) **megawatt** = one million watts; ***power plants with a capacity larger than 50 megawatts***

megalo- ['megələʊ] *prefix* meaning abnormally large

◇ **megalopolis** [megə'lɒpəlɪs] *noun* agglomeration of several small towns into one huge urban area

meiosis *or US* **miosis** [maɪ'əʊsɪs] *noun* process of cell division which results in two pairs of cells each with only one set of chromosomes; *compare* MITOSIS

melanin ['melənɪn] *noun* dark pigment which gives colour to skin and hair

◇ **melanism** ['melənɪzm] *noun* abnormal deposits of dark pigment on the skin; **industrial melanism** = phenomenon where certain animals (such as butterflies and moths) become darker in colour in industrial areas allowing them to match the trees and leaves which are covered with soot

> COMMENT: because dark animals are better camouflaged in the sooty environment, they can avoid predators better than lighter ones. They therefore increase in number and eventually outnumber the lighter-coloured animals

melanoma [melə'nəʊmə] *noun* tumour formed of dark pigment cells, caused by exposure to sunlight; ***cases of melanoma could rise by between 5 and 7 per cent for each percentage decrease in ozone in the atmosphere***

melt [melt] *verb* (i) to heat a solid so that it becomes liquid; (ii); *(of solid)* to become liquid after being heated; ***the gradual rise in air temperature melted the glaciers*** *or* ***made the glaciers melt;*** **melting point** = temperature at which a solid turns to liquid; ***the melting point of ice is 0°C*** *see also* MOLTEN

◇ **meltdown** ['meltdaʊn] *noun* point in an accident in a nuclear reactor, where the fuel overheats and the core melts while the nuclear reaction is still in progress

◇ **meltwater** ['meltwɔːtə] *noun* water from melting ice, especially from a glacier *or* from winter snow

membrane ['membreɪn] *noun* thin layer of tissue covering an organ

Mendel's laws ['mendəlz 'lɔːz] *plural noun* laws governing heredity

> COMMENT: the two laws set out by Georg Mendel following his experiments growing peas, were (in modern terms): that genes for separate genetic characters assort independently of each other and that the genes for a pair of genetic characters are carried by different gametes

menstruation [menstruː'eɪʃn] *noun* bleeding from the uterus which occurs in a woman each month when the lining of the womb is shed because no fertilized egg is implanted in it

◇ **menstrual** ['menstruəl] *adjective* referring to menstruation; **menstrual cycle** = period (usually about 28 days) during which a woman ovulates, then the walls of the uterus swell and bleeding takes place if the ovum has not been fertilized

Mercalli [mɜː'kæli] *see* MODIFIED MERCALLI SCALE

mercury ['mɜːkjuri] *noun* poisonous liquid metal, used in thermometers and electric batteries; **mercury barometer** = barometer made of a glass tube containing mercury: one end of the tube is sealed, the other is open, resting in a bowl of mercury; as the atmospheric pressure changes, so the column of mercury in the tube rises or falls; **mercury poisoning** = poisoning by eating *or* drinking mercury *or* mercury compounds *or* by inhaling mercury vapour

◇ **mercurous chloride** **($HgCl_2$)** ['mɜːkjurəs 'klɔːraɪd] *noun* calomel *or* poisonous substance used to kill moss on lawns and to treat pinworms in the intestine (NOTE: chemical symbol of mercury is **Hg;** atomic number is **80)**

meridian [me'rɪdiən] *noun* line of longitude *or* imaginary line on the surface of the earth running from the North Pole to the South Pole, at right angles to the equator; **Greenwich meridian** *or* **prime meridian** = line of longitude situated at 0° and passing through Greenwich, England

meridional [mə'rɪdiənəl] *adjective* going from the North Pole to the South Pole *or* from the South Pole to the North Pole; **meridional airstream** = airstream blowing from north to south *or* from south to north

◇ **meridionality** [merɪdiə'nælɪti] *noun* phenomenon of air which blows from north to south *or* from south to north

mes- *or* **meso-** ['mesəʊ] *prefix* meaning middle

mesa ['meɪsə] *noun* high plateau (in the south west of the USA)

mesobenthos [mesəʊ'benθɒs] *noun* animals *or* plants living on the seabed, between 250 and 1,000 metres below the surface

◇ **mesoclimate** [mezəʊ'klaɪmət] *noun* variant of climate only found in a certain locality, no more than several kilometres in radius

◇ **mesohaline** [mezəʊ'hælaɪn] *adjective* (water) which is partly salt, but less so than polyhaline water

◇ **mesopause** [mezəʊ'pɔːz] *noun* thin layer of atmosphere between the mesosphere and the thermosphere

◇ **mesophyll** ['mezəʊfɪl] *noun* tissue inside a leaf

◇ **mesophyte** ['mezəʊfaɪt] *noun* plant which needs a normal amount of water to survive

◇ **mesoplankton** [mezəʊ'plæŋktən] *noun* organisms which take the form of plankton for part of their life cycle

◇ **mesosaprobic** [mezəʊsə'prəʊbɪk] *adjective* (organism) which can survive in moderately polluted water

◇ **mesosphere** ['mezəʊsfiə] *noun* zone of the earth's atmosphere between the stratosphere and the thermosphere (i.e. between 50 and 80 kilometres above the surface, with the stratopause at the bottom and the mesopause at the top: the air temperature falls steadily as one rises through the mesosphere)

◇ **mesothelium** [mezəʊ'θiːliəm] *noun* layer of cells lining a membrane; *see also* ENDOTHELIUM, EPITHELIUM

◇ **mesotherm** ['mezəʊθɜːm] *noun* plant which grows in warm conditions

◇ **mesotrophic** [mezəʊ'trɒfɪk] *adjective* (water) containing a moderate amount of nutrients; **mesotrophic lake** = lake which is between the oligotrophic and the richer eutrophic state and which has a moderate amount of nutrients in its water

meta- ['metə] *prefix* meaning which changes *or* which follows

◇ **metabolism** [me'tæbəlɪzm] *noun* chemical processes which are continually taking place in organisms and which are essential to life; **basal metabolism** = amount of energy used by a body in exchanging oxygen and carbon dioxide when at rest (i.e. energy needed to keep the body functioning and the temperature normal)

◇ **metabolic** [metə'bɒlɪk] *adjective* referring to metabolism; **basal metabolic rate (BMR)** = BASAL METABOLISM; **metabolic cycle** = cycle by which plants absorb sunlight, transform it into energy by photosynthesis and create carbon compounds; **metabolic waste** = substance produced by metabolism, such as carbon dioxide, which is not needed by the organism which produces it

◇ **metabolize** [me'tæbəlaɪz] *verb* to change the nature of something by metabolism; ***the liver metabolizes proteins and carbohydrates***

> COMMENT: metabolism covers all changes which take place in an organism: the building of tissue (anabolism); the breaking down of tissue (catabolism); the conversion of nutrients into tissue; the elimination of waste matter; the action of hormones, etc.

metal ['metəl] *noun* material (either an element *or* a compound) which can carry heat and electricity. (Some metals are essential for life)

◇ **metallic** [me'tælɪk] *adjective* like a metal *or* referring to a metal; **metallic element** = chemical element which is a metal

metaldehyde [met'ældɪhaɪd] *noun* substance, sold in small blocks, used to light fires *or* to kill slugs and snails

metalimnion [metə'lɪmniən] *noun* middle layer of water in a lake, between the epilimnion and the hypolimnion

metamorphic [metə'mɔːfɪk] *adjective* (rock, such as marble) which has changed because of external influences (such as pressure from other rocks, temperature changes, etc.)

◇ **metamorphism** [metə'mɔːfɪzm] *noun* creation of metamorphic rock

metamorphosis [metə'mɔːfəsɪs] *noun* change, especially the change of a larva into an adult insect

metamorphose [metə'mɔːfəʊz] *verb* to change; to undergo metamorphosis *or* metamorphism

metastasis [me'tæstəsɪs] *noun* spreading of a malignant disease from one part of the body to another through the bloodstream *or* the lymph system (NOTE: plural is **metastases**)

meteor ['mi:tiɔ:] *noun* solid body which enters the earth's atmosphere from outer space, usually burning up as it does so

◇ **meteoric** [mi:ti'ɒrɪk] *adjective* (rainwater) which has reached the water table

◇ **meteorite** ['mi:tiəraɪt] *noun* solid body which falls from outer space onto the earth's surface

COMMENT: outer space contains many millions of small solid bodies which sometimes come into contact with the earth. Large meteorites can create craters and form dust clouds when they hit the earth and it is believed that very large meteorites hitting the earth at some time during its past history may have been responsible for major climatic changes

meteorology [mi:tiə'rɒlədʒi] *noun* science of studying the weather and the atmosphere

◇ **meteorological** [mi:tiərə'lɒdʒɪkl] *adjective* referring to meteorology *or* to the climate; **Meteorological Office** = central government office which analyses weather reports and forecasts the weather

◇ **meteorologist** [mi:tiə'rɒlədʒɪst] *noun* scientist who specializes in the study of the weather and the atmosphere

meter ['mi:tə] **1** *noun* **(a)** device for counting *or* measuring; ***hundreds of power workers will be asked to carry radiation meters;*** **electric meter** *or* **gas meter** *or* **water meter** = device which records how much electricity *or* gas *or* water has been used **(b)** *US* = METRE **2** *verb* to count *or* measure with a meter

methaemoglobin [meθhi:mə'gləʊbɪn] *noun* dark brown substance formed from haemoglobin which develops during illness *or* following treatment with certain drugs

COMMENT: methaemoglobin cannot transport oxygen round the body and so causes cyanosis

◇ **methaemoglobinaemia** [meθhi:məgləʊbɪ'ni:miə] *noun* presence of methaemoglobin in the blood

methane (CH_4) ['mi:θeɪn] *noun* marsh gas *or* colourless gas, produced naturally from rotting organic waste and also found in coal mines where it is called firedamp; **methane converter** = process which takes the gas produced by rotting waste in landfill sites and processes it into a usable form; **methane fermentation** = breaking down of food in the gut of ruminant animals, especially cattle, producing methane which is eliminated from the animal's body

◇ **methanation** *or* **methanisation** [meθə'neɪʃn *or* meθənaɪzeɪʃn] *noun* process of converting a mixture into methane

◇ **methanogenesis** [meθənəʊ'dʒenəsɪs] *noun* generation of methane

COMMENT: methane is produced naturally from rotting vegetation in marshes, where it can sometimes catch fire, creating the phenomenon called will o' the wisp, a light flickering over a marsh. It occurs more widely as the product of animal excreta (in cattle farming). Excreta from cattle can be passed into tanks where methane is extracted leaving the slurry which is then used as fertilizer. The methane can be used for heating or as a power source. Methane is also a greenhouse gas, and it has been suggested that methane from rotting vegetation, from cattle excreta, from water in paddy fields, and even from termites' nests, all contribute to the greenhouse effect

QUOTE: the principal sources of methane are - enteric fermentation in livestock and insects, ricefields and natural wetlands, biomass burning, landfills and gas and coal fields
Nature

QUOTE: people are worried about the possible escapes of methane gas should the tip be established
Environment Now

methanol ['meθənɒl] *noun* alcohol (CH_3OH) manufactured from coal, natural gas *or* waste wood, which can be used as fuel *or* solvent

COMMENT: methanol can be used as a fuel in any type of burner; its main disadvantage is that it is less efficient than petrol and can cause pollution if it escapes into the environment, as it mixes easily with water. Production of methanol from coal or natural gas does not help fuel conservation, since it depletes the earth's fossil fuel resources

methionine ['meθɪəni:n] *noun* essential amino acid

methyl alcohol (CH_3OH) ['meθɪl 'ælkəhɒl] *noun* methanol *or* wood alcohol

◇ **methylated spirits** ['meθɪleɪtɪd 'spɪrɪts] *noun* almost pure alcohol (which has wood alcohol and colouring added to make it unfit for human consumption)

◇ **methylene blue** ['meθɪli:n 'blu:] *noun* blue dye; **methylene blue test** = test to see if a sample of effluent has the ability to remain in an oxidized condition (i.e. stable), by treating it with methylene blue over a period of time. The effluent is considered stable if it retains the blue colour of the dye throughout the testing period

◇ **methyl isocyanate (MIC)** ['meθɪl aɪsəʊ'saɪəneɪt] *noun* compound used in the production of insecticides

COMMENT: MIC is a very lethal gas and was the gas which leaked at Bhopal in India in 1984

metre *or US* **meter** ['mi:tə] *noun* base SI unit of measurement length (NOTE: with figures usually written **m: 1.3m)**

◇ **metric system** ['metrɪk 'sɪstəm] *noun* decimal measuring system (such as the SI system), which is calculated in units of ten

metropolis [met'rɒpəlɪs] *noun* very large town, usually the capital of a country

◇ **metropolitan** [metrə'pɒlɪtən] *adjective* referring to a large town; ***smog covered the whole metropolitan area***

Mg *chemical symbol for* magnesium

mg = MILLIGRAM

MHD = MAGNETOHYDRODYNAMICS

MIC = METHYL ISOCYANATE

mica ['maɪkə] *noun* silicate mineral which splits into thin transparent flakes, used as an insulator in electrical appliances

micro- ['maɪkrəʊ] *prefix* meaning very small (NOTE: opposite is **macro-** *or* **mega-** *or* **megalo-**. Note also that the prefix **micro-** is used in front of SI units to indicate a one millionth part) **microsecond** = one millionth of a second

microbe ['maɪkrəʊb] *noun* microorganism *or* very small organism which can only be seen with a microscope. (Viruses, bacteria, protozoa and fungi are all forms of microbe)

◇ **microbial** [maɪ'krəʊbiəl] *adjective* referring to microbes; **microbial disease** = disease caused by a microbe; **microbial ecology** = study of the way in which microbes develop in nature; **microbial fermentation** = breaking down of a substance caused by the action of microbes

◇ **microbiological** [maɪkrəʊbaɪə'lɒdʒɪkl] *adjective* referring to microbiology

◇ **microbiologist** [maɪkrəʊbaɪ'ɒlədʒɪst] *noun* scientist who specializes in the study of microorganisms

◇ **microbiology** [maɪkrəʊbaɪ'ɒlədʒi] *noun* scientific study of microorganisms

QUOTE: nitrates and organic matter from dead and decaying vegetation and from animal droppings, stimulate microbial activity in the soil
New Scientist

microclimate [maɪkrəʊ'klaɪmət] *noun* climate covering a very small area, such as a pond *or* a tree *or* a field; *compare* MACROCLIMATE

◇ **microeconomic** [maɪkrəʊi:kə'nɒmɪk] *adjective* concerning microeconomics

◇ **microeconomics** [maɪkrəʊi:kə'nɒmɪks] *noun* branch of economics dealing with particular aspects, such as one business *or* commodity, etc.; *compare* MACROECONOMICS

◇ **microenvironment** [maɪkrəʊɪn'vaɪrənmənt] *noun* = MICROHABITAT

◇ **microfauna** ['maɪkrəʊfɔ:nə] *noun* (i) very small animals, which can only be seen with a microscope; (ii) animals living in a microhabitat

◇ **microhabitat** [maɪkrəʊ'hæbɪtæt] *noun* single small area, such as the bark of a tree, where fauna and/or flora live

◇ **micrometer** [maɪ'krɒmɪtə] *noun* **(a)** instrument for taking very small measurements, such as measuring the width *or* thickness of very thin pieces of metal, etc. **(b)** *US* = MICROMETRE

◇ **micrometre** *or* **micron** ['maɪkrəʊmi:tə *or* 'maɪkrɒn] *noun* unit of measurement of thickness (= one thousandth of a millimetre) (NOTE: usually written μ with figures: **25μ)**

◇ **micronutrient** [maɪkrəʊ'nju:triənt] *noun* nutrient which an organism uses in very small quantities, such as iron, zinc, copper, etc.

microorganism [maɪkrəʊ'ɔ:gənɪzm] *noun* microbe *or* very small organism which can only be seen with a microscope. (Viruses, bacteria, protozoa and fungi are all forms of microbe)

microplankton [maɪkrəʊ'plæŋktən] *noun* very small plankton

◇ **micropollutant** [maɪkrəʊpə'lu:tənt] *noun* pollutant which exists in very small traces in water

microscope ['maɪkrəskəʊp] *noun* scientific instrument with lenses, which makes very small objects appear larger; ***the tissue was examined under the microscope; under the microscope it was possible to see the pollen particles;*** **electron microscope (EM)** = microscope which uses a beam of electrons instead of light

COMMENT: in an ordinary or light microscope, the image is magnified by lenses; an electron microscope uses a beam of electrons instead of light, and so achieves much greater magnification

◇ **microscopic** [maɪkrə'skɒpɪk] *adjective* so small that it can only be seen through a microscope

◇ **microscopy** [maɪ'krɒskəpi] *noun* science of the use of microscopes

microtherm [maɪkrəʊ'θɜːm] *noun* plant which grows in cool regions

mid- [mɪd] *prefix* meaning middle; **mid-latitudes** = areas halfway between the poles and the equator; ***in the mid-latitudes, global warming would produce dry hot summers and mild winters;*** **mid-ocean ridge** = ridge running down the middle of an ocean, such as the Atlantic, caused by the upward movement of magma

migrate [maɪ'greɪt] *verb* **(a)** *(of an animal or a bird)* to move from one place to another according to the season; ***as winter approaches, the herds of deer migrate south*** **(b)** to move to another place; ***waste materials are allowed to migrate from landfill sites into the surrounding soil***

◇ **migrant** ['maɪgrənt] *noun* **(a)** animal *or* bird which moves from one place to another according to the season **(b)** worker who moves from country to another to find work; *compare* NOMAD

◇ **migration** [maɪ'greɪʃn] *noun* movement of an animal *or* a bird from one place to another according to the season; ***the islands lie along one of the main migration routes from Siberia to Australia***

◇ **migratory** ['maɪgrətəri] *adjective* (animal *or* bird) which moves from one place to another according to the season; ***estuaries are important feeding grounds for migratory birds, and are also important for the passage of migratory fish like salmon***

> COMMENT: some examples of migration: birds (such as swallows) which breed in Northern Europe but which fly south for the winter; fish (such as salmon or eels) which spawn in one place, often a river, and then migrate to the sea after spawning

mildew ['mɪldjuː] *noun* fungus which produces a fine powdery film on the surface of an organism

mile [maɪl] *noun* unit of measurement of distance (one statute mile = 1.609 kilometres, one nautical mile = 1.852 kilometres)

◇ **mileage** ['maɪlɪdʒ] *noun* distance in miles; total number of miles

milk teeth ['mɪlk 'tiːθ] *plural noun* first teeth of a mammal which are gradually replaced by permanent teeth

mill [mɪl] *noun* factory where a substance is crushed to make powder, especially one for making flour from the dried seeds of corn; **paper mill** = factory for making paper out of crushed wood pulp; **water mill** *or* **windmill** = mill driven by water *or* wind; **mill race** = channel of water which turns the wheel of a water mill

◇ **millwheel** ['mɪlwiːl] *noun* large wheel (with wooden bars) which is turned by the force of water

millet ['mɪlɪt] *noun* common cereal crop *(Panicum miliaceum)* grown in many of the hot, dry regions of Africa and Asia, where it is a staple food

> COMMENT: the two most important species are finger millet and bulrush millet. Millet grains are used in various types of food. It can be boiled and eaten like rice; made into flour for porridge, pasta or chapatis; mixed with wheat flour to make bread. Millet can also be malted to make beer. Millets are also grown as forage crops, and the seed is used as a poultry feed

milli- ['mɪli] *prefix* meaning one thousandth

◇ **millibar** ['mɪlɪbɑː] *noun* unit of measurement of atmospheric pressure, used in meteorology and in drawing weather maps; *see also* ISOBAR

◇ **milligauss** ['mɪlɪgaʊs] *noun* unit of measurement of magnetic flux density; ***a person living under a low-voltage power line is exposed to 20 milligauss of radiation***

◇ **milligram** ['mɪlɪgræm] *noun* unit of measurement of weight (= one thousandth of a gram) (NOTE: with figures usually written **mg)**

◇ **millilitre** *or US* **milliliter** ['mɪlɪliːtə] *noun* unit of measurement of liquid (= one thousandth of a litre) (NOTE: with figures usually written **ml)**

◇ **millimetre** *or US* **millimeter** ['mɪlɪmiːtə] *noun* unit of measurement of length (= one thousandth of a metre) (NOTE: with figures usually written **mm)**

◇ **millisievert** ['mɪlɪsiːvət] *noun* unit of measurement of radiation (NOTE: with figures usually written **mSv)** **millisievert/year (mSv/year)** = number of millisieverts per year

> QUOTE: radiation limits for workers should be cut from 50 to 5 millisieverts, and those for members of the public from 5 to 0.25
> **Guardian**

million tonnes of coal equivalent (MTCE) *noun* measure of energy from a source which is not coal; ***renewable energy can provide up to 50 MTCE per annum***

mimic ['mɪmɪk] **1** *noun* animal which imitates another; ***starlings are excellent mimics*** **2** *verb* to imitate (another animal); ***the starling mimicked the call of the thrush***

◇ **mimicry** ['mɪmɪkri] *noun* situation where an animal imitates another, to prevent itself from being attacked

> COMMENT: some animals mimic others which are unpleasant or poisonous so that predators will not try to eat them (Batesian mimicry); other animals mimic animals

which have an equally unpleasant taste as themselves (Mullerian mimicry). In some animals, mimicry is a form of camouflage: insects mimic sticks or leaves so that predators cannot see them clearly

Minamata disease [mɪnə'mɑːtə dɪ'ziːz] *noun* form of mercury poisoning from eating polluted fish, found first in Japan

mine [maɪn] **1** *noun* hole dug in the ground to extract a mineral; **coal mine** *or* **gold mine** = hole dug in the ground to extract coal *or* gold **2** *verb* to dig into the ground to extract a mineral; **open-cast mining** *or* **open-cut mining** *or* **strip mining** *or* **surface mining** = form of mining where the mineral is dug from the surface instead of digging underground to get it

COMMENT: open-cast mining damages the environment. The top layer of soil and rock is pushed away from the surface of the ground to expose the mineral without digging underground, destroying the natural vegetation of the mined area and its surroundings. Unless the site is filled in and replanted when mining is completed the whole area remains devastated

mineral ['mɪnərəl] *noun* inorganic solid substance which is found in nature; **mineral deposits** = deposits of rocks containing useful minerals; **mineral matter** = the solid part of the soil composed of stones, sand, silt and clay (as opposed to the vegetable matter, formed from dead or decaying plants); **mineral nutrients** = nutrients (except carbon, hydrogen and oxygen) which are inorganic and are absorbed by plants from the soil; **mineral oil** = (i) oil which derives from petroleum and is made up of hydrocarbons; (ii); *US* oil distilled from petroleum and used as a laxative (NOTE: British English is **liquid paraffin**) **mineral resources** = supply of minerals and metals which are necessary for the running of a country's economy, but not including natural resources such as wood *or* food plants; **mineral water** = water taken from a natural spring and sold in bottles

COMMENT: the most important minerals required by the body are: calcium (found in cheese, milk and green vegetables) which helps the growth of bones and encourages blood clotting; iron (found in bread and liver) which helps produce red blood cells; phosphorus (found in bread and fish) which helps in the growth of bones and the metabolism of fats; and iodine (found in fish) which is essential to the functioning of the thyroid gland

◇ **mineralization** [mɪnərəlaɪ'zeɪʃn] *noun* breaking down of organic waste into its non-organic chemical components

minimum ['mɪnɪməm] **1** *adjective* smallest possible; ***the law provides only the minimum protection for workers;*** **minimum lethal dose** = smallest amount of a substance needed to kill an organism **2** *noun* smallest possible amount (NOTE: plural is **minimums** *or* **minima**)

◇ **minimal** ['mɪnɪməl] *adjective* very small; **minimal area** = smallest area for sampling in which specimens of all species can be found

minute [maɪ'njuːt] *adjective* very small

miosis [maɪ'əʊsɪs] *noun US* = MEIOSIS

mire [maɪə] *noun* wet land saturated with water, like bog *or* marsh

mist [mɪst] *noun* fine water droplets suspended in the air, which reduce visibility; **sea mist** = mist which forms over the sea

◇ **misty** ['mɪsti] *adjective* covered in mist; ***a misty autumn morning***

COMMENT: usually mists form at night, when the temperature falls because the sky is clear. If visibility falls below 1,000 metres, the mist becomes a fog

mistral ['mɪstrɑːl] *noun* strong, cold wind from the north which blows down the Rhone valley into the Mediterranean

misuse [mɪs'juːz] *verb* to use wrongly

QUOTE: 50% of farms misuse the land or keep it fallow
Environmental Action

mite [maɪt] *noun* tiny animal of the spider family which may be free-living in the soil *or* parasitic on animals *or* plants

◇ **miticide** ['mɪtɪsaɪd] *noun* substance which kills mites

mitochondrion [maɪtə'kɒndrɪən] *noun* tiny rod-shaped part of a cell's cytoplasm responsible for cell respiration (NOTE: plural is **mitochondria**)

mitosis [maɪ'təʊsɪs] *noun* process of cell division, whereby a cell divides into two identical cells; *compare* MEIOSIS

mixed [mɪkst] *adjective* made up of different elements, types, sexes, etc.; **mixed cropping** *or* **mixed culture** = growing more than one species of plant on the same piece of land at the same time (NOTE: opposite is **monoculture**) **mixed economy** = economic system which contains both nationalized industries and private enterprise; **mixed farming** = farming involving arable and dairy farming; **mixed woodland** = woodland containing conifers and deciduous trees

ml = MILLILITRE

mm = MILLIMETRE

Mn *chemical symbol for* manganese

Mo *chemical symbol for* molybdenum

moder ['məʊdə] *noun* humus which is partly acid mor and partly neutral mull

moderate ['mɒdəreɪt] *verb* to slow down the speed at which neutrons move in a nuclear reactor, so that the neutrons are more likely to hit a fissile atom

◇ **moderator** ['mɒdəreɪtə] *noun* substance (such as graphite *or* heavy water) which is used to slow down the speed of the neutrons in a nuclear reactor

COMMENT: using a moderator allows fuel enriched with uranium-235 to be used in a reactor

Modified Mercalli Scale ['mɒdɪfaɪd mɜː'kælɪ 'skeɪl] *noun* scale for measuring the damage caused by an earthquake

COMMENT: the scale runs as follows: 1: shock not felt; 2: shock felt by people on the top floors of houses, and light fittings swing; 3: noticed by many people, especially on the top floors of houses; 4: shocks very noticeable, with plates rattling in cupboards; 5: plaster on walls cracks and small delicate items get broken; 6: more damage, furniture moves, people run outside for safety; 7: ordinary buildings are slightly damaged, weak buildings are badly damaged; 8: even very strong buildings are damaged and chimneys fall down; 9: cracks appear in the ground, most buildings are badly damaged; 10: floods and landslides occur, power lines are brought down, railways and roads damaged; 11: bridges and most buildings destroyed; 12: ground moves so strongly that objects are thrown into the air and the whole area is completely devastated

Mohorovičić discontinuity [məʊhə'rəʊvɪtʃɪtʃ dɪskəntɪ'njuːɪti] *noun* boundary layer in the interior of the earth between the crust and the mantle, below which seismic shocks move more rapidly (NOTE: sometimes called the **Moho** for short)

moist [mɔɪst] *adjective* wet *or* damp; **moist tropical forest** = forest which does not receive as much rain as other types of tropical forest

◇ **moisture** ['mɔɪstʃə] *noun* water *or* other liquid; **moisture content** = amount of water *or* other liquid which a substance contains; **moisture meter** = device for measuring the amount of water *or* other liquid which a substance contains

mol [məʊl] = MOLE (a)

mold [məʊld] *US* = MOULD

mole [məʊl] *noun* **(a)** base SI unit of measurement of the amount of a substance (NOTE: with figures usually written **mol**) **(b)** small grey mammal which lives underground and which eats worms and insects

◇ **molehill** ['məʊlhɪl] *noun* small heap of earth pushed up to the surface by a mole as it makes its tunnel

molecule ['mɒlɪkjuːl] *noun* smallest unit of a substance with at least two atoms

◇ **molecular** [mə'lekjuːlə] *adjective* referring to a molecule; **molecular biology** = study of the molecules which form the structure of living matter; **molecular weight** *or* **relative molecular mass** = (i) sum of all the atomic weights of the atoms in a molecule; (ii) ratio of the average mass of one molecule of a substance to one twelfth of the mass of an atom of carbon-12

mollusc ['mɒlʌsk] *noun* any of many animals with soft bodies, usually living in shells; ***slugs and snails are molluscs, as are oysters and other shellfish***

◇ **molluscicide** [mə'lʌskɪsaɪd] *noun* substance used to kill molluscs such as snails which spread Bilharzia

molt [məʊlt] *US* = MOULT

molten ['məʊltən] *adjective* which has become liquid with heat; **molten lava** = liquid rock and minerals which flows out of an erupting volcano; *see also* MELT

molybdenum [mə'lɪbdənəm] *noun* metallic trace element, essential to biological life and used to make electric wiring (NOTE: chemical symbol is **Mo**; atomic number is **42**)

monitor ['mɒnɪtə] **1** *noun* screen (like a TV screen) on a computer; **cardiac monitor** = instrument which checks the functioning of the heart in an intensive care unit **2** *verb* to check *or* to examine regularly and record the progress of something; ***scientists have been monitoring air pollution for the last ten years***

◇ **monitoring** ['mɒnɪtrɪŋ] *noun* regular checking *or* examining and recording of the progress of something; ***scientists have set up a monitoring programme to check the changes in the sun's radiation***

QUOTE: extensive environmental and human health monitoring must be carried out
Guardian

mono- ['mɒnəʊ] *prefix* meaning single *or* one

◇ **monocline** ['mɒnəʊklaɪn] *noun* rock formation where sedimentary rock slopes sharply on one side of a fold

◇ **monocotyledon** [mɒnəʊkɒtə'liːdən] *noun* plant such as grass *or* the lily which has a single cotyledon *or* seed leaf: one of the two classifications of plants; *compare* COTYLEDON, DICOTYLEDON

◇ **monocotyledenous** [mɒnəʊkɒtə'liːdənəs] *adjective* referring to monocotyledons

◇ **monocropping** *or* **monocrop system** *or* **monoculture** ['mɒnəkrɒpɪŋ *or* 'mɒnəkʌltʃə] *noun* system of cultivation where only one species is grown on the same piece of land over a period of years; **conifer monoculture** = system of afforestation where only one type of conifer is grown (NOTE: the opposite is **mixed culture, polyculture)**

◇ **monogamy** [mə'nɒgəmi] *noun* breeding arrangement where a male and female mate for life; *compare* POLYGAMY

◇ **monoxide** [mə'nɒksaɪd] *see* CARBON

◇ **monophagous** [mɒ'nɒfəgəs] *adjective* (organism) which feeds on only one kind of food

◇ **monosodium glutamate (MSG)** [mɒnə'səʊdiəm 'glu:təmeɪt] *noun* substance (E621) added to processed food to enhance the flavour, but causing a reaction in hypersensitive people

> QUOTE: the conventional method of monocrop planting at 2.5m ignores all the opportunities of working with the diversity and the succession and stacking principles of natural systems
> **Permaculture Magazine**

monozygotic twins [mɒnəʊzaɪ'gɒtɪk 'twɪnz] *plural noun* two offspring born at the same time and from the same ovum, so exactly the same in appearance and sex

monsoon [mɒn'su:n] *noun* **(a)** season of wind and rain in tropical countries; **monsoon forest** = tropical rainforest in an area where rain falls during the monsoon season **(b)** wind which blows in opposite directions according to the season, especially the wind blowing north from the Indian Ocean in the summer

moon [mu:n] *noun* satellite of the earth, which orbits the earth every 27 days; **full moon** = middle point of a lunar cycle, when the whole face of the moon is lit by the sun; **new moon** = beginning of a lunar phase, when the moon's face is not lit by the sun

> COMMENT: the moon always shows the same face to the earth and shines with light reflected from the sun. As it moves round the earth, the face of the moon gradually becomes completely lit by the sun and is then said to be full. See also LUNAR PHASES. The moon exerts a gravitational pull on the earth and influences the tides

moor [mʊə] *noun* high land which is not cultivated, formed of acid soil covered with grass and low shrubs such as heather; **turf moor** = area of land where peat is found

◇ **moorland** ['mʊəlænd] *noun* area of land covered with moor

mor [mɔ:r] *noun* type of humus found under coniferous forests, which is acid and contains few nutrients; *compare* MULL

moraine [mə'reɪn] *noun* deposit of gravel and sand left by a glacier

> COMMENT: there are various types of moraine: ground moraine, which is a deposit left under a glacier; terminal moraine, which is the heap of soil and sand pushed by a glacier and left behind when it melts; lateral moraines are deposits left at the sides of a glacier as it moves forward

moratorium [mɒrə'tɔ:riəm] *noun* period when everyone agrees to stop a certain activity; ***the government has called for a moratorium on nuclear testing; they voted to impose a ten-year moratorium on whale catching; the conference rejected a motion calling for a moratorium on nuclear reprocessing***

> QUOTE: nothing but a moratorium on commercial peat cutting would be acceptable
> **Natural World**

morbidity rate [mɔ:'bɪdɪti 'reɪt] *noun* number of cases of a disease per hundred thousand of population

morph [mɔ:f] *noun* organism and its particular shape

morphology [mɔ:'fɒlədʒi] *noun* study of the structure and shape of living organisms; *see also* GEOMORPHOLOGY

mortality [mɔ:'tælɪti] *noun* death; ***the population count in spring is always lower than that in the autumn because of winter mortality;*** **mortality rate** = number of deaths per year, shown per thousand of the population

mortar ['mɔ:tə] *noun* mixture of sand, cement and water, used to bind bricks together when building a wall; *see also* DRY-STONE WALL

mosaic [mə'zeɪɪk] *noun* virus disease of plants which makes spots on the leaves and can seriously affect some crops

mosquito [mɒs'ki:təʊ] *noun* insect which sucks blood and passes viruses or parasites into the bloodstream; **mosquito control** = way of reducing the number of mosquitoes, usually by the use of pesticides

◇ **mosquitocide** [mɒs'ki:təsaɪd] *noun* substance which kills mosquitoes

> COMMENT: in tropical countries dengue, filariasis, malaria and yellow fever are transmitted by the mosquito. Mosquitoes breed in water and they spread rapidly in lakes or canals created by dams and other irrigation schemes. Because irrigation is more widely practised in tropical countries, mosquitoes are increasing and diseases such as malaria are spreading

moss [mɒs] *noun* very small plant without roots, which grows in damp places and forms mats of vegetation; **Irish moss** = CARRAGEEN; **moss peat** = dried and sterilized peat formed from the remains of mosses, sold in bags for horticultural purposes

moth [mɒθ] *noun* insect of the order Lepidoptera, which flies generally at night; **house moth** = small moth which sometimes lives in houses and whose larvae can destroy clothes and blankets, etc. kept in cupboards

COMMENT: moths are similar to butterflies, but are dull in colour and lay their wings over their bodies when at rest

mother cell ['mʌðə 'sel] *noun* original cell which splits into daughter cells during mitosis

mother-of-pearl ['mʌðə əv 'pɜːl] *noun* hard substance, mostly made up of calcium carbonate, which forms the inner layer of some shells, such as the shell of the oyster

mould *or US* **mold** [məʊld] *noun* **(a)** any of various plants of the Kingdom Fungi, especially mildew, a fungus which produces a fine powdery film on the surface of an organism **(b)** soft earth; **leaf mould** = soft fibrous material formed of rotten and broken-down dead organic matter such as leaves

moult [məʊlt] **1** *noun* losing feathers *or* hair at a certain period of the year; **moult plumage** = small feathers which remain on a bird when it is moulting **2** *verb* to lose feathers *or* hair at a certain period of the year; ***most animals moult at the beginning of summer;*** **moulting season** = time of year when a bird *or* animal moults

mountain ['maʊntən] *noun* **(a)** mass of rock rising higher than a hill; **mountain plant** = plant which grows in *or* comes originally from a mountain region; **mountain sheep** = breeds of sheep which live in *or* come originally from a mountain region; **mountain sickness** *or* **altitude sickness** = condition where a person suffers from oxygen deficiency at a high altitude (as on a mountain) where the level of oxygen in the air is low **(b)** **commodity mountain** = surplus of a certain agricultural product produced in the EU, for example the 'butter mountain'; *see also* LAKE

◇ **mountainous** ['maʊntənəs] *adjective* (area of land) where there are high mountains

◇ **mountainside** ['maʊntənsaɪd] *noun* side of a mountain

mouth [maʊθ] *noun* opening where a river joins the sea

MSG = MONOSODIUM GLUTAMATE

mSv = MILLISIEVERT

MTCE = MILLION TONNES OF COAL EQUIVALENT

mud [mʌd] *noun* mixture of soil and water; **blue mud** = deposit found on the seabed in the Pacific Ocean and elsewhere, containing decaying organic matter and iron sulphide, the latter giving it its blue colour; **drilling mud** = mixture of clay, water and minerals, used as a lubricant and coolant when drilling into rock; **green mud** = deposit found on the seabed off the south-eastern US and elsewhere, containing fine clay and glauconite, the latter giving it its green colour; **red mud** = deposit found on the seabed in the China Sea and elsewhere, containing dust and iron oxide, the latter giving it its red colour; **mud volcano** = heap of hot mud thrown up round a hot spring

◇ **muddy** ['mʌdi] *adjective* containing mud *or* covered with mud

◇ **mud flat** ['mʌd 'flæt] *noun* area of flat mud, usually in river estuaries

◇ **mudslide** ['mʌdslaɪd] *noun* fall of a large amount of mud down a slope

muffler ['mʌflə] *noun US* device attached to a car exhaust which reduces the sound emitted (NOTE: British English is **silencer**)

mulch [mʌltʃ] **1** *noun* organic material (such as dead leaves *or* straw) used to spread over the surface of the soil to prevent evaporation *or* erosion **2** *verb* to spread organic material (such as dead leaves *or* straw) over the surface of the soil to prevent evaporation *or* erosion

COMMENT: black plastic sheeting is often used by commercial horticulturists, but the commonest mulches are organic. Apart from preventing evaporation, mulches reduce weed growth and encourage worms

mull [mʌl] *noun* type of humus found under deciduous forests, with rotted leaves and many nutrients; *compare* MOR

Mullerian mimicry [mʌ'leəriən 'mɪmɪkri] *noun* form of mimicry where a animal mimics another animal which has an equally unpleasant taste

multicellular [mʌltɪ'seljuːlə] *adjective* composed of several *or* many cells

multi-cropping ['mʌltɪ 'krɒpɪŋ] *noun* multiple cropping, growing more than one crop on the same piece of land in one year; ***wet rice is often multi-cropped***

municipal [mjuː'nɪsɪpəl] *adjective* referring to a town; **municipal dump** = place where a town's refuse is disposed of after it has been collected; **municipal refuse** *or* **municipal waste** = refuse collected in a town

mushroom ['mʌʃrʊm] *noun* common edible fungus, often grown commercially

muskeg ['mʌskeg] *noun (mainly in Canada)* bog in high cold plateau regions

mutant ['mju:tənt] *noun & adjective* (i) gene in which mutation has occurred; (ii) organism carrying a gene in which mutation has occurred

◇ **mutagen** ['mju:tədʒen] *noun* agent which causes mutation

◇ **mutagenicity** [mju:tədʒe'nɪsɪti] *noun* ability to make genes mutate

◇ **mutate** [mju:'teɪt] *verb* to undergo a change in a gene *or* chromosome; ***bacteria can mutate suddenly and become increasingly able to infect***

◇ **mutation** [mju:'teɪʃn] *noun* change in a gene *or* chromosome

myc- *or* **myco-** ['maɪkəʊ] *prefix* referring to fungus

◇ **mycelium** [maɪ'si:liəm] *noun* mass of threads which forms the main part of a fungus

◇ **Mycobacterium** [maɪkəʊbæk'ti:əriəm] *noun* one of a group of bacteria, including those which cause leprosy and tuberculosis

◇ **mycology** [maɪ'kɒlədʒi] *noun* study of fungi

myiasis ['maɪəsɪs] *noun* infestation of animals by the larvae of flies

Myrtaceae [mɜ:'teɪʃɪi] *noun* family of Australian plants, including the eucalyptus

Mysticeti [mɪstɪ'seti] *noun* baleen whales, including blue whales, humpbacks, etc.

myxomatosis [mɪksəmə'təʊsɪs] *noun* usually fatal virus disease affecting rabbits, transmitted by fleas

Nn

N 1 *chemical symbol for* nitrogen **2** = NEWTON

> QUOTE: in an attempt to reduce nitrate leaching and minimize nitrogen starvation, beet growers might be able to apply N as a foliar spray
> **Farmers Weekly**

Na *chemical symbol for* sodium

nacreous clouds ['neɪkriəs 'klaʊdz] *noun* thin clouds, possibly made of ice crystals, which form a layer about 25km above the earth and look like mother-of-pearl

nanoplankton ['nænəʊplæŋktən] *noun* very small plankton

nastic response ['næstɪk rɪ'spɒns] *noun* response of plants and flowers to a stimulus which is not connected to the direction from which the stimulus comes, such as the closing of flowers at night

◇ **-nasty** ['næsti] *suffix* referring to a nastic response; **photonasty** = response of plants to light (without turning towards the light source); **thermonasty** = response of plants to heat (NOTE: opposed to **-tropism** and **-taxis** which are responses of plants which move or turn towards the stimulus)

natality [nə'tælɪti] *noun* birth; **natality rate** = birth rate *or* number of births per year, shown per thousand of the population

national ['næʃnl] *adjective* concerning a particular country; belonging to the people of a particular country; **national income** *or* **national revenue** = total amount of money received during a certain period by all the people of a particular country and made up of wages, rents, interest, etc.; **National Nature Reserve** = *(in the UK)* nature reserve designated by the Nature Conservancy Council for the protection of plants and animals living in it; **national park** = large area of unspoilt land, owned and managed by the government for recreational use by the public; **National Radiological Protection Board (NRPB)** = UK agency which monitors radiation risks to the British population; **National Rivers Authority (NRA)** = UK statutory body responsible for water management, flood defence and the regulation of water quality; **National Trust** = organization in Britain which preserves historic buildings and parks and special areas of natural beauty

native ['neɪtɪv] *adjective* which belongs to a place; ***the tiger is native to Asia;*** **native element** = element which exists in a pure state in nature (such as gold *or* carbon); **native species** = species which exists naturally in a certain place (as opposed to an introduced species)

nature ['neɪtʃə] *noun* **(a)** (i) essential quality of something; (ii) kind *or* sort; (iii) plants, animals and their environment in general; **Nature Conservancy Council (NCC)** *or* **English Nature** = official body in the UK, established in 1973, which takes responsibility for the conservation of fauna and flora. Since April 1991 the branch of the Council dealing with England has also been called English Nature; **nature conservation** = active management of the earth's natural resources and environment to ensure their quality is maintained and that they are wisely used; **nature reserve** = special area where the wildlife is protected (National Nature Reserves are designated by the NCC); **nature study** = learning about plant and animal life at school; **nature trail** = path through the countryside with signs to draw attention to important and interesting features **(b)** **human nature** = general characteristics of humans

◇ **natural** ['nætʃrəl] *adjective* **(a)** normal *or* not surprising; ***the animal's behaviour was quite natural; it's natural for wild animals to be frightened of people*** **(b)** not made by people; (thing) which comes from nature; **natural background** = surrounding level of radiation in a particular location; **natural childbirth** = childbirth where the mother is not given pain-killing drugs but is encouraged to give birth to the baby with as little medical assistance as possible; **natural environment** *or* **natural habitat** = type of environment in which an organism lives and which is not made by people; **natural evaporation** = evaporation of moisture from lakes and rivers caused by the wind or the sun; **natural historian** = scientist who specializes in the study of natural history; **natural history** = study of nature; **natural immunity** = immunity from disease inherited by newborn offspring from birth, acquired in the womb *or* from the mother's milk; **natural increase** = increase in population when births exceed deaths; **natural pollutant** = polluting substance (such as ash from a volcano) which occurs naturally; **natural resources** = part of the environment which can be used commercially (such as coal); **natural science** = science (such as biology, chemistry, geology, physics) which studies the physical world, as opposed to the theoretical sciences, such as mathematics, philosophy and sociology; **natural selection** = evolution of a species, whereby characteristics which help individual organisms to survive and reproduce are passed on to the offspring and those characteristics which do not help are not passed on; **natural vegetation** = plants existing in a natural state (such as a rainforest) and not planted *or* managed by people

◇ **natural gas** ['nætʃrəl 'gæs] *noun* gas found underground and not manufactured, brought to towns for domestic use

◇ **naturalist** ['nætʃrəlɪst] *noun* person who is interested in and studies natural history

◇ **naturalize** ['nætʃrəlaɪz] *verb* to introduce a species into an area where it has not lived before so that it becomes established as part of the ecosystem

> COMMENT: natural gas is often found near petroleum deposits, although it can occur without petroleum. It is mainly formed of methane but also contains small amounts of butane and propane. It contains no sulphur (coal gas contains sulphur) and since it mixes with air it burns completely, creating very little carbon monoxide

nauplius ['nɔːpliəs] *noun* swimming larva of a crustacean (NOTE: plural is **nauplii**)

nautical ['nɔːtɪkl] *adjective* referring to ships and the sea; **nautical mile** = unit of measurement of distance, used at sea and in the air (= 1.852 kilometres)

NCC = NATURE CONSERVANCY COUNCIL

neap tide ['niːp 'taɪd] *noun* tide which occurs at the first and last quarters of the moon, when the difference between high and low water is less than normal; *compare* SPRING TIDE

Nearctic Region [niː'ɑːktɪk 'riːdʒən] *noun* biogeographical region (part of Arctogea) comprising North America and Greenland

need [niːd] **1** *noun* something that is necessary *or* required; ***we must pay more attention to the needs of Third World farmers;*** **in need** = requiring something, especially financial aid *or* food aid **2** *verb* to require; ***we need more bottle banks in our town***

needle ['niːdl] *noun* **(a)** thin hard leaf of a conifer **(b)** thin pointer on a dial such as a compass

> COMMENT: pine needles are covered with a waxy coating which prevents water loss and also stops diseases attacking the tree. Conifers reduce the water content in their needles as winter approaches

NEF = NOISE EXPOSURE FORECAST

negative feedback ['negətɪv 'fiːdbæk] *noun* situation where the result of a process inhibits the action which caused the process

> QUOTE: the increases in the concentration of methane and carbon dioxide may have caused significant warming already, if negative feedback mechanisms such as cloud cover are ineffective
> **Nature**

neighbourhood noise ['neɪbəhʊd 'nɔɪz] *noun* general noise from a local source (such as the noise of a factory) which is disturbing to people living in the area

nekton ['nektən] *noun* swimming marine animals (such as fish), as opposed to floating *or* drifting animals like plankton

nematode ['nemətəʊd] *noun* type of roundworm, some of which are parasites of animals, such as hookworms, while others live in the roots of plants

◇ **nematicide** [ne'mætɪsaɪd] *noun* substance which kills nematode worms

> QUOTE: root-knot nematodes are among the most widespread pests that limit agricultural productivity. Almost all plants that account for the majority of the world's food crop - including the potato - are susceptible to these pests
> **Appropriate Technology**

neo- ['ni:əʊ] *prefix* meaning new

◇ **neo-Darwinism** [ni:əʊ'dɑ:wɪnɪzm] *noun* revised form of Darwin's theory of evolution which includes Mendel's laws of genetics and other more recent discoveries

◇ **Neogea** *or* **Neotropical Region** [ni:əʊ'dʒi:ə *or* ni:əʊ'trɒpɪkl 'ri:dʒən] *noun* one of the main biogeographical regions of the earth comprising Central and South America together with the islands in the Caribbean; *see also* ARCTOGEA, NOTOGEA

◇ **neonatal** [ni:əʊ'neɪtl] *adjective* referring to the first few weeks after birth; **neonatal death rate** = number of newborn babies who die, shown per thousand babies born

neon ['ni:ɒn] *noun* inert gas found in very small quantities in the atmosphere and used in illuminated signs (NOTE: chemical symbol is **Ne**; atomic number is **10**)

neptunium [nep'tju:niəm] *noun* natural radioactive element (NOTE: chemical symbol is **Np**; atomic number is **93**)

neritic [ne'rɪtɪk] *adjective* (animal *or* plant) which lives in the shallow sea over the continental shelf; **neritic facies** = appearance of sedimentary rocks laid down in shallow water, where the ripples marks made by waves are clearly visible; **neritic zone** = area of warm shallow water at the edge of a lake *or* sea, the habitat of plants and other organisms

nerve [nɜ:v] *noun* bundle of fibres in a body which takes impulses from one part of the body to another; **nerve cell** *or* **neurone** = cell in the nervous system which transmits nerve impulses; **nerve centre** = point at which nerves come together; **nerve ending** = terminal at the end of a nerve fibre, where a nerve cell connects with another nerve *or* with a muscle; **nerve fibre** = fibre leading from a nerve cell, carrying the impulses; **nerve gas** = gas which attacks the nervous system and can cause death; **nerve impulse** = electrical impulse transmitted by nerve cells

◇ **nervous system** ['nɜ:vəs 'sɪstəm] *noun* all the nerve cells of a body, including the spinal cord

nest [nest] **1** *noun* construction built by birds and some fish for their eggs; construction made by some social insects, such as ants and bees, for the colony to live in; ***the birds have built a nest in the chimney; when the little birds are several weeks old they leave the nest;*** **nest builder** = animal which build a nest **2** *verb* to build a nest (in a place); ***the robins have nested in an old kettle;*** **nesting bird** = bird which is sitting on its eggs to incubate them; **nesting site** = place where a bird may build a nest

◇ **nestling** ['nestlɪŋ] *noun* very small bird still in the nest

net [net] *adjective* (amount) remaining after all deductions have been made; **net national product** = gross national product after an allowance for the depreciation of capital goods has been deducted; **net profit** = actual gain after expenses, tax, wages, etc. have been paid

network ['netwɜ:k] *noun* interconnecting system of lines, roads, pipes, cables, veins, etc.; ***a network of fine blood vessels;*** **railway network** = interconnecting system of railway lines

neurone ['njʊərəʊn] *noun* nerve cell, a cell in the nervous system which transmits nerve impulses; **sensory neurone** = nerve cell which transmits impulses relating to sensations from the receptor to the central nervous system

neurotoxin [nju:rəʊ'tɒksɪn] *noun* natural substance (such as the poison of a snake *or* insect) which prevents the victim's nerve impulses from working

neuston ['nju:stən] *noun* organisms like plankton which float *or* swim in the surface film of a body of water

neutral ['nju:trəl] *adjective* **(a)** neither acid nor alkali; ***a pH factor of 7 is neutral;*** **neutral soil** = soil which is neither acid nor alkaline, that is, where the pH value is neutral **(b)** neither positive nor negative

◇ **neutralize** ['nju:trəlaɪz] *verb* **(a)** to make an acid neutral; ***acid in drainage water can be neutralized by limestone; ammonia is produced in the nose to neutralize the effects of acid in the air*** **(b)** to counteract the effect of something; *(in bacteriology)* to make a toxin harmless by combining it with the correct amount of antitoxin; ***alkali poisoning can be neutralized by applying acid solution***

◇ **neutralizing** ['nju:trəlaɪzɪŋ] *adjective* (substance) which counteracts the effect of something

> QUOTE: alkaline soils, such as soils rich in limestone, can neutralize acid directly
> **Scientific American**

neutron ['nju:trɒn] *noun* particle with no electric charge in the nucleus of an atom; *compare* ION

névé ['neveɪ] *noun* spring snow on high mountains which becomes harder and more like ice during the summer

newton ['nju:tən] *noun* SI unit of measurement of force (NOTE: usually written **N** after figures: **the muscle exerted a force of 5N)**

> COMMENT: 1 newton is the force required to move 1 kilogram at the speed of 1 metre per second

NGO = NON-GOVERNMENTAL ORGANIZATION

Ni *chemical symbol for* nickel

niacin ['naɪəsɪn] *noun* nicotinic acid *or* vitamin of the vitamin B complex, found in milk, meat, liver, yeast, beans, peas and bread

niche [ni:ʃ] *noun* place in an ecosystem which a species is specially adapted to fit; **ecological niche** = all the characters (chemical, physical, biological) that determine the position of an organism *or* species in an ecosystem (commonly called the 'role' *or* 'profession' of an organism, e.g. an aquatic predator, a terrestrial herbivore)

nickel ['nɪkl] *noun* metallic element, used in computer wiring (NOTE: chemical symbol is **Ni;** atomic number is **28)**

nicotine ['nɪkəti:n] *noun* toxic substance in tobacco, also used as an insecticide

◇ **nicotinic acid** [nɪkə'ti:nɪk 'æsɪd] *noun* = NIACIN

nid- [nɪd] *prefix* meaning nest

◇ **nidicolous** [nɪ'dɪkələs] *adjective* (baby bird) which is helpless and remains in the nest for some time

◇ **nidifugous** [nɪ'dɪfjʊgəs] *adjective* (baby bird) which is so well developed when hatched that it can leave the nest immediately

night soil ['naɪt 'sɔɪl] *noun* human excreta, collected and used for fertilizer in some parts of the world

NIH = NOT INVENTED HERE phrase used to criticize something which has not been locally manufactured and therefore may not be suitable for the region

nimbostratus [nɪmbəʊ'streɪtəs] *noun* grey mass of cloud with precipitation in the form of rain *or* snow about 1,000m above the ground

NIMBY ['nɪmbi] = NOT IN MY BACKYARD phrase used to describe people who encourage the development of agricultural land for building houses *or* factories, provided it is not near where they themselves are living

El Niño [el 'ni:njəʊ] *noun* **(a)** warm current flowing south along the Pacific coast of South America in December **(b)** phenomenon occurring every few years in the Pacific Ocean, where a mass of warm water moves from west to east, rising as it moves, giving very high tides along the pacific coast of South America

> COMMENT: El Niño not only brings high tides but also influences the rainfall patterns all around the Pacific basin, with most rain falling along the Pacific coast of South America, in South India and in the Pacific Islands, but making North India and Australia drier than usual. The phenomenon seems to occur as a cycle, every two to five years, and seems to be caused by fluctuations in the water of the Pacific Ocean itself, although some scientists have suggested it is connected with heavy snowfalls in Central Asia

> QUOTE: El Niño events are associated with general tropical tropospheric warming and occurrences of drought over large areas
>
> **New Scientist**

> QUOTE: during an El Niño, which occurs roughly twice per decade but with extremely variable intensity, complex interactions of ocean currents, atmosphere and shifting precipitation belts produce droughts and other meteorological anomalies throughout the world for several months or even longer
>
> **Natural History**

nitrate ['naɪtreɪt] *noun* (i) ion with the formula NO_3; (ii) chemical compound containing the nitrate ion, such as sodium nitrate; **nitrate-sensitive area (NSA)** = region of the country where nitrate pollution is likely and where the use of nitrate fertilizers is strictly controlled

> COMMENT: nitrates are a source of nitrogen for plants; they are used as fertilizers but can poison babies if they get into drinking water

◇ **nitrification** [nɪtrɪfɪ'keɪʃn] *noun* process by which bacteria found in the soil break down nitrogen compounds and form nitrates which plants can absorb

◇ **nitrify** ['nɪtrɪfaɪ] *verb* to convert nitrogen *or* nitrogen compounds into nitrates; **nitrifying bacteria** = bacteria which convert nitrogen into nitrates

◇ **nitrite** ['naɪtraɪt] *noun* (i) ion with the formula NO_2; (ii) chemical compound containing the nitrite ion, such as sodium nitrite

> COMMENT: nitrites are formed by bacteria from nitrogen as an intermediate stage in the formation of nitrates

nitric ['naɪtrɪk] *adjective* containing nitrogen; **nitric acid (HNO_3)** = corrosive acid used in the manufacture of explosives and fertilizers; **nitric oxide (NO)** = colourless gas which forms red fumes of nitrogen dioxide in air

nitrogen ['naɪtrədʒən] *noun* chemical element, a gas which is essential to biological life and which is the main component of air and an essential part of protein; **atmospheric**

nitrogen = nitrogen as found in the atmosphere; **nitrogen cycle** = process by which nitrogen enters living organisms. The nitrogen is absorbed into green plants in the form of nitrates, the plants are then eaten by animals and the nitrates are returned to the ecosystem through the animal's excreta or when an animal or a plant dies; **nitrogen deficiency** = lack of nitrogen in soil, found where organic matter is low; it results in thin, weak growth of plants, especially grasses, cereals, kales and cabbages; **nitrogen dioxide (NO_2)** = poisonous gas, one of the pollutant gases produced by car engines; **nitrogen fertilizers** = many straight and compound fertilizers contain nitrogen, but it is quickly changed by bacteria in the soil to the nitrate form; **nitrogen fixation** = process by which nitrogen in the air is converted by bacteria in certain plants into nitrogen compounds: when the plants die the nitrogen is released into the soil and acts as a fertilizer; **nitrogen-fixing bacteria** = bacteria in the soil which convert nitrogen in the air into nitrogen compounds by means of the process of nitrogen fixation in plants; **nitrogen-fixing plants** = plants, such as lucerne or beans, which form an association with bacteria which convert nitrogen from the air into nitrogen compounds which pass into the soil (NOTE: chemical symbol is **N**; atomic number is **7**)

COMMENT: nitrogen is taken into the body by digesting protein-rich foods; excess nitrogen is excreted in urine. When the intake of nitrogen and the excretion rate are equal, the body is in nitrogen balance or protein balance. The nitrogen cycle is the series of processes by which nitrogen is converted into nitrates which are absorbed into green plants. The plants are then eaten by animals, which themselves are eaten by other animals. The nitrates are returned to the ecosystem in excreta or when animals or plants die

QUOTE: the global nitrogen cycle has been altered by human activity to such an extent that more nitrogen is fixed annually by humanity (primarily for nitrogen fertilizer, also by legume crops and as a byproduct of fossil fuel combustion) than by all natural pathways combined
Ecology

nitrogen oxide (NO_x) ['naɪtrədʒən 'ɒksaɪd] *noun* oxide (such as nitric oxide (NO) *or* nitrogen dioxide (NO_2) formed when nitrogen is oxidized

◇ **nitrous oxide (N_2O)** ['naɪtrəs 'ɒksaɪd] *noun* colourless gas with a sweet smell, used as an anaesthetic in dentistry and surgery and as an aerosol propellant

COMMENT: in general nitrogen oxides form the major part of air pollution, though nitric oxide (produced by burning fossil fuel) is not directly dangerous to humans. Nitrogen dioxide is produced by car engines and is toxic. Nitrogen oxides are also produced when farmland is sprayed with fertilizers; the bacteria in the soil feed on the fertilizer and produce the gas

QUOTE: NO_x and hydrocarbons can react in sunlight to form ozone
Environment Now

NNI = NOISE AND NUMBER INDEX

noble gases ['nəʊbl 'gæsɪz] *noun* inert gases *or* gases (helium, neon, argon, krypton, xenon and radon) which do not react chemically with other substances; **noble metal** = metal such as gold *or* silver which resists corrosion and does not form compounds with non-metals (as opposed to base metals)

node [nəʊd] *noun* point on the stem of a plant where a leaf is attached

◇ **nodule** ['nɒdju:l] *noun* small node found on the roots of leguminous crops; the nodules contain types of bacteria which can convert nitrogen from the air into nitrogen compounds

noise [nɔɪz] *noun* any unpleasant sound; **background noise** = (i); *(in the environment)* general level of noise which is always there (ii); *(in an electronic instrument)* unwanted interference noise; **neighbourhood noise** = general noise from a local source (such as the noise of a factory) which is disturbing to people living in the area; **Noise Abatement Society** = association of people who try to influence others to reduce levels of noise; **noise and number index (NNI)** = way of measuring noise from aircraft; **noise charge** = fee paid by a company to be allowed to make a certain amount of noise in the course of its business; **noise criteria** = levels of noise which are acceptable to people who hear them; **noise exposure forecast (NEF)** = forecast of the effect which industrial *or* aircraft noise will have on people; **noise level** = loudness of a noise which can be measured; ***the factory has announced plans to keep noise levels down to a minimum;*** **noise nuisance** = noise which is annoying *or* disturbing *or* unpleasant; **noise pollution** = unpleasant sounds which cause discomfort; **noise pollution level (NPL)** = loudness of unpleasant noise which can be measured; **noise zones** = areas which are classified according to the amount of noise

◇ **noisy** ['nɔɪzi] *adjective* (engine, etc.) which makes a loud noise; (place) where there is a lot of noise

nomad ['nəʊmæd] *noun* person whose home is a large area of land, who moves from place to place in it without settling in any one spot (e.g. a person who hunts game *or* a herdsman who drives his animals from place to place to find grazing)

◇ **nomadic** [nəʊ'mædɪk] *adjective* referring to nomads; ***the herdsmen in the area lead a nomadic existence; nomadic pastoralists move their livestock around to feed on available grazing***

◇ **nomadism** [nəʊmæ'dɪzm] *noun* habit of certain animals which move around from place to place without having a fixed habitat

non- [nɒn] *prefix* meaning not

◇ **Non-Attainment Area** [nɒnə'teɪnmənt 'eəriə] *noun* area of the USA which does not meet the standards of clean air laid down in the Clean Air Act of 1970

◇ **non-biodegradable** *or* **non-degradable** [nɒnbaɪəʊdɪ'greɪdəbl *or* nɒndɪ'greɪdəbl] *adjective* (substance, like lead) which is not broken down in nature; **non-degradation** = preventing pollution of clean air

◇ **non-disposable** [nɒndɪs'pəʊzəbl] *adjective* which is not thrown away after use, but which can be returned for recycling

◇ **non-ferrous** ['nɒnferəs] *adjective* which does not contain iron

◇ **non-governmental organization (NGO)** [nɒngʌvən'mentəl ɔ:gənaɪ'zeɪʃn] *noun* organization (such as a pressure group *or* charity *or* voluntary agency) which is not funded by a government and which works on a local *or* national *or* international level

◇ **non-indigenous** [nɒnɪn'dɪdʒənəs] *adjective* which is not native to a place

◇ **non-nucleated** [nɒn'nju:klɪeɪtɪd] *adjective* (cell) with no nucleus

◇ **non-organic** [nɒnɔ:'gænɪk] *adjective* (substance) which does not come from plants *or* animals; (compound) which does not contain carbon

◇ **non-persistent** [nɒnpɜ:'sɪstənt] *adjective* (pesticide) which does not remain toxic for long so does not enter the food chain

◇ **non-renewable resources** [nɒnrɪ'nju:əbl rɪ'sɔ:sɪz] *plural noun* natural resources (such as coal *or* oil) which cannot be replaced if they are consumed

◇ **non-resistant** [nɒnrɪ'zɪstənt] *adjective* (strain of organisms) which is affected by a disease *or* antibiotic *or* pesticide *or* herbicide, etc.

◇ **non-selective weedkiller** [nɒnsɪ'lektɪv 'wi:dkɪlə] *noun* weedkiller which kills all plants

◇ **non-smoker** [nɒn'sməʊkə] *noun* person who does not smoke

◇ **non-toxic** [nɒn'tɒksɪk] *adjective* (substance) which is not poisonous *or* harmful

north [nɔ:θ] *adjective, adverb & noun* one of the directions on the earth's surface, the direction facing away from the sun at midday and towards the North Pole; ***the wind is blowing from the north; the river flows north into the ocean;*** **North Atlantic Drift** = warm current which travels northwards along the east coast of North America and then crosses the Atlantic to hit the British Isles; **North Pole** = point which is furthest north on the earth, through which the lines of longitude run; **north wind** = wind which blows from the north

◇ **northerly** ['nɔ:ðəli] **1** *adjective* to *or* from the north; ***the cyclone moved in a northerly direction towards the coast*** **2** *noun* wind which blows from the north

◇ **northern** ['nɔ:ðən] *adjective* in the north; towards the north; ***the herds spend summer on the northern plains;*** **Northern Lights** = Aurora Borealis, spectacular illumination of the sky in the Northern Hemisphere caused by ionized particles striking the atmosphere

◇ **North Sea** ['nɔ:θ 'si:] *noun* sea to the north of the Netherlands and Germany, east of the UK and west of Scandinavia; ***effluent is discharged directly into the North Sea;*** **North Sea oil** *or* **North Sea gas** = oil *or* gas extracted from the rocks under the North Sea

notify ['nəʊtɪfaɪ] *verb* to inform someone officially; ***the local doctor notified the Health Service of the case of cholera*** (NOTE: you notify someone **of** something)

◇ **notifiable disease** [nəʊtɪ'faɪəbl dɪ'zi:z] *noun* **(a)** serious infectious disease which in Great Britain has to be reported by a doctor to the Department of Health and Social Security so that steps can be taken to stop it spreading **(b)** serious infectious disease of animals and poultry which in Great Britain has to be reported to the police when an outbreak is confirmed on a farm; diseases and pests of plants which must be notified to the Ministry of Agriculture

> COMMENT: the following include notifiable diseases of humans: cholera, diphtheria, dysentery, encephalitis, food poisoning, jaundice, meningitis, ophthalmia neonatorum, paratyphoid, plague, poliomyelitis, relapsing fever, scarlet fever, smallpox, tuberculosis, typhoid and typhus. The following are some of the notifiable diseases of animals: anthrax; foot and mouth disease; Newcastle disease; rabies; sheep pox; sheep scab; swine fever

Notogea ['nəʊtədʒiə] *noun* Australasian Region, one of the main biogeographical regions of the earth, comprising Australia, New Zealand and the Pacific Islands; *see also* ARCTOGEA, NEOGEA

NO_x = NITROGEN OXIDE

noxious ['nɒkʃəs] *adjective* harmful (drug *or* gas *or* fumes)

noy [nɔɪ] *noun* unit of measurement of perceived noise

Np *chemical symbol for* neptunium

NPK nitrogen, phosphorus and potassium used together as a fertilizer

NPL = NOISE POLLUTION LEVEL

NRA = NATIONAL RIVERS AUTHORITY

NRC = NUCLEAR REGULATORY COMMISSION

NRPB = NATIONAL RADIOLOGICAL PROTECTION BOARD

NSA = NITRATE-SENSITIVE AREA

nuclear ['nju:kliə] *adjective* **(a)** referring to nuclei, especially referring to the fission *or* fusion of nuclei; **nuclear bomb** = bomb whose destructive power is produced by nuclear fission *or* fusion; **nuclear energy** = energy created during a nuclear reaction, either fission *or* fusion, which, in a nuclear power station, produces heat which warms water and forms steam which runs a turbine to generate electricity; **nuclear fission** = splitting of the nucleus of an atom (such as uranium-235) into several small nuclei which then releases energy and neutrons; **nuclear fuel** = substance which is fissile (such as uranium-238) and can be used to create a controlled reaction in a nuclear reactor; **nuclear fuel cycle** = series of processes by which uranium ore is extracted, processed to make uranium oxide, then enriched until it is ready for use in a reactor; **nuclear fusion** = joining together of several nuclei to form a single large nucleus, creating energy (as in a hydrogen bomb); **nuclear power** = electricity generated by a nuclear power station; power generated by a nuclear reactor; **nuclear power station** *or* **nuclear power plant** = power station in which nuclear reactions are used to provide energy to run turbines which generate electricity; **nuclear-powered** = operated by nuclear power; **nuclear reaction** = physical reaction of the nucleus of an atom, which when bombarded by radiation particles creates an isotope; **nuclear reactor** = device which creates heat and energy by starting and controlling atomic fission; *see also* REACTOR *US* **Nuclear Regulatory Commission (NRC)** = US government body which regulates and licenses nuclear plants; **nuclear reprocessing plant** = place where spent nuclear fuel is subjected to chemical processes which produce further useful materials (such as plutonium); *see also* REPROCESSING; **nuclear test** = test on a nuclear weapon; **nuclear test ban** = ban on testing of nuclear weapons; **nuclear war** = war using nuclear weapons; **nuclear waste** = radioactive waste from a nuclear reactor (including spent fuel rods and coolant); **nuclear weapon** = bomb *or* missile whose destructive power is produced by nuclear fission *or* fusion; **nuclear winter** = period which scientists believe will follow after a nuclear war, when there would be no warmth and light because dust particles would obscure the sun and most life would be affected by radiation **(b)** referring to a central group; **nuclear family** = main family group formed of two parents and their offspring; *compare* EXTENDED FAMILY

◇ **nuclear-free** ['nju:kliə 'fri:] *adjective* with no nuclear reactors; ***while some countries remain nuclear-free, nuclear reactors supply about 15% of all electricity generated in the world;*** **nuclear-free zone** = area which will not allow the use of nuclear reactors *or* the use of nuclear weapons

> COMMENT: fusion is the opposite process to fission, fission being currently used in all nuclear power stations. Fusion is used to create bombs such as the hydrogen bomb and research is being carried out into ways of using it in power stations

nucleating agent ['nju:klɪeɪtɪŋ 'eɪdʒənt] *noun* substance (solid carbon dioxide) scattered on clouds to make them release rain

◇ **nucleic acid** [nju:'kleɪɪk 'æsɪd] *noun* any of several organic acids combined with proteins (DNA *or* RNA) which exist in the nucleus and protoplasm of all cells

◇ **nucleus** ['nju:kliəs] *noun* **(a)** central core of an atom, formed of neutrons and protons **(b)** central body in a cell, containing DNA and RNA, and controlling the function and characteristics of the cell **(c)** centre, round which something gathers; **condensation nucleus** = particle on which moisture condenses, forming a raindrop; ***moisture readily condenses on an existing surface and sulphate particles are ideal condensation nuclei*** (NOTE: plural is **nuclei**)

nuée ardente ['nu:eɪ ɑ:'dɒnt] *noun* cloud of burning gas which flows downhill during a volcanic eruption

nuisance ['nju:səns] *noun* something annoying *or* disturbing *or* unpleasant; **noise nuisance** = noise which is annoying *or* disturbing *or* unpleasant; **odour nuisance** = smell which is annoying *or* unpleasant; **nuisance threshold** = point at which something, such as a noise *or* smell, becomes annoying *or* unpleasant

nuke [nju:k] *noun informal* = NUCLEAR BOMB, NUCLEAR WEAPON, NUCLEAR POWER, NUCLEAR POWER STATION; **nuke waste** = NUCLEAR WASTE

nursery ['nɜːsəri] *noun* place where plants are grown until they are large enough to be planted in their final positions

nut [nʌt] *noun* hard seed case of certain trees

nutrient ['njuːtriənt] *noun* (i) substance (such as protein *or* fat *or* vitamin) in food which is necessary to provide energy or to help the body grow, repair and maintain itself; (ii) substance which a plant needs to allow it to grow and produce seed (such as carbon, hydrogen, oxygen, nitrogen, phosphorus, potassium, calcium, magnesium and sulphur); **nutrient cycle** = process in which nutrients from living organisms are transferred into the physical environment and back to the organisms: this process is essential for organic life to continue; **nutrient requirement** = type and amount of nutrients needed by an organism; **nutrient stripping** = removal of nutrients, as, for example, from sewage in order to prevent eutrophication of water in reservoirs

> QUOTE: in the rainforest, despite the profusion of plants and trees, the underlying soils are poor, almost all the nutrients being bound up in the vegetation. Once the forests have been cut down, those few nutrients that remain in the soil are quickly washed away, transforming the land into a barren wasteland
>
> **Ecologist**

> QUOTE: poor farmers burn patches of rainforest to release some of the nutrients locked in the biomass back into the soil, but those nutrients that are released are often exhausted after only one year of intensive agriculture
>
> **New Scientist**

nutrition [njuː'trɪʃn] *noun* nourishment *or* food

◇ **nutritional** [njuː'trɪʃənəl] *adjective* referring to nutrition; ***the nutritional quality of meat;*** **nutritional disorder** = disorder (such as obesity) related to food and nutrients; **nutritional requirement** = type and amount of food needed by an organism

◇ **nutritionist** [njuː'trɪʃənɪst] *noun* dietitian, person who specializes in the study of nutrition and advises on diets

◇ **nutritious** [njuː'trɪʃəs] *adjective* providing nourishment

◇ **nutritive** ['njuːtrətɪv] *adjective* (i) providing nourishment; (ii) referring to nutrition

nyct- [nɪkt] *prefix* referring to night

◇ **nyctinasty** ['nɪktɪnæsti] *noun* response of flowers (such as the closing of petals) to darkness at night

nymph [nɪmf] *noun* insect (such as an immature dragonfly) at the stage in its development between the larval stage and adulthood

◇ **nymphal** ['nɪmfəl] *adjective* referring to the stage in the development of certain insects (such as the dragonfly) between the larval stage and adulthoood

Oo

O *chemical symbol for* oxygen

oak [əʊk] *noun* common hardwood tree of the genus *Quercus* found in temperate regions; it provides valuable timber

◇ **oak apple** *or* **oak gall** ['əʊkæpl *or* 'əʊk 'gɔːl] *noun* small hard round growth found on oak trees, caused by a parasitic wasp

◇ **oakwood** ['əʊkwʊd] *noun* wood of an oak tree

oasis [əʊ'eɪsɪs] *noun* place in an arid desert where the water table is near the surface and where vegetation can grow; in the oases of the hot desert regions, the date palm forms an important food supply; **oasis effect** = loss of water from an irrigated area due to hot dry air coming from an unirrigated area nearby

oats [əʊts] *noun* hardy cereal crop *(Avena sativa)* grown in most types of soil in cool wet northern temperate regions

obliterative shading *or* **obliterative countershading** [ə'blɪtərɪtɪv 'ʃeɪdɪŋ] *noun* grading of the colour of an animal which minimizes relief and gives a flat appearance (as when the back of the animal is dark shading towards a light belly)

OC = ORGANIC CARBON

occidental [ɒksɪ'dentəl] *adjective* referring to the west

occluded front *or* **occlusion** [ə'kluːdɪd 'frʌnt *or* ə'kluːʒn] *noun* front where warm and cold air masses meet and mix together, with the warm air rising away from the surface of the ground

occupation [ɒkjuːˈpeɪʃn] *noun* job *or* work

◇ **occupational** [ɒkjuːˈpeɪʃənl] *adjective* referring to work; **occupational disease** = disease which is caused by the type of work someone does *or* the conditions in which someone works (such as disease caused by dust *or* chemicals in a factory)

ocean [ˈəʊʃn] *noun* large area of sea. (There are four oceans: the Atlantic, the Pacific, the Indian and the Arctic); **ocean circulation** = movement of surface water between oceans, caused by wind, temperature *or* salinity; **ocean current** = movement of the surface water of an ocean, caused by wind, temperature *or* salinity; **ocean dumping** = discharging waste (solid *or* liquid *or* radioactive) into the ocean; **Ocean Thermal Energy Conversion (OTEC)** = process which is still being researched whereby the difference in temperature between the upper and lower layers of water in tropical seas is used to generate electricity and fresh water. Warmer water from the upper layer is converted to steam to drive turbines and then condensed to provide fresh water

◇ **oceanarium** [əʊʃəˈneəriəm] *noun* type of large saltwater aquarium where marine animals are kept

◇ **oceanic** [əʊʃɪˈænɪk] *adjective* referring to an ocean, especially referring to deep sea water beyond the continental shelf; (organism) which lives in the ocean; **oceanic crust** = part of the earth's crust beneath the ocean; **oceanic farm** = place where fish *or* shellfish *or* marine plants are grown; **oceanic trench** = long deep valley in the floor of the ocean, where two tectonic plates meet, usually associated with volcanic activity

◇ **oceanography** [əʊʃənˈɒgrəfi] *noun* study of all the physical aspects of the ocean, including the fauna and flora living there

◇ **oceanology** [əʊʃəˈnɒlədʒi] *noun* study of the geographical distribution of the ocean's economic resources

octane (C_8H_{18}) [ˈɒkteɪn] *noun* hydrocarbon found in petrol; ***a car which runs on low-octane petrol;*** **octane rating** *or* **octane number** = classification of the quality and performance of petrol, according to the amount of hydrocarbon in it

> COMMENT: petrol without the addition of hydrocarbons will make the engine knock. Hydrocarbons, such as octane or aromatic hydrocarbons, or lead tetraethyl, can be added to the petrol to give better performance, while increasing the octane rating. Unleaded petrol has a relatively low octane rating and leaded petrol (which contains an antiknock additive) has a high rating. Leaded petrol produces more atmospheric pollution than unleaded

Odontoceti [ədɒntəˈseti] *noun* toothed whales

odour *or US* **odor** [ˈəʊdə] *noun* smell, especially an unpleasant smell; **odour nuisance** = smell which is annoying *or* unpleasant

ODP = OZONE-DEPLETING POTENTIAL

oe- (NOTE: words beginning with **oe-** are written **e-** in American English)

OECD = ORGANIZATION FOR ECONOMIC COOPERATION AND DEVELOPMENT

oestrogen [ˈiːstrədʒən] *noun* hormone produced in the ovaries of animals, which stimulates the reproductive system; also produced in small amounts by the testis

> COMMENT: synthetic oestrogens are used in most oral contraceptives and are also used in animal feeds and in the treatment of menstrual and menopausal disorders

offend [əˈfend] *verb* to be unpleasant to someone

◇ **offensive** [əˈfensɪv] *adjective* unpleasant *or* which offends; **offensive industry** *or* **offensive trade** = trade (such as preparing leather) which causes very unpleasant smells

offpeak [ɒfˈpiːk] *adjective* (time) when electricity consumption is less; ***offpeak electricity costs less; by using thermal storage we can move 50% of electricity demand into offpeak hours***

offshore [ɒfˈʃɔː] *adjective & adverb* **(a)** in sea water near the coast; **offshore island** *or* **offshore oil platform** = island *or* platform situated a short distance (up to 20 miles) from the coast **(b)** away from the coast; **offshore wind** = wind which blows from the coast towards the sea

oil [ɔɪl] *noun* **(a)** liquid which cannot be mixed with water (there are three types: vegetable *or* animal oils, essential volatile oils and mineral oils); **cod liver oil** = oil from the liver of codfish, which is rich in calories and in vitamins A and D; **corn oil** = vegetable oil obtained from maize grains, used for cooking and as a salad oil; **essential oil** *or* **volatile oil** = concentrated oil from a scented plant used in cosmetics and as an antiseptic; **fixed oil** = oil which is liquid at 20°C; **mineral oil** = oil which derives from petroleum and is made up of hydrocarbons; **palm oil** = edible oil produced from the seed or fruit of an oil palm; **sunflower oil** = oil extracted from sunflower seeds **(b)** specifically, oil extracted from underground deposits, used to make petrol and other petroleum products; **crude oil** = oil before it is refined and processed into petrol and other products; **oil industry** = industry which extracts and processes oil; **oil installation** = plant for processing and handling oil; **oil platform** = large construction which is positioned over an oil well in the sea, containing living quarters for workers and the pumping and drilling equipment; **oil pollution** = pollution with oil, such as pollution of the sea by oil from a damaged oil tanker; **oil pool** = reservoir of oil found under rock in the earth's surface; **oil regeneration plant** = place where waste oil is reprocessed into high-grade oil; **oil rig** = large metal construction containing the drilling and pumping equipment for an oil well; **oil sand** = geological formation of sand *or* sandstone containing bitumen, which can be extracted and processed to give oil; **oil shale** = geological formation of sedimentary rocks containing oil *or* bitumen which can be extracted by crushing and heating the rock; **oil slick** = oil which has escaped into water and floats on the surface; **oil spill** = escape of oil into the environment (as from a tanker which hits rocks *or* from a ruptured pipeline); **oil tanker** = large ship specially constructed for carrying oil

◇ **oil-bearing** ['ɔɪl 'beərɪŋ] *adjective* (rock *or* sand *or* shale) which contains oil

◇ **oil-exporting country** [ɔɪlɪk'spɔːtɪŋ 'kʌntri] *noun* country which produces enough oil for its own use and to sell to other countries

◇ **oilfield** ['ɔɪlfiːld] *noun* area of rock under which one *or* more pools of oil lie, which can be exploited; ***the search is on for new oilfields to replace fields which have been exhausted***

◇ **oil-fired** ['ɔɪlfaɪəd] *adjective* (power station) which produces electricity using oil as fuel; (boiler *or* central heating) which uses oil as fuel

◇ **oil-importing country** [ɔɪlɪm'pɔːtɪŋ 'kʌntri] *noun* country which buys oil from other countries for home consumption

◇ **oily** ['ɔɪli] *adjective* containing oil

COMMENT: oil is made up of varying types of hydrocarbon together with sulphur compounds and usually occurs in combination with natural gas or water, and when these are removed it is called crude oil or crude petroleum. Refined crude oil gives various products such as petrol, LPG, diesel oil, paraffin wax, tar, etc. Crude oil is found in geological deposits, mainly in the Middle East, in the North Sea, Central America and Asia.

QUOTE: oil, because of its use for transport, industrial applications and power generation, already contributes more to greenhouse gas emissions than coal

Green Magazine

okta ['ɒktə] *noun* unit of measurement of cloud cover, meaning one eighth of the sky area

COMMENT: to measure cloud cover, the sky is divided into imaginary sections, each covering one eighth of the total. A cloudless sky is 'zero oktas', and sky which is completely covered with clouds is 'eight oktas' or 'eight eighths'

olefin *or* **olefine** ['əʊlɪfiːn] *noun* an aliphatic hydrocarbon

olig- *or* **oligo-** ['ɒlɪgəʊ] *prefix* meaning few *or* little

◇ **oligohaline** [ɒlɪgəʊ'hælaɪn] *adjective* (water) which has traces of salt in it

◇ **oligosaprobic** [ɒlɪgəʊsə'prəʊbɪk] *adjective* (animal *or* organism, such as the trout) which cannot survive in polluted water

◇ **oligotrophic** [ɒlɪgəʊ'trɒfɪk] *adjective* (water) containing few nutrients; **oligotrophic lake** = lake which has a balance between decaying vegetation and living organisms, where the lowest layer of water never loses its oxygen and where the water contains few nutrients but sustains a fish population. (This situation is typical of a young lake; over a period of time the lake will eutrophy and become richer in nutrients and also in algae); *compare* EUTROPHIC

QUOTE: input of phosphates and other nutrients has been restricted for some time, with the aim of keeping the lake oligotrophic

Nature

olive ['ɒlɪv] *noun* Mediterranean tree *(Olea europaea),* with fruit from which an edible oil can be produced; a considerable quantity of fruit is grown for direct consumption

ombrogenous [ɒm'brɒdʒənəs] *adjective* which receives rain; (bog) which gets its water only from rain; (plant) which gets its nourishment only from rain

ombudsman ['ɒmbədsmən] *noun* Parliamentary Commissioner *or* official who investigates complaints by the public against government departments *or* other large organizations

COMMENT: the main ombudsman is the Parliamentary Commissioner but there are other ombudsmen, such as the Health Service Commissioner, who investigates complaints against the Health Service, and the Local Ombudsman who investigates complaints against local authorities. Although an ombudsman will make his recommendations to the department concerned and may make his recommendations public, he has no power to enforce them. The ombudsman may only investigate complaints which are addressed to him through an MP; the member of the public first brings his complaint to his MP and if the MP cannot get satisfaction from the department against which the complaint is made, then the matter is passed to the ombudsman

omnivore *or* **omnivorous animal** ['ɒmnɪvɔː *or* ɒm'nɪvərəs 'ænɪməl] *noun* animal which eats anything, both vegetation and meat; *compare* CARNIVORE, HERBIVORE

onshore [ɒn'ʃɔː] *adjective & adverb* **(a)** situated on land; **onshore oil installation** = plant for processing and handling oil, built on land (as opposed to offshore facilities) **(b)** towards the coast; **onshore wind** = wind which blows from the sea towards the coast

ooze [uːz] **1** *noun* soft mud (especially at the bottom of a lake *or* the sea); **diatom ooze** = ooze formed from the fossil remains of diatoms; **ooze mud** = OOZE **2** *verb (of liquid)* to flow slowly

opacity [ə'pæsɪti] *noun* not allowing light to pass through

◇ **opaque** [əʊ'peɪk] *adjective* not transparent; (substance) which does not allow light *or* other rays to pass through it; **radio-opaque dye** = liquid which is introduced into soft organs (such as the kidney) so that they show up clearly on an X-ray photograph

OPEC ['əʊpek] = ORGANIZATION OF PETROLEUM EXPORTING COUNTRIES

open ['əʊpən] *adjective* not closed; **open country** *or* **open land** = area of land which is not built over *or* which does not have many trees *or* high mountains

◇ **open-air** [əʊpən'eə] *adjective* referring to the environment outside buildings

◇ **open burning** [əʊpən'bɜːnɪŋ] *noun* burning of waste matter in the open air creating pollution with smoke

◇ **open dump** ['əʊpən 'dʌmp] *noun* place where waste is left on the ground and not buried in a hole

open-cast mining *or* **open-cut mining** ['əʊpənkɑːst] *noun* form of mining where the mineral is dug from the surface instead of digging underground to get it (NOTE: also called **strip mining**)

COMMENT: open-cast mining damages the environment. The top layer of soil and rock is pushed away from the surface of the ground to expose the mineral without digging underground, destroying the natural vegetation of the mined area and its surroundings. Unless the site is filled in and replanted when mining is completed the whole area remains devastated

operate ['ɒpəreɪt] *verb* to work *or* cause to work; **this machine is operated by electricity** = is driven by *or* works by electricity

◇ **operating costs** ['ɒpəreɪtɪŋ 'kɒsts] *plural noun* amount of money required to keep a machine *or* a factory *or* a business working

◇ **operator** ['ɒpəreɪtə] *noun* person who works a machine

opportunity [ɒpə'tjuːnɪti] *noun* chance *or* situation where something can be done successfully

◇ **opportunist** *or* **opportunistic** [ɒpə'tjuːnɪst *or* ɒpətjuː'nɪstɪk] *adjective* (organism) which quickly colonizes an available habitat

orbit ['ɔːbɪt] **1** *noun* curved path of a planet *or* satellite round the earth **2** *verb* to go round a planet in a curved path

◇ **orbital** ['ɔːbɪtl] *adjective* which moves in an orbit *or* which goes round; **orbital road** = road which goes right round a town, at some distance from the built-up areas; *compare* RING ROAD

order ['ɔːdə] *noun* classification of animals *or* plants, formed of several families (NOTE: orders of animals have names ending in **-a;** orders of plants have names ending in **-ales**)

ore [ɔː] *noun* mineral found in the ground containing a metal which can be extracted from it; ***iron ore deposits were found in the mountains; the ore is heated to a high temperature to extract the metal;*** **high-grade ore** = ore which contains a large percentage of metal; **low-grade ore** = ore which contains a small percentage of metal; **ore body** = mass of ore which can be dug and processed

◇ **ore-bearing** ['ɔːbeərɪŋ] *adjective* (rock) which contains ore

organ ['ɔːgən] *noun* part of the body which is distinct from other parts and has a particular function (such as the liver *or* an eye *or* the ovaries, etc.); **critical organ** = part of the body which is particularly sensitive to radiation; **organ transplant** = taking an organ from one person and putting it into another

◇ **organic** [ɔː'gænɪk] *adjective* **(a)** referring to organs in the body; **organic disorder** = disorder caused by changes in body tissue *or* in an organ **(b)** (substance) which comes from an animal *or* a plant; (substance) which contains carbon; **organic carbon** = carbon which comes from an animal *or* plant; **organic chemistry** = branch of chemistry dealing with compounds which contain carbon **(c)** (food) which has been cultivated naturally, without any chemical fertilizers *or* pesticides; **organic farming** = method of farming which does not involve using chemical fertilizers *or* pesticides; ***organic farming may become more economic than conventional farming;*** **organic**

fertilizers = plant nutrients which are returned to the soil from dead or decaying plant matter and animal wastes, such as compost farmyard manure and bone meal

◇ **organically** [ɔː'gænɪkli] *adverb* (food) grown using natural fertilizers and not chemicals

COMMENT: organic farming uses natural fertilizers and rotates stock farming (i.e. raising of animals) with crop farming. Organic farming may produce lower yields than traditional or intensive farming, but the lower yields may be offset by the high cost of the chemical fertilizers used in intensive farming. In areas of overproduction, organic farming has the advantage of reducing crop production without loss of quality and without taking land out of agricultural use

organism ['ɔːgənɪzm] *noun* any single living plant, animal, bacterium *or* fungus

Organization for Economic Cooperation and Development (OECD) *noun* international intergovernmental association set up in 1961 to coordinate the economic policies of member nations

Organization of Petroleum Exporting Countries (OPEC) *noun* association set up in 1960 to represent the interests of the major oil-exporting nations, to fix the price of oil and the amounts which can be produced

organochlorine [ɔːgænəʊ'klɔːriːn] *noun* chlorinated hydrocarbon, any of several chemical compounds containing chlorine, used as an insecticide (such as aldrin, dieldrin, etc.)

COMMENT: organochlorine insecticides are very persistent, with a long half-life of up to 15 years, while organophosphorous insecticides have a much shorter life. Chlorinated hydrocarbon insecticides not only kill insects, but also enter the food chain and kill small animals and birds which feed on the insects

organophosphorous compound [ɔːgənəʊ'fɒsfərəs 'kɒmpaʊnd] *noun* an organic compound containing phosphorus

◇ **organophosphate** *or* **organophosphorous insecticide** [ɔːgənəʊ'fɒsfeɪt *or* ɔːgənəʊ'fɒsfərəs ɪn'sektɪsaɪd] *noun* any of several synthetic chemical insecticides (such as malathion), based on chemical compounds including phosphate, which attack the nervous system but are not as persistent as the organochlorines

COMMENT: organophosphates are not persistent and so do not enter the food chain. They are, however, very toxic and need to be handled carefully, as breathing in their vapour may be fatal

organotherapy [ɔːgənəʊ'θerəpi] *noun* treatment of a disease by using an extract from the organ of an animal (such as using liver extract to treat anaemia)

organo-tin paint [ɔːgənəʊ'tɪn 'peɪnt] *noun* antifouling paint (such as TBT) which is based on tin and is extremely toxic to organisms in the sea

oriental [ɔːri'entəl] *adjective* referring to the east; **Oriental Region** = biogeographical region (part of Arctogea) comprising the Indian subcontinent, Southeast Asia, Indonesia and the Philippines

origin ['ɒrɪdʒɪn] *noun* beginning; where something *or* someone comes from; **the fruit must be marked with its country of origin** = with the name of the country where it was grown

◇ **original** [ə'rɪdʒɪnəl] *adjective* from the beginning; ***the original inhabitants of the village left many years ago***

ornithology [ɔːnɪ'θɒlədʒi] *noun* study of birds

◇ **ornithological** [ɔːnɪθə'lɒdʒɪkl] *adjective* referring to ornithology

◇ **ornithologist** [ɔːnɪ'θɒlədʒɪst] *noun* scientist who studies birds

orographic effects [ɒrəʊ'græfɪk ɪ'fekts] *noun* atmospheric disturbances that are caused by, or relate to, the existence of mountains or other high land

oscillation [ɒsɪ'leɪʃn] *noun* regular movement from side to side; **Southern Oscillation** = regular cycle by which air is exchanged between the Pacific basin and the Indian Ocean, occurring every two to five years and linked to changes in the sea temperature and to the El Niño effect

-osis ['əʊsɪs] *suffix* referring to disease

osmosis [ɒz'məʊsɪs] *noun* movement of a solution from one part through a semi-permeable membrane to another part, where enough of the molecules in solution pass through the membrane to make the two solutions balance

◇ **osmoreceptor** [ɒzməʊrɪ'septə] *noun* cell in the hypothalamus which checks the level of osmotic pressure in the blood and regulates the amount of water in the blood

◇ **osmoregulation** [ɒzməʊregjuː'leɪʃn] *noun* process by which cells and simple organisms maintain a balance with the fluid in their surroundings

◇ **osmotic pressure** [ɒz'mɒtɪk 'preʃə] *noun* pressure required to stop the flow of a solvent through a membrane

ost- *or* **osteo-** ['ɒstiəʊ] *prefix* referring to bone

◇ **osteomalacia** [ɒstiəʊmə'leɪʃə] *noun* condition where the bones become soft because of lack of calcium or phosphate

◇ **osteophony** [ɒsti'ɒfəni] *noun* bone conduction *or* conduction of sound waves to the inner ear through the bones of the skull (as opposed to air conduction)

OTEC = OCEAN THERMAL ENERGY CONVERSION

outback ['aʊtbæk] *noun (in Australia)* large area of wild or semi-wild land in the centre of the continent; ***many wild animals which used to live in the outback are becoming rare as the outback is reclaimed for farming***

◇ **outbreak** ['aʊtbreɪk] *noun* series of cases of a disease which start suddenly; ***there is an outbreak of typhoid fever*** *or* ***a typhoid outbreak in the town***

◇ **outbreeding** ['aʊtbri:dɪŋ] *noun* breeding between individuals who are not related; *compare* INBREEDING

◇ **outcrop** ['aʊtkrɒp] **1** *noun* area of rock which stands out above the surface of the soil **2** *verb (of rock)* to stand out above the surface of the soil

outer ['aʊtə] *adjective* further outside *or* further away from the centre; ***the outer coating of the seed is very hard;*** **outer continental shelf** = edge of the continental shelf which is furthest away from the shore; **outer space** = area outside the earth's atmosphere in which the sun, stars and planets move; **outer suburbs** = residential part of a town, furthest away from the centre, but still within the built-up area

outfall (sewer) ['aʊtfɔ:l] *noun* pipe which takes sewage (either raw *or* treated) and discharges it into a river *or* lake *or* the sea; ***27% of outfalls discharge totally untreated sewage; sewage should be discharged through outfall pipes which are situated far enough from the shore***

> QUOTE: objectors claim a sea outfall pipe into the bay would spoil local shellfish, harm tourism and would be liable to damage by strong tidal action
> **Environment Now**

outlet ['aʊtlət] *noun* opening *or* channel through which something can go out

outlier ['aʊtlaɪə] *noun* (i) organism that occurs naturally some distance away from the principal area in which its species is found; (ii) area in which younger rocks are completely surrounded by older rocks

output ['aʊtpʊt] *noun* amount of energy *or* of work *or* of a substance *or* of manufactured goods, etc. produced

outwash ['aʊtwɔ:ʃ] *noun* water which flows from a melting glacier and creates deposits of silt called outwash fans *or* outwash deposits

ova ['əʊvə] *see* OVUM

over- ['əʊvə] *prefix* too much

◇ **overburden** [əʊvə'bɜ:dən] *noun* soil and rock lying on top of a coal seam *or* a mineral vein which is dug away from the surface of the land in strip mining to expose the coal *or* mineral below; **overburden pressure** = force of the soil *or* rock pressing down onto a coal seam *or* mineral vein

◇ **overcome** [əʊvə'kʌm] *verb* **(a)** to fight something and win; ***they failed to overcome public opposition to the construction of a nuclear power station*** **(b)** to make someone lose consciousness; ***two people were overcome by smoke in the fire***

◇ **overcropping** [əʊvə'krɒpɪŋ] *noun* growing too many crops on poor soil, which has the effect of weakening the soil still further

◇ **overcultivated** [əʊvə'kʌltɪveɪtɪd] *adjective* (land) which has been too intensively cultivated and so is exhausted

◇ **overexploit** [əʊvəɪk'splɔɪt] *verb* to cultivate soil *or* to work mineral deposits, etc. so much that they become exhausted; ***we overexploit land in the same way as we overexploit the sea; irrigated land given over to cash crops can be overexploited***

◇ **overexploitation** [əʊvəeksplɔɪ'teɪʃn] *noun* cultivating soil *or* working mineral deposits, etc. so much that they become exhausted; ***overexploitation has reduced herring stocks by half***

◇ **over-fertilization** [əʊvəfɜ:tɪlaɪ'zeɪʃn] *noun* putting too much fertilizer on land. (The runoff from over- fertilization can cause water pollution)

◇ **overfish** [əʊvə'fɪʃ] *verb* to catch too many fish (in the sea *or* in a river, etc.) so that the fish become rare

◇ **overfishing** [əʊvə'fɪʃɪŋ] *noun* catching too many fish; ***herring stocks have been reduced by overfishing***

◇ **overflow 1** *noun* ['əʊvəfləʊ] excess liquid which flows over the edge of a container; **overflow pipe** = pipe attached to the top of a container to channel away excess liquid **2** *verb* [əʊvə'fləʊ] to flow over the edge of a container; ***the floods made the reservoir overflow; the river overflowed its banks and flooded hundred of hectares of farmland***

◇ **overgraze** [əʊvə'greɪz] *verb* to graze a pasture so much that it loses nutrients and is no longer rich enough to provide food for livestock

◇ **overgrazing** [əʊvə'greɪzɪŋ] *noun* grazing a pasture so much that it loses nutrients and is no longer rich enough to provide food for livestock; ***overgrazing has led to soil erosion and desertification***

◇ **overland flow** ['əʊvəlænd 'fləʊ] *noun* movement of rainwater *or* meltwater over the surface of the ground in a broad thin layer

◇ **overproduction** [əʊvəprə'dʌkʃn] *noun* producing too much; ***beef mountains and wine lakes are caused by overproduction in the EU***

> QUOTE: of the almost 500 million hectares of land that has been degraded by human action throughout Africa, about half has resulted primarily from livestock overgrazing
> **Environmental Conservation**

overshot wheel [əʊvə'ʃɒt 'wi:l] *noun* type of waterwheel where the water falls on the wheel from above. (It is more efficient than the undershot wheel, where the water flows underneath the wheel)

◇ **overstorey** *or* **overwood** ['əʊvəstɔ:ri *or* 'əʊvəwʊd] *noun* topmost vegetation layer in a forest formed by the highest trees

overuse 1 *noun* [əʊvə'ju:s] using something too much; ***the overuse of pesticides is contaminating the rivers*** **2** *verb* [əʊvə'ju:z] to use something too much; ***farmers are warned against overusing chemical fertilizers***

overwinter [əʊvə'wɪntə] *verb* (i) to spend winter in a particular place; (ii) to remain alive though the winter; ***the herds overwinter on the southern plains; geraniums will not overwinter in the garden in this part of the country***

ovicide ['əʊvɪsaɪd] *noun* substance, especially an insecticide, which kills eggs

oviparous [əʊ'vɪpərəs] *adjective* (animal) which carries and lays eggs; *compare* VIVIPAROUS

ovulate ['ɒvju:leɪt] *verb* to release a mature ovum from the ovary

◇ **ovulation** [ɒvju:'leɪʃn] *noun* release of a mature ovum from the ovary

◇ **ovum** ['əʊvəm] *noun* egg (NOTE: plural is **ova**)

> COMMENT: at regular intervals (in the human female, once a month) ova, or unfertilized eggs, leave the ovaries and move down the Fallopian tubes to the uterus; at the point where the Fallopian tubes join the uterus an ovum may be fertilized by a sperm cell

ox-bow lake ['ɒksbəʊ 'leɪk] *noun* curved lake, formed when a meander of a river becomes cut off from the main body of the river by silt

oxidant ['ɒksɪdənt] *noun* = OXIDIZING AGENT

oxidase ['ɒksɪdeɪz] *noun* enzyme which encourages oxidation by removing hydrogen

◇ **oxidation** [ɒksɪ'deɪʃn] *noun* action of forming an oxide by the reaction of oxygen with another chemical substance; **oxidation ditch** *or* **oxidation pond** = ditch *or* pond where sewage is purified by allowing biochemical reactions to take place in it over a period of time

◇ **oxidation-reduction** [ɒksɪ'deɪʃn rɪ'dʌkʃn] *noun* reversible chemical reaction between two substances where one is oxidized (gains oxygen atoms) and the other is reduced (loses oxygen atoms)

◇ **oxide** ['ɒksaɪd] *noun* chemical compound formed with oxygen (such as carbon dioxide, nitric oxide, etc.)

◇ **oxidizability** [ɒksɪdaɪzə'bɪlɪti] *noun* ability of a substance to oxidize

◇ **oxidizable matter** [ɒksɪ'daɪzəbl 'mætə] *noun* substance which can oxidize

◇ **oxidize** ['ɒksɪdaɪz] *verb* to form an oxide by the reaction of oxygen with another chemical substance; ***under oxidizing conditions which exist in the tropical atmosphere, methane is converted into carbon dioxide***

◇ **oxidizing agent** ['ɒksɪdaɪzɪŋ 'eɪdʒənt] *noun* substance that forms an oxide with another substance

> COMMENT: carbon compounds form oxides when metabolized with oxygen in the body, producing carbon dioxide

> QUOTE: decaying vegetable matter, like raw sewage, does break down in water through the natural process of oxidation
> **Environment Now**

oxychlorination [ɒksɪklɔ:rɪ'neɪʃn] *noun* process of neutralizing bacteria in water intended for drinking

oxygen ['ɒksɪdʒən] *noun* chemical element, a common colourless gas which is present in the air, essential to biological life and is widespread in rocks (NOTE: chemical symbol is **O**; atomic number is **8**)

◇ **oxygenate** ['ɒksɪdʒəneɪt] *verb* to treat (blood) with oxygen; to become filled with oxygen

◇ **oxygenation** [ɒksɪdʒə'neɪʃn] *noun* treating blood with oxygen; becoming filled with oxygen

> COMMENT: oxygen is an important constituent of living matter, as well as water and air. Oxygen is absorbed from the air

into the bloodstream through the lungs and is carried to the tissues along the arteries; it is essential to normal metabolism. Oxygen is formed by plants from carbon dioxide in the atmosphere during photosynthesis and released back into the air

oxyhaemoglobin [ɒksɪhiːmə'gləʊbɪn] *noun* compound of haemoglobin and oxygen, which is the way oxygen is carried in arterial blood from the lungs to the tissues; *see also* HAEMOGLOBIN

◇ **oxyphobe** ['ɒksɪfəʊb] *noun* plant which cannot survive in acid soil

◇ **oxyphyte** ['ɒksɪfaɪt] *noun* plant which lives on acid soil

◇ **oxytocin** [ɒksɪ'təʊsɪn] *noun* hormone produced by the pituitary gland, which controls the contractions of the uterus and encourages the flow of milk

COMMENT: an extract of oxytocin is used as an injection to start contractions of the uterus

ozone (O_3) ['əʊzəʊn] *noun* poisonous form of oxygen found naturally in the atmosphere, which is toxic to humans at concentrations above 0.1 parts per million; **ozone hole** = thin part *or* gap in the ozone layer, which forms over Antarctica each year at the end of winter; **ozone layer** = ozonosphere *or* layer of ozone in the atmosphere between 20 and 50km above the surface of the earth; **ozone monitoring device** = device which measures the levels of ozone in the atmosphere

◇ **ozone-depleting potential (ODP)** [əʊzəʊndɪ'pliːtɪŋ pə'tenʃl] *noun* measurement of the effect of a substance on ozone in the atmosphere; ***CFC-12 has an ODP of one***

◇ **ozonize** ['əʊzəʊnaɪz] *verb* to convert (oxygen) into ozone; to treat (a substance) with ozone

◇ **ozonosphere** [əʊ'zəʊnəsfiːə] *noun* ozone layer *or* layer of ozone in the atmosphere between 20 and 50km above the surface of the earth

COMMENT: ozone is created in the stratosphere by the effect of ultraviolet radiation from the sun on oxygen. Ozone also splits and becomes oxygen again as part of a continuous cycle of chemical change. It is destroyed by reaction with nitric oxide (created by burning fossil fuel) or water or chlorine compounds (from chlorofluorocarbons used in aerosols and packaging). The reduction of ozone in the stratosphere by any of these reactions creates a thin area or 'hole' in the ozone layer. The ozone layer in the stratosphere acts as a protection against the harmful effects of the sun's radiation, and the destruction or reduction of the layer has the effect of allowing more radiation to pass through the atmosphere with harmful effects (such as skin cancer) on humans. The first ozone hole was detected over Antarctica, but it is now suspected that others may be developing in other parts of the stratosphere. In the lower atmosphere, ozone is created by the effect of sunlight on hydrocarbons and nitrogen oxides released from burning fossil fuels. It is a major constituent of smog. The maximum amount of ozone which is considered safe for humans to breathe is 80 parts per billion. Even in lower concentrations it irritates the throat, makes people cough and gives headaches and asthma attacks similar to hay fever. At these lower levels ozone has a harmful effect on green plants and can kill trees such as conifers

QUOTE: the ozone layer filters out ultraviolet radiation reaching earth from the sun. In the summer, the sun's elevation is high and the amount of ultraviolet light that reaches the ground through the ozone hole will be much higher than in the early days of spring when the hole first forms
New Scientist

QUOTE: while the natural layer of ozone in the upper atmosphere is critical to the preservation of earth-bound life, at ground level ozone forms when hydrocarbons and nitrogen oxides emitted by vehicles, factories, power plants and hundreds of other sources chemically react with sunlight
Environmental Action

Pp

P *chemical symbol for* phosphorus

pack [pæk] *noun* **(a)** group of predatory animals (especially of the dog family, such as wolves, but also referring to animals such as killer whales which live and hunt together) **(b)** **pack ice** = large area of ice floating at sea consisting of a mixture of ice of various sizes and ages crushed together so that there is little or no open water

package ['pækɪdʒ] *noun* object wrapped in paper, plastic, cardboard, etc.

◇ **packaged** ['pækɪdʒd] *adjective* wrapped in paper, plastic, cardboard, etc.

◇ **packaging** ['pækɪdʒɪŋ] *noun* paper, plastic, cardboard, etc. used to wrap an object; ***many of the packaging materials we used to throw away may now be re-used to save energy and raw materials***

paddy *or* **padi** ['pædi] *noun* growing rice crop; **paddy (field)** = field filled with water, in which rice is grown; ***paddies are breeding grounds for mosquitoes; scientists believe that paddy fields contribute to the methane in the atmosphere***

PAH = POLYCYCLIC AROMATIC HYDROCARBON

paint [peɪnt] *noun* liquid substance put on a surface to give it colour *or* to protect it; **antifouling (paint)** = special pesticide painted onto the bottom of a ship to prevent organisms growing on the hull and which may be toxic enough to pollute sea water; **lead paint** = paint containing lead, which makes it more durable, but which is forbidden on children's toys and furniture

palae- *or* **palaeo-** *or US* **paleo-** *or* **pale-** ['pæliəʊ] *prefix* meaning ancient *or* prehistoric; **palaeobotany** = study of fossil plants; **palaeoclimatology** = scientific study of the climate of the geological past; **palaeoecology** = study of the ecology of fossils; **palaeontology** = study of fossil organisms; **palaeozoology** = study of fossil animals

◇ **Palaearctic Region** [pælɪ'ɑːktɪk 'riːdʒən] *noun* biogeographical region (part of Arctogea) covering Europe, North Asia and North Africa

◇ **palaeomagnetism** [pæliəʊ'mægnɪtɪzm] *noun* study of the magnetism of ancient rocks

> COMMENT: rocks indicate the direction (and therefore the position) of the magnetic pole (which slowly changes its position over thousands of years) at the time the rocks were formed

palm (tree) [pɑːm] *noun* large tropical plant like a tree with branching fern-like leaves, producing fruits which give oil and other foodstuffs; **coconut palm** = palm which produces coconuts; **date palm** = palm which produces dates; **palm oil** = edible oil produced from the seed *or* fruit of a palm; **palm kernel oil** = vegetable oil produced from the kernels of the oil palm nut

paludism ['pæljuːdɪzm] *noun* malaria, a tropical disease caused by a parasite *Plasmodium* which enters the body after a bite from a mosquito

> COMMENT: paludism is a recurrent disease which produces regular periods of shivering, vomiting, sweating and headaches as the parasites develop in the body; the patient also develops anaemia

palynology [pælɪ'nɒlədʒi] *noun* scientific study of pollen, especially of pollen found in peat and coal deposits

pampas ['pæmpəs] *noun* wide grassy plains found in South America; **pampas grass** = type of tall feathery grass found on the plains of South America and grown elsewhere for ornament

> COMMENT: the eastern pampas, which has a higher rainfall, has a natural covering of tall coarse grass, known as pampas grass. Vast areas of the pampas are now cultivated, and cattle and sheep are reared on them

pan [pæn] *verb* to search for precious minerals, such as gold, by passing sandy deposits through a sieve in running water, allowing the smaller particles to be washed away

PAN = PEROXYACETYL NITRATE

pan- [pæn] *prefix* meaning generalized *or* affecting everything

◇ **pandemic** [pæn'demɪk] *adjective or noun* epidemic disease which affects many parts of the world; *compare* ENDEMIC, EPIDEMIC

panemone ['pænɪməʊn] *noun* type of windmill, where flat surfaces spin round a vertical axis

panicle ['pænɪkl] *noun* inflorescence where several small flowers branch from the same stem

paper ['peɪpə] *noun* substance made from the pulp of wood, rags, etc., rolled into sheets for writing on, wrapping, etc.; **paper industry** = business of manufacturing paper; **paper mill** =

factory where paper is manufactured; **paper pulp** = wet mixture of pulverized fibres of wood, rags, etc. from which paper is made

para- ['pærə] *prefix* meaning (i) similar to *or* near; (ii) changed *or* beyond

paraffin ['pærəfɪn] *noun* **(a) paraffin (oil)** = kerosene *or* oil produced from petroleum, used as a fuel in aircraft engines, for domestic heating and lighting and as a solvent **(b) paraffin (wax)** = white insoluble solid which melts between 50 and 60°C, used to make candles, as a waterproofing and as the base of some ointments **(c)** *(in chemistry)* saturated aliphatic hydrocarbon

parallel ['pærəlel] *noun* imaginary line running round the earth, linking points at an equal distance from the equator

COMMENT: the parallels are numbered and some of them act as national boundaries: the 49th parallel marks most of the border between the USA and Canada

parameter [pə'ræmɪtə] *noun* characteristic of a population (such as mean height *or* mean wing length) for which a measurement is attempted

paraquat ['pærəkwɒt] *noun* non-selective contact herbicide; it becomes quite inert on contact with the soil

parasite ['pærəsaɪt] *noun* plant *or* animal which lives on or inside another organism (the host) and derives its nourishment and other needs from it; **social parasite** = parasite which benefits from the host's normal behaviour (such as the cuckoo, which lays its eggs in the nest of another bird who brings up the cuckoo's young as if it were its own); *see also* ECTOPARASITE, ENDOPARASITE

◇ **parasitic** [pærə'sɪtɪk] *adjective* referring to parasites; ***a parasitic plant;*** **parasitic life** = organisms which are parasites; **parasitic disease** = disease caused by a parasite

◇ **parasiticide** [pærə'sɪtɪsaɪd] *noun & adjective* (substance) which kills parasites

◇ **parasitism** ['pærəsaɪtɪzm] *noun* state in which one organism (the parasite) lives on or inside another organism (the host) and derives its nourishment and other needs from it

◇ **parasitize** ['pærəsɪtaɪz] *verb* to live as a parasite on (another organism); ***sheep are parasitized by flukes***

◇ **parasitoid** ['pærəsaɪtɔɪd] *noun* animal which is a parasite only at one stage in its development

◇ **parasitology** [pærəsaɪ'tɒlədʒi] *noun* scientific study of parasites

COMMENT: the commonest parasites affecting animals are lice on the skin and various types of worms in the intestines. Many diseases of humans (such as malaria and amoebic dysentery) are caused by infestation with parasites. Viruses are parasites on animals, plants and even on bacteria

QUOTE: individual cuckoos specialize on particular hosts: some parasitize only reed warblers, others prefer pied wagtails
BBC Wildlife

parathion [pærə'θaɪən] *noun* organophosphorous insecticide used to control insects

parcel ['pɑːsəl] *noun* **(a)** quantity of wood, either growing in a forest *or* felled, which is sold **(b) parcel of air** = large mass of air

parent ['peɪərənt] *noun* mother *or* father; **parent cell** *or* **mother cell** = original cell which splits into daughter cells during mitosis; **parent plant** = plant from which others are produced

park [pɑːk] *noun* area of open land used as a place of recreation; **business park** *or* **industrial park** *or* **science park** = area of land with buildings specially designed and constructed for business premises *or* light industries *or* science; **country park** = area in the countryside set aside for the public to visit and enjoy; **marine park** = natural park created on the bottom of the sea (as on a tropical reef) where visitors go into observation chambers under the sea to look at the fish and plant life; *compare* OCEANARIUM; **national park** = large area of unspoilt land, owned and managed by the government for recreational use by the public

part [pɑːt] *noun* piece, one of the sections which make up a whole; **parts per billion (ppb)** *or* **parts per hundred million (pphm)** *or* **parts per million (ppm)** *or* **parts per thousand million (pptm)** = measure of the concentration of a substance in a gas, liquid *or* solid; ***the concentration of methane in the atmosphere now averages 1.56 ppm***

parthenogenesis [pɑːθənəʊ'dʒenəsɪs] *noun* reproduction by unfertilized ova (as in aphids)

partial ['pɑːʃəl] *adjective* not complete *or* affecting only part of something; **partial eclipse** = eclipse where only part of the sun *or* moon is hidden

particle ['pɑːtɪkl] *noun* **(a)** very small piece of a substance; ***particles of volcanic ash were carried into the upper atmosphere*** **(b)** *(in physics)* part of an atom; **alpha particle** = nucleus of the same composition as a helium atom, which is emitted by the nuclei of some radioactive elements, such as radon and which, when emitted, will pass through gas but not through solids; **beta particle** = electron which will pass through thin substances such as metal and can harm living tissue; *see also* RADIATION

◇ **particulate** [pɑː'tɪkjʊlət] **1** *adjective* (i) referring to particles; (ii) made up of separate particles; ***particulate matter in the atmosphere forms the nuclei around which raindrops form*** **2** *noun* tiny solid particle of pollutant; **primary particulates** = particles of matter sent into the air from fires *or* from industrial processes *or* from volcanic eruptions *or* from sandstorms, etc.; **secondary particulates** = particles of matter formed in the air by chemical reactions such as smog

> COMMENT: the finest particulates are the most dangerous as they are easily inhaled into the bronchioles in the lungs. Fine particulates from volcanic eruptions can enter the stratosphere and have a cooling effect by preventing the heat from the sun reaching the earth's surface (i.e. the opposite of the greenhouse effect)

passage ['pæsɪdʒ] *noun* (i) long narrow channel; (ii) moving from one place to another; **air passage** = tube which takes air to the lungs; **bird of passage** = bird which migrates from one area to another, stopping for a short time before moving on

passerines *or* **Passeriformes** ['pæsəraɪnz *or* pæsərɪ'fɔːmiːz] *noun* very large order of birds which perch in branches

passive ['pæsɪv] *adjective* not active; **passive margin** = area at the edge of a continental mass where there is no volcanic activity; *compare* ACTIVE MARGIN; **passive smoking** = inhaling smoke exhaled by other people, even if you are not a smoker yourself; ***dogs with short noses such as bulldogs are at the highest risk from passive smoking***

> COMMENT: even though a person may never have smoked a cigarette, it is possible for him to develop lung cancer through being continuously in the presence of other people who are smoking

> QUOTE: from a scientific view there is now a consensus that passive smoking causes some cases of lung cancer
>
> **New Scientist**

Pasteurella [pɑːstju'relə] *noun* genus of parasitic bacteria, one of which causes the plague

◇ **pasteurization** [pɑːstʃəraɪ'zeɪʃn] *noun* heating of food *or* food products to destroy bacteria

◇ **pasteurize** ['pɑːstʃəraɪz] *verb* to heat food *or* food products to destroy bacteria; ***the government is telling people to drink only pasteurized milk***

> COMMENT: Pasteurization is carried out by heating food for a short time at a lower temperature than that used for sterilization: the two methods used are heating to 72°C for fifteen seconds (the high-temperature-short-time method) or to 65°C for half an hour, and then cooling rapidly. This has the effect of killing tuberculosis bacteria

pastoral ['pɑːstərəl] *adjective (of land)* available for pasture

◇ **pastoralist** ['pɑːstərəlɪst] *noun* farmer who keeps grazing animals on pasture; ***the people most affected by the drought in the Sahara are nomadic pastoralists***

pasture ['pɑːstʃə] **1** *noun* land covered with grass *or* other small plants, used by farmers as a feeding place for animals; ***the cows are in the pasture;*** **to put cattle to pasture** = to put them onto land covered with grass; **alpine pasture** = grass fields in high mountains which are used by cattle farmers in the summer; **pasture agronomist** = person who specializes in selecting and growing varieties of grasses **2** *verb* to put animals onto land covered with grass *or* other small plants; ***their cows are pastured in fields high in the mountains***

◇ **pastureland** ['pɑːstʃəlænd] *noun* land covered with grass *or* other small plants, used by farmers as a feeding place for animals

> QUOTE: scarcity of pastureland which cattle need in abundant quantities often leads to deforestation and hill erosion
>
> **Environmental Action**

path- *or* **patho-** [pæθ] *prefix* referring to disease

◇ **pathogen** ['pæθədʒən] *noun* germ, virus, etc.; microorganism which causes a disease

◇ **pathogenesis** [pæθə'dʒenəsɪs] *noun* origin *or* production *or* development of a disease

◇ **pathogenetic** [pæθədʒə'netɪk] *adjective* referring to pathogenesis

◇ **pathogenic** [pæθə'dʒiːnɪk] *adjective* which can cause *or* produce a disease; **pathogenic bacteria** *or* **organisms** = bacteria *or* organisms responsible for transmitting a disease

◇ **pathogenicity** [pæθədʒə'nɪsəti] *noun* ability of a pathogen to cause a disease

◇ **pathological** [pæθə'lɒdʒɪkl] *adjective* referring to a disease *or* which is caused by a disease; which indicates a disease; **pathological report** = report on tests carried out to find the cause of a disease; **pathological waste** = waste (such as waste from a hospital) which may contain pathogens and which could cause disease

◇ **pathologist** [pə'θɒlədʒɪst] *noun* **(a)** doctor who specializes in the study of diseases and the changes in the body caused by disease **(b)** doctor who examines dead bodies to find out the cause of death

◇ **pathology** [pə'θɒlədʒi] *noun* study of diseases and the changes in structure and function which diseases cause in the body; **clinical pathology** = study of disease as applied to treatment of patients; **pathology report** = report on tests carried out to find the cause of a disease

Pb *chemical symbol for* lead

PBB = POLYBROMINATED BIPHENYL

PCB = POLYCHLORINATED BIPHENYL

peak [pi:k] *noun* (i) top of a mountain; (ii) high point on a graph *or* in a series of figures; ***the annual rainfall has seasonal peaks and troughs;*** **concentration peak** = largest amount of a substance in a solution *or* in a certain volume; **peak period** = time when most travellers are travelling *or* when most consumers are using electricity, etc.; *see also* OFFPEAK; **peak traffic** = largest number of vehicles within any given period

peasouper [pi:'su:pə] *noun (informal)* type of thick dark yellow fog caused by sulphur particles from burning coal, formerly common in London in the winter, but not seen since the 1960s

peat [pi:t] *noun* wet, partly-decayed mosses and other plants which form the soil of a bog; **peat bog** = soft wet land where peat has formed

◇ **peat-free** ['pi:t 'fri:] *adjective* (material, such as coir) which does not contain peat and can be used in horticulture as an alternative to peat

◇ **peatland** ['pi:tlænd] *noun* area of land covered with peat bog

◇ **peaty** ['pi:ti] *adjective* containing peat

COMMENT: peat can be cut and dried in blocks, which can then be used as fuel. In some countries there are peat-fired power stations. It is in fact an early form of coal, as coal is peat which has been compressed by layers of sediment until it becomes rock

pebble ['pebl] *noun* small piece of rock, less than 64mm in diameter

pelagic [pə'lædʒɪk] *adjective* referring to the top and middle layers of sea water; **pelagic deposits** *or* **pelagic sediment** = material that has fallen to the floor of the pelagic zone; **pelagic organism** = organism which lives in open water in the sea *or* in a lake, but not on the bottom, or near the shore; **pelagic zone** = part of the sea which is open, not near the shore and not immediately above the seabed; *compare* DEMERSAL

penalty ['penəlti] *noun* fine *or* other action taken to punish someone for doing something; ***the government has imposed penalties on industrial concerns which emit too much pollution***

peneplain ['pi:nɪpleɪn] *noun* plain formed after mountains have been completely eroded

Penicillium [penɪ'sɪliəm] *noun* fungus from which penicillin is derived

◇ **penicillin** [penɪ'sɪlɪn] *noun* common antibiotic produced from a fungus

COMMENT: penicillin is effective against many microbial diseases, but some people can be allergic to it and this fact should be noted on medical record cards

peninsula [pe'nɪnsjʊlə] *noun* long narrow piece of land, surrounded on three sides by sea

◇ **peninsular** [pe'nɪnsjʊlə] *adjective* referring to a peninsula

pentad ['pentæd] *noun* five-day period (used especially in meteorological forecasting and recording)

pepsin ['pepsɪn] *noun* enzyme in the stomach which breaks down the proteins in food

per [pɜ:] *preposition* for every; ***we travelled at 90 kilometres per hour; the machine works at 2,000 revolutions per minute; the cost will be $20 per person;*** **per annum** = for *or* in every year; ***245,000 sq km of land per annum is cleared of trees;*** **per capita** = of *or* for each person; **per capita income** = amount of money received by each person; **per head** = for each person; ***average annual consumption has increased to 29 litres per head***

perceived noise level (PNL) [pə'si:vd 'nɔɪz 'levəl] *noun* measurement of the loudness of a sound as heard by the human ear; **perceived noise level in decibels (PNdB)** = measurement of sound pressure in decibels

percentage [pə'sentɪdʒ] *noun* proportion *or* rate in every hundred *or* for every hundred; ***developing countries possess by far the largest percentage of the world's rainforest;*** **per cent** = in *or* for every hundred; ***only one per cent of all prosecutions for water pollution were successful;*** **there has been a five per cent increase in applications** = the number of applications has gone up by five in every hundred (usually written % with figures: **we need to increase output by 5%**)

percolate ['pɜ:kəleɪt] *verb* to trickle slowly through a quantity of solid particles; **percolating filter** *or* **trickling filter** = filter bed through which liquid sewage is passed to purify it

> QUOTE: rainfall is limited in this part of Florida, and what rain does fall percolates rapidly through the white sand to a depth beyond the roots of the plants
>
> **Natural History**

perennial [pə'reniəl] **1** *adjective* which lasts for many years; **perennial agriculture** = system of agriculture (as in the tropics) where there is no winter and several crops can be grown on the same land each year **2** *adjective & noun* (plant) which lives for a long time, flowering each year without dying; *compare* ANNUAL, BIENNIAL

period ['pi:riəd] *noun* interval of geological time, the subdivision of an era, and divided into epochs

◇ **periodic** [pi:ri'ɒdɪk] *adjective* referring to periods *or* which occurs from time to time; **periodic table** = table of elements listed according to their atomic mass

peripheral [pə'rɪfərəl] *adjective* at the edge; **peripheral nerves** = pairs of motor and sensory nerves which branch out from the brain and spinal cord; **peripheral nervous system (PNS)** = all the nerves in different parts of the body which are linked and governed by the central nervous system

permafrost ['pɜ:məfrɒst] *noun* ground which is permanently frozen, as in the Arctic regions, where the top layer of soil melts and softens in the summer while the soil beneath remains frozen

permanent ['pɜ:mənənt] *adjective* which exists always; **permanent grassland** *or* **permanent pasture** = land which remains solely as grassland over a long period of time and is not ploughed; **permanent hardness** = hardness of water, caused by calcium and magnesium, which remains even after the water has been boiled

permeability [pɜ:miə'bɪləti] *noun (of a rock)* ability to allow water to pass through; *(of a membrane)* ability to allow fluid containing chemical substances to pass through

◇ **permeable** ['pɜ:miəbl] *adjective* (rock) which allows water to pass through; (membrane) which allows fluid containing chemical substances to pass through (NOTE: US English is **pervious)**

peroxyacetyl nitrate (PAN) [pərɒksɪə'setɪl 'naɪtreɪt] *noun* substance contained in photochemical smog, which is extremely harmful to plants

persist [pə'sɪst] *verb* to continue for some time; to remain active for a period of time; ***some substances persist in toxic forms in the air for weeks***

◇ **persistence** [pə'sɪstəns] *noun* ability of a pollutant to remain active for a period of time

◇ **persistent** [pə'sɪstənt] *adjective* which continues for some time; which remains active for a period of time; **persistent insecticide** *or* **pesticide** = insecticide *or* pesticide that remains toxic (either in the soil *or* in the body of an animal) and is passed from animal to animal through the food chain (NOTE: the opposite is **non-persistent)**

> QUOTE: the persistence of very low ozone levels into the southern hemisphere's summer could have serious ecological effects
>
> **New Scientist**

pervious ['pɜ:viəs] *adjective US* (rock) which allows water to pass through; (membrane) which allows fluid containing chemical substances to pass through (NOTE: GB English is **permeable)**

pest [pest] *noun* animal *or* plant which is troublesome *or* harmful to people, such as farmers; ***a spray to remove insect pests;*** **pest control** = keeping down the number of pests (by killing them *or* preventing them from attacking)

> COMMENT: the word is a relative term: a pest to one person may not be a pest to another, so foxes are pests to chicken farmers, but not to naturalists

pesticide ['pestɪsaɪd] *noun* substance which kills pests; **pesticide residue** = amount of pesticide that remains in the environment after spraying

> COMMENT: there are three basic types of pesticide. 1. organochlorides, which have a high persistence in the environment of up to about 15 years (DDT, dieldrin and aldrin); 2. organophosphates, which have an intermediate persistence of several months (parathion, carbaryl and malathion); 3. carbamates, which have a low persistence of around two weeks (Tenik, Zectran and Zineb). Most pesticides are broad-spectrum, that is they kill all insects in a certain area and may kill other animals like birds and small mammals. Pesticide residue levels in food in the UK are generally low. Pesticide residues have been found in bran products, bread and baby foods, as well as in milk and meat. Where pesticides are found, the levels are low and rarely exceed international maximum residue levels

PET = POLYETHYLENE TEREPHTHALATE

petal ['petl] *noun* outer coloured part of the corolla of a flower

petrify ['petrɪfaɪ] *verb* to turn into stone; **petrified forest** = remains of trees which have been petrified and are found in other rocks

◇ **petrifaction** [petrɪ'fækʃn] *noun* turning into stone

petrochemical [petrəʊ'kemɪkl] *noun* chemical derived from petroleum *or* natural gas; **petrochemical industry** = industry which processes petroleum *or* natural gas and produces petrochemicals

◇ **petrochemistry** [petrəʊ'kemɪstrɪ] *noun* (i) scientific study of the chemical composition of petroleum and substances derived from it; (ii) scientific study of the chemical composition of rocks

petrol ['petrəl] *noun* liquid, made from petroleum, used as a fuel in internal combustion engines; ***the car is very economic on petrol; we are looking for a car with a low petrol consumption;*** **petrol engine** = engine which uses petrol as a fuel

◇ **petrol-engined** ['petrəl 'endʒɪnd] *adjective* (motor vehicle) which uses petrol as a fuel (NOTE: US English is **gasoline**)

> COMMENT: in a petrol engine the petrol is mixed with air making it more combustible. It is then sprayed or injected into the cylinders, where it is ignited by an electric spark. Petrol is made of a mixture of several hydrocarbons, such as butane and benzene. It also contains additives, in particular tetraethyl lead, which prevent an engine from knocking. The use of petrol in vehicle engines is responsible for a high level of pollutants in the atmosphere, as the engines emit carbon monoxide and various lead compounds

petroleum [pə'trəʊliəm] *noun* mineral oil found in the ground; **crude petroleum** = petroleum before it is refined and processed into petrol and other products; **petroleum-exporting country** = country which produces enough oil for its own use and to sell to other countries; **petroleum gas** = natural gas occurring in combination with petroleum; **petroleum industry** = industry which makes products such as petrol, soap, paint, etc. from crude petroleum; **petroleum products** = products (such as petrol, soap, paint, etc.) which are made from crude petroleum; **petroleum revenue** = income from selling oil

> COMMENT: petroleum is made up of varying types of hydrocarbon together with sulphur compounds and usually occurs in combination with natural gas or water, and when these are removed it is called crude oil or crude petroleum. Refined crude petroleum gives various products such as petrol, LPG, diesel oil, paraffin wax, tar, etc. Crude petroleum is found in geological deposits, mainly in the Middle East, in the North Sea, Central America and Asia.

petrology [pə'trɒlədʒi] *noun* study of rocks and minerals (NOTE: not connected with **petrol**)

PFBC = PRESSURIZED FLUIDIZED-BED COMBUSTION

pH [piː'eitʃ] *noun* measure of the concentration of hydrogen ions in a solution, which shows how acid *or* alkaline it is; ***there was a sudden increase in pH; polluted water with a low pH value is more acid and corrosive;*** **pH factor** *or* **number** *or* **value** = number which indicates how acid *or* alkaline a solution is; **pH meter** = meter which measures how acid *or* alkaline a solution is; **pH test** = test to see how acid *or* alkaline a solution is

> COMMENT: the pH factor is shown as a number. A value of 7 is neutral. Lower values indicate increasing acidity and higher values indicate increasing alkalinity. So 0 is most acid and 14 is most alkaline. Acid rain has been known to have a pH of 2 or less, making it as acid as lemon juice. Most freshwater fish cannot survive even slightly acid conditions. Salmon and trout cannot stand a pH value of 6 and only pike can survive in water at less than pH 4

> QUOTE: the sulphuric and nitric acids in cloud droplets can give them an extremely low pH. Water collected near the base of clouds in the eastern US during the summer typically has a pH of about 3.6 but values as low as 2.6 have been recorded. In the greater Los Angeles area the pH of fog has fallen as low as 2
>
> **Scientific American**

> QUOTE: lack of fermentation means that the pH will often remain high - about 5, rather than dropping to 4 as in clamp silage. This can be dangerous when there is soil contamination because listeria bacteria in the soil will thrive at this pH level
>
> **Farmers Weekly**

Phaeophyta [fiːəʊ'fiːtə] *noun* brown algae such as brown seaweed

phag- *or* **phago-** ['fægəʊ] *prefix* referring to eating

◇ **-phage** [feɪdʒ] *suffix* which eats

◇ **phagocyte** ['fægəʊsaɪt] *noun* cell (such as a white blood cell) which can surround and destroy other cells, such as bacteria cells

◇ **phagocytic** [fægə'sɪtɪk] *adjective* (i) referring to phagocytes; (ii) (cell) which destroys other cells

phase [feɪz] *noun* stage *or* period of development; ***the larval phase of an insect;*** **phases of the moon** = various stages which the moon's face appears to pass through every 29 days: first quarter, full moon, last quarter and new moon

phenology [fe'nɒlədʒi] *noun* science which studies the effect of climate on annually recurring phenomena such as animal migration, plant budding, etc.

phenomenon [fə'nɒmɪnən] *noun* event that exists and is experienced; ***chemical reactions in the atmosphere generate phenomena such as acid rain*** (NOTE: plural is **phenomena**)

phenotype ['fi:nətaɪp] *noun* physical composition of an organism, produced by the genes, such as brown eyes, height, red feathers, etc.; *compare* GENOTYPE

phenylalanine [fi:nɪl'æləni:n] *noun* essential amino acid

-philia ['fɪliə] *suffix* meaning attraction *or* liking for something

pheromone ['ferəməʊn] *noun* chemical substance produced and released into the environment by an animal, influencing the behaviour of another individual of the same species

philoprogenitive [fɪləʊprəʊ'dʒenətɪv] *adjective* which produces many offspring

phloem ['fləʊəm] *noun* tissue in a plant which carries organic substances from the leaves to the rest of the plant, and is formed of living cells (as opposed to xylem which is formed of dead cells)

phosphate ['fɒsfeɪt] *noun* salt of phosphoric acid, which is an essential plant nutrient and which is formed naturally by weathering of rocks

> COMMENT: phosphates escape into water from sewage, especially in waste water containing detergents, and encourage the growth of algae by eutrophication. Natural organic phosphates are provided by guano and fishmeal; otherwise phosphates are mined. Artificially produced phosphates are used in agriculture and are known as superphosphates because they are highly concentrated

phosphorescence [fɒsfə'resəns] *noun* production of light with no heat, either caused by oxidation of phosphorus (as in sea water) *or* generated by some animals such as glow-worms

◇ **phosphorescent** [fɒsfə'resənt] *adjective* producing light with no heat

phosphoric acid [fɒs'fɒrɪk] *noun* acid which forms phosphates

◇ **phosphorous** ['fɒsfərəs] *adjective* containing phosphorus

◇ **phosphorus** ['fɒsfərəs] *noun* toxic chemical element which is essential to biological life, being present in bones and nerve tissue. (Pure phosphorus will cause burns if it touches the skin and can poison if swallowed); **phosphorus cycle** = cycle by which phosphorus atoms are circulated through living organisms (NOTE: chemical symbol is **P**; atomic number is **15**)

> COMMENT: phosphorus is an essential part of DNA and RNA, and when an organism dies the phosphorus contained in its tissues returns to the soil and is taken up by plants in the phosphorus cycle

phot- *or* **photo-** [fəʊt] *prefix* referring to light

◇ **photic zone** ['fəʊtɪk 'zəʊn] *noun* euphotic zone, top layer of water in the sea *or* a lake, which sunlight can penetrate and in which photosynthesis takes place; *compare* APHOTIC

photochemistry [fəʊtəʊ'kemɪstri] *noun* study of chemical changes brought about by light and other forms of radiation

◇ **photochemical** [fəʊtəʊ'kemɪkl] *adjective* (chemical reaction) which is caused by light; ***gases rise into the upper atmosphere and undergo photochemical change;*** **photochemical oxidant** = substance (such as ozone) which is caused by chemical reactions with rays of light; **photochemical pollution** = pollution caused by the action of light; **photochemical smog** = smog caused by the action of sunlight on polluting gases

> COMMENT: when the atmosphere near ground level is polluted with nitrogen oxides from burning fossil fuels together with hydrocarbons, ultraviolet light from the sun sets off a series of reactions that result in photochemical smog, containing, among other substances, ozone

photoconverter [fəʊtəʊkən'vɜ:tə] *noun* device which converts energy from light into electric energy

◇ **photodecomposition** [fəʊtəʊdɪkɒmpə'zɪʃn] *noun* breaking down of a substance by the action of light

◇ **photoelectric cell** [fəʊtəʊɪ'lektrɪk 'sel] *noun* cell in which light that falls on the cell is converted to electricity

◇ **photogenic** [fəʊtəʊ'dʒi:nɪk] *adjective* (i) which is produced by the action of light; (ii) which produces light

◇ **photograph** ['fəʊtəgrɑ:f] *noun* picture taken with a camera, using the chemical action of light on sensitive film; ***aerial photographs of the area showed changes in land use***

◇ **photography** [fə'tɒgrəfi] *noun* taking pictures with a camera

◇ **photonastic** [fəʊtəʊ'næstɪk] *adjective* referring to photonasty

◇ **photonasty** ['fəʊtəʊnæsti] *noun* response of plants to light (without turning towards the light source)

◊ **photo-oxidant** [fəʊtəʊ'ɒksɪdənt] *noun* chemical compound produced by the action of sunlight on nitrogen oxides and hydrocarbons

◊ **photo-oxidation** [fəʊtəʊɒksɪ'deɪʃn] *noun* changing the chemical constitution of a compound by the action of sunlight; ***photo-oxidation breaks down polluted air and converts the gases to sulphur dioxide***

◊ **photoperiod** [fəʊtəʊ'pi:riəd] *noun* period in every 24 hours when an organism is exposed to daylight

◊ **photoperiodicity** [fəʊtəʊpi:riə'dɪsɪti] *noun* way in which plants and animals react to changes in the length of the period of daylight from summer to winter

◊ **photoperiodism** [fəʊtəʊ'pi:riədɪzm] *noun* response of an organism (such as growth) to the amount of daylight it receives in every 24 hours

◊ **photophilic** *or* **photophilous** [fəʊtəʊ'fɪlɪk *or* fə'tɒfɪləs] *adjective* (organism) which grows best in strong light

◊ **photorespiration** [fəʊtəʊrespɪ'reɪʃn] *noun* respiration in some types of plant which is activated by changes in light

◊ **photosensitive** [fəʊtəʊ'sensɪtɪv] *adjective* (skin *or* lens) which is sensitive to light *or* which is stimulated by light

◊ **photosensitivity** [fəʊtəʊsensɪ'tɪvɪti] *noun* being sensitive to light

photosynthesis [fəʊtəʊ'sɪnθəsɪs] *noun* process by which green plants convert carbon dioxide and water into sugar and oxygen using sunlight as energy

◊ **photosynthesize** [fəʊtəʊ'sɪnθəsaɪz] *verb* to carry out photosynthesis; ***acid rain falling on trees reduces their ability to photosynthesize***

> QUOTE: light powers photosynthesis, the process by which a plant converts carbon dioxide and water into sugars, starch and oxygen
> **Scientific American**

phototrophic [fəʊtəʊ'trɒfɪk] *adjective* (organism) which obtains energy from sunlight

◊ **phototropic** [fəʊtəʊ'trɒpɪk] *adjective* (plant *or* cell) which turns *or* grows towards a stimulus of light

◊ **phototropism** [fəʊtəʊ'trɒpɪzm] *noun* turning *or* growing towards a stimulus of light, found in certain plants

photovoltaic [fəʊtəʊvɒl'teɪɪk] *adjective* which converts the energy from electromagnetic radiation (e.g. light) into electricity; **photovoltaic cell** *or* **panel** = device which converts the energy from light into electricity

> COMMENT: as light strikes the cell the electrons in it become mobile and create electricity

phreatic [fri:'ætɪk] *adjective* referring to the water table; **phreatic gas** = gas produced when water comes into contact with magma; **phreatic water** = water in the layers of soil beneath the water table

◊ **phreatophyte** [fri:'ætəfaɪt] *noun* plant whose roots go down into the water table

phycology [faɪ'kɒlədʒi] *noun* scientific study of algae

phylogenesis *or* **phylogeny** [faɪləʊ'dʒenəsɪs *or* faɪ'lɒdʒəni] *noun* evolution of a taxon of organisms

◊ **phylum** ['faɪləm] *noun* major subdivision in the classification of organisms, below kingdom (NOTE: plural is **phyla**)

physi- *or* **physio-** ['fɪziəʊ] *prefix* referring to (i) physiology; (ii) physical

physics ['fɪzɪks] *noun* scientific study of matter, including electricity, radiation, magnetism and other phenomena which do not change the chemical composition of matter

◊ **physical** ['fɪzɪkl] *adjective* referring to matter *or* to the body; **physical geography** = geomorphology *or* study of the physical features of the earth's surface, their development and how they are related to the core beneath; **physical medicine** = branch of medicine which deals with physical disabilities *or* with treatment of disorders after they have been diagnosed

◊ **physically** ['fɪzɪkli] *adverb* referring to the body; ***an organism adapts physically to changes in temperature***

physiological [fɪziə'lɒdʒɪkl] *adjective* referring to physiology *or* to the functions of living organisms; **physiological specialization** = phenomenon whereby some members of a population appear to be identical in appearance but differ biochemically from each other

◊ **physiologist** [fɪzɪ'ɒlədʒɪst] *noun* scientist who specializes in the study of the functions of living organisms

◊ **physiology** [fɪzɪ'ɒlədʒi] *noun* scientific study of the functions of living organisms; **human physiology** = study of the functions of the human body; **plant physiology** = study of the functions of plants

phyt- *or* **phyto-** ['faɪtəʊ] *prefix* referring to plants *or* coming from plants

◊ **phytobenthos** [faɪtəʊ'benθɒs] *noun* part of a stream *or* lake covered in plants

◊ **phytochemistry** [faɪtəʊ'kemɪstri] *noun* study of the chemistry of plants

◊ **phytogeography** [faɪtəʊdʒi'ɒgrəfi] *noun* study of plants and their geographical distribution

◊ **phytome** ['faɪtəʊm] *noun* plant community

◊ **phytophagous** [faɪ'tɒfəgəs] *adjective* (animal) which is herbivorous *or* which eats plants

phytoplankton [faɪtəʊ'plæŋktən] *noun* microscopic plants which float in the sea *or* in a lake, etc.; **phytoplankton bloom** = large mass of plankton which develops regularly at different periods of the year and floats on the surface of the sea *or* of a lake, etc.

◊ **phytoplanktonic** [faɪtəʊplæŋk'tɒnɪk] *adjective* referring to phytoplankton

◊ **phytoplankter** [faɪtəʊ'plæŋktə] *noun* single microscopic plant which floats in the sea *or* in a lake, etc.

COMMENT: phytoplankton are formed mainly of diatoms and carry out photosynthesis as other plants do under the influence of the sunlight which penetrates the surface of the water. Phytoplankton are the basis of the food chain of all aquatic animals

QUOTE: excess ultraviolet light is killing off phytoplankton, which is the primary source of the oceans' ecosystem
Guardian

phytotoxic [faɪtəʊ'tɒksɪk] *adjective* (substance) which is poisonous to plants

◊ **phytotoxicant** [faɪtəʊ'tɒksɪkənt] *noun* substance which is poisonous to plants

◊ **phytotoxin** [faɪtəʊ'tɒksɪn] *noun* poisonous substance produced by a plant

PIC = PRODUCT OF INCOMPLETE COMBUSTION

piezometric *or* **piezometrical** [pi:zəʊ'metrɪk *or* pi:zəʊ'metrɪkl] *adjective* (level) reached by water under its own pressure in a borehole *or* in a piezometer

◊ **piezometer** [pi:'zɒmɪtə] *noun* instrument for measuring the pressure of a liquid

COMMENT: to measure the pressure of water in the earth, a tube is inserted into the ground and readings taken from it

pig iron ['pɪg 'aɪən] *noun* impure iron produced in a blast furnace and used to make purer forms of iron *or* steel

pigment ['pɪgmənt] *noun* substance which gives colour to a part of an organism such as blood *or* skin *or* hair *or* a plant's leaves; **blood pigment** = HAEMOGLOBIN; **respiratory pigment** = blood pigment which can carry oxygen collected in the lungs and release it in tissues; **pigment cell** = cell in an organism which contains pigment

◊ **pigmentary** ['pɪgməntri] *adjective* referring to *or* producing pigment

◊ **pigmentation** [pɪgmən'teɪʃn] *noun* colouring of the body, especially that produced by deposits of pigment

COMMENT: the body contains several substances which control colour: melanin gives dark colour to the skin and hair; bilirubin gives yellow colour to bile and urine; haemoglobin in the blood gives the skin a pink colour.

pilot project ['paɪlət 'prɒdʒekt] *noun* small-scale project carried out to see if a large-scale project will work; ***they are running a pilot project in the area for three months before deciding on the next stage***

pine (tree) [paɪn] *noun* evergreen tree of the genus *Pinus* growing in temperate latitudes; ***the north of the country is covered with pine forests;*** **pine cone** = hard case containing the seeds of a pine; **pine oil** = essential oil of the pine, used as a perfume

◊ **pinewood** ['paɪnwʊd] *noun* forest of pine trees

pipeline ['paɪplaɪn] *noun* long pipe, often underground, carrying oil, natural gas, etc.

pisciculture ['pɪsɪkʌltʃə] *noun* fish farming *or* breeding edible fish in special pools for sale as food

pit [pɪt] *noun* large hole in the ground, for example, a hole excavated for the extraction of minerals, especially coal; **gravel pit** = area of land from which gravel is extracted

pitch [pɪtʃ] *noun* **(a)** frequency of a sound: a low-pitched sound has a low frequency and a high-pitched sound has a high frequency **(b)** dark sticky substance obtained from tar, used to make objects watertight; **mineral pitch** = ASPHALT

pituitary body *or* **pituitary gland** [pɪ'tju:ɪtəri] *noun* main endocrine gland in the body

COMMENT: the pituitary gland is about the size of a pea and hangs down from the base of the brain. The front lobe of the gland secretes several hormones which stimulate the adrenal and thyroid glands, and stimulate the production of sex hormones, melanin and milk. The posterior lobe of the gland secretes oxytocin. The pituitary gland is the most important gland in the body because the hormones it secretes control the functioning of the other glands

placenta [plə'sentə] *noun* tissue which grows inside the uterus in mammals during pregnancy and links the embryo to the mother

◇ **placental** [plə'sentəl] *adjective* referring to the placenta

COMMENT: the placenta allows an exchange of oxygen and nutrients to be passed from the mother to the embryo. It does not exist in marsupials

plagioclimax [pleɪdʒiə'klaɪmæks] *noun* stage in the development of a plant ecosystem where the system is stable because of outside interference, as in managed woodlands

plague [pleɪg] *noun* **(a)** infectious disease which occurs in epidemics where many organisms are killed; **bubonic plague** = fatal disease caused by *Pasteurella pestis* in the lymph system transmitted to humans by fleas from rats **(b)** widespread infestation by a pest; ***a plague of locusts has invaded the region and is destroying crops***

COMMENT: bubonic plague was the Black Death of the Middle Ages; its symptoms are fever, delirium, vomiting and swelling of the lymph nodes

plain [pleɪn] *noun* level country with few trees

plan [plæn] **1** *noun* **(a)** organized way of doing something; **contingency plan** = plan which will be put into action if something unexpected happens; **development plan** = plan drawn up by a government *or* local council to show how an area should be developed over a long period **(b)** drawing which shows how something is arranged *or* how something will be built; **street plan** *or* **town plan** = map of a town showing streets and buildings **2** *verb* to organize carefully how something should be done

◇ **planned** [plænd] *adjective* **planned economy** = economic system where the government plans all business activity

◇ **planner** ['plænə] *noun* **(a)** person who plans; **the government's economic planners** = people who plan the future economy of the country for the government; **town planner** = person who supervises the design of a town and the way the streets and buildings in a town are laid out and developed **(b) desk planner** *or* **wall planner** = book *or* chart which shows days *or* weeks *or* months so that the work of an office can be shown by diagrams

◇ **planning** ['plænɪŋ] *noun* **(a)** organizing carefully how something should be done; **economic planning** = planning the future financial state of the country for the government; **family planning** = using contraception to control the number of children in a family; **manpower planning** = planning to get the right number of workers in each job **(b)** organizing how land and buildings are to be used; **town planning** *or US* **city planning** = designing a town *or* city, including the way the streets and buildings are laid out and developed; **country planning** *or* **rural planning** = organizing how land is to be used in the country and the amount and type of building there will be; **planning authority** = local body which gives permission for changes to be made to existing buildings *or* for new use of land; **planning controls** = legislation used by local authority to control building; **planning department** = section of a local government office which deals with requests for planning permission; **planning inquiry** = hearing before a government inspector relating to a decision of a local authority in planning matters; **planning permission** = official document allowing a person *or* company to plan new buildings on empty land *or* to alter existing buildings; **outline planning permission** = general permission to build a property on a piece of land, but not the final approval because there are no details given; ***they have been refused planning permission; we are waiting for planning permission before we can start building; the land is to be sold with planning permission***

QUOTE: buildings are closely regulated by planning restrictions
Investors Chronicle

plane tree ['pleɪn triː] *noun Platanus acerifolia*, a common temperate deciduous hardwood tree, frequently grown in towns because of its resistance to air pollution

planet ['plænɪt] *noun* large body in the solar system, such as the earth, Mars, Mercury, etc.

◇ **planetary** ['plænɪtri] *adjective* referring to a planet

plankton ['plæŋktən] *noun* microscopic animals and plants which live and drift in water and are eaten by many aquatic animals (NOTE: the word is plural and a single organism is a **plankter**)

◇ **planktivorous** [plæŋk'tɪvərəs] *adjective* (organism) which eats plankton

◇ **planktonic** [plæŋk'tɒnɪk] *adjective* referring to plankton; ***blooms are population explosions of planktonic plants***

COMMENT: plankton are divided into two groups: zooplankton, which are microscopic animals and phytoplankton which are microscopic plants capable of photosynthesis. Plankton float near the surface of the water and provide food for many fish and other marine animals

plant [plɑːnt] *noun* **(a)** organism containing chlorophyll with which it carries out photosynthesis; ***botanists discovered several new species of plant in the jungle;***

herbaceous plant = plant without perennial stems above the ground; **host plant** = plant on which a parasite lives; **indicator plant** = plant which is very sensitive to changes in the environment such as increased pollution; **plant biology** = study of the structure, reproduction, growth, distribution, etc. of plants; **plant breeder** = person who breeds new forms of plants; **plant community** = group of plants living together in an area; **plant cover** = number of plants growing on a certain area of land; **plant-eater** = animal which eats plants; **plant ecology** = study of the relationship between plants and their environment; **plant-feeder** = PLANT-EATER; **plant kingdom** = category of all organisms classed as plants; **plant nutrients** = minerals whose presence in the soil is essential for the healthy growth of plants; **plant physiology** = study of the functions of plants; **plant plankton** = phytoplankton *or* microscopic plants which float in the sea *or* in a lake, etc.; **plant population** = the number of plants in an area, such as the number of plants per hectare **(b)** very large factory *or* installation; ***a nuclear power plant is to be built near the town; the area was sealed off after leaks from the chemical plant;*** **chemical plant** = factory where chemicals are produced; **nuclear power plant** = power station in which nuclear reactions are used to provide energy to run turbines which generate electricity; **power plant** = (i) factory *or* installation where electricity is generated; (ii) machine *or* equipment (such as an engine) producing power to work something; **purification plant** = installation where impurities are removed from water

plantation [plɑːn'teɪʃn] *noun* **(a)** estate, especially one in the tropics on which large-scale production of cash crops takes place; plantations specialize in the production of a single crop, and crops such as cocoa, coffee, cotton, tea and rubber are typical plantation crops; **coffee plantation** = plantation of coffee bushes **(b)** area of land planted with trees for commercial purposes (NOTE: plantations of conifers are often by mistake called **forests)**

> QUOTE: extensive cash crop plantation schemes such as rubber and oil palm estates form a vital part of the economy
>
> **Ecologist**

plasma ['plæzmə] *noun* yellow watery liquid which makes up the main part of blood; **plasma cell** = white blood cell which produces a certain type of antibody

Plasmodium [plæz'məʊdiəm] *noun* type of parasite which infests red blood cells and causes malaria

plastic ['plæstɪk] **1** *noun* artificial, usually organic, material made from petroleum and used to make many objects **2** *adjective* which can take on different shapes when under stress and does not return to its original shape when the stress is no longer there

> COMMENT: plastics are moulded by heating a substance under pressure and retain their shape after being formed. Thermoplastics are heated while being shaped and can be heated and shaped again for re-use. Thermosetting plastics are heated while being shaped but cannot be reheated for recycling. Waste plastics containing chlorine can produce hydrogen chloride when incinerated. Plastics formed from ethylene or propylene (i.e. polyethylene and polypropylene) are not degradable and must be recycled or destroyed by incineration

plate [pleɪt] *noun* large area of solid rock in the earth's crust, which floats on the mantle and moves very slowly; **plate tectonics** = theory that the earth's crust is made up of a series of large plates of solid rock which float on the mantle and move very slowly, the places where they meet being subject to earthquakes and volcanic eruptions

plateau ['plætəʊ] *noun* area of high flat land (NOTE: plural is **plateaux)**

platinum ['plætɪnəm] *noun* metallic element, an important rare metal which does not corrode (NOTE: chemical symbol is **Pt;** atomic number is **78)**

Pleistocene ['plaɪstəʊsiːn] *noun* period in the earth's evolution which lasted until about ten thousand years ago, including the main Ice Ages

plough *or US* **plow** [plaʊ] **1** *noun* agricultural implement used to turn over the surface of soil in order to cultivate crops **2** *verb* to turn over the soil with a plough; **deep ploughing** = ploughing very deep into the soil (used when reclaiming previously virgin land for agricultural purposes)

plug [plʌg] *noun* round block of igneous rock forming the central vent of an old volcanic opening

plumbism ['plʌmbɪzm] *noun* lead poisoning *or* poisoning caused by taking in lead salts

plume [pluːm] *noun* **(a)** (i) long cloud of smoke *or* gas from a factory chimney *or* volcano; (ii) cloud of powdered snow blowing from a mountain crest; ***build a tall chimney stack so that the gas plume will be carried further by the wind; plumes from power stations can increase the levels of acid in clouds; the plume of radioactive material covered most of the country*** **(b)** *(of a bird)* feather

plutonium [pluːˈtəʊniəm] *noun* natural radioactive element, also made in nuclear reactors; **plutonium-239** = extremely poisonous and carcinogenic isotope of plutonium which is formed from uranium-238 and is itself used as a fuel in nuclear reactors; **enriched plutonium** = plutonium to which uranium-235 has been added and which is used as a fuel in nuclear reactors (NOTE: chemical symbol is **Pu**; atomic number is **94**)

PNdB = PERCEIVED NOISE LEVEL IN DECIBELS

pneum- *or* **pneumo-** [ˈnjuːməʊ] *prefix* referring to air *or* to the lungs *or* to breathing

◇ **pneumoconiosis** [njuːməʊkəʊnɪˈəʊsɪs] *noun* lung disease where fibrous tissue forms in the lungs because the patient has inhaled particles of stone *or* dust over a long period of time

PNL = PERCEIVED NOISE LEVEL

PNS = PERIPHERAL NERVOUS SYSTEM

Po *chemical symbol for* polonium

poach [pəʊtʃ] *verb* to catch animals *or* birds *or* fish illegally on someone else's land

◇ **poacher** [ˈpəʊtʃə] *noun* person who catches animals *or* birds *or* fish illegally on someone else's land

◇ **poaching** [ˈpəʊtʃɪŋ] *noun* catching animals *or* birds *or* fish illegally on someone else's land

QUOTE: the move comes against a background of growing international concern over the fate of the tiger, whose numbers are dropping rapidly in the face of poaching, principally for tiger bone for medicinal purposes

Guardian

pod [pɒd] *noun* legume *or* casing for several seeds (as for peas or beans)

podocarpus [pɒdəʊˈkɑːpəs] *noun* common softwood tree in Australasia

podsol *or* **podzol** [ˈpɒdsɒl *or* ˈpɒdzɒl] *noun* type of acid soil where oxides have been leached from the light-coloured top layer into a darker lower layer which is impervious and contains little organic matter; *compare* CHERNOZEM

◇ **podsolic** *or* **podzolic** [pɒdˈsɒlɪk *or* pɒdˈzɒlɪk] *adjective* referring to *or* concerning podsol *or* podzol

◇ **podzolic soil** *or* **podzolized soil** [pɒdˈzɒlɪk *or* ˈpɒdzəlaɪzd ˈsɔɪl] *noun* soil from which iron and aluminium oxides have been leached from the topsoil in moist cool climates

◇ **podzolization** [pɒdzəlaɪˈzeɪʃn] *noun* process by which oxides leach from the top layer of soil making it acid

COMMENT: on the whole podzols make poor agricultural soils, due to their low nutrient status and the frequent presence of an iron pan; large areas of the coniferous forest regions of Canada and Russia are covered with podzols

poikilo- [pɔɪˈkiːləʊ] *prefix* meaning irregular *or* varied

◇ **poikilosmotic** [pɔɪkiːlɒzˈmɒtɪk] *adjective* (aquatic animal) whose body fluids change by osmosis depending on the composition of the surrounding water

◇ **poikilotherm** [pɔɪˈkiːləʊθɜːm] *noun* animal which has cold blood *or* cold-blooded animal; *compare* HOMOIOTHERM

COMMENT: the body temperature of cold-blooded animals changes with the outside temperature

point [pɔɪnt] *noun* **(a)** sharp end, such as the end of a piece of land jutting into the sea **(b)** dot used to show the division between whole numbers and parts of numbers in a decimal; *5.7; 14.61* **(c)** particular mark *or* place in a series of numbers; **critical point** = moment at which a substance undergoes a change in temperature, volume or pressure; **dew point** = temperature at which dew forms on grass *or* leaves, etc.; **freezing point** = temperature at which a liquid becomes solid; **frost point** = temperature at which moisture in saturated air turns to ice; **melting point** = temperature at which a solid turns to liquid; ***what's the freezing point of water?; the melting point of ice is 0°C*** **(d) point quadrat** = device for measuring the leaf cover of ground, where a piece of wood with many holes in it is placed over a quadrat and rods passed through the holes, counting the number of leaves which they touch

poison [ˈpɔɪzən] **1** *noun* substance which can kill *or* harm when eaten, drunk, breathed in *or* touched; **poison ivy** *or* **poison oak** = North American plants whose leaves can cause a painful rash if touched **2** *verb* to give someone *or* something a poison; ***the workers were poisoned by toxic fumes; the wound was poisoned by bacterial infection; hundreds of fish were poisoned by the runoff***

◇ **poisoning** [ˈpɔɪznɪŋ] *noun* killing *or* harming an organism with a poison; **blood poisoning** = condition where bacteria are present in blood and cause illness; **lead poisoning** = plumbism *or* poisoning caused by taking in lead salts; **salmonella poisoning** = illness caused by eating food which is contaminated with Salmonella bacteria which develop in the intestines; **staphylococcal poisoning** = poisoning by Staphylococci which have spread in food

◇ **poisonous** [ˈpɔɪznəs] *adjective* (substance) which is toxic *or* full of poison *or* which can kill or harm; ***some mushrooms are good to eat and some are poisonous;*** **poisonous gas** = gas which can kill *or* which can make someone ill

COMMENT: the commonest poisons, of which even a small amount can kill, are arsenic, cyanide and strychnine. Many common foods and drugs can be poisonous if taken in large doses. Common household materials such as bleach, glue and insecticides can also be poisonous. Some types of poisoning, such as Salmonella, can be passed to other people through lack of hygiene

polar ['pəʊlə] *adjective* referring to the North *or* South Pole; **polar vortex** *or* **circumpolar vortex** = circular movement of air around one of the poles

polder ['pəʊldə] *noun* piece of low-lying land which has been reclaimed from the sea and is surrounded by earth banks

pole [pəʊl] *noun* **(a)** end of an axis; one of two opposite points which are linked **(b)** one of two points (the North and South Poles) where longitudinal lines meet and which are the most northerly *or* southerly points on the earth; **geomagnetic pole** *or* **magnetic pole** = one of the two poles of the earth (near to, but not identical with, the geographical poles) which are the centres of the earth's magnetic field and to which a compass points

policy ['pɒlɪsi] *noun* plan for dealing with something *or* way of dealing with someone; **Common Agricultural Policy (CAP)** = agreement between members of the EU to protect farmers by paying subsidies to fix prices of farm produce; **agricultural policy** = government's way of dealing with all matters concerning agriculture; **environmental policy** = plan for dealing with all matters affecting the environment on a national *or* local scale; **foreign policy** = government's way of dealing with foreign countries

politics ['pɒlɪtɪks] *noun* practice *or* study of how to govern a country

◇ **political** [pə'lɪtɪkl] *adjective* concerning politics

◇ **politician** [pɒlɪ'tɪʃən] *noun* person actively involved in politics, especially a member of a parliament *or* assembly

pollard ['pɒlɑːd] **1** *noun* tree of which the branches have been cut back to a height of about two metres above the ground **2** *verb* to cut back the branches on a tree every year *or* every few years to a height of about two metres above the ground; *compare* COPPICE

COMMENT: pollarding allows new shoots to grow as in coppicing, but high enough above the ground to prevent them being eaten by animals. Willow trees are often pollarded

pollen ['pɒlən] *noun* cells in flowers which convey the male gametes; **pollen analysis** = scientific study of pollen, especially of pollen found in peat *or* coal deposits; *see also* PALYNOLOGY; **pollen count** = measurement of the amount of pollen in a sample of air

◇ **pollinate** ['pɒlɪneɪt] *verb* to transfer pollen from male to female reproductive organs in a flower

◇ **pollination** [pɒlɪ'neɪʃn] *noun* action of pollinating a flower; **insect pollination** = pollination of a flower by an insect

◇ **pollinator** ['pɒlɪneɪtə] *noun* organism such as a bee *or* bird *or* other plant which helps pollinate a plant; ***some apple trees need to be planted with pollinators as they are not self-fertile; birds are pollinators for many types of tropical plant***

◇ **pollinosis** [pɒlɪ'nəʊsɪs] *noun* hay fever, inflammation of the nose and eyes caused by an allergic reaction to pollen *or* fungus spores *or* dust in the atmosphere

COMMENT: tree pollen is most prevalent in spring, followed by the pollen of flowers and grasses during the summer months and fungal spores in the early autumn. The pollen is released by the stamens of a flower and floats in the air until it finds a female flower. Pollen in the air is a major cause of hay fever. It enters the nose and eyes and chemicals in it irritate the mucus and force histamines to be released by the sufferer, causing the symptoms of hay fever to appear

pollute [pə'luːt] *verb* to discharge harmful substances in abnormally high concentrations into the environment, often done by people but can occur naturally; ***polluting gases react with the sun's rays; polluted soil must be removed and buried***

◇ **pollutant** [pə'luːtənt] *noun* substance *or* agent which pollutes; ***discharge pipes take pollutants away from the coastal area into the sea;*** **air pollutant** *or* **atmospheric pollutant** = substance which pollutes the air *or* the atmosphere, such as gas *or* smoke

◇ **polluter** [pə'luːtə] *noun* person *or* company which causes pollution; ***certain industries are major polluters of the environment;*** **polluter pays principle** = principle that if pollution occurs, the person *or* company responsible should be required to pay for the consequences of the pollution and for avoiding it in future

◇ **pollution** [pə'luːʃn] *noun* presence of abnormally high concentrations of harmful substances in the environment, often put there by people; ***in terms of pollution, gas is by far the cleanest fuel; pollution of the atmosphere***

has increased over the last 50 years; soil pollution round mines poses a problem for land reclamation; **air pollution** *or* **atmospheric pollution** = polluting of the air by gas *or* smoke, etc.; **environmental pollution** = polluting of the environment; **noise pollution** = unpleasant sounds which cause discomfort; **water pollution** = polluting of the sea, rivers, lakes, canals; **pollution charges** = cost of repairing *or* stopping environmental pollution; **pollution control** = means of limiting pollution

COMMENT: pollution is caused by natural sources or by human action. Pollution can be caused by a volcanic eruption or by a nuclear power station. Pollutants are not only chemical substances, but can be a noise or an unpleasant smell (as from a grinding works or from a sewage farm)

QUOTE: the principle culprits of the area's serious pollution are the more than 2.5 million automobiles that discharge an estimated 80% of the 5 million tons of air pollutants that enter the city's skies each year

Environment

QUOTE: In November, a £50 million package of grants was announced by the Agriculture Minister for stopping farm pollution. Farmers can claim up to half the cost of building or improving facilities for the storage, treatment and disposal of slurry and silage effluent

Guardian

polonium [pə'ləʊniəm] *noun* natural radioactive element (NOTE: chemical symbol is **Po**; atomic number is **84**)

poly- ['pɒli] *prefix* meaning many *or* much *or* touching many organs

polybrominated biphenyl (PBB) [pɒlɪ'brɒmɪneɪtɪd baɪ'fenɪl] *noun* highly toxic aromatic compound containing benzene and bromine, used in plastics and fire-retardant materials and thought to be carcinogenic

polychlorinated biphenyl *or* **polychlorobiphenyl (PCB)** [pɒlɪ'klɔ:rɪneɪtɪd baɪ'fenɪl *or* pɒlɪklɔ:rəʊ'baɪfenɪl] *noun* one of a group of compounds produced by chlorination of biphenyl

COMMENT: PCBs are stable compounds and were extensively used in electrical fittings and paints. Although they are no longer manufactured they are extremely persistent and remain in huge quantities in the atmosphere and in landfill sites. They are not water-soluble and float on the surface of water where they are eaten by aquatic animals and so enter the food chain. PCBs are fat-soluble, and are therefore easy to take into the system, but difficult to excrete

polycondensed plastic [pɒlɪkən'densd 'plæstɪk] *noun* type of plastic (such as nylon) which can be recycled

◇ **polyculture** ['pɒlɪkʌltʃə] *noun* rearing or growing of more than one species of plant or animal on the same area of land at the same time

◇ **polycyclic aromatic hydrocarbon (PAH)** [pɒlɪ'sɪklɪk ærə'mætɪk haɪdrəʊ'kɑ:bən] *noun* one of a group of chemical compounds which are carcinogenic

polyethylene [pɒlɪ'eθɪli:n] *noun* = POLYTHENE

◇ **polyethylene terephthalate (PET)** [pɒlɪ'eθɪli:n terɪf'θæleɪt] *noun* type of plastic used to make artificial fibres and plastic bottles, which can be recycled

polygamy [pɒ'lɪgəmi] *noun* breeding arrangement where a male has several mates; *compare* MONOGAMY

polyhaline [pɒlɪ'hælaɪn] *adjective* (water) which is very salt, containing almost as much salt as sea water

polymer ['pɒlɪmə] *noun* natural *or* artificial chemical compound whose large molecules are made up of smaller molecules combined in repeated groups

◇ **polymerization** [pɒlɪmərаɪ'zeɪʃn] *noun* chemical reaction *or* process in which a polymer is formed

polymorphism [pɒlɪ'mɔ:fɪzm] *noun* (i) existence of different forms during the life cycle of an organism (such as a butterfly which exists as a caterpillar, then a pupa, before becoming a butterfly); (ii) existence of different forms of an organism (such as bees, which exist as workers, queens and drones)

◇ **polymorphic** *or* **polymorphous** [pɒlɪ'mɔ:fɪk *or* pɒlɪ'mɔ:fəs] *adjective* (organism) which exists in different forms

polyp ['pɒlɪp] *noun* sedentary form of aquatic animals, such as sea anemones

polyphagous [pə'lɪfəgəs] *adjective* (animal) which eats more than one type of food

polypropylene [pɒlɪ'prəʊpɪli:n] *noun* thermoplastic used to make artificial fibres, bottles, pipes, etc., which is not degradable and must be recycled *or* destroyed by incineration

polysaprobe [pɒlɪ'sæprəʊb] *noun* organism which can survive in heavily polluted water

◇ **polysaprobic** [pɒlɪsæ'prəʊbɪk] *adjective* (organism) which can survive in heavily polluted water

◇ **polystyrene** [pɒlɪ'staɪri:n] *noun* thermoplastic which can be made into hard lightweight foam by blowing air *or* gas into it (expanded polystyrene), used as an insulating and packaging material

◇ **polythene** *or* **polyethylene** ['pɒlɪθi:n *or* pɒlɪ'eθɪli:n] *noun* thermoplastic used to make artificial fibres, packaging, boxes, etc.

◇ **polyunsaturated fat** [pɒlɪʌn'sætʃəreɪtɪd 'fæt] *noun* fatty acid capable of absorbing more hydrogen (typical of vegetable and fish oils)

◇ **polyvinylchloride (PVC)** [pɒlɪvɪnɪl'klɔ:raɪd] *noun* thermoplastic that is not biodegradable, used for floor coverings, clothes, shoes, pipes, etc.

pome [pəʊm] *noun* type of fruit which develops from the axis of the flower

> COMMENT: common fruit like apples and pears are pomes, in which the fleshy part develops from the receptacle of a flower and not from the ovary as in other fruit

◇ **pomology** [pə'mɒlədʒi] *noun* study of growing fruit

pond [pɒnd] *noun* small area of still water; **maturation pond** = pond used in the final stages of sewage treatment; **settling pond** = tank where liquid is held until the solid particles in it have fallen to the bottom; *see also* DEW POND; **pond life** = organisms which live in a pond

◇ **ponding** ['pɒndɪŋ] *noun (of liquid)* collecting into a pond *or* puddle

◇ **pondweed** ['pɒndwi:d] *noun* type of plant which lives in a pond

pool [pu:l] *noun* **(a)** small area of water; **rock pool** = pool of salt water left in rocks by the sea **(b)** group *or* combination of things *or* people; **gene pool** = genetic information carried by the sex cells of all the individual organisms in a population **(c)** pool of oil *or* gas which collects in porous sedimentary rock

poor [pɔ:] *adjective* not very good; **poor soil** = soil with few useful nutrients and so less suitable for plants

population [pɒpju'leɪʃn] *noun* number of organisms of the same species living and breeding in a certain area; number of people living in a country *or* town; ***the fish population has been severely reduced; population statistics show that the birth rate is slowing down; the government has decided to screen the whole population of the area; population growth is a major threat to conservation efforts;*** **population control** = limiting the number of organisms living in a certain area; policy of limiting the number of human beings in poor and densely populated areas of the world by the use of contraception and sterilization; **population cycle** = regular changes in the population of a species, usually a cycle in which the population gradually increases and then falls away again; **population decrease** = reduction in the number of organisms living in a certain area; **population density** = number of organisms living in a certain area; **population dispersion** = pattern of organisms found over a wide area; **population dynamics** = study of the changes in the number of organisms living in a certain area; **population ecology** = study of the size, growth and distribution of populations; **population equilibrium** = state where the population stays at the same level because the number of deaths is the same as the number of births; **population explosion** = rapid increase in the number of organisms, especially of human beings, living in a certain area; **zero population growth** = state when the numbers of births and deaths in a population are equal and so the size of the population remains the same; **population pyramid** = graphical representation showing the distribution of a population according to age, sex, etc.

◇ **populate** ['pɒpjuleɪt] *verb* to fill an area with organisms; ***starlings introduced into the USA in the 19th century soon populated the whole eastern seaboard***

pore [pɔ:] *noun* (i) tiny hole in the skin through which sweat passes; (ii) tiny space in a rock formation

◇ **porosity** [pɔ:'rɒsɪti] *noun* degree to which a substance is porous; ***clay has a lower porosity than lighter soils***

◇ **porous** ['pɔ:rəs] *adjective* (i) containing pores; (ii) (rock) which has many small pores in it and can absorb water; (iii) (tissue) which allows fluid to pass through it; ***porous rock is not necessarily permeable***

positive ['pɒzɪtɪv] *adjective* which shows the presence of something; **positive feedback** = situation where the result of a process stimulates the process which caused it (as when a change at one trophic level affects all levels above it)

◇ **positively** [pɒzɪtɪvli] *adverb* in a positive way

post- [pəʊst] *prefix* meaning after *or* later

◇ **postclimax** ['pəʊstklaɪmæks] *noun* climax community still existing in a place where the environmental conditions are no longer suitable for it

potamology [pɒtə'mɒlədʒi] *noun* scientific study of rivers

◇ **potamoplankton** [pɒtəməʊ'plæŋktən] *noun* plankton which live in rivers

◇ **potamous** ['pɒtəməs] *adjective* (animal) which lives in rivers

potash ['pɒtæʃ] *noun* general term used to describe potassium salts; potash salts are crude minerals and contain much sodium chloride; **potash fertilizers** = fertilizers based on potassium

potassium [pə'tæsiəm] *noun* metallic element, essential to biological life (NOTE: chemical symbol is **K;** atomic number is **19)**

◇ **potassium permanganate** [pə'tæsiəm pɜː'mæŋgənət] *noun* purple-coloured salt, used as a disinfectant

> COMMENT: potassium is one of the three major soil nutrients needed by growing plants (the others are nitrogen and phosphorus)

potential [pə'tenʃl] **1** *adjective* possible *or* probable *or* which may happen; ***this is a potential site for the new nature reserve;*** **potential energy** = energy of a body *or* system as a result of its position in an electric, magnetic *or* gravitational field **2** *noun* **(electric) potential** = work required to transfer a unit of positive electric charge from an infinite distance to a given point; **biotic potential** = estimate of the maximum rate of increase of a species, assuming no competition from predators *or* parasites: the rate is never achieved in reality because of natural selection; **redox potential** = measure of the ability of the natural environment to bring about an oxidation *or* a reduction process (e.g. involving sulphur)

◇ **potentiate** [pə'tenʃieɪt] *verb (of two substances)* to increase each other's probable toxic effects

◇ **potentiation** [pətenʃi'eɪʃn] *noun* degree of probable increased damage caused by the synergism of toxic substances

poverty ['pɒvəti] *noun* lack of nutrients in the soil

powder ['paʊdə] *noun* fine dry particles like dust

◇ **powdered** ['paʊdəd] *adjective* crushed so that it forms a fine dry dust

◇ **powdery** ['paʊdəri] *adjective* which looks like powder; ***mildew forms as a powdery layer on leaves***

power ['paʊə] *noun* energy, especially electricity, which makes something move; ***the storm cut off power supplies to the town;*** **fossil-fuel power** = power, especially electricity, generated from fossil fuels (such as coal, oil *or* gas); **nuclear power** = power generated by a nuclear reaction; electricity generated by a nuclear power station; **tidal power** *or* **wave power** = energy obtained from the movement of the tides *or* waves; electricity produced by turbines driven by the force of the tides *or* waves; **wind power** = power generated by using wind to drive a machine (as in a windmill) *or* to drive a turbine which creates electricity; **power cables** *or* **power lines** = wires which take high-tension electric current from a power station, carried across the countryside on pylons; **power raising** = action of starting up a power station

◇ **-powered** ['paʊəd] *suffix* meaning driven by; ***a wind-powered pump; a nuclear-powered submarine***

> COMMENT: it is believed by some scientists that living near or under high-tension power lines can affect people's health, causing headaches, depression and possibly cancer

power plant *or* **power station** ['paʊə plɑːnt *or* 'paʊə steɪʃən] *noun* building with machines which make electricity; **coal-fired** *or* **oil-fired power station** = station which produces electricity using coal *or* oil as fuel; **nuclear power station** *or* **nuclear power plant** = power station in which nuclear reactions are used to provide energy to run turbines which generate electricity

> COMMENT: a power station makes electricity by using steam to turn turbines which themselves drive generators. High-pressure steam is heated by burning coal or oil, or by energy from a nuclear reactor. The steam passes over the turbines as a high-pressure jet, making the turbines rotate. The steam is produced from fresh water which is heated in a boiler, then passed through the furnace again to be superheated. Superheated steam drives the first turbine and then is returned to the boiler again for reheating and then passing over the second and third turbines. The steam is then condensed into water which then passes back through the system again. Although the steam used in power stations is heated several times and is condensed into reusable water at the end of the cycle, much of the energy generated to heat the water and create steam is not translated into electric power and is therefore wasted

ppb = PARTS PER BILLION

pphm = PARTS PER HUNDRED MILLION

ppm = PARTS PER MILLION

pptm = PARTS PER THOUSAND MILLION

PPP = POLLUTER PAYS PRINCIPLE

prairie ['preəri] *noun* area of grass-covered plain in North America, mainly treeless, with many different species of grasses and herbs; the prairie lands of the United States and Canada are responsible for most of the North America's wheat production (NOTE: in Europe, the equivalent is the **steppe) long-grass prairie** *or* **tall-grass prairie** = area in the east of the North American prairies where mainly varieties of tall grasses grow; **short-grass prairie** = area in the west of the North American prairies where mainly varieties of short grasses grow

> QUOTE: within a lifetime, most of Canada's wild prairies have been lost or damaged by extensive ranching of cattle and urban or industrial development
> **Guardian**

pre- [priː] *prefix* meaning before *or* in front of; ***nothing was known of this effect pre-1964***

precipitate 1 *noun* [prɪˈsɪpɪtət] solid particles which separate from a solution during a chemical reaction **2** *verb* [prɪˈsɪpɪteɪt] to make solid dissolved particles separate from a solution

◊ **precipitant** [prɪˈsɪpɪtənt] *noun* substance added to a solution to make solid dissolved particles precipitate

◊ **precipitation** [prɪsɪpɪˈteɪʃn] *noun* **(a)** action of forming solid particles in a solution; **precipitation tank** = sewage tank in which a chemical is added to the sewage before it passes to the sedimentation tanks **(b)** water which falls from clouds as rain, snow, hail, etc.; ***precipitation in the mountain areas is higher than in the plains;*** **acid precipitation** = rain *or* snow *or* hail, etc. which contains a higher level of acid than normal; *see also* ACID RAIN

◊ **precipitator** [prɪˈsɪpɪteɪtə] *noun* **electrostatic precipitator** = device for collecting minute particles of dust suspended in gas by charging the particles as they pass through an electrostatic field

precise [prɪˈsaɪs] *adjective* very exact *or* correct; ***the instrument can give precise measurements of changes in atmospheric pressure***

precursor [prɪˈkɜːsə] *noun* substance *or* cell from which another substance *or* cell is developed; ***the biggest share of ozone precursors comes from emissions from vehicles***

> QUOTE: methane is a precursor to reactions that destroy ozone in the stratosphere
> **New Scientist**

predator [ˈpredətə] *noun* animal which kills and eats other animals; ***the larvae are predators of aphids*** *see also* PREY

◊ **predatory** [ˈpredətəri] *adjective* referring to a predator; ***predatory animals, such as lions and tigers***

◊ **predation** [prɪˈdeɪʃn] *noun* killing and eating other animals

predominate [priːˈdɒmɪneɪt] *verb* to be more powerful than others; ***a cold northerly airstream predominates during the winter; dark forms of the moths predominate in industrial areas***

◊ **predominance** [priːˈdɒmɪnəns] *noun* being more powerful than others

◊ **predominant** [priːˈdɒmɪnənt] *adjective* which is more powerful than others; ***the predominant airstream is from the west***

preferendum [prefəˈrəndəm] *noun* area where a species flourishes best

preserve [prɪˈzɜːv] **1** *noun* area of land where wild animals are kept to be hunted and killed for sport **2** *verb* to keep something in the same state *or* to stop something from changing *or* rotting

◊ **preservation** [prezəˈveɪʃn] *noun* keeping something in the same state *or* stopping something from changing *or* rotting; ***food preservation allows some types of perishable food to be eaten during the winter months when fresh food is not available;*** **preservation order** = order from a local government department which stops a building from being demolished; **tree preservation order (TPO)** = order from a local government department which stops a tree from being felled; ***they have put a preservation order on this building***

◊ **preservative** [prɪˈzɜːvətɪv] *noun* substance added to food to preserve it by slowing natural decay caused by bacteria, fungi, etc. (In the EU preservatives are given E numbers E200 - 297)

pressure [ˈpreʃə] *noun* **(a)** (i) action of squeezing *or* of forcing; (ii) force of something on its surroundings; **atmospheric pressure** = normal pressure of the air on the surface of the earth; **atmospheric pressure zones** = bands of high and low pressure round the earth: high pressure near the poles, low pressure areas between 40 and 70 degrees latitude north and south, high pressure along the 30 degrees latitude line in the subtropics and then low pressure again round the equator; **osmotic pressure** = pressure required to stop the flow of a solvent through a membrane; **pressure gradient** = change in atmospheric pressure from one place to another on the ground (as shown on a map by isobars); **pressure vessel** = container that houses the core, coolant and moderator in a nuclear reactor **(b)** force *or* strong influence to make someone change his mind *or* change his actions; **pressure group** = group of people with similar interests who try to influence politicians; ***the environmental association set up a pressure group to lobby parliament***

◊ **pressurize** [ˈpreʃəraɪz] *verb* to alter the atmospheric pressure in a container; **pressurized fluidized-bed combustion (PFBC)** = economic method of burning low-grade coal in a furnace in which air is blown upwards through the burning fuel; **pressurized water reactor (PWR)** = type of nuclear reactor in which water is heated to steam under high pressure to turn turbines

prevailing wind [prɪ'veɪlɪŋ 'wɪnd] *noun* wind which usually blows from a certain direction

prevalent ['prevələnt] *adjective* common (in comparison to something) *or* occurring frequently; ***the disease is prevalent in some African countries; a fungus which is more prevalent in deciduous forests***

◇ **prevalence** ['prevələns] *noun* frequency of occurrence; ***the prevalence of malaria in some tropical countries; the prevalence of cases of malnutrition in large towns***

prevent [prɪ'vent] *verb* to stop something happening; ***the pressure group tried to prevent the construction of the new airport***

◇ **prevention** [prɪ'venʃn] *noun* stopping something happening; **accident prevention** = taking steps to stop accidents happening

◇ **preventive** *or* **preventative** [prɪ'ventɪv *or* prɪ'ventətɪv] *adjective* which stops something happening; **preventive measure** = step taken to stop something happening; **preventive medicine** = medical action to stop a disease from occurring

COMMENT: preventive measures against disease include immunization, vaccination and quarantine

prey [preɪ] **1** *noun* animal which is killed and eaten by another; ***small mammals are the prey of owls;*** **bird of prey** = carnivorous bird which kills and eats animals **2** *verb* to prey on = to kill and eat (another animal); ***water snakes prey on frogs and small fish***

primary ['praɪməri] *adjective* (something) which is first and leads to another (secondary); **primary colour** = one of the three main colours in the spectrum (red, green and blue) from which other colours are formed; **primary commodities** = basic raw materials *or* food; **primary consumer** = animal (such as a herbivore) which eats plants (which are producers in the food chain); **primary coolant** = substance used to cool a nuclear reactor, which then passes to a heat exchanger to transfer its heat to another coolant which is used to turn the turbines; **primary energy** = energy required to produce other forms of energy such as heat *or* electricity; **primary forest** = forest which originally covered a region before changes in the environment brought about by people; **primary industry** = industry dealing with raw materials (such as coal, food, farm produce, wood); **primary mineral** = mineral formed initially from cooling magma and which has remained unchanged; **primary particulates** = particles of matter sent into the air from fires *or* from industrial processes *or* from volcanic eruptions *or* from sandstorms, etc.; **primary product** = product (such as wood, milk, fish) which is a basic raw material; **primary production** = amount of organic matter formed by photosynthesis; **primary productivity** = (i) rate at which plants produce organic matter through photosynthesis; (ii) amount of organic matter produced in a certain area over a certain period of time (such as a crop during a growing season); **primary sere** = first plant community which develops in a place where no plants have grown before (as on cooled lava from a volcano); **primary succession** = ecological community which develops in a place where nothing has lived before (as on cooled lava from a volcano); **primary treatment** = first stage in the treatment of sewage in which suspended solids are removed; *compare* SECONDARY

QUOTE: farmers are convinced that primary industry no longer has the capacity to meet new taxes
Australian Financial Times

QUOTE: because it needs such large trees, the woodpecker is confined to primary forest - the evergreen broadleaved forest that once covered all the hills
New Scientist

QUOTE: the primary production of the oceans is estimated to be on the order of 190 billion tons per year of microscopic phytoplankton and other marine plants
Natural Resources Forum

primates ['praɪmeɪts] *noun* order of mammals containing monkeys, apes and human beings

COMMENT: primates are the highest order of animals, with brains which are proportionately larger for their size than those of other animals

primitive ['prɪmətɪv] *adjective* referring to very early *or* prehistoric times or to an early stage in an organism's development; **primitive area** = area of undeveloped land, such as a forest, which is set aside and protected as a national park, etc.; **primitive rock** = rock formed in *or* before the Palaeozoic era

primordial [praɪ'mɔːdiəl] *adjective* in a very early stage of development

probe [prəʊb] **1** *noun* device inserted into something to investigate the inside *or* to obtain information; **space probe** = device sent into space to obtain information about conditions in space **2** *verb* to investigate the inside of something

process ['prəʊses] **1** *noun* technical *or* scientific action; ***a new process for extracting***

oil from shale; **Bessemer process** = method of making steel; **breakdown process** = action of separating into elements *or* of decomposing; **chemical process** = reaction which happens when two *or* more chemicals come into contact; **process chemical** = chemical which is manufactured by industrial process **2** *verb* **(a)** to make a substance undergo a chemical reaction; to produce something by treating a raw material in a factory; **fish processing plant** = factory which takes fish and produces fertilizers, etc. from them **(b)** to examine *or* to test samples; ***the core samples are being processed by the laboratory*** *see also* REPROCESSING

COMMENT: the Bessemer process involves heating molten metal and blowing air into it at the same time; this is done in a type of furnace called a Bessemer converter. The process is used to remove phosphorus and carbon from pig iron. The air forms iron oxide, which removes impurities from the molten metal, including carbon monoxide which burns off. Finally, manganese is added to the metal to remove the iron oxide

produce [prə'dju:s] *verb* to make; ***a factory producing agricultural machinery; a drug which increases the amount of milk produced by cows***

◇ **producer** [prə'dju:sə] *noun* **(a)** person *or* company which produces something **(b)** organism (such as a green plant) which takes energy from outside an ecosystem and channels it into the system (the first level in the food chain); **producer gas** = mixture of carbon monoxide and nitrogen made by passing air over hot coke and used as a fuel

◇ **product** ['prɒdʌkt] *noun* (i) thing which is produced; (ii) result *or* effect of a process; **product of incomplete combustion (PIC)** = compound formed when combustion does not destroy all the waste being incinerated

◇ **production** [prə'dʌkʃn] *noun* **(a)** act of manufacturing; **production line** = system of manufacturing a product, where the item moves slowly through the factory with new pieces being added to it as it goes along; **production platform** = oil rig in the sea where oil from several wells is collected; **production residues** = waste left after a production process **(b)** act of producing something by organisms; amount of heat *or* energy produced by the biomass in an area; **primary production** = amount of organic matter formed by photosynthesis; **gross primary production** = rate at which a biomass assimilates organic matter; **production ecology** = study of groups of organisms from the point of view of the food which they produce; **gross production rate** = rate at which a biomass assimilates organic matter; **production ration** = quantity of food needed to make a farm animal produce meat, milk, eggs, etc. (It is always more than the basic maintenance ration)

◇ **productive** [prə'dʌktɪv] *adjective* which produces a lot; **productive soil** = soil which is very fertile and produces large crops

◇ **productivity** [prɒdʌk'tɪvɪti] *noun* rate at which something is produced; ***with new strains of rice, productivity per hectare can be increased;*** **net productivity** = difference between the amount of organic matter produced by photosynthesis and the amount of organic matter used by plants in their growth; **primary productivity** = (i) rate at which plants produce organic matter through photosynthesis; (ii) amount of organic matter produced in a certain area over a certain period of time (such as a crop during a growing season)

QUOTE: rising global temperatures could completely alter the face of the earth. Many of the world's most fertile regions are likely to become drier and less productive, whilst regions like India and the Middle East are expected to become wetter and more fertile

Ecologist

profession [prə'feʃn] *noun* all the characters (chemical, physical, biological) that determine the position of an organism *or* species in an ecosystem (such as an aquatic predator, a terrestrial herbivore)

profundal zone [prə'fʌndəl 'zəʊn] *noun* area of water in a lake below the limnetic zone

programme ['prəʊgræm] *noun* planned course of action

prohibit [prə'hɪbɪt] *verb* to forbid *or* to say that something should not be done; ***the US has prohibited the use of CFCs in aerosols; smoking is prohibited in this area***

project 1 *noun* ['prɒdʒekt] plan *or* scheme of work; ***the land reclamation project will cost several million dollars;*** **development project** = plan drawn up by a business *or* government *or* local council to build on an area of land; **housing project** = plan drawn up by a business *or* local council to build houses on an area of land; **pilot project** = small-scale project carried out to see if a large-scale project will work; ***they are running a pilot project in the area for three months before deciding on the next stage*** **2** *verb* [prə'dʒekt] to protrude *or* to stick out

QUOTE: rural health projects enable vaccines and other essential health technologies to reach more of the population. Most rural electrification projects involve the government providing electricity supplies at little or no cost

Appropriate Technology

promontory ['prɒməntri] *noun* area of high land which sticks out into the sea

promote [prəˈməʊt] *verb* to encourage something to take place; ***growth-promoting hormones are used to increase the weight of beef cattle***

◇ **promoter** [prəˈməʊtə] *noun* **growth promoter** = substance which makes an organism grow

◇ **promotion** [prəˈməʊʃn] *noun* encouraging something to take place; ***growth promotion may alter the nutritional quality of wheat***

propagate [ˈprɒpəgeɪt] *verb* to produce new plants

◇ **propagation** [prɒpəˈgeɪʃn] *noun* producing new plants

◇ **propagator** [ˈprɒpəgeɪtə] *noun* device in which seed can be sown *or* cuttings taken to produce new plants

> COMMENT: the commonest of the various forms of plant propagation are: growing from seed or from tubers (such as potatoes); growing clones (as by taking cuttings, grafting, budding roses, layering)

propane [ˈprəʊpeɪn] *noun* gas found in petroleum, which is sold commercially in liquid form under pressure as LPG

propellant [prəˈpelənt] *noun* substance used to make something move; gas used in an aerosol can to make the spray come out

> QUOTE: many companies have announced that they will stop using CFCs as propellants
> **Environment Now**

propene [ˈprəʊpiːn] *noun* = PROPYLENE

property [ˈprɒpəti] *noun* **(a)** distinctive characteristic of a substance; ***we use the energy-producing properties of uranium isotopes in nuclear reactors*** **(b)** something which belongs to someone; ***the farmland is the property of a large corporation and they have applied to use it for industrial construction***

prophylaxis [prɒfɪˈlæksɪs] *noun* prevention of a disease (as by spraying to kill insects)

◇ **prophylactic** [prɒfɪˈlæktɪk] *adjective & noun* (substance) which helps to prevent the development of a disease (such as antimalaria tablets)

propylene **(C_3H_6)** [ˈprɒpɪliːn] *noun* substance obtained from petroleum, used in the manufacture of plastics and chemicals

protease [ˈprəʊtieɪz] *noun* digestive enzyme which breaks down proteins in food

protect [prəˈtekt] *verb* to keep something safe from harm; ***the population must be protected against the spread of the virus***

◇ **protection** [prəˈtekʃn] *noun* thing which protects; ***children are vaccinated as a protection against disease;*** **environmental protection** = act of making sure that the environment is not harmed by regulating the discharge of waste, the emission of pollutants, and other human activities

◇ **protective** [prəˈtektɪv] *adjective* which protects; **protective coloration** = pattern of colouring which protects an animal from attack

protein [ˈprəʊtiːn] *noun* nitrogen compound which is present in and is an essential part of all living cells in the body, formed by the condensation of amino acids; **protein balance** = situation when the nitrogen intake in protein is equal to the excretion rate (in the urine); **protein deficiency** = lack of enough proteins in the diet

◇ **proteolysis** [prəʊtɪˈɒlɪsɪs] *noun* breaking down of proteins in food by proteolytic enzymes

◇ **proteolytic** [prəʊtiəʊˈlɪtɪk] *adjective* referring to proteolysis; **proteolytic enzyme** *or* **protease** = digestive enzyme which breaks down proteins in food

> COMMENT: proteins are necessary for growth and repair of the body's tissue; they are mainly formed of carbon, nitrogen and oxygen in various combinations as amino acids. Certain foods (such as beans, meat, eggs, fish and milk) are rich in protein

> QUOTE: the global fish catch provides some 16% of the world's supply of animal protein
> **Appropriate Technology**

protest [ˈprəʊtest] **1** *noun* act of showing disagreement; ***they refused to buy from the supermarket in protest against the company's policy of not stocking organic food*** **2** *verb* **to protest against something** = to show you disagree with something; ***crowds protested against the killing of seals*** (NOTE: in US English, you **protest something** but in British English you protest **against** something)

prothrombin [prəʊˈθrɒmbɪn] *noun* protein in blood which helps blood to coagulate and which needs vitamin K to be effective

protium [ˈprəʊtiəm] *noun* commonest light isotope of hydrogen

proto- [ˈprəʊtəʊ] *prefix* meaning first *or* at the beginning

protocol [ˈprəʊtəkɒl] *noun* formal statement (of decision *or* results *or* ruling, etc.)

proton [ˈprəʊtɒn] *noun* particle with a positive charge found in the nucleus of an atom

protoplasm ['prəʊtəʊplæzm] *noun* substance like a jelly which makes up the largest part of each cell

◇ **protoplasmic** [prəʊtəʊ'plæzmɪk] *adjective* referring to protoplasm

◇ **protoplast** ['prəʊtəʊplɑ:st] *noun* basic cell unit in a plant formed of a nucleus and protoplasm

Protozoa [prəʊtəʊ'zəʊə] *plural noun* tiny simple organisms like amoebae, with a single cell *or* no cell at all, but with a nucleus (NOTE: singular is **protozoon)**

◇ **protozoan** [prəʊtəʊ'zəʊən] **1** *adjective* referring to the Protozoa **2** *noun* = PROTOZOON

COMMENT: parasitic Protozoa can cause several diseases, such as amoebiasis, malaria and other tropical diseases

pseud- *or* **pseudo-** ['sju:dəʊ] *prefix* meaning false *or* similar to something, but not the same

◇ **pseudokarst** ['sju:dəʊkɑ:st] *noun* terrain like karst, but not in limestone country and not due to weathering

psych- *or* **psycho-** ['saɪkəʊ] *prefix* referring to the mind

◇ **psychological** [saɪkə'lɒdʒɪkl] *adjective* referring to psychology; caused by a mental state

◇ **psychology** [saɪ'kɒlədʒi] *noun* study of human behaviour and mental processes

Pt *chemical symbol for* platinum

Pu *chemical symbol for* plutonium

public ['pʌblɪk] *adjective* concerning *or* available to all people in general; **Public Health Inspector** = official of a local authority who examines the environment and tests for air pollution *or* bad sanitation *or* noise pollution, etc.; **public transport** = system of buses, trains, aircraft, trams, boats, etc. which all people may use; **public utility** = essential service such as electricity, gas, water, telephone, railway, etc. which is available to all people in general

puddingstone ['pʊdɪŋstəʊn] *noun* type of conglomerate, stone which is formed from other stones fused together

pulp [pʌlp] *noun* any soft wet matter, usually inside a harder exterior; **paper pulp** = wet mixture of pulverized fibres of wood, rags, etc. from which paper is made; **wood pulp** = softwood which has been pulverized into small fibres and mixed with water, used to make paper; **pulp and paper industry** = business of making pulp to manufacture paper; **pulpwood** = softwood used for making paper

pulse [pʌls] *noun* **(a)** pressure wave which can be felt in an artery each time the heart beats to pump blood; **to take someone's pulse** = to place fingers on an artery to feel the pulse and count the number of beats per minute **(b)** (i) any regularly recurring variation in quantity; (ii) sudden change in quantity; **acid pulse** = sudden increase in acidity in rainwater *or* river water **(c) pulses** = seeds of leguminous plants (lentils, beans and peas) used as food; ***pulses provide a large amount of protein***

pulverize ['pʌlvəraɪz] *verb* to reduce (something) to small particles like powder

◇ **pulverization** [pʌlvəraɪ'zeɪʃn] *noun* stage in the treatment of waste where it is reduced to small particles

pumice (stone) ['pʌmɪs] *noun* light glass-like substance formed from foam at the edge of a lava flow

pump [pʌmp] **1** *noun* machine which forces liquid *or* air into or out of something; **heat pump** = device which cools or heats by transferring heat from cold areas to warm ones (used in refrigerators or for heating large buildings); **wind pump** = pump driven by the wind, which raises water out of the ground **2** *verb* to force liquid *or* air into or out of something; **pumped-storage system** = hydroelectric system, where electricity is generated at times of peak demand and water is pumped up to a high reservoir during offpeak periods; **pumping station** = works where water *or* sewage *or* gas, etc. is pumped along a pipe *or* out of the ground *or* up to a storage tank, etc.

COMMENT: a pumped-storage turbine acts as an electricity generator when water pressure is high and becomes a water pump when water pressure is low, pumping water back up to the reservoir

pupa ['pju:pə] *noun* chrysalis *or* stage in the development of certain insects (such as butterflies) where the larva becomes encased in a hard shell (NOTE: plural is **pupae** ['pju:pi:])

◇ **pupal** ['pju:pəl] *adjective* referring to a pupa

◇ **pupate** [pju:'peɪt] *verb (of insect)* to move from the larval to the pupal stage

purchase ['pɜ:tʃəs] **1** *noun* act of buying; something bought **2** *verb* to buy; **purchasing power** = (of person *or* country, etc.) wealth *or* ability to buy things; (of currency) ability to buy a certain quantity

pure [pjʊə] *adjective* very clean *or* not mixed with other substances *or* not containing any other substances; **pure alcohol** *or* **alcohol BP** = alcohol with 5% water; **pure strain of plants** = plants bred by self-fertilization whose characteristics remain always the same; **pure**

tone = sound formed of a single frequency containing no harmonics

◊ **purebred** ['pjʊəbred] *adjective* (animal) which is the offspring of parents which are themselves the offspring of parents of the same breed

◊ **purification** [pjʊrɪfɪ'keɪʒn] *noun* action of making pure *or* removing impurities; ***activated sludge speeds up the purification process;*** **water purification** = removing impurities from water; **purification plant** = installation where impurities are removed from water

◊ **purify** ['pju:rɪfaɪ] *verb* to make pure *or* to remove impurities

◊ **purity** ['pju:rɪtɪ] *noun* state of being pure

putrefaction [pju:trɪ'fækʃn] *noun* decomposition of dead organic substances by bacteria

◊ **putrefy** ['pju:trɪfaɪ] *verb* to decompose *or* to rot

◊ **putrescibility** [pju:tresɪ'bɪlɪti] *noun* ability (of waste matter) to decompose *or* rot

◊ **putrescible** [pju:'tresɪbl] *adjective* (waste matter) which can decompose *or* rot

PVC = POLYVINYLCHLORIDE

PWR = PRESSURIZED WATER REACTOR

pylon ['paɪlən] *noun* tall metal construction for carrying high-tension electric cables; ***pylons can be unsightly and can be avoided by putting the cables they carry underground***

pyr- *or* **pyro-** ['paɪrəʊ] *prefix* referring to burning

pyramid ['pɪrəmɪd] *noun* chart *or* graphical representation showing the structure of an ecosystem, measured in terms of number, biomass *or* energy; **pyramid of biomass** = chart showing the different amounts of biomass at each trophic level, with the highest biomass at producer level and the lowest at secondary consumer level. (The biomass at each level is about ten per cent of that of the level beneath); **pyramid of energy** = chart showing the amounts of energy consumed at each trophic level; **pyramid of numbers** *or* **biotic pyramid** *or* **ecological pyramid** = chart showing the structure of an ecosystem in terms of who eats what: the base is composed of producer organisms (usually plants), then herbivores, then carnivores; **food pyramid** = chart of a food chain showing the number of predators at each level

pyrethrum [paɪ'ri:θrəm] *noun* organic pesticide, developed from a form of chrysanthemum, which is not very toxic and is non-persistent

pyridoxine [pɪrɪ'dɒksi:n] *noun* vitamin B_6

> COMMENT: pyridoxine is present in meat, cereals and treacle. Lack of pyridoxine causes vomiting and convulsions in babies

pyrite **(FeS_2)** *or* **pyrites** ['paɪraɪt *or* paɪ'raɪti:z] *noun* iron sulphide *or* yellow mineral found in igneous and metamorphic rocks

pyrolysis [paɪ'rɒlɪsɪs] *noun* conversion of a substance into another one by heat

Qq

Q unit of energy equal to 10^{18}Btu

quad [kwɒd] unit of energy equal to 10^{15}Btu

quadrat ['kwɒdrət] *noun* area of land measuring one square metre, chosen as a sample for research into plant populations; ***the vegetation of the area was sampled using the quadrat method***

quake [kweɪk] = EARTHQUAKE

quality ['kwɒlɪti] *noun* what something is like *or* how good or bad something is; ***good quality*** *or* ***bad quality;*** **high quality** *or* **top quality** = very best quality; **we sell only quality farm produce** = we sell only farm produce of the best quality; **quality of life** = good points about the type of life which people live *or* hope to live; ***the aim of the ban on emitting polluting gases is to improve the quality of life in the surrounding regions***

◊ **qualitative** ['kwɒlɪtətɪv] *adjective* referring to quality; **qualitative analysis** = analysis of what elements are present in a chemical compound; **qualitative inheritance** = inheritance of a major characteristic which distinguishes individual specimens of a species

quango ['kwæŋgəʊ] *noun* official body, set up by a government to investigate or deal with a special problem

quantify ['kwɒntɪfaɪ] *verb* **to quantify the effect of something** = to show the effect of something in figures; ***it is impossible to quantify the effect of the new legislation on pollution levels***

◇ **quantifiable** [kwɒntɪ'faɪəbl] *adjective* which can be quantified; ***the effect of the change in the waste disposal systems is not quantifiable***

quantity ['kwɒntɪti] *noun* amount *or* number of items; **to carry out a quantity survey** = to estimate the amount of materials and the cost of the labour required for a construction project

◇ **quantitative** ['kwɒntɪtətɪv] *adjective* referring to quantity; **quantitative analysis** = analysis of the quantity of a chemical element present in a compound; **quantitative inheritance** = inheritance of a characteristic which can vary slightly from specimen to specimen in a species

quarantine ['kwɒrənti:n] **1** *noun* period (originally forty days) when an animal *or* person *or* ship just arrived in a country has to be kept separate in case it carries a serious disease, to allow the disease time to develop and so be detected; ***the animals were put in quarantine on arrival at the port; a ship in quarantine shows a yellow flag called the quarantine flag*** **2** *verb* to put a person *or* animal in quarantine

COMMENT: animals coming into Britain are quarantined for six months because of the danger of rabies. People who are suspected of having an infectious disease can be kept in quarantine for a period which varies according to the incubation period of the disease. The main diseases concerned are cholera, yellow fever and typhus

quarry ['kwɒri] **1** *noun* place where rock is removed from the ground for commercial purposes; **slate quarry** = place where slate is removed from the ground for commercial purposes **2** *verb* to remove rock from the ground for commercial purposes

quartile ['kwɔ:taɪl] *noun* one of three figures below which 25%, 50% and 75% of a total falls

quartz [kwɔ:ts] *noun* mineral form of silica, found often as crystals in igneous rocks. (Pure quartz is known as rock crystal)

QUOTE: quartz is resistant to weathering and lacks the metals needed for cation exchange and so percolation through quartz does little to buffer acid
Scientific American

quasi- ['kweɪzaɪ] *prefix* almost *or* which seems like; ***a quasi-official body***

quaternary era [kwə'tɜ:nəri 'ɪərə] *noun* geological era which is still currently in existence

quench layer ['kwenʃ 'leɪə] *noun* deposit of unburnt hydrocarbons on the walls of cylinders in an internal combustion engine

quick-freeze [kwɪk'fri:z] *verb* to preserve food products by freezing them rapidly

quicklime ['kwɪklaɪm] *noun* calcium compound (calcium oxide) made from burnt limestone, used to spread on soil to reduce acidity and add calcium, in the composition of cement and in many industrial processes

quiescent [kwaɪ'esənt] *adjective* (volcano) which is inactive; (seed) which is not germinating because the conditions are unsatisfactory

quinine ['kwɪni:n] *noun* alkaloid drug made from the bark of a South American tree (the cinchona); **quinine poisoning** = illness caused by taking too much quinine

◇ **quininism** *or* **quinism** ['kwɪni:nɪzm *or* 'kwɪnɪzm] *noun* quinine poisoning

COMMENT: quinine was formerly used to treat the fever symptoms of malaria, but is not often used now because of its side effects. Symptoms of quinine poisoning are dizziness and noises in the head. Small amounts of quinine have a tonic effect and are used in tonic water

quota ['kwəʊtə] *noun* fixed amount of something which is allowed; ***the government has imposed a quota on the fishing of herring;*** **quota system** = system where imports *or* supplies are regulated by fixing maximum *or* minimum amounts

quotient ['kwəʊʃənt] *noun* result when one number is divided by another; **intelligence quotient (IQ)** = ratio of the result of an intelligence test shown as a relationship of the mental age to the actual age of the person tested (the average being 100); **respiratory quotient (RQ)** = ratio of the amount of carbon dioxide taken into the alveoli of the lungs from the blood to the amount of oxygen which the alveoli take from the air

Rr

R = ROENTGEN

R-value ['ɑ:'vælju:] *noun* unit of measurement of resistance to the flow of heat

COMMENT: an insulated outside wall has an R-value of R-11, an internal ceiling R-19

Ra *chemical symbol for* radium

raceme ['ræsi:m] *noun* inflorescence on which flowers are borne on stalks (as opposed to a panicle where several small flowers branch from the same stem)

rad [ræd] *noun* former unit of measurement of absorbed radiation dose: gray is now used to mean one hundred rads; *see also* GRAY, BECQUEREL

radar ['reɪdɑ:] *noun* method of finding objects by sending out high-frequency radio waves and detecting returning radio waves reflected by the object

COMMENT: radar can detect storm clouds and is used in meteorology as well as by aircraft to avoid flying through storm clouds

radiant ['reɪdiənt] *adjective* which is sent out in the form of rays; **radiant heat** = heat which is transmitted by infrared rays from hot bodies, e.g. an electric fire sends out radiant heat from a hot wire coil

radiate ['reɪdieɪt] *verb* **(a)** to spread out in all directions from a central point **(b)** to send out rays; ***heat radiates from the body; beta rays are radiated from a radioactive isotope***

◇ **radiation** [reɪdɪ'eɪʃn] *noun* **(a)** spreading out in all directions from a central point; **adaptive radiation** = development of a species from a single ancestor in such a way that different forms evolve to fit different environmental conditions **(b)** waves of energy which are given off when heat is transferred; **thermal radiation** = emission of radiant heat; **ultraviolet radiation** = invisible rays which have very short wavelengths and are beyond the violet end of the spectrum, forming the tanning and burning element in sunlight; **radiation fog** = fog which forms when the air just above ground level is cooled as the land surface immediately beneath it cools at night due to radiation **(c)** waves of energy which are given off by a radioactive substance; **background radiation** = radiation which comes from rocks *or* the earth *or* the atmosphere and not from a single man-made source which can be traced; **high-level radiation** = radiation from a very radioactive substance; **ionizing radiation** = radiation (such as alpha particles *or* X-rays) which produces atoms with electrical charges as it passes through a medium; **low-level radiation** = radiation from a substance which is less radioactive; **radiation burn** = burning of the skin caused by exposure to radiation from a radioactive substance; **radiation enteritis** = enteritis caused by X-rays; **radiation injury** = injury caused by exposure to radiation from a radioactive substance; **radiation pollution** = contamination of the environment by radiation; **radiation sickness** = illness caused by exposure to radiation from a radioactive substance; **radiation treatment** = RADIOTHERAPY; **radiation zone** = area that is contaminated by radiation and which people are not allowed to enter

COMMENT: prolonged exposure to ionizing radiation from various sources can be harmful. Nuclear radiation from fallout from nuclear weapons or from power stations, background radiation from substances naturally present in the soil, exposure to X-rays (either as a patient being treated or as a radiographer) can cause radiation sickness. First symptoms of the sickness are diarrhoea and vomiting, but radiation can also be followed by skin burns and loss of hair. Massive exposure to radiation can kill quickly and any person exposed to radiation is more likely to develop certain types of cancer than other members of the population. The main radioactive pollutants are strontium-90, caesium-137, iodine-131 and plutonium-239

QUOTE: radiation fog is the most commonly occurring type of fog in the populated lowlands of the UK. In such regions, fog is present on typically 5-10% of early mornings in late autumn and early winter

Weather

radio- ['reɪdiəʊ] *prefix* referring to (i) radiation; (ii) radioactive substances

radioactive [reɪdiəʊ'æktɪv] *adjective* (substance) whose nucleus disintegrates and gives off energy in the form of radiation which can pass through other substances; **radioactive decay** = gradual disintegration of the nucleus of radioactive matter; **radioactive isotope** = natural *or* artificial isotope which sends out radiation, used in radiotherapy; **radioactive waste** = used radioactive materials produced by nuclear power stations, industrial plants, hospitals, etc.

◇ **radioactivity** [reɪdɪəʊæk'tɪvəti] *noun* energy in the form of radiation emitted by a radioactive substance; **environmental**

radioactivity = energy in the form of radiation emitted into the environment by radioactive substances

COMMENT: the commonest naturally radioactive substances are radium and uranium. Other substances can be made radioactive for industrial or medical purposes by making their nuclei unstable, so forming radioactive isotopes. Radioactive wastes are classified as low-level (which are not very dangerous), intermediate, and high-level (which emit dangerous levels of radiation and cause disposal problems)

radiobiologist [reɪdiəʊbaɪ'ɒlədʒɪst] *noun* scientist who specializes in radiobiology

◇ **radiobiology** [reɪdiəʊbaɪ'ɒlədʒi] *noun* scientific study of radiation and its effects on living things

◇ **radiocarbon** [reɪdiəʊ'kɑːbən] *noun* radioactive isotope of carbon, whose presence is used to date ancient objects; **radiocarbon dating** = determining the age of an object by the amount of radiocarbon that has decayed

◇ **radiodermatitis** [reɪdiəʊdɜːmə'taɪtɪs] *noun* inflammation of the skin caused by exposure to radiation

◇ **radiogenic heat** [reɪdiəʊ'dʒenɪk 'hiːt] *noun* heat generated by the decay of a radioactive substance

◇ **radiograph** ['reɪdiəʊgrɑːf] *noun* X-ray photograph

◇ **radiographer** [reɪdi'ɒgrəfə] *noun* person specially trained to operate a machine to take X-ray photographs *or* radiographs

◇ **radiography** [reɪdi'ɒgrəfi] *noun* examining the internal parts of a patient by taking X-ray photographs

radioisotope [reɪdiəʊ'aɪsətəʊp] *noun* isotope of a chemical element which has been made radioactive

COMMENT: radioisotopes are used in medicine to provide radiation for radiation treatment. Radioisotopes of iodine are used to investigate thyroid activity

radiologist [reɪdi'ɒlədʒɪst] *noun* doctor who specializes in radiology

◇ **radiology** [reɪdi'ɒlədʒi] *noun* use of radiation to diagnose disorders (as in the use of X-rays or radioactive tracers) *or* to treat diseases such as cancer

◇ **radioscopy** [reɪdi'ɒskəpi] *noun* examining an X-ray photograph on a fluorescent screen

◇ **radiosensitive** [reɪdiəʊ'sensɪtɪv] *adjective* (cancer cell) which is sensitive to radiation and can be treated by radiotherapy

◇ **radiosensitivity** [reɪdiəʊsensə'tɪvəti] *noun* sensitivity (of a cancer cell) to radiation

◇ **radiosonde** ['reɪdiəʊsɒnd] *noun* radio transmitter sent into the atmosphere attached to a balloon to report readings of altitude, pressure, temperature, etc.

◇ **radiotherapy** [reɪdiəʊ'θerəpi] *noun* treating a disease by exposing the affected part to radioactive rays such as X-rays *or* gamma rays

COMMENT: many forms of cancer can be treated by directing radiation at the diseased part of the body

radium ['reɪdiəm] *noun* natural radioactive metallic element (NOTE: chemical symbol is **Ra**; atomic number is **88**)

radon ['reɪdɒn] *noun* inert natural radioactive gas, formed from the radioactive decay of radium (NOTE: chemical symbol is **Rn**; atomic number is **86**)

COMMENT: radon occurs naturally in soil, in construction materials and even in ground water. It can seep into houses and causes radiation sickness

rain [reɪn] *noun* water which falls from clouds as small drops; **acid rain** = rain which contains a higher level of acid than normal; *see also* ACID; **artificial rain** = rain which is made by 'seeding' clouds with crystals of salt, carbon dioxide and other substances; **rain-bearing cloud** = clouds which carries moisture in droplet form which can fall as rain; **rain shadow** = reduction of rainfall on the lee side of a mountain

◇ **rainbow** ['reɪnbəʊ] *noun* natural phenomenon which occurs when light strikes water droplets, especially when sunlight hits rain *or* spray from a waterfall, creating a semicircle of rings of each of the colours of the spectrum

◇ **raindrop** ['reɪndrɒp] *noun* drop of water which falls from a cloud

◇ **rainfall** ['reɪnfɔːl] *noun* amount of water which falls as rain on a certain area over a certain period

◇ **rainforest** ['reɪnfɒrɪst] *noun* thick tropical forest which grows in regions where the rainfall is very high; ***poor farmers have cleared hectares of rainforest to grow cash crops***

◇ **rainmaking** ['reɪnmeɪkɪŋ] *noun* attempting to create rain by 'seeding' clouds with crystals of salt, carbon dioxide and other substances

◇ **rainout** ['reɪnaʊt] *noun* process where particles in the atmosphere act as centres round which water can form drops which then fall as rain; *compare* WASHOUT

◇ **rains** [reɪnz] *plural noun* season when heavy rain falls

◇ **rainstorm** ['reɪnstɔːm] *noun* heavy rain accompanied by wind

◇ **rainwash** ['reɪnwɒʃ] *noun* erosion of soil by rain

◇ **rainwater** ['reɪnwɔːtə] *noun* water which falls as rain from clouds

◇ **rainy** ['reɪni] *adjective* with a lot of rain; **rainy season** = period in some countries when a lot of rain falls (as opposed to the dry season); ***the rainy season lasts from April to August***

COMMENT: rain is normally slightly acid (about pH 5.6) but becomes more acid when pollutants from burning fossil fuels are released into the atmosphere

QUOTE: the Amazon Basin houses the Earth's greatest rainforest, a symbol of international determination to bring deforestation to a halt
Times

raise [reɪz] *verb* **(a)** to increase; to make higher; **raised beach** = beach of sand left higher than sea level because the level of the sea has fallen; **raised bog** = bog where the dead moss has accumulated without decomposing, so raising the level of the bog above the surrounding land **(b)** (i) to make plants germinate and nurture them as seedlings; (ii) to breed livestock; ***the plants are raised from seed in special seedbeds***

ramet ['ræmɪt] *noun* single cloned organism

ranch [rɑːnʃ] *noun* **cattle ranch** = very large grassland farm where cattle are raised; ***governments have encouraged the conversion of rainforest to livestock ranches***

◇ **ranching** ['rɑːnʃɪŋ] *noun* raising cattle on ranches

QUOTE: plantations and ranching have laid waste millions of hectares of forest. In Central America cattle ranching is responsible for the clearance of almost two-thirds of the forests
Ecologist

random ['rændəm] *adjective* not specially selected; **at random** = in a way which is not specially selected; **random sample** = sample for testing taken without any special selection

range [reɪnʒ] *noun* **(a)** series of different but similar things; **mountain range** = series of mountains running in a line for many miles **(b)** (i) difference between lowest and highest values in a series of data; (ii) area within two or more points; ***the temperature range is over 50°C; the geographical range of the plant is from the Arctic Circle to Southern Europe*** **(c)** large area of grass-covered farmland used for raising cattle *or* sheep

◇ **ranger** ['reɪndʒə] *noun* person in charge of the management and protection of a forest *or* park *or* reserve

rape [reɪp] *noun* **oil-seed rape** = plant of the cabbage family *(Brassica napus)* extensively grown to provide oil, also used as green manure

rapid ['ræpɪd] **1** *adjective* quick *or* happening during a short space of time; ***the rapid growth of grasses after the rains came; the rapid beating of the wings of a hummingbird*** **2** *noun* **rapids** = part of a river where the water flows rapidly over large rocks

raptor ['ræptə] *noun* bird of prey

◇ **raptorial** [ræp'tɔːriəl] *adjective* referring to raptors

rare [reə] *adjective* not common *or* (disease) of which there are very few cases; **rare earth elements** = metallic elements (numbers 57 to 71 on the periodic table); **rare gases** = inert gases *or* gases (helium, neon, argon, krypton, xenon and radon) which do not react chemically with other substances

◇ **rarity** ['reərɪti] *noun* state of being rare

COMMENT: species are classified internationally into several degrees of rarity: rare, means that a species is not numerous and is confined to small local populations, but is not necessarily likely to become extinct. Vulnerable means that a species has a small population and that population is declining. Endangered means that the species has such a small population that it is likely to become extinct

rate [reɪt] *noun* **(a)** amount *or* proportion of something compared to something else; **rate of natural increase** = difference between the crude birth rate and the death rate; **rate of population growth** = increase in population in a certain area divided by the initial population; **birth rate** = number of births per year, shown per thousand of the population; **death rate** *or* **mortality rate** = number of deaths per year, shown per thousand of the population; **fertility rate** = number of births per year, shown, in humans, per thousand females aged between 15 and 44 **(b)** number of times something happens; **heart rate** = number of times the heart beats per minute

COMMENT: worldwide, the current human birth rate is 34 per 1,000 and the death rate is 15 per 1,000, showing that the world's population is increasing. There are, however, substantial differences between rates in different countries

rating ['reɪtɪŋ] *noun* value of something; **antiknock rating** = classification of petrol, showing how likely it is to cause knocking; **octane rating** = classification of the quality and performance of petrol, according to the amount of hydrocarbon in it

ratio ['reɪʃiəʊ] *noun* number which shows a proportion *or* which is the result of one number divided by another; ***an IQ is the ratio of the person's mental age to his chronological age;*** **efficiency ratio** = number which shows the proportion of work done *or* energy produced by a machine *or* engine, etc., to the energy supplied to it (usually expressed as a percentage); **energy efficiency ratio (EER)** = measure of the efficiency of a heating *or* cooling system (such as a heat pump *or* an air-conditioning system) shown as the ratio of the output in Btu per hour to the input in watts

ration ['ræʃən] *noun* amount of food given to an animal; **maintenance ration** = quantity of food needed to keep a farm animal healthy but not productive; **production ration** = quantity of food needed to make a farm animal produce meat, milk, eggs, etc. (It is always more than the basic maintenance ration)

raw [rɔː] *adjective* **(a)** (food) which has not been cooked **(b)** (sewage, water, etc.) which has not been treated

◇ **raw material** ['rɔː mə'tɪəriəl] *noun* substance which is used to manufacture something, such as ore for making metals, wood for making furniture, etc.

ray [reɪ] *noun* line of light *or* radiation *or* heat; **infrared rays** = long invisible rays, below the visible red end of the colour spectrum, which form part of the warming radiation which the earth receives from the sun; **ultraviolet rays (UV rays)** = short invisible rays, beyond the violet end of the colour spectrum, which form the tanning and burning element in sunlight; *see also* X-RAY

RBE = RELATIVE BIOLOGICAL EFFECTIVENESS

RDF = REFUSE-DERIVED FUEL

react [ri'ækt] *verb* **(a) to react to something** = to act because of something else *or* to act in response to something; ***the government reacted slowly to the news of the disaster*** **(b)** *(of a chemical substance)* **to react with something** = to change because of the presence of another substance; ***ozone is produced as a result of oxides reacting with sunlight***

◇ **reaction** [rɪ'ækʃn] *noun* (i) action which takes place because of something which has happened earlier; (ii) effect produced by a stimulus; (iii) chemical change when two substances come into contact and cause each other to change; **chain reaction** = (i) nuclear reaction where a neutron hits a nucleus, makes it split, and so releases further neutrons; (ii) chemical reaction where each stage is started by a chemical substance which reacts with another, producing further substances which can continue to react; **nuclear reaction** = physical reaction of the nucleus of an atom, which when bombarded by radiation particles creates an isotope; **reaction turbine** = turbine where the blades on the turbine adjust to the angle at which the jets of water hit them

◇ **reactive** [ri'æktɪv] *adjective* (chemical) which reacts easily with other substances

◇ **reactivity** [riːæk'tɪvəti] *noun* ability of something to react

> QUOTE: if ozone gets into the lower atmosphere, its greater reactivity over oxygen makes it highly toxic to plants
>
> **New Scientist**

reactor [ri'æktə] *noun* **breeder reactor** = nuclear reactor which produces more fissile material than it consumes; **fast breeder reactor (FBR)** *or* **fast (neutron) reactor** = nuclear reactor which produces more fissile material than it consumes, using fast-moving neutrons and making plutonium-239 from uranium-238, thereby increasing the reactor's efficiency; **nuclear reactor** = device which creates heat and energy by starting and controlling atomic fission

> COMMENT: nuclear reactors are used (a) to produce energy in power stations or motors; (b) to produce plutonium-239 for use in weapons; (c) for research purposes. A nuclear reactor needs fuel and the most common fuel used is uranium dioxide, which contains both fissionable uranium-235 and fertile uranium-238 isotopes. The fuel in a reactor is in the form of hundreds of rods, each separated from the others by cladding. The fission of the fuel produces fast neutrons which have to be slowed down by a moderator such as water (H_2O), heavy water (D_2O) or graphite. All nuclear power stations produce electricity from steam-driven generators: the heat required to produce the steam is produced by nuclear fission and transferred to the water in several ways. In boiling water reactors (BWRs), light water is heated to boiling point and the steam produced drives a turbine. In pressurized water reactors (PWRs), the water is heated under pressure to a very high temperature, and this pressurized hot water is passed through a water tank heating the water in the tank until it becomes steam to drive the turbines. In the advanced gas-cooled reactors (AGRs), carbon dioxide or helium is heated to a very high temperature and then passed through water to create steam. Pollution from nuclear reactors is possible due to the escape of radiation into the surrounding environment. This may take the form of discharge of the coolant, radiation from

the disposal of spent fuel, or by accidental overheating which ruptures pipes

reading ['ri:dɪŋ] *noun* note taken of figures, especially of degrees on a scale; ***readings were taken over an area of 60 square miles round the nuclear power station; temperature readings over the last decade show a gradual rise in surface temperature***

reafforest [ri:ə'fɒrɪst] *verb* to plant trees again in an area which was once covered by forest

◇ **reafforestation** [ri:æfɒrɪs'teɪʃn] *noun* planting trees again in an area which was once covered by forest

> QUOTE: the first priority in any programme of ecological recovery must be a worldwide programme of reafforestation. Even the most degraded lands in the dry tropics can be restored to forest, successful reafforestation schemes having been implemented in Costa Rica and India
> **Ecologist**

reagent [ri'eɪdʒənt] *noun* chemical substance which reacts with another substance (especially when used to detect the presence of the second substance)

rear [rɪə] *verb* to look after young animals until they are old enough to look after themselves

receptor (cell) [rɪ'septə 'sel] *noun* nerve ending which senses a change in the surrounding environment *or* in the body (such as cold, heat, pressure, pain) and reacts to it by sending an impulse through the central nervous system to the brain; *see also* CHEMORECEPTOR, EXTEROCEPTOR, INTEROCEPTOR

recessive [rɪ'sesɪv] *adjective* (gene *or* genetic trait) which is weaker than and hidden by a dominant gene

◇ **recessiveness** [rɪ'sesɪvnəs] *noun* (of a gene *or* genetic trait) being recessive

> COMMENT: since each physical characteristic is governed by two genes, if one gene is dominant and the other recessive, the resulting trait will be that of the dominant gene. Traits governed by recessive genes will appear if genes from both parents are recessive

reclaim [rɪ'kleɪm] *verb* **(a)** to take virgin land *or* marshland *or* a waste site *or* land which has already been developed, etc. and make it available for agricultural *or* commercial purposes **(b)** to recover useful materials from waste; ***they reclaimed the tiles from the roof of a derelict factory***

◇ **reclamation** [reklə'meɪʃn] *noun* **(a)** (i) act of reclaiming land; (ii) land which has been reclaimed; ***the authority is studying the costs of the land reclamation scheme in the city centre*** **(b)** recovering useful materials from waste; **heat reclamation** = collecting heat from substances heated during a process and using it to heat further substances, so as to avoid heat loss

recommend [rekə'mend] *verb* to suggest that something should be done; ***the report recommended choosing a different site for the disposal of radioactive waste***

◇ **recommendation** [rekəmen'deɪʃn] *noun* suggesting that something should be done; ***the council accepted the committee's recommendations about disposal of waste***

reconstitute [ri:'kɒnstɪtju:t] *verb* to put something back into its original state; ***the government has insisted that the mining company should reconstitute the site after the open-cast mining operation has closed down***

record 1 *verb* [rɪ'kɔ:d] to note information; ***the chart records the variations in the rainfall pattern*** **2** *noun* ['rekɔ:d] piece of information about something; ***the winter was the wettest since records were started in 1896***

recover [rɪ'kʌvə] *verb* **(a)** to get better; to get back to a previous state; ***the fish populations may never recover from overfishing*** **(b)** to obtain (metals *or* other useful materials) from waste by separating and purifying it

◇ **recovery** [rɪ'kʌvəri] *noun* **(a)** getting better; getting back to a previous state; ***the area has been overfished to such an extent that the recovery of the fish population is impossible*** **(b)** obtaining metals *or* other useful materials from waste; **heat recovery** = collecting heat from substances heated during a process and using it to heat further substances, so as to avoid heat loss

recycle [ri:'saɪkl] *verb* to process waste material so that it can be used again; ***the glass industry recycles tonnes of waste glass each year; switching to using recycled paper can be cost-effective;*** **recycled paper** = paper made from old waste paper

◇ **recyclable** [ri:'saɪkləbl] *adjective* (waste) which can be processed so that it can be used again

◇ **recycling** [ri:'saɪklɪŋ] *noun* processing waste material so that it can be used again

> COMMENT: many waste items can be recycled. In particular, precious metals (lead, copper, silver, etc.) can be recovered from old batteries or computers; paper can be recycled from old newspapers and packaging materials (after deinking); glass can be manufactured from cullet from old bottles

red [red] *adjective & noun* (of) a colour like the colour of blood; **red algae** = Rhodophyta *or* type of very small algae *or* phytoplankton, mainly found on the seabed and which cause the phenomenon called red tide; **Red Data Book** = catalogue published by the IUCN, listing species which are rare or in danger of becoming extinct; **red lead (Pb_3O_4)** = red poisonous oxide of lead, used as a colouring in paints; **red-lead ore** = CROCOITE; **red sea** = RED TIDE; **red snow** = snow coloured red by the presence of algae; **red tide** = phenomenon where the sea becomes red, caused by Rhodophyta, a type of very small algae *or* phytoplankton

◇ **redness** ['rednəs] *noun* being red *or* red colour

◇ **redwood** ['redwʊd] *noun* American conifer, often growing to a very large size

redevelop [ri:dɪ'veləp] *verb* to demolish the buildings on an area of land and build new ones; ***the company has put forward a plan to redevelop the area around the railway station***

◇ **redevelopment** [ri:dɪ'veləpmənt] *noun* act of redeveloping an area of land; ***the council's planning committee has approved the redevelopment plan for the town centre***

redistribution of land [ri:dɪstrɪ'bju:ʃn əv 'lænd] *noun* taking land from large landowners and splitting it into smaller plots for peasant owners

redox potential (rH) ['ri:dɒks pə'tenʃl] *noun* measure of the ability of the natural environment to bring about an oxidation *or* a reduction process (e.g. involving sulphur)

reduce [rɪ'dju:s] *verb* **(a)** to make *or* become less *or* smaller *or* lower **(b)** *(in chemistry)* to add electrons *or* hydrogen to a substance

◇ **reducer** [rɪ'dju:sə] *noun* decomposer *or* organism (such as the earthworm, a fungus *or* bacteria) which breaks down dead organic matter

◇ **reduction** [rɪ'dʌkʃn] *noun* **(a)** making *or* becoming less *or* smaller *or* lower; ***the key to controlling acid rain must be the reduction of emissions from fossil-fuelled power stations*** **(b)** *(in chemistry)* adding electrons *or* hydrogen to a substance

redundant [rɪ'dʌndənt] *adjective* no longer useful; **redundant farmland** = land which is currently being used for farming but which will not be in the future

reed [ri:d] *noun* aquatic plant, growing near the shores of lakes, used to make thatched roofs

◇ **reedbed** ['ri:dbed] *noun* mass of reeds growing together

reef [ri:f] *noun* low rocks *or* coral near the surface of the sea; **barrier reef** = long coral reef lying along the shore and enclosing a lagoon

refine [rɪ'faɪn] *verb* to process something to remove impurities; ***a by-product of refining oil***

◇ **refinery** [rɪ'faɪnəri] *noun* plant where raw material, such as ore, oil *or* sugar, is processed to remove impurities

reflect [rɪ'flekt] *verb* to send back light *or* heat *or* sound towards it source or in a different direction; ***some solar radiation is reflected back by clouds or by the earth's surface***

◇ **reflection** [rɪ'flekʃn] *noun* light *or* sound *or* heat which is sent back towards its source; ***bats find their way by sending out high frequency sounds and listening to the reflection of the sound from objects in their way;*** **reflection seismology** = study of how pressure waves from seismic movements are reflected by rock structures in the earth

reflex ['ri:fleks] *noun* automatic reaction to a stimulus; **accommodation reflex** = reaction of the pupil when the eye focuses on an object which is close; **light reflex** = reaction of the pupil of the eye which changes size according to the amount of light going into the eye; **reflex action** = automatic reaction on the part of an animal to a stimulus (such as a sneeze after breathing in dust)

reflux ['ri:flʌks] *noun (of a liquid)* flowing backwards in the opposite direction to normal flow; *(in distillation)* collecting vapour and condensing it so that it can return to the boiler again

reforest [ri:'fɒrɪst] *verb* = REAFFOREST

◇ **reforestation** [ri:fɒrɪs'teɪʃn] *noun* = REAFFORESTATION

refract [rɪ'frækt] *verb* to make light rays change direction as they go from one medium (such as air) to another (such as water)

◇ **refraction** [rɪ'frækʃn] *noun* (i) change of direction of light rays as they enter a different medium; (ii) measurement of the angle at which the light rays bend

refrigerate [rɪ'frɪdʒəreɪt] *verb* to make something cold

◇ **refrigerant** [rɪ'frɪdʒərənt] *noun* substance (such as liquid ammonia) which makes other substances cold

◇ **refrigeration** [rɪfrɪdʒə'reɪʃn] *noun* making something cold

◇ **refrigerator** [rɪ'frɪdʒəreɪtə] *noun* machine which keeps things cold

refuge *or* **refugium** ['refju:dʒ *or* rɪ'fju:dʒiəm] *noun* safe place where species can escape environmental change and continue to exist as before (NOTE: plural is **refugia**)

refuse ['refju:s] *noun* rubbish *or* waste matter; **domestic refuse** *or* **household refuse** = waste material from houses; **refuse collection and disposal** = gathering waste together and getting rid of it; **refuse-derived fuel (RDF)** = fuel which is made *or* processed from refuse; **refuse dump** = place where refuse is thrown away

regenerate [rɪ'dʒenəreɪt] *verb* to grow again; ***a forest takes about ten years to regenerate after a fire***

◇ **regenerative** [rɪ'dʒenərətɪv] *adjective* which allows new growth to replace damaged tissue; **regenerative properties** = ability of something, such as an ecosystem, to recover from pollution

◇ **regeneration** [rɪgenə'reɪʃn] *noun* growing again (of vegetation on land which has been cleared); ***grazing by herbivores prevents the regeneration of forests destroyed by fire***

regime [reɪ'ʒi:m] *noun* general pattern *or* system; ***the two rivers have very different flow regimes: one has very rapid flow down from high mountains, the other is slower and mainly crosses fertile plains***

region ['ri:dʒən] *noun* area

register ['redʒɪstə] *verb* to note a piece of data; ***several lakes have registered a steady increase in acidity levels until their natural fish populations have started to decline; the earthquake registered 6.2 on the Richter scale***

regolith ['regəlɪθ] *noun* layer of weathered rock fragments which covers most of the Earth's land area

regress [rɪ'gres] *verb* to return to a more primitive earlier state

◇ **regression** [rɪ'greʃn] *noun* returning to a more primitive earlier state, as when cultivated land returns to a wild state

◇ **regressive** [rɪ'gresɪv] *adjective* (water level) which is getting lower

regulate ['regjuleɪt] *verb* (i) to make something work (in a regular way); (ii) to change *or* maintain something by law; ***planning permission is regulated by local authorities***

◇ **regulation** [regju'leɪʃn] *noun* **(a)** making something work (in a regular way); **osmotic regulation** *or* **osmoregulation** = process by which cells and simple organisms maintain a balance with the fluid in their surroundings **(b)** rule made by a government *or* club *or* society; ***the safety regulations have not been complied with; the pamphlet lists the regulations concerning visits to nature reserves***

◇ **regulator** ['regjuleɪtə] *noun* thing which controls development; **growth regulator** = a chemical used to control the growth of certain plants; mainly used for weed control in cereals and grassland

◇ **regulatory** [regju'leɪtəri] *adjective* (organization) which makes sure something works according to the rules; **Nuclear Regulatory Commission (NRC)** = US government body which regulates and licenses nuclear plants

rehabilitation [ri:həbɪlɪ'teɪʃn] *noun* reclaiming and redeveloping land and buildings which have been abandoned

reheat [rɪ'hi:t] *verb* to heat something again; ***water which has been cooled and condensed after passing through the boilers is sent back to the boilers for reheating***

◇ **reheater** [rɪ'hi:tə] *noun* section of a power station where steam which has been used to turn the first turbine is heated again to create enough pressure to turn the second turbine

relate [rɪ'leɪt] *verb* to connect to

◇ **-related** [rɪ'leɪtɪd] *suffix* connected to

◇ **relationship** [rɪ'leɪʃnʃɪp] *noun* way in which someone *or* something is connected to another

◇ **relative** ['relətɪv] *adjective* which is compared to something else; **relative abundance** = number of individual specimens of an animal *or* plant seen over a certain period of time in a certain place; **relative biological effectiveness (RBE)** = measure of radiation used as a protection; **relative humidity** = ratio between the amount of water vapour in air and the amount which would be present if the air was saturated (shown as a percentage); **relative transpiration** = rate at which water transpires from the surface of a plant

release [rɪ'li:s] **1** *noun* act of letting something out *or* of letting something go free; ***the report points out the danger of radiation releases from nuclear power stations*** **2** *verb* to let something out *or* to let something go free; ***acid rain leaches out nutrients from the soil and releases harmful substances such as lead into the soil; using an aerosol spray releases CFCs into the atmosphere***

◇ **releaser** [rɪ'li:sə] *noun* stimulus which provokes a reaction in an animal (such as the sight of a hawk *or* the sound of a gun)

relevé ['rələveɪ] *noun* list of plant species growing in a certain area

reliability [rɪlaɪə'bɪlɪti] *noun* degree to which a person *or* thing can be trusted

◇ **reliable** [rɪ'laɪəbl] *adjective* (person *or* thing) who *or* which can be trusted; ***it is not a very reliable method of measuring the depth***

relict ['relɪkt] *noun* species which is still in existence, even though the environment in which it originally developed is no longer present

relief [rɪ'li:f] *noun* differences in height between points on the earth's surface; **relief map** = map with contour lines which shows differences in height between points on the earth's surface

rely on [re'laɪ 'ɒn] *verb* to trust

rem = ROENTGEN EQUIVALENT MAN

remote sensing [rɪ'məʊt 'sensɪŋ] *noun* getting information about the physical aspects of the earth (such as the location of mineral deposits, the movement of water *or* pests, etc.) from satellite observation and aerial photography

remove [rɪ'mu:v] *verb* to take something out; to extract (a mineral) from the ground

◇ **removal** [rɪ'mu:vl] *noun* taking something out; extracting (a mineral) from the ground

renew [rɪ'nju:] *verb* to replace *or* replenish; to make something new again

◇ **renewable** [rɪ'nju:əbl] *adjective* (natural resource) which can be replaced *or* replenish itself; ***herring stocks are a renewable resource if they are not overfished;*** **renewable (sources of) energy** = energy from the sun *or* wind *or* waves *or* tides *or* from geothermal deposits *or* from burning waste, none of which uses up fossil fuel reserves (NOTE: opposite is **finite**)

> QUOTE: today's primary sources of energy are mainly non-renewable: natural gas, coal, peat and conventional nuclear power. There are also renewable sources, including wood, plants, dung, falling water, geothermal sources and solar, tidal, wind and wave energy. Nuclear reactors that produce their own fuel (breeders) and eventually fusion reactors, are also in this category
> **Brundtland report**

> QUOTE: even though renewable energy technologies may be commercially viable, they rarely achieve significant market penetration because of a wide range of constraints on their recognition and use as viable alternatives to existing, conventional energy sources
> **Appropriate Technology**

repel [rɪ'pel] *verb* to drive away

◇ **repellant** [rɪ'pelənt] *noun* = REPELLENT

◇ **repellent** [rɪ'pelənt] *adjective & noun* (substance) which repels; ***the coating of wax on leaves acts as a repellent to pollutants;*** **insect repellent** = chemical which protects by repelling insects

replant [ri:'plɑ:nt] *verb* to plant again; ***after the trees were felled the land was cleared and replanted with mixed conifers and broad-leaved species***

replication [replɪ'keɪʃn] *noun* process in the division of a cell where the DNA makes copies of itself

report [rɪ'pɔ:t] **1** *noun* official document giving an account of something *or* stating what action has been taken *or* what the current state is *or* what the results are from a test, etc. **2** *verb* to give an account of something; **reportable disease** = disease (such as asbestosis *or* hepatitis *or* anthrax) which may be caused by working conditions or may infect other workers and must be reported to the health authority

repository [rɪ'pɒzɪtəri] *noun US* place where nuclear waste can be stored

repower [rɪ'paʊə] *verb* to rebuild an old power station, converting it to a more modern combustion system, such as pressurized fluidized-bed combustion

reprocess [rɪ'prəʊses] *verb* to process again

◇ **reprocessing** [rɪ'prəʊsesɪŋ] *noun* processing something again, such as taking spent nuclear fuel and subjecting it to chemical processes which produce further useful materials (such as plutonium)

> COMMENT: reprocessing plants are the cause of much controversy; some countries do not have facilities for reprocessing spent nuclear fuel and therefore export it to those countries with suitable plants. Unfortunately, reprocessing plants are just as likely to cause radioactive waste as nuclear power stations; the transport of the spent material from the power station to the processing plant is also a potential radiation hazard

reproduce [rɪprə'dju:s] *verb* **(a)** *(of an animal or plant, etc.)* to produce another individual like itself; *(of bacteria, etc.)* to produce new cells **(b)** to do a test again in exactly the same way

◇ **reproduction** [ri:prə'dʌkʃn] *noun* producing another individual *or* new cells, etc.; **organs of reproduction** = REPRODUCTIVE ORGANS; **asexual reproduction** = reproduction by taking cuttings of plants *or* by cloning

◇ **reproductive** [ri:prə'dʌtɪv] *adjective* referring to reproduction; **reproductive organs** = parts of an organism which are involved in reproduction; **reproductive system** = (in an animal) arrangement of organs and ducts which produce spermatozoa and ova; **reproductive tract** = (in an animal) series of tubes and ducts which carry spermatozoa and ova from one part of the body to another

> COMMENT: in the male animal, the testes form the spermatozoa which pass out of the body through the urethra and penis on

ejaculation; in the female, ova are produced by the ovaries and pass through the Fallopian tubes where they are fertilized by spermatozoa from the male. The fertilized ovum moves down into the uterus where it develops into an embryo

reptile ['reptaɪl] *noun* member of the Reptilia, the class of cold-blooded animals (such as crocodiles, tortoises, snakes) which lay eggs and have scaly skins

requirement [rɪ'kwaɪəmənt] *noun* something that is needed; **energy requirements** = amount of energy needed

research [rɪ'sɜːtʃ] **1** *noun* scientific study which investigates something new; ***a research borehole found water at a depth of 2,000 feet;*** **field research** = scientific study made in the open air as opposed to in a laboratory; **research area** = piece of ground *or* area of knowledge in which a scientific study is being carried out **2** *verb* to carry out scientific study; ***he is researching the causes of the ozone hole***

reserve [rɪ'zɜːv] *noun* **(a)** area of unspoilt land where no commercial exploitation is allowed; **game reserve** = area of land where wild animals are kept to be hunted and killed for sport; **nature reserve** = special area where the wildlife is protected **(b)** amount stored *or* kept back for future use; **energy reserves** = amount of energy stored; often refers to the stocks of non- renewable fuel, such as oil, which a nation, for example, possesses

◇ **reservation** [rezə'veɪʃn] *noun* (i) area of land set aside for a special purpose; (ii); *(in North America)* area of land set aside for native Indian tribes to live in

reservoir ['rezəvwɑː] *noun* **(a)** artificial *or* natural area of water, used for storing water for domestic *or* industrial use; ***the town's water supply comes from reservoirs in the mountains; after two months of drought the reservoirs were beginning to run dry*** **(b)** natural hole in rock which contains liquid (such as water *or* oil) *or* gas; **reservoir rock** = underground layer of rock containing liquid (such as water *or* oil) *or* gas

residence time ['rezɪdəns 'taɪm] *noun* amount of time during which something remains in the same place or in the same state until it is lost or transformed into something else

resident ['rezɪdənt] *noun & adjective* (animal) which lives in a place; ***the introduced species wiped out the resident population of flightless birds***

residual [re'zɪdjuəl] **1** *adjective* remaining *or* which is left behind; **residual oil** = oil which is left after crude oil has been through various refining processes; **residual spraying** = spraying insecticide onto walls so that it will stay there and kill insects which come into contact with it **2** *noun* = RESIDUE

◇ **residue** ['rezɪdjuː] *noun* material left after a process has taken place *or* after a material has been used; **chemical residue** = waste left after a chemical process has taken place; **consumption residues** = waste matter left after manufactured goods are used; **pesticide residue** = amount of pesticide that remains in the environment after spraying

resilience [rɪ'zɪliəns] *noun* strength to stand up to shocks, especially the ability of an ecosystem to return to its normal state after being disturbed

resin ['rezɪn] *noun* **(a)** sticky oil which comes from some types of conifer **(b)** **(synthetic) resin** = solid *or* liquid organic compound, a polymer used in the making of plastic

◇ **resinous** ['rezɪnəs] *adjective* referring to resin

resist [rɪ'zɪst] *verb* to be strong enough to avoid being killed *or* attacked by a disease *or* by a pesticide

◇ **resistance** [rɪ'zɪstəns] *noun* **(a)** (i) ability of an organism not to get a disease; (ii) ability of a germ to be unaffected by antibiotics; (iii) ability of a pest *or* weed to be unaffected by a pesticide *or* herbicide; ***the bacteria have developed a resistance to certain antibiotics; after living in the tropics his resistance to colds was low; increasing insect resistance to chemical pesticides is a major problem;*** **selective resistance** = ability of an organism to be unaffected by certain poisons, pollutants, pesticides, herbicides, etc. **(b) environmental resistance** = ability to withstand all the pressures like predation, competition, weather, food availability, which inhibit the potential growth of a population

◇ **resistant** [rɪ'zɪstənt] *adjective* unaffected by something; ***the rabbits are becoming resistant to myxomatosis; the bacteria are resistant to some antibiotics;*** **resistant strain** = group of organisms which is unaffected by a certain disease *or* antibiotic *or* pesticide *or* herbicide, etc.

◇ **-resistant** [rɪ'zɪstənt] *suffix* meaning which is unaffected by something; ***DDT-resistant strain of insects; fungus-resistant genetic material was found in genetic stocks from Mexico; a new strain of virus-resistant rice***

COMMENT: resistant strains develop quite rapidly after application of the treatment. Some strains of insect have developed which are resistant to DDT. The resistance develops as non-resistant strains die off, leaving only individuals which possess a slightly different and resistant chemical makeup

resorption [rɪ'zɔːpʃn] *noun* absorbing again of a substance already produced back into an organism; melting of solid rock which falls into magma

resource [rɪ'zɔːs] *noun* anything in the environment which can be used by an organism; ***muddy estuaries are among the most productive and important wildlife resources;*** **finite resources** = resources which will in the end be used up (such as reserves of coal *or* oil *or* gas); **mineral resources** = supply of minerals and metals which are necessary for the running of a country's economy, but not including natural resources such as wood *or* food plants; **natural resources** = part of the environment which can be used commercially (such as coal); **resource management** = system of controlling the use of resources in such a way as to avoid waste and to use them to the best effect

respiration [respɪ'reɪʃn] *noun* action of breathing; **external respiration** = part of respiration concerned with oxygen in the air being exchanged in the lungs for carbon dioxide from the blood; **internal respiration** = part of respiration concerned with the passage of oxygen from the blood to the tissues and the passage of carbon dioxide from the tissues to the blood

◇ **respiratory** [rɪ'spɪrətəri] *adjective* referring to breathing; **respiratory disorder** = illness which affects the patient's breathing; **respiratory pigment** = blood pigment which can carry oxygen collected in the lungs and release it in tissues; **respiratory quotient (RQ)** = ratio of the amount of carbon dioxide taken into the alveoli of the lungs from the blood to the amount of oxygen which the alveoli take from the air; **respiratory system** = series of organs and passages which take air into the lungs and exchange oxygen for carbon dioxide

COMMENT: respiration includes two stages: breathing in (inhalation) and breathing out (exhalation). Air is taken into the respiratory system through the nose or mouth and goes down into the lungs through the pharynx, larynx and windpipe. In the lungs, the bronchi take the air to the alveoli (air sacs) where oxygen in the air is passed to the bloodstream in exchange for waste carbon dioxide which is then breathed out

response [rɪs'pɒns] *noun* reaction of an organism to a stimulus such as a pollutant

responsible [rɪs'pɒnsɪbl] *adjective* answerable for something; causing something; ***the action of sunlight on gases is responsible for the formation of ozone; they accused the power station of being responsible for deaths from cancer in the area***

◇ **responsibility** [rɪspɒnsɪ'bɪlɪti] *noun* being answerable for something; looking after something; ***the general feeling is that large-scale industrial companies are only governed by commercial requirements and have no social responsibility***

rest [rest] *verb* **to rest land** = to let land lie fallow, without growing any crops

restock [riː'stɒk] *verb* to provide another supply of something that has been used up; ***so many animals died that they had to restock the farm in the spring***

restore [rɪ'stɔː] *verb* to give back *or* to get back; ***by letting the land lie fallow for a couple of years, farmers hope to restore some of the natural nutrients which have been removed from the soil***

◇ **restoration** [restə'reɪʃn] *noun* giving back *or* getting back; **land restoration** = bringing a piece of land back to its former state, such as bringing back into productive use a piece of land which was not usable before (e.g. a waste site)

QUOTE: the plan to revert the arable to downland will be the biggest restoration scheme ever carried out by any voluntary group in the country
Natural World

result [rɪ'zʌlt] **1** *noun* figures at the end of a calculation *or* at the end of a test; ***what was the result of the test?*** **2** *verb* to happen because of something; ***the cancer resulted from exposure to radiation at work; ozone results from the reaction of chemicals in the atmosphere to sunlight***

retina ['retɪnə] *noun* inside layer of the eye which is sensitive to light

retinol ['retɪnɒl] *noun* vitamin A, vitamin which is soluble in fat and can be formed in the body, but which is mainly found in food such as liver, vegetables, eggs and cod liver oil

returnable [rɪ'tɜːnəbl] *adjective* (bottle, etc.) which can be taken back to the place where it was bought and which can then be reused

reverse [rɪ'vɜːs] **1** *adjective* going backwards; **reverse osmosis** = method of water purification, where water is forced through a membrane which removes impurities **2** *verb* to change a decision *or* to do the opposite of what was done before; ***the decision of the planning committee was reversed and the planning application went ahead***

rH = REDOX POTENTIAL

rhizome ['raɪzəʊm] *noun* plant stem which lies under the ground and contains leaf buds

(as opposed to a root which lies under the ground but is not a stem)

rhizosphere ['raɪzəʊsfiə] *noun* soil surrounding the roots of a plant

Rhodophyta [rəʊdə'fi:tə] *noun* red algae, mainly found on the seabed

ria ['rɪə] *noun* valley which has been filled by the sea

ribbon development ['ribən dɪ'veləpmənt] *noun* building of houses in an uninterrupted row along a main road

riboflavine [raɪbəʊ'fleɪvɪn] *noun* vitamin B_2, a vitamin found in eggs, liver, green vegetables, milk and yeast and also used as an additive (E101) in processed food. Lack of riboflavine will affect a child's growth and can cause anaemia and inflammation of the mouth and tongue

ribonuclease [raɪbəʊ'nju:klɪeɪz] *noun* enzyme which breaks down RNA

ribonucleic acid (RNA) ['raɪbəʊnju:kli:ɪk 'æsɪd] *noun* one of the nucleic acids in the nucleus of all living cells, which takes coded information from DNA and translates it into specific enzymes and proteins; *see also* DNA

ribosomal [raɪbə'səʊməl] *adjective* referring to ribosomes

◇ **ribosome** ['raɪbəsəʊm] *noun* tiny particle in a cell, containing RNA and protein, where protein is synthesized

rice [raɪs] *noun* cereal grass *(Oryza sativa)* the most important cereal crop, and the staple food of half the population of the world; **wild rice** = species of grass which is found naturally in North America and which is similar to rice

rich [rɪtʃ] *adjective* **(a)** (soil) which has many useful nutrients and in which plants grow well; **rich in** = having a lot of something; ***green vegetables are rich in minerals; the doctor has prescribed a diet which is rich in protein; the forests are rich in mosses and other forms of moisture-loving plants*** **(b)** (food) which has high calorific value **(c)** (mixture of fuel and air) which has a relatively high ratio of fuel to air

◇ **-rich** [rɪtʃ] *suffix* meaning having a lot of; ***a nutrient-rich detergent; a protein-rich diet; many wild plants have oil-rich seeds which can help in the manufacture of detergents***

◇ **richness** ['rɪtʃnəs] *noun* number of species found in a certain area

> QUOTE: the seas in the region are rich in krill, plankton and fish
> **New Scientist**

Richter scale ['rɪktə 'skeɪl] *noun* scale of measurement of the force of an earthquake; ***there were no reports of injuries after the quake which hit 5.2 on the Richter scale***

> COMMENT: the scale, devised by Charles Richter, has values from zero to ten, with the strongest earthquake ever recorded being 8.9. Earthquakes of 5 or more on the Richter scale cause damage. The Richter scale measures the force of an earthquake; the damage caused is measured on the Modified Mercalli scale

> QUOTE: on 9 February 1971 there was an earthquake registering 6.4 points on the Richter scale. Its epicentre was situated 14 kms from the dam
> **Ecologist**

ridge [rɪdʒ] *noun* **(a)** long raised section of ground, occurring as part of a mountain range and on the ocean floor **(b)** long narrow band of high pressure leading away from the centre of an anticyclone; ***a ridge of high pressure is lying across the country*** (NOTE: the opposite is **trough**)

rift valley ['rɪft 'væli] *noun* long valley with steep walls, formed when land between two fault lines sinks or possibly when a fault widens as plates forming the earth's crust move apart

rig [rɪg] *noun* **drilling rig** = large metal construction containing the drilling and pumping equipment for extracting oil *or* gas; **oil rig** = large metal construction containing the drilling and pumping equipment for an oil well

rill [rɪl] *noun* little stream

rime [raɪm] *noun* feathery ice formed when freezing fog settles on surfaces

ring [rɪŋ] **1** *noun* circle which goes round something; **annual ring** *or* **tree ring** = ring of new wood formed each year in the trunk of a tree and which can easily be seen when the tree is cut down; **ring road** = road which goes right round a town, usually quite close to built-up areas, allowing traffic to bypass the town and not go through the centre **2** *verb* to attach a numbered ring to the leg of a bird so that its movements can be recorded

> COMMENT: ringing is a very common methods of tracing bird movements and providing information about bird's ages. It can also cause stress to the birds

Rio Declaration ['ri:əʊ deklə'reɪʃn] *noun* 27-page document, approved at the 1992 Earth Summit, laying down broad principles of environmentally-sound development

riparian [rɪ'peəriən] *noun* referring to the bank of a river; **riparian fauna** = animals which live on the banks of rivers

ripple ['rɪpl] *noun* little wave on the surface of water; ***ripple marks can be seen in some sedimentary rocks where the sand was marked by water*** *see also* NERITIC FACIES

riptide *or* **rip current** [rɪp 'taɪd *or* rɪp 'kʌrənt] *noun* (i) area of rough water in the sea where currents meet; (ii) current which flows against the flow of the incoming waves

river ['rɪvə] *noun* large flow of water, running from mountains *or* hills down to the sea; **river authority** = official body which manages the rivers in an area; **river basin** = large, low-lying area of land, drained by a river; **river capture** = incorporation of smaller streams into a large river by a process of erosion; **river profile** = slope along the bed of a river, expressed as a graph of distance-from-source against height; **river system** = series of small streams and rivers which connect with each other; **river terrace** = flat plain left when a river cuts more deeply into the bottom of a valley

◇ **riverine** ['rɪvəraɪn] *adjective* referring to a river; ***the dam has destroyed the riverine fauna and flora for hundreds of kilometres***

Rn *chemical symbol for* radon

RNA [ɑːen'eɪ] = RIBONUCLEIC ACID; **messenger RNA** = type of RNA which transmits information from DNA to form enzymes and proteins

rock [rɒk] *noun* solid mineral substance which forms the outside crust of the earth; **basaltic rock** = rock which contains basalt; **igneous rock** = rock, such as basalt *or* granite, formed from molten lava; **metamorphic rock** = rock, such as marble, which has changed because of external influences (such as pressure from other rocks, temperature changes, etc.); **mother rock** *or* **parent rock** = main layer of rock; **sedimentary rock** = rock which has been formed from silt, broken down from older rocks, deposited as sediment at the bottom of lakes or the sea, and then subjected to pressure; **volcanic rock** = rock formed from lava; **rock crystal** = pure quartz

rod [rɒd] *noun* one of the two types of light-sensitive cell in the retina of the eye (the rods are sensitive to poor light); *see also* CONE

rodent ['rəʊdənt] *noun* order of mammals including rats and mice, which have sharp teeth for gnawing

roentgen *or* **röntgen** ['rɒntgən] *noun* unit of measurement of the amount of exposure to X-rays *or* gamma rays; **roentgen equivalent man (rem)** = unit of measurement of ionizing radiation equivalent to the effect of absorbing one roentgen (now replaced by the sievert); **roentgen rays** = X-rays *or* gamma rays which can pass through tissue and leave an image on a photographic film

◇ **roentgenology** *or* **röntgenology** [rɒntgə'nɒlədʒi] *noun* = RADIOLOGY

role [rəʊl] *noun* all the characters (chemical, physical, biological) that determine the position of an organism *or* species in an ecosystem (such as an aquatic predator, a terrestrial herbivore)

root [ruːt] **1** *noun* part of a plant which is normally under the ground and absorbs water and nutrients from the surrounding soil; **root cutting** = piece of root cut from a living plant and put in soil where it will sprout; **root crops** = plants which store edible material in a root, corm or tuber; root crops used as food vegetables or fodder include carrots, parsnips, swedes and turnips; starchy root crops include potatoes, cassavas and yams; **root system** = all the roots of a plant **2** *verb* to produce roots; ***the cuttings root easily in moist sand;*** **rooting compound** = powder containing plant hormones (auxins) into which cuttings can be dipped to encourage the formation of roots

◇ **rootstock** ['ruːtstɒk] *noun* **(a)** rhizome *or* plant stem which lies under the ground and contains leaf buds, as opposed to a simple root which is not a stem **(b)** plant with roots onto which a piece (the scion) of another plant is grafted; **dwarfing rootstock** = plant which is normally low-growing so making the grafted plant grow smaller than it would otherwise

Rossby wave ['rɒsbɪ 'weɪv] *noun* huge, side-to-side swing in air and ocean currents caused by the Coriolis force

rot [rɒt] *verb (of organic tissue)* to decay *or* to become putrefied (because of bacterial *or* fungal action)

rotate [rəʊ'teɪt] *verb* to move in a circle

◇ **rotation** [rəʊ'teɪʃn] *noun* moving in a circle; **rotation of crops** *or* **crop rotation** = system of cultivation where three different crops needing different nutrients are planted in three consecutive growing seasons to prevent the nutrients in the soil being totally used up. The land is then allowed to lie fallow for the fourth year

> COMMENT: the advantages of rotating crops are, firstly, that pests particular to one crop are discouraged from spreading and, secondly, that some crops actually benefit the soil. Legumes (peas and beans) increase the nitrogen content of the soil if their roots are left in the soil after harvesting

rotenone ['rəʊtənəʊn] *noun* insecticide derived from derris

rotor ['rəʊtə] *noun* **(a)** central shaft of a generator, which turns inside the stator **(b)** rapidly turning mass of air, surrounded by clouds

rough [rʌf] *adjective* not smooth; **rough fish** = fish which are not used for sport *or* food

◇ **roughage** ['rʌfɪdʒ] *noun* dietary fibre, fibrous matter in food, which cannot be digested

COMMENT: roughage is found in cereals, nuts, fruit and some green vegetables. It is believed to be necessary to help digestion and avoid developing constipation, obesity and appendicitis

roundworm ['raʊndwɜːm] *noun* type of worm with a round body, some of which are parasites of animals, others of roots of plants

Royal Society for Nature Conservation (RSNC) British society formed in 1981 from earlier societies, whose aim is the conservation of fauna and flora and which is the umbrella organization for a network of regional trusts for nature conservation

Royal Society for the Protection of Birds (RSPB) British society whose aim is the conservation of birds and their habitats

RQ = RESPIRATORY QUOTIENT

RSNC ROYAL SOCIETY FOR NATURE CONSERVATION

RSPB = ROYAL SOCIETY FOR THE PROTECTION OF BIRDS

rubber ['rʌbə] *noun* material which can be stretched and compressed, made from a thick white fluid (latex) from a tropical tree

ruderal ['ruːdərəl] *adjective* (plant) which grows in rubbish *or* on wasteland

ruminant ['ruːmɪnənt] *noun* animal (such as a cow *or* sheep) which chews cud

◇ **rumination** [ruːmɪ'neɪʃn] *noun* process by which food taken to the stomach of a ruminant is returned to the mouth, chewed again and then swallowed

COMMENT: ruminants have stomachs with four sections. They take in foodstuffs into the upper chamber where it is acted upon by bacteria. The food is then regurgitated into their mouths where they chew it again before passing it to the last two sections where normal digestion takes place

runoff ['rʌnɒf] *noun* **(a)** removing of water (as from a river in flood) by opening sluices **(b)** flow of rainwater *or* melted snow from the surface of land into streams and rivers; **stream runoff** = rainwater *or* melted snow which flows into streams; **runoff water** = rainwater *or* melted snow which flows into rivers and streams **(c)** flow of excess fertilizer *or* pesticide from farmland into rivers; ***nitrate runoff causes pollution of lakes and rivers; fish are extremely susceptible to runoff of organophosphates; fertilizer runoff, such as runoff of phosphates, can cause problems of eutrophication;*** **runoff rate** = amount of excess fertilizer *or* pesticide from farmland which flows into rivers

rural ['ruːrəl] *adjective* referring to the country, as opposed to the town; ***many rural areas have been cut off by floods; the government is planning an electrification project for rural areas;*** **rural migration** = movement of people away from the country and into the towns in order to find work

rust [rʌst] **1** *noun* **(a)** reddish powder (iron oxide) which forms on the surface of iron and iron compounds on contact with damp air **(b)** type of fungus disease which gives plants a powdery surface **2** *verb* to become covered with reddish powder through contact with damp air

rye [raɪ] *noun* hardy cereal crop grown in temperate areas

Ss

S *chemical symbol for* sulphur

saccharin ($C_7H_5NO_3S$) ['sækərɪn] *noun* substance used as a substitute for sugar

safe [seɪf] *adjective* not likely to hurt *or* cause damage; ***technicians are trying to make the damaged reactor safe; it is not safe to drink the water here;*** **safe dose** = amount of radiation which can be absorbed without causing harm to someone

◇ **safely** ['seɪfli] *adverb* without danger *or* without being hurt *or* without causing damage; ***low-level waste can be safely disposed of by burying***

◇ **safety** ['seɪfti] *noun* being safe *or* without danger; **safety belt** = belt worn in a car *or* aircraft to help to prevent a passenger being hurt if there is an accident; **safety cab** = protective cab fitted to a tractor to prevent injury to the driver if the tractor turns over; **to take safety precautions** = to do certain things which make your actions *or* condition safe; **safety rod** = tube inserted into a nuclear reactor in order to alter the speed of the reaction; **safety zone** = area in which people are not at risk from radiation

Sahara [sə'hɑ:rə] *noun* large desert region running across North Africa

◇ **Saharan** [sə'hɑ:rən] *adjective* referring to the Sahara; **sub-saharan** = referring to the area south of the Sahara; ***rural woodfuel supplies are falling in many countries of sub-saharan Africa***

Sahel [sə'hel] *noun* semi-desert region south of the Sahara in the process of being desertified

◇ **Sahelian** [sə'hi:liən] *adjective* referring to the Sahel; ***a drought in the Sahelian regions***

saline ['seɪlaɪn] *adjective* referring to *or* containing salt; ***salt marshes have saline soil;*** **saline lake** = low-lying saltwater lake which is not connected with the sea but where the water contains a lot of salt because of evaporation and lack of incoming fresh water; **saline solution** = salt solution made of distilled water and sodium chloride (such as that used in medicine which is introduced into the body intravenously through a drip)

◇ **salinity** [sə'lɪnɪti] *noun* proportion of salt (sodium chloride) present in a given amount of water *or* soil; ***rising salinity has now become a severe health hazard***

◇ **salination** *or* **salinization** [sælɪ'neɪʃn *or* sælɪnaɪ'zeɪʃn] *noun* process by which soil *or* water becomes more salty, found especially in hot countries where irrigation is practised

◇ **salinized** ['sælɪnaɪzd] *adjective* (land) where evaporation leaves salts as a crust on the dry surface

◇ **salinometer** [sælɪ'nɒmɪtə] *noun* instrument for measuring the amount of salt in a saline solution *or* in sea water

salmon ['sæmən] *noun* large fish that spawns in fresh water and swims down to the ocean to develop into an adult

◇ **salmonid** ['sælmənɪd] *noun* family of fish, including trout, which are very susceptible to pollution in water and whose presence in water indicates that it is pure

Salmonella [sælmə'nelə] *noun* genus of bacteria in the intestines, which are acquired by eating contaminated food and which cause food poisoning and typhoid fever; it is found in eggs from infected hens, but can be destroyed by heating; **salmonella poisoning** = illness caused by eating food which is contaminated with Salmonella bacteria which develop in the intestines

salt [sɔ:lt] **1** *noun* **(a) common salt** = sodium chloride (NaCl), white crystals used to make food, especially meat, fish and vegetables, taste better; **salt dome** = area where salt from beneath the surface of the soil has risen to form a small hill; **salt lick** = naturally occurring deposit of salt which animals lick *or* block of salt given to animals to lick **(b)** chemical compound formed from an acid and a metal **2** *adjective* tasting of salt; ***the soup was very salt*** **3** *verb* to preserve food by keeping it in salt or in salt water; cabbage, gherkins, ham and many types of fish are salted for preservation

◇ **salt lake** ['sɔ:lt 'leɪk] *noun* low-lying saltwater lake which is not connected with the sea but where the water contains a lot of salt because of evaporation and lack of incoming fresh water

◇ **salt marsh** ['sɔ:lt mɑ:ʃ] *noun* marsh near the sea, formed with sea water; ***salt marshes are covered by salt water at high tide***

◇ **saltpan** ['sɔ:ltpæn] *noun* area where salt from beneath the soil surface rises to form crystals on the surface

◇ **salt water** ['sɔ:lt 'wɔ:tə] *noun* water which contains salt, such as sea water (as opposed to fresh water in rivers and lakes)

◇ **saltwater** ['sɔ:ltwɔ:tə] *adjective* (lake) containing salt water; (animal) living in salt water

◇ **salty** ['sɔ:lti] *adjective* full of salt; tasting of salt; ***excess minerals in fertilizers combined with naturally saline ground to make the land so salty that it can no longer produce crops***

> COMMENT: salt forms a necessary part of diet as it replaces salt lost in sweating and helps to control the water balance in the body. It also improves the working of the muscles and nerves. Salt is used on roads to clear ice and snow and has a corrosive effect on metal, ruining car bodies

sample ['sæmpl] **1** *noun* small quantity of something used for testing; ***soil samples were taken to test for alkalinity*** **2** *verb* to take a small quantity of something to test; ***ecologists used sampling to determine the age of the vegetation in the ecosystem***

sanctuary ['sæŋktjuəri] *noun* special area where the wildlife is protected; ***a bird sanctuary has been created on the island***

sand [sænd] *noun* fine grains of weathered rock, usually round grains of quartz, found especially on beaches and in the desert; **oil sand** *or* **tar sand** *or* **bituminous sand** = geological formation of sand *or* sandstone containing bitumen, which can be extracted and processed to give oil

◇ **sand bar** ['sænd 'bɑ:] *noun* long bank of sand in shallow water either in a river *or* the sea

◇ **sand dunes** ['sænd 'dju:nz] *plural noun* areas of sand blown by the wind into small hills *or* ridges, often crescent-shaped in the desert *or* sometimes covered with sparse grass when near the sea

◇ **sandflea** ['sændfli:] *noun* jigger, a tropical insect, whose larvae enter the skin between the toes and travel under the skin, causing intense irritation

◇ **sandfly** ['sændflaɪ] *noun* insect of the genus *Phlebotomus* which transmits sandfly fever, a virus infection like influenza, common in the Middle East

◇ **sandpit** ['sændpɪt] *noun* place where sand is extracted from the ground

◇ **sandstone** ['sændstəʊn] *noun* type of sedimentary rock, formed of round particles of quartz

◇ **sandstorm** ['sændstɔ:m] *noun* high wind in the desert, which carries large amounts of sand with it

◇ **sandy soils** ['sændi 'sɔɪlz] *noun* soil containing a high proportion of sand particles

sanitary ['sænɪtəri] *adjective* (i) clean; (ii) referring to hygiene *or* to health; ***they did not have the sanitary conditions necessary to perform the operation;*** **sanitary landfill** *or* **sanitary landfilling** = disposal of waste in specially dug holes in the ground, as opposed to throwing it away anywhere

◇ **sanitation** [sænɪ'teɪʃn] *noun* being hygienic, especially referring to public hygiene; ***poor sanitation in crowded conditions can result in the spread of disease;*** **sanitation control** = measures taken to protect public hygiene, such as spraying oil onto the surface of water to prevent insects such as mosquitoes from breeding

sap [sæp] *noun* liquid carrying nutrients which flows inside a plant

◇ **sapling** ['sæplɪŋ] *noun* young tree

◇ **sappy** ['sæpi] *adjective* (wood) which is full of sap

◇ **sapwood** ['sæpwʊd] *noun* outer layer of wood on the trunk of a tree, which is younger than the heartwood inside and carries the sap; *see also* HEARTWOOD

sapele [sə'pi:li] *noun* fine African hardwood, formerly widely exploited but now becoming rarer

sapro- ['sæprəʊ] *prefix* meaning decay *or* rotting

◇ **saprobe** ['sæprəʊb] *noun* bacterium that lives in rotting matter

◇ **saprobic** [sə'prəʊbɪk] *adjective* referring to the classification of organisms according to the way in which they tolerate pollution; *see also* MESOSAPROBIC, OLIGOSAPROBIC, POLYSAPROBIC

◇ **saprogenic** *or* **saprogenous** [sæprəʊ'dʒenɪk *or* sə'prɒdʒənəs] *adjective* (organism) which grows on decaying organic matter

◇ **saprophagous** [sə'prɒfəgəs] *adjective* (organism) which feeds on decaying organic matter

◇ **saprophyte** ['sæprəfaɪt] *noun* organism (such as a fungus) which lives and feeds on dead *or* decaying organic matter

◇ **saprophytic** [sæprə'fɪtɪk] *adjective* (organism) which lives and feeds on dead *or* decaying organic matter

◇ **saproplankton** [sæprəʊ'plæŋktən] *noun* plankton which live and feed on dead *or* decaying organic matter

sarcoma [sɑ:'kəʊmə] *noun* cancer of connective tissue such as bone, muscle *or* cartilage

Sargasso Sea [sɑ:'gæsəʊ 'si:] *noun* area of still water in the north Atlantic Ocean, which is surrounded by currents and contains drifting waste matter

satellite ['sætəlaɪt] *noun* **(a)** body which orbits round a larger body in space; ***the Moon is the Earth's only satellite*** **(b)** man-made device that orbits the earth, receiving, processing and transmitting signals and generating images such as weather pictures

saturate ['sætʃəreɪt] *verb* to fill with liquid *or* with the maximum amount of a substance which can be absorbed; ***nitrates leached from forest soils, showing that the soils are saturated with nitrogen;*** **saturated fat** = fat which has the largest amount of hydrogen possible

COMMENT: animal fats such as butter and fat meat are saturated fatty acids and contain large amounts of hydrogen. It is known that increasing the amount of unsaturated and polyunsaturated fats (mainly vegetable fats and oils and fish oil) and reducing saturated fats in the food intake help reduce the level of cholesterol in the blood

saturation [sætʃə'reɪʃn] *noun* being filled with liquid *or* with the maximum amount of a substance which can be absorbed; **degree of saturation** = amount of a substance which can be absorbed; **saturation point** = level at which no more of a substance can be absorbed; **the moisture in the air reached saturation point and fell as rain** = the air could absorb no more water

COMMENT: air has different saturation levels at different temperatures. The hotter the air temperature, the more moisture the air can absorb

saturnism ['sætənɪzm] *noun* lead poisoning

savanna *or* **savannah** [sə'vænə] *noun* dry grass-covered plain with few trees (usually referring to the grasslands of South America *or* Africa); growth is abundant during the rainy season but vegetation dies back during the dry season

scale [skeɪl] *noun* **(a)** flake of tissue as on the skins of reptiles and fish; **scale insect** = flat parasitic insect that secretes a protective scale around itself and which lives on plants **(b)** calcium deposit which forms in pipes and kettles, caused by the hardness of the water **(c)** series of degrees of measurement; ***the Celsius temperature scale runs from 0° to 100°;*** **Modified Mercalli scale** = scale for measuring the damage caused by an earthquake; **Richter scale** = scale of measurement of the force of an earthquake, measured from zero to ten, with the strongest earthquake ever recorded being 8.9

◊ **scaly** ['skeɪli] *adjective* covered in flakes of tissue, like a reptile *or* fish

scar [skɑː] **1** *noun* mark left on the surface after a wound has healed; **scar tissue** = fibrous tissue which forms a scar **2** *verb* to leave a mark on the surface of something; ***the landscape was scarred by open-cast mines***

scarce [skeəs] *adjective* uncommon *or* rare *or* in short supply

◊ **scarcity** ['skeəsɪti] *noun* rareness *or* shortage *or* lack

QUOTE: a senior UN official predicted that water scarcity will reach crisis proportions
Green Magazine

scavenge ['skævɪndʒ] *verb* **(a)** to feed on dead and decaying matter **(b)** to remove (impurities *or* pollutants) from a gas

◊ **scavenger** ['skævɪndʒə] *noun* (i) detritovore *or* organism such as a bacterium, fungus *or* worm which feeds on dead matter like leaf litter *or* refuse; (ii) mammal *or* bird which feeds on animals which have been killed by lions *or* other predators; (iii) generally, any organism which feeds on dead animals, dead plants *or* refuse left unconsumed by other organisms

◊ **scavenging** ['skævɪndʒɪŋ] *noun* **(a)** eating organic matter *or* dead animals; ***vultures and hyenas sometimes feed by scavenging*** **(b)** removing impurities *or* pollutants from a gas; **precipitation scavenging** = removing particles of polluting substances from the air in the form of acid rain

QUOTE: in flue gas desulphurization wet limestone is sprayed into the plant's hot exhaust gases where it scavenges as much as 90% of the sulphur dioxide
Scientific American

QUOTE: a few species in the community are carnivores and scavengers, and eat the animals that live off the bacteria
New Scientist

scenery ['siːnəri] *noun* appearance of the landscape

scent [sent] *noun* (i) pleasant smell; (ii) cosmetic substance which has a pleasant smell; (iii) smell given off by a substance which stimulates the sense of smell; ***the scent of flowers makes some people sneeze***

schist [ʃɪst] *noun* type of metamorphic rock with flakes which make it split

Schistosoma [ʃɪstə'səʊmə] *noun* Bilharzia, a fluke which enters the patient's bloodstream and causes schistosomiasis

◊ **schistosomiasis** [ʃɪstəsəʊ'maɪəsɪs] *noun* bilharziasis, a tropical disease caused by flukes in the intestine *or* bladder

COMMENT: the larvae of the fluke enter the skin through the feet and lodge in the walls of the intestine or bladder. They are passed out of the body in stools or urine and return to water, where they lodge and develop in the water snail, the secondary host, before going back into humans. Patients suffer from fever and anaemia

school [skuːl] *noun* group of aquatic animals, such as fish *or* porpoises, which all move together and keep an equal distance apart

scion ['saɪən] *noun* piece of a plant which is grafted onto a stock

sclerophyll ['sklerəfɪl] *noun* woody plant which grows in hot dry regions, with thick leathery evergreen leaves which lose very little moisture

-scope [skəʊp] *suffix* referring to an instrument for examining by sight; **microscope** = scientific instrument with lenses, which makes very small objects appear larger

scorbutus [skɔː'bjuːtəs] *noun* scurvy, a disease caused by lack of vitamin C *or* ascorbic acid which is found in fruit and vegetables

◊ **scorbutic** [skɔː'bjuːtɪk] *adjective* referring to scurvy; *see note at* SCURVY

Scots pine ['skɒts 'paɪn] *noun* *Pinus sylvestris,* common commercially grown European conifer

Scottish Natural Heritage branch of the Nature Conservancy Council responsible for the conservation of fauna and flora in Scotland

scrap [skræp] **1** *noun* waste material, especially metal; ***50% of steel is made from recycled scrap; the old computers have been sold for scrap; the value of scrap metal is higher than you might think; municipal dumps have special containers for scrap paper*** (NOTE: no plural) **2** *verb* to demolish *or* to destroy something because it is no longer useful

scrapie ['skreɪpi] *noun* brain disease of sheep and goats; *see note at* BOVINE SPONGIFORM ENCEPHALOPATHY

scree [skri:] *noun* loose rocks and stones covering the side of a mountain

screen [skri:n] **1** *noun* hedge *or* row of trees grown to shelter other plants *or* to protect something from the wind *or* to prevent something from being seen **2** *verb* to examine someone to test for the presence of a disease; ***the population of the village was screened for meningitis***

◇ **screening** ['skri:nɪŋ] *noun* examining someone to test for the presence of a disease

screwworm ['skru:wɜ:m] *noun* *Cochliomya hominivorax,* a fly similar to the bluebottle, but dark green in colour. It is common in Central and South America; it devastated cattle in the USA in the 1950s, but has now been eradicated there

scrub [skrʌb] **1** *noun* (i) small trees and bushes; (ii) area of land covered with small trees and bushes **2** *verb* to remove sulphur and other pollutants from waste gases produced by power stations

◇ **scrubber** ['skrʌbə] *noun* device for removing sulphur and other pollutants from waste gases

◇ **scrubland** ['skrʌblænd] *noun* land covered with small trees and bushes

◇ **scrub typhus** ['skrʌb 'taɪfəs] *noun* tsutsugamushi disease, a severe form of typhus found in Southeast Asia, caused by the Rickettsia bacterium and transmitted to humans by mites

> QUOTE: the vegetation is dominated by stands of pine with a thick understorey of evergreen scrub oaks and densely branched shrubs. This is the true scrub
>
> **Natural History**

scurvy ['skɜ:vi] *noun* scorbutus, a disease caused by lack of vitamin C *or* ascorbic acid which is found in fruit and vegetables

> COMMENT: scurvy causes general weakness and anaemia, with bleeding from the gums, joints and under the skin. In severe cases the teeth drop out. Treatment consists of vitamin C tablets and a change of diet to include more fruit and vegetables

Se *chemical symbol for* selenium

sea [si:] *noun* body of salt water, smaller than an ocean, which covers a large part of the earth; **sea breeze** = light wind which blows from the sea towards the land, for example, in the evening when the land cools; **sea current** = flow of water in the sea; **sea farming** = systematic production of food from the sea; **sea fog** = type of advection fog which forms along the coast *or* over sea; **sea salt** = (i) crystals of sodium chloride, extracted from sea water; (ii) salt blown into the air from the sea which has an effect on plants and buildings near the sea

◇ **seabed** ['si:bed] *noun* bottom of the sea; ***fish which feeds on minute debris on the seabed***

◇ **seabird** ['si:bɜ:d] *noun* bird (such as a gull) which lives near the sea and lives on fish

◇ **seaboard** ['si:bɔ:d] *noun* area of land along a stretch of sea; ***the eastern seaboard of the USA***

◇ **seagull** ['si:gʌl] *noun* general name for a number of species of birds which live near the sea and have a stout build, rather hooked bills and webbed feet

sea level ['si: levl] *noun* altitude of the land at the edge of sea

> COMMENT: taken as the base for references to altitude; a mountain 300m high is three hundred metres above sea level; the Dead Sea is 395m below sea level. Sea levels in general have risen over the past hundred years and much more rapid rises are forecast if the greenhouse effect results in the melting of the polar ice caps

sea loch ['si: lɒk] *noun* fjord *or* long inlet of the sea in Scotland

◇ **sea water** ['si: 'wɔ:tə] *noun* water in the sea

◇ **seaweed** ['si:wi:d] *noun* general name for several species of large algae, growing in the sea and usually rooted to a surface

seam [si:m] *noun* layer of mineral in rock beneath the earth's surface; ***the coal seams are two metres thick; the gold seam was worked out some years ago***

season ['si:zn] **1** *noun* **(a)** one of the four parts into which a year is divided (spring, summer, autumn and winter) **(b)** time of year when something happens; **breeding season** = time of year when organisms produce offspring; **dry season** = time of year in some countries when very little rain falls (as opposed to the rainy season); **grazing season** = time of year when animals can feed outside on grass; **growing season** = time of year when a plant grows; **hunting season** = time of year when people are allowed to hunt; **rainy season** *or* **wet season** = time of year in some countries when a lot of rain falls (as opposed to the dry season) **2** *verb* to allow the sap in timber to dry so that the wood can be used; ***the construction industry needs large quantities of properly seasoned wood***

◇ **seasonal** ['si:zənəl] *adjective* referring to a season; occurring at a season; ***seasonal changes in temperature; plants grow according to a seasonal pattern;*** **tropical seasonal forest** = biome where plants grow at a certain time during the year

> QUOTE: during the breeding season, if weather conditions are favourable, the average territory provides enough food to sustain two adults and a brood of six young
> **Natural History**

second ['sekənd] **1** *noun* base SI unit of measurement of time equal to one sixtieth of a minute; unit of measurement of the circumference of a circle equal to one sixtieth of a degree **2** *adjective* coming after the first

◇ **secondary** ['sekəndri] *adjective* (i) which comes after the first; (ii) (condition) which develops from another condition (the primary condition); **secondary consumer** = animal (such as a carnivore) which eats other consumers in the food chain; **secondary industry** = industry which uses basic raw materials to make manufactured goods; **secondary mineral** = mineral formed after chemical reactions *or* weathering have taken place in magma; **secondary particulates** = particles of matter formed in the air by chemical reactions such as smog; **secondary substances** = chemical substances found in plant leaves, believed to be a form of defence against herbivores; **secondary succession** = ecological community which develops in a place where a previous community has been removed (as by fire, flooding *or* cutting down of trees); *compare* PRIMARY

secrete [sɪ'kri:t] *verb (of a gland)* to produce a substance such as a hormone *or* oil *or* enzyme

◇ **secretion** [sɪ'kri:ʃn] *noun* substance produced by a gland

sediment ['sedɪmənt] *noun* solid particles, usually insoluble, which fall to the bottom of a liquid; **marine sediment** = solid particles which fall to the seabed

◇ **sedimentary** [sedɪ'mentəri] *adjective* which falls to the bottom of a liquid; **sedimentary cycle** = process by which sediment falls to the bottom of water, becomes rock, then is weathered to form sediment again; **sedimentary deposit** = solid particles which have fallen to the bottom of a liquid; **sedimentary rock** = rock which has been formed from silt, broken down from older rocks, deposited as sediment at the bottom of lakes or the sea, and then subjected to pressure

◇ **sedimentation** [sedɪmen'teɪʃn] *noun* (i) formation of sedimentary rock; (ii) action of solid particles falling to the bottom of a liquid, as in the treatment of sewage; **sedimentation basin** = area of land where the rocks have been formed from matter carried there by wind and water; **sedimentation tank** = tank in which sewage is allowed to stand so that solid particles can sink to the bottom; **sedimentation rate** = speed with which solid particles fall to the bottom of a liquid

seed [si:d] **1** *noun* fertilized ovule which forms a new plant on germination; **seed bank** = (i) all the seeds existing in the soil; (ii) collection of seeds from plants, kept for research purposes; **seed corn** = cereal grown to give grain which is used as seed; **seed leaf** = cotyledon *or* first leaf of a plant as the seed sprouts **2** *verb* **(a)** to sow seeds in an area; ***the area of woodland was cut and then seeded with pines;*** **to seed itself** = *(of plant)* to propagate itself by dropping seed which germinates and produces plants in following seasons; ***the poppies seeded themselves all over the garden; several self-seeded poppies have come up in the vegetable garden*** **(b)** to encourage rain to fall by flying over a cloud and dropping crystals of salt, carbon dioxide and other substances onto it

◇ **seedcase** ['si:dkeɪs] *noun* hard outside cover which protects the seeds in certain plants

◇ **seedling** ['si:dlɪŋ] *noun* very small plant which has just sprouted from a seed

◇ **seed tree** ['si:d 'tri:] *noun* tree left standing when others are cut down, to allow it to seed the cleared land

seep [si:p] *verb (of a liquid)* to flow slowly through a substance; ***water seeped through the rock; chemicals seeped out of the container***

◇ **seepage** ['si:pɪdʒ] *noun* action of flowing slowly through a substance; **seepage tank** = tank attached to a septic tank, into which the liquids from the septic tank are drained

seiche [seɪʃ] *noun* tide in a lake, caused usually by the wind *or* by movements in water level

seism [saɪzm] *noun* earthquake

◇ **seismic** ['saɪzmɪk] *adjective* referring to earthquakes; **seismic shock** *or* **seismic wave** = shock wave which spreads out from the centre *or* focus of an earthquake

◇ **seismograph** ['saɪzməgrɑ:f] *noun* instrument for measuring earthquakes; *see also* RICHTER

◇ **seismological** [saɪzmə'lɒdʒɪkl] *adjective* referring to the study of earthquakes

◇ **seismologist** [saɪz'mɒlədʒɪst] *noun* scientist who studies earthquakes

◇ **seismology** [saɪz'mɒlədʒi] *noun* scientific study of earthquakes; **reflection seismology** = study of how pressure waves from seismic movements are reflected by rock structures in the earth

◇ **seismonasty** [saɪzmə'nɑːsti] *noun* response of plants to a shock (as when the leaves of *Mimosa pudica* or sensitive plant, fold up when touched)

select [sɪ'lekt] *verb* to make a choice *or* to choose one thing but not others; ***the strongest plants are selected for further breeding***

◇ **selection** [sɪ'lekʃn] *noun* act of choosing one thing but not others; **artificial selection** = selection by people of individual animals *or* plants from which to breed further generations because the animals *or* plants have useful characteristics; **natural selection** = evolution of a species, whereby characteristics which help individual organisms to survive and reproduce are passed on to the offspring and those characteristics which do not help are not passed on

◇ **selective** [sɪ'lektɪv] *adjective* which chooses one thing but not others; **selective herbicide** *or* **selective weedkiller** = weedkiller which is supposed to kill only certain plants and not others

> QUOTE: a species of small mammal has evolved selective resistance to venom
> **BBC Wildlife**

selenium [sə'liːniəm] *noun* non-metallic trace element (NOTE: chemical symbol is **Se**; atomic number is **34**)

self- [self] *prefix* referring to oneself; **self-fertile** = (plant) which fertilizes itself by male gametes from its own flowers; **self-fertility** = ability of a plant to fertilize itself; **self-fertilization** *or* **selfing** = action of fertilizing itself; **self-pollination** = pollination of a plant by pollen from its own flowers; *compare* CROSS-POLLINATION; **self-purification** = ability of water to clean itself of polluting substances

◇ **self-regulating** [self'regjuːleɪtɪŋ] *adjective* (ecosystem) which keeps itself in balance; ***most tropical rainforests are self-regulating environments***

◇ **self-sterile** [self'steraɪl] *adjective* (plant) which cannot fertilize itself from its own flowers

◇ **self-sterility** [selfstə'rɪlɪti] *noun* inability of a plant to fertilize itself

sell-by date ['selbaɪdeɪt] *noun* date on the label of a food product, which is last date on which the product should be sold and be guaranteed as of good quality

selva ['selvə] *noun* tropical rain forest in the Amazon Basin

semen ['siːmən] *noun* thick pale fluid containing spermatozoa, produced by the testes and ejaculated from the penis

semi- ['semi] *prefix* meaning half; **semi-arid** = (region) which has very little rain; **semi-deciduous forest** = forest in which some trees lose their leaves *or* needles at some point during the year; **semi-desert** = area of land which has very little rain; **semi-parasitic** = (plant) which lives as a parasite but also undergoes photosynthesis; (plant *or* animal) which usually lives as a parasite but is also capable of living on dead *or* decaying organic matter; **semi-permeable** = (membrane) which allows a liquid to pass through but not substances dissolved in the liquid

◇ **semiconductor** [semɪkən'dʌktə] *noun* material that has conductive properties between those of a conductor (such as a metal) and an insulator

sensitive ['sensɪtɪv] *adjective* **(a)** which can respond to stimuli; ***the leaves of the plant are sensitive to frost;*** **sensitive plant** = *Mimosa pudica,* the leaves of which fold up when touched **(b)** (apparatus) which can record very small changes; ***the earthquake was a small one and only registered on the most sensitive equipment***

◇ **sensitivity** [sensɪ'tɪvɪti] *noun* being able to respond to stimuli

◇ **sensitize** ['sensɪtaɪz] *verb* to make someone sensitive to a drug *or* allergen

◇ **sensory** ['sensəri] *adjective* referring to the senses; **sensory adaptation** = any alteration in a receptor because of increased, decreased *or* prolonged stimulation; **sensory organ** = part of an organism which receives stimuli, such as a nerve

sepal ['sepəl] *noun* green outer part surrounding a flower

separate ['sepərət] *adjective* (thing) which is apart *or* not linked; **separate system** = drainage system where the sewage and rainwater are not collected together but drain through different channels

◇ **separator** ['sepəreɪtə] *noun* device which separates one substance from another, such as steam from water

septic ['septɪk] *adjective* referring to the process of decomposition of organic matter; **septic sludge** = solid part of sewage undergoing the process of purification by decomposition; **septic tank** = (i) tank for household sewage, constructed in the ground near a house which is not connected to the main drainage system, and in which the waste is purified by the action of anaerobic bacteria; (ii) tank at a sewage treatment works in which sewage is collected to begin its treatment by anaerobic bacteria

COMMENT: the sewage collects in the tank, the solids settle and are decomposed and the liquid, purified by bacterial action, drains off into the soil or into a seepage tank

sere [sɪə] *noun & suffix* series of plant communities which succeed one another; **primary sere** = first plant community which develops in a place where no plants have grown before (as on cooled lava from a volcano); **hydrosere** = series of plant communities growing in water *or* in wet conditions; *see also* CLISERE, LITHOSERE, XEROSERE

sessile ['sesaɪl] *adjective* (i) attached directly (not by a stalk) to a branch *or* stem; (ii) permanently attached to a surface

set aside ['set ə'saɪd] *verb* to use a piece of formerly arable land for something else, such as allowing it to lie fallow *or* using it as woodland *or* for recreation, etc.

◇ **set-aside** ['setəsaɪd] *noun* using a piece of formerly arable land for something else, such as allowing it to lie fallow *or* using it as woodland *or* for recreation, etc. (NOTE: the US equivalent is **the acreage reduction program**)

QUOTE: the scheme will offer subsidies to farmers if they set aside - i.e. leave fallow - 20 per cent of their land for five years
Sunday Times

QUOTE: agriculture is changing, in the form of set-aside policies for the reduction of overproduction and conservation of important habitats and landscape. Set-aside land must either be managed with green cover crops like clover and kept in good agricultural condition, or put to forestry, or used for non-farming purposes
Environment Now

QUOTE: he will be setting aside some whole fields on poorer-yielding chalk soils, but mainly 15m headland strips adjacent to woodland which suffer from shading
Farmers Weekly

QUOTE: British farmers have already entered 650,000 hectares of fields into these set-aside schemes
New Scientist

settle ['setl] *verb* **(a)** to stop moving and stay in one place; ***settled cultivators live in the south of the country; early nomadic herdsmen eventually settled as farmers*** **(b)** *(of sediment)* to fall to the bottom of a liquid

◇ **settlement** ['setlmənt] *noun* **(a)** action of staying in one place **(b)** *(of sediment)* falling to the bottom of a liquid **(c)** small village *or* colony where people have settled

Seveso [sə'veɪzəʊ] *noun* town in Italy, scene of a disaster in 1976 when tetrachlorodibenzoparadioxin gas escaped from a chemical factory during the manufacture of 2,4,5-T

sewer ['suːə] *noun* large pipe *or* tunnel which takes waste water and refuse away from buildings; **main sewer** = large sewer which collects sewage from an area; **storm sewer** = specially large pipe for taking away large amounts of rainwater

◇ **sewage** ['suːɪdʒ] *noun* waste water and other refuse such as faeces, carried away in sewers; **domestic sewage** = sewage from houses; **industrial sewage** = liquid *or* solid waste from industrial processes; **raw sewage** = sewage which has not been treated in a sewage farm; ***the municipality was accused of discharging raw sewage into the lake;*** **sewage disposal** = removing sewage from buildings and passing it into a river *or* lake, etc.; **sewage effluent** = liquid *or* solid waste carried away in sewers; **sewage farm** = place where sewage is treated, especially to make it safe to be used as fertilizer; **sewage farming** = treating sewage, especially to make it safe to be used as a fertilizer; **sewage gas** = gas (methane mixed with carbon dioxide) which is given off by sewage; **sewage treatment plant** *or* **sewage works** = place where sewage is treated to make it safe to be pumped into a river *or* the sea

◇ **sewerage** ['suːərɪdʒ] *noun* system of pipes and treatment plants which collect and dispose of sewage in a town

COMMENT: modern sewage systems treat sewage in special plants. Some systems still remove sewage from houses and discharge it directly into the sea through outfall pipes. Apart from the health hazards to bathers and the possibility of contamination of shellfish, raw sewage helps the eutrophication of the sea, increasing the numbers of plankton and algae. Sewage can be safely treated by dilution in large bodies of water provided there is sufficient water available. Otherwise, a modern sewage treatment plant works by passing the sewage through a series of processes. It is first screened to remove large particles, then passed into sedimentation tanks where part of the solids remaining in the sewage settle. The sewage then continues into an aerator which adds air to activate the bacteria. The sewage then settles in a second sedimentation tank before being discharged into a river or the sea. Sludge which settles at the bottom of the sedimentation tanks is treated in digestion tanks where it is digested anaerobically by bacteria

> QUOTE: Britain's coastal waters are now some of the worst for sewage pollution in the whole of Europe. Industry tips vast quantities of waste fluids into the domestic sewerage system. In some areas the industrial content of sewage is as high as 20%
> **Environment Now**

sex [seks] *noun* one of two groups (male and female) into which animals and plants can be divided; ***the relative numbers of the two sexes in the population are not equal, more males being born than females;*** **sex chromatin** = chromatin which is found only in female cells and which can be used to identify the sex of a baby before birth; **sex chromosome** = chromosome which determines if an organism is male or female; **sex determination** = way in which the sex of an individual organism is fixed by the number of chromosomes which make up its cell structure; **sex organs** = organs which are associated with reproduction and sexual intercourse

> COMMENT: in mammals, individuals have either a pair of identical XX chromosomes (and so are female) or have one X and one Y chromosome and are male. Out of the twenty-three pairs of chromosomes in each human cell, only two are sex chromosomes. Females have a pair of X chromosomes and males have a pair consisting of one X and one Y chromosome. The sex of a baby is determined by the father's sperm. While the mother's ovum only carries X chromosomes, the father's sperm can carry either an X or a Y chromosome. If the ovum is fertilized by a sperm carrying an X chromosome, the embryo will contain the XX pair and so be female

sex-linkage [seks'lɪŋkɪdʒ] *noun* existence of characteristics which are transmitted through the X chromosomes

◇ **sex-linked** [seks'lɪŋkt] *adjective* (i) (gene) which is linked to X chromosomes; (ii) (characteristic, such as colour blindness) which is transmitted through the X chromosomes

◇ **sexology** [sek'sɒlədʒɪ] *noun* study of sex and sexual behaviour

◇ **sexual** ['seksjuəl] *adjective* referring to sex; ***a study of the sexual behaviour of moths;*** **sexual reproduction** = reproduction in which gametes from two individuals fuse together

shade [ʃeɪd] *noun* **(a)** place sheltered from direct sunlight; **shade plant** = plant which prefers to grow in the shade; **shade-tolerant tree** = tree (such as beech) which will grow in the shade of a larger tree; **shade-intolerant tree** = tree (such as Douglas fir) which will not grow in the shade of other trees **(b)** relative darkness of a colour; ***the new leaves are a light shade of green***

◇ **shading** ['ʃeɪdɪŋ] *noun* **(a)** action of cutting off the light of the sun; ***parts of the field near tall trees suffer from shading*** **(b)** changing the strength of colours; **obliterative shading** = grading of the colour of an animal which minimizes relief and gives a flat appearance (as when the back of the animal is dark shading towards a light belly)

shale [ʃeɪl] *noun* sedimentary rock formed from clay, which cracks along horizontal straight lines; **oil shale** = geological formation of sedimentary rocks containing oil, which can be extracted by crushing and heating the rock

shear waves ['ʃɪə 'weɪvz] *noun* type of slow seismic waves which alter direction as they pass through different types of rock (used in reflection seismology)

sheep [ʃi:p] *noun* group of mammals related to goats and cattle, farmed for their wool and meat; **sheep farming** = raising sheep on a farm; **sheep station** = very large sheep farm in Australia

sheet lightning ['ʃi:t 'laɪtnɪŋ] *noun* lightning where the flash cannot be seen, but where the clouds are lit by it

shelf [ʃelf] *noun* layer of rock *or* ice which juts out; **continental shelf** = seabed surrounding a continent and covered with shallow water: usually taken to be the area between the shore and the 183m deep line

◇ **shelf-life** ['ʃelflaɪf] *noun* number of days or weeks which a product can stay on the shelf of a shop and still be good to use

shell [ʃel] *noun* hard outer covering of an animal *or* egg *or* seed

◇ **shellfish** ['ʃelfɪʃ] *noun* sea animals which live in shells

shelter ['ʃeltə] **1** *noun* protection (from wind *or* sun, etc.); **shelter belt** = row of trees planted to protect crops, etc. from wind **2** *verb* to protect something (from wind *or* sun, etc.)

◇ **shelterwood** ['ʃeltəwʊd] *noun* large trees left standing when others are cut, to act as shelter for seedling trees

shield [ʃi:ld] **1** *noun* large area of very old rocks (such as the Canadian Shield) **2** *verb* to protect; ***the plants grow in crevices shielded from the wind; ozone in the stratosphere shields the earth from solar ultraviolet radiation***

shifting cultivation ['ʃɪftɪŋ 'kʌltɪ'veɪʃn] *noun* agricultural practice using the rotation of fields rather than crops, short cropping periods followed by long fallows and the maintenance of fertility by the regeneration of vegetation; *see also* FALLOW

shingle ['ʃɪŋgl] *noun* pebbles found on beaches, between about 1 and 7cm in diameter; **shingle beach** = beach covered with pebbles

shock [ʃɒk] *noun* sudden movement caused by pressure; **shock wave** = wave of high pressure which radiates out from a central source (such as an explosion); *see also* AFTERSHOCK

shore [ʃɔː] *noun* land at the edge of the sea *or* a lake; **shore terrace** = flat strip of land on a sloping shore

◇ **shorebird** ['ʃɔːbɜːd] *noun* bird which lives and nests on the shore

◇ **shoreline** ['ʃɔːlaɪn] *noun* line of land at the edge of the sea *or* a lake

COMMENT: the shore is divided into different zones: the upper shore is the area which is only occasionally covered by sea water at the very highest tides; the middle shore is the main area of shore which is covered and uncovered by the sea at each tide; and the lower shore which is very rarely uncovered and only at the lowest tides

shortage ['ʃɔːtɪdʒ] *noun* lack; ***there is a shortage of qualified geologists for the project;*** **food shortage** = lack of food

short-day plant [ʃɔːt'deɪ 'plɑːnt] *noun* plant (such as the chrysanthemum) which flowers as the days get shorter in the autumn

shower ['ʃaʊə] *noun* brief fall of rain

shrivel ['ʃrɪvəl] *verb* to become dry and wrinkled

shrub [ʃrʌb] *noun* low plant like a short tree

◇ **shrubby** ['ʃrʌbi] *adjective* (plant) that grows like a shrub

shut down ['ʃʌt 'daʊn] *verb* to stop working

◇ **shutdown** ['ʃʌtdaʊn] *noun* action of stopping something working; **emergency shutdown** = stopping a nuclear reactor working when it seems that something dangerous may happen

Si *chemical symbol for* silicon

SI *abbreviation for* Système International, the international system of metric measurements; **SI units** = international system of units for measuring physical properties (such as weight *or* speed *or* light, etc.); ***lumen is the SI unit of light per second***

sib *or* **sibling** ['sɪblɪŋ] *noun* brother *or* sister; one of the offspring of the same parents; ***brothers and sisters are all siblings;*** **sibling species** = species which look alike, but which cannot interbreed

sick [sɪk] *adjective* ill *or* not well; ***half the forests in Germany are sick;*** **sick building syndrome** = condition where many people working in a building feel ill *or* have headaches, caused by blocked air-conditioning ducts in which stale air is recycled round the building, often carrying allergenic substances *or* bacteria

-side [saɪd] *suffix* meaning the side *or* edge of something; ***the eroded hillsides are scarred with gullies; waterside plants thrive in shade***

side effects ['saɪd ɪ'fekts] *plural noun* secondary and undesirable effect of a drug, etc.; ***draining the marsh has had several unexpected side effects***

sievert ['siːvət] *noun* unit of measurement of the absorbed dose of radiation (calculated as the amount of radiation from one milligram of radium at a distance of one centimetre for one hour) (NOTE: usually written **Sv** with figures)

significant [sɪg'nɪfɪkənt] *adjective* important *or* considerable *or* notable; ***there has been no significant reduction in the amount of raw sewage being released into the sea***

◇ **significantly** [sɪg'nɪfɪkəntli] *adverb* to any considerable *or* notable degree; ***the amount of raw sewage released into the sea has not been significantly reduced***

silage ['saɪlɪdʒ] *noun* food for cattle, formed of grass and other green plants, cut and stored in silos; **silage additive** = substance containing bacteria and/or chemicals, used to speed up *or* improve the fermentation process in silage *or* to increase the amount of nutrients in it; **silage effluent** = acidic liquid produced by the silage process which can be a serious pollutant, especially if it drains into a watercourse

COMMENT: silage is made by fermenting a crop with a high moisture content under anaerobic conditions

silencer ['saɪlənsə] *noun* device attached to a car exhaust which reduces the sound emitted (NOTE: US English is **muffler**)

silica ['sɪlɪkə] *noun* silicon dioxide, mineral which forms quartz and sand, and is used to make glass

◇ **silicate** ['sɪlɪkeɪt] *noun* (i) chemical compound of silicon and oxygen, the most widespread form of mineral being found in most rocks and soils; (ii) particles of silica found in clay

◇ **silicon** ['sɪlɪkən] *noun* chemical element, which is found naturally in various compounds such as silica and is manufactured to produce silicon chips, used in the electronics industry; **silicon chip** = electronic component made of silicon; **silicon dioxide** = SILICA (NOTE: chemical symbol is **Si**; atomic number is **14**)

◇ **silicosis** [sɪlɪ'kəʊsɪs] *noun* kind of pneumoconiosis *or* disease of the lungs caused by inhaling silica dust from mining *or* stone-crushing operations

COMMENT: this is a serious disease which makes breathing difficult and can lead to emphysema and bronchitis

silk [sɪlk] *noun* thread produced by the larvae of a moth to make its cocoon and used by man to make a fabric; **silk culture** = rearing the larvae needed to produce silk

◇ **silkworm** ['sɪlkwɜːm] *noun* larva of a moth, which produces silk

silo ['saɪləʊ] *noun* large container for storing grain *or* silage

silt [sɪlt] **1** *noun* **(a)** particles of mineral intermediate between clay and sand (with a diameter of between 0.002 and 0.06mm in the UK), generally formed of fine quartz **(b)** *(in general)* soft mud which settles at the bottom of water **2** *verb* **to silt up** = *(of a harbour or river)* to become full of silt, so that boats can no longer sail

◇ **siltation** *or* **silting** [sɪl'teɪʃn *or* 'sɪltɪŋ] *noun* action of depositing silt at the bottom of water; ***increased sedimentation and siltation in backwaters***

silver ['sɪlvə] *noun* white-coloured metallic element which is not corroded by exposure to air (NOTE: chemical symbol is **Ag;** atomic number is **47)**

silvi- ['sɪlvi] *prefix* referring to trees

◇ **silvicide** ['sɪlvɪsaɪd] *noun* substance which kills trees

◇ **silvicolous** [sɪl'vɪkələs] *adjective* living *or* growing in woodland

◇ **silvicultural** [sɪlvɪ'kʌltʃərəl] *adjective* referring to the cultivation of trees

◇ **silviculture** ['sɪlvɪkʌltʃə] *noun* cultivation of trees

similarity [sɪmɪ'lærəti] *noun* being alike; **similarity coefficient** = degree to which two areas of vegetation are alike

sink [sɪŋk] **1** *verb* to fall to the bottom of water **2** *noun* **(a)** place into which a substance passes to be stored *or* to be absorbed; chemical *or* physical process which removes *or* absorbs a substance; ***the commonest ozone sink is the reaction with nitric oxide to form nitrogen dioxide and oxygen;*** **carbon sink** = part of the biosphere (such as a tropical forest) which absorbs carbon, as opposed to animals which release carbon into the atmosphere in the form of carbon dioxide; **heat sink** = place which can absorb extra heat; ***the oceans may delay global warming by acting as a heat sink;*** **oxygen sink** = part of a plant which stores oxygen **(b) sink** *or* **sinkhole** low-lying piece of land where water collects to form a pond

QUOTE: the main sink for CH_4 (methane) is reaction with atmospheric hydroxyl radicals
Nature

QUOTE: if the oceans act as a huge sink for CO_2 estimates about how much the gas contributes to the greenhouse effect may have to be rethought
Farmers Weekly

Sirenians [saɪ'riːniənz] *noun* large marine animals of the order Sirenia, such as the manatee *or* sea cow, living in warm estuaries

sirocco [si'rɒkəʊ] *noun* dry wind blowing from the desert northwards in North Africa

site [saɪt] **1** *noun* place *or* position of something *or* place where something happened; ***the area around the nuclear test site is closed to the public; hazardous chemicals found on the site include arsenic, lead mercury and cyanide;*** **Site of Special Scientific Interest (SSSI)** = small area of land which has been noted as particularly important by the Nature Conservancy Council, and which is preserved for its fauna, flora *or* geology; **nesting site** = place where a bird may build a nest **2** *verb* to put something *or* to be in a particular place; ***the nesting area is sited on the west side of the cliff***

◇ **siting** ['saɪtɪŋ] *noun* putting something *or* being in a particular place

QUOTE: after farming, recreation is the biggest source of damage to Britain's sites of special scientific interest
New Scientist

Sitka spruce ['sɪtkə 'spruːs] *noun Picea sitchensis,* a fast-growing temperate softwood tree, used for making paper

skeleton ['skelɪtən] *noun* all the bones which make up a body

◇ **skeletal** ['skelɪtəl] *adjective* referring to a skeleton; **skeletal muscle** *or* **voluntary muscle** = muscle which is attached to a bone and which makes a limb move

sky [skaɪ] *noun* space above the earth, which appears blue during the daytime

◇ **skyline** ['skaɪlaɪn] *noun* line of hills *or* buildings seen against the sky; ***the need to protect mountain skylines from planting of conifers***

COMMENT: the sunlight that strikes the atmosphere is scattered by the particles which it hits. The scattering affects short light waves most, hence the short blue light waves colour the sky

slack water ['slæk 'wɔːtə] *noun* part of the tidal cycle occurring between the ebb and flood tides at the point when the flows are reversing direction

slag [slæg] *noun* waste matter which floats on top of the molten metal during smelting, used to lighten heavy soils, such as clay, and also for making cement; **basic slag** = calcium phosphate, waste from blast furnaces, used as a fertilizer because of its phosphate content; **slag heap** = large pile of waste material from an industrial process such as smelting *or* from coal mining

slaked lime ['sleɪkt 'laɪm] *noun* calcium hydroxide, mixture of calcium oxide and water, used to spread on soil to reduce acidity and add calcium

slash [slæʃ] *verb* to make a long cut with a knife; **slash and burn agriculture** = swidden farming *or* form of agriculture where forest is cut down and burnt to create open space for growing crops; the space is abandoned after several crops have been grown and further forest is cut down

slate [sleɪt] *noun* hard metamorphic rock which splits easily along cleavage lines, used especially for making roofs

slaughter ['slɔːtə] **1** *noun* killing of a large number of animals; ***protesters tried to stop the annual slaughter of seals*** **2** *verb* **(a)** to kill animals for food **(b)** to kill large numbers of animals

◇ **slaughterhouse** ['slɔːtəhaʊs] *noun* building where animals are slaughtered and the carcasses prepared for sale for human consumption

sleeping sickness ['sliːpɪŋ sɪknəs] *noun* African disease, spread by the tsetse fly, where trypanosomes infest the blood

sleet [sliːt] **1** *noun* ice and rain mixed together **2** *verb* to fall as ice and rain mixed together

slick [slɪk] *noun* **(oil) slick** = oil which has escaped into water and floats on the surface; ***the slick contaminated over 40 miles of coast***

> QUOTE: as the slick moved down river, hundreds of thousands of residents were without water, schools and businesses closed
> **Environmental Action**

slime [slaɪm] *noun* algal substance, such as green algae, which forms on damp stone surfaces

slip-off slope ['slɪpɒf 'sləʊp] *noun* more gently sloping bank on the inside of a meandering river

slough [slʌf] **1** *noun* dead tissue (especially dead skin) which has separated from healthy tissue **2** *verb (of snake)* to lose dead skin which falls off

sludge [slʌdʒ] *noun* the solid part of sewage; **activated sludge** = solid sewage containing active microorganisms and air, which is used to mix with untreated sewage to speed up the purification process; **raw sludge** = solid sewage before it is treated, which falls to the bottom of a sedimentation tank; **sewage sludge** = the solid part of sewage; ***sewage sludge is dumped directly into the North Sea;*** **sludge digestion** = final treatment of sewage when it is digested anaerobically by bacteria; **sludge gas** = sewage gas (methane mixed with carbon dioxide) which is given off by sewage; **sludge gulper** = large truck which removes sludge from cesspits and septic tanks; **sludge processing** = treating the solid part of sewage so that it may safely be dumped *or* used as a fertilizer

> QUOTE: sewage sludge dumping operations are conducted by the Thames Water Authority in the outer Thames estuary. Four purpose-built vessels dump up to 12,000 wet tonnes of treated sewage sludge daily
> **London Environmental Bulletin**

sluice [sluːs] *noun* channel for water, especially channel for water through a dam

slum [slʌm] *noun* area of a city where the buildings are in bad condition and often where a lot of poor people live very closely together

slurry ['slʌri] *noun* **(a)** liquid waste from animals, stored in tanks and treated to be used as fertilizer; it may also be stored in a lagoon, from which it can either be piped to the fields or transferred to tankers and then distributed; **slurry spreader** = machine which spreads slurry **(b)** **lime slurry** = mixture of lime and water added to hard water to make it softer

> COMMENT: slurry can be spread on the land using pumps, a pipeline system and slurry guns; more often, slurry spreaders are used, which can load, transport and spread the material. New regulations to control the pollution from slurry were introduced in 1989. These require the base and walls of silage clamps to be impervious. An artificial embankment must be constructed around slurry tanks

smallholding ['smɔːlhəʊldɪŋ] *noun* small agricultural unit under 20 hectares in area; often run as a family concern, and with some form of specialization, such as market gardening, raising goats, etc.

◇ **smallholder** ['smɔːlhəʊldə] *noun* person who farms a smallholding

smelt [smelt] *verb* to extract metal from ore by heating it

◇ **smelting** ['smeltɪŋ] *noun* process of extracting metal from ore by heating

◇ **smelter** ['smeltə] *noun* plant where ore is heated and metal extracted from it

smog [smɒg] *noun* pollution of the atmosphere in towns, caused by warm damp air combined with exhaust fumes from cars; **photochemical smog** = smog caused by the action of sunlight on polluting gases

COMMENT: smog originally meant smoke and fog, and the London 'peasoupers' were caused by coal smoke in foggy weather in winter. Today's smog is usually caused by car exhaust fumes and can occur in sunny weather. When the atmosphere near ground level is polluted with nitrogen oxides from burning fossil fuels together with hydrocarbons, ultraviolet light from the sun sets off a series of reactions that result in photochemical smog, containing, among other substances, ozone. Temperature inversion (where the air temperature at ground level is colder than the air above) helps to form smog by making it impossible for the pollutants in the air to rise.

smoke [sməʊk] **1** *noun* white, grey *or* black product formed of small particles, given off by something which is burning; ***the room was full of cigarette smoke; several people died from inhaling toxic smoke;*** **smoke control area** = area of a town where it is not permitted to produce smoke from chimneys *or* fires **2** *verb* **(a)** to emit smoke; ***a smoking volcano*** **(b)** to breathe in smoke from a cigarette *or* cigar *or* pipe; ***she was smoking a cigarette; he smokes a pipe; doctors are trying to persuade people to stop smoking*** **(c)** to preserve food by hanging it in the smoke from a fire (used now mainly for fish, but also for some bacon and cheese)

◇ **smokeless** ['sməʊkləs] *adjective* (place) where there is no smoke *or* where smoke is not allowed; (fuel) which does not produce smoke when it is burned; **smokeless** *or* **smoke-free area** = part of a public place, such as a restaurant *or* aircraft, etc., where smoking is not allowed; **smokeless zone** = area of a town where it is not permitted to produce smoke from chimneys *or* fires

◇ **smoker** ['sməʊkə] *noun* person who smokes tobacco; **smoker's cough** = dry asthmatic cough, often found in people who smoke large numbers of cigarettes

◇ **smokestack** ['sməʊkstæk] *noun* very tall chimney, as in a factory, usually containing several flues

◇ **smoky** ['sməʊki] *adjective* (place) which is full of smoke; (chimney) which makes a lot of smoke

COMMENT: By banning the use of fuels which create smoke, the air above towns has become much cleaner than it was some years ago. This has the disadvantage of allowing the sunlight to penetrate and cause photochemical reactions to take place, increasing the acidity of the atmosphere, with the result that apparently clean air can be as dangerous as smog

smoking ['sməʊkɪŋ] *noun* action of smoking tobacco; ***smoking can damage your health;*** **no smoking area** = part of a public place, such as a restaurant *or* aircraft, etc., where smoking is not allowed; **passive smoking** = inhaling smoke exhaled by other people, even if you are not a smoker yourself; ***dogs with short noses such as bulldogs are at the highest risk from passive smoking***

COMMENT: the connection between smoking tobacco, especially cigarettes, and lung cancer has been proved to the satisfaction of the British government, which prints a health warning on all packets of cigarettes. Smoke from burning tobacco contains nicotine and other substances which stick in the lungs and can, in the long run, cause cancer. Even though a person may never have smoked a cigarette, it is possible for him or her to develop lung cancer through being continuously in the presence of other people who are smoking

QUOTE: from a scientific view there is now a consensus that passive smoking causes some cases of lung cancer

New Scientist

smut [smʌt] *noun* **(a)** black flake of oily carbon emitted from a fire; ***smuts from the oil depot fire covered the town*** **(b)** disease of cereal plants, caused by a fungus, which covers the plant with black spots

Sn *chemical symbol for* tin

snow [snəʊ] **1** *noun* water which falls as white flakes of ice crystals in cold weather; **snow blindness** = temporary painful blindness caused by bright sunlight shining on snow; **snow line** = line on a high mountain above which there is permanent snow **2** *verb* to fall as snow; ***it snowed heavily during the night***

◇ **snowfall** ['snəʊfɔːl] *noun* quantity of snow which comes down at any one time; ***a heavy snowfall blocked the main roads***

◇ **snowflake** ['snəʊfleɪk] *noun* small piece of snow formed from a number of ice crystals

◇ **snow-melt** ['snəʊmelt] *noun* melting of snow in spring, often the cause of floods

◇ **snowstorm** ['snəʊstɔːm] *noun* heavy fall of snow accompanied by wind

◇ **snowy** ['snəʊi] *adjective* (weather) when a lot of snow falls

COMMENT: snow is formed when the atmospheric temperature is cold enough to make the moisture in clouds freeze into ice crystals

soak [səʊk] *verb* to put something in liquid, so that it absorbs some of it; ***the newspaper is soaked in water, then formed into bricks and dried***

◇ **soakaway** ['səʊkəweɪ] *noun* channel in the ground filled with gravel, which takes rainwater from a downpipe *or* liquid sewage from a septic tank and allows it to be absorbed into the surrounding soil

◇ **soak up** ['səʊk 'ʌp] *verb* to absorb a liquid

soar [sɔː] *verb (of bird)* using updraughts of warm air, to fly higher *or* stay airborne without any movement of the wings

social ['səʊʃəl] *adjective* referring to society *or* to a group of animals *or* people; **social animal** = animal which lives in an organized society (such as ants *or* bees); **social group** = several people *or* animals living together in an organized way; group of people with a similar position in society; **social medicine** = medicine as applied to treatment of diseases which occur in certain social groups; **social parasite** = parasite which benefits from the host's normal behaviour (such as the cuckoo, which lays its eggs in the nest of another bird who brings up the cuckoo's young as if it were its own); **social responsibility** = caring about the effects upon society and people's health of environmental pollution and industrial processes

society [sə'saɪəti] *noun* (i) group of animals which live together in an organized way; (ii) group of plants within a larger community; (iii) group of people who live together and have the same laws and customs

sociology [səʊsɪ'ɒlədʒi] *noun* study of human societies; **plant sociology** = study of communities of plants

soda lake ['səʊdə 'leɪk] *noun* salt lake with a high proportion of sodium in the water; *see also* CAUSTIC SODA

sodium ['səʊdiəm] *noun* chemical element which is the basic substance in salt; **sodium balance** = balance maintained in the body between salt lost in sweat and urine, and salt taken in from food; the balance is regulated by aldosterone; **sodium chlorate** = herbicide which is taken up into the plant through the roots; **sodium chloride** = common salt; **sodium hydroxide (NaOH)** = caustic soda *or* compound of sodium and water which is used to make soap and to clear blocked drains (NOTE: chemical symbol is **Na;** atomic number is **11)**

soft [sɒft] *adjective* not hard; **soft water** = water which does not contain calcium and other minerals which are found in hard water and which is easily able to make soap lather

◇ **soften** ['sɒfn] *verb* to make *or* become soft

◇ **softwood** ['sɒftwʊd] *noun* (i) wood from pine trees and other conifers; (ii) any pine tree *or* conifer which produces such wood; **softwoods** = (forests of) pine trees and other conifers (as opposed to hardwoods)

soil [sɔɪl] *noun* earth in which plants grow; **poor soil** = soil with few useful nutrients and so less suitable for plants; **rich soil** = soil with many useful nutrients in which plants grow well; **soil conservation** = using methods such as irrigation, mulching, etc., to prevent soil from being eroded *or* overcultivated; **soil creep** = slow movement of soil downhill; **soil depletion** = reduction of the soil layer by erosion; **soil drainage** = flowing of water from soil (either naturally *or* by putting pipes and drainage channels into the soil); **soil erosion** = removal of soil by the effect of rain, wind *or* sea; **soil fertility** = potential capacity of soil to grow plants; **soil flora** = minute plants (such as fungi and algae) which live in the soil; **soil horizon** = layer of soil which is of a different colour *or* texture from the rest: the topsoil containing humus, the subsoil containing minerals leached from the topsoil, etc.; **soil improvement** = making the soil more fertile; **soil map** = map showing the different types of soil found in an area; **soil profile** = vertical section through the soil showing the different layers; **soil salinity** = measurement of the quantity of mineral salts found in a soil; **soil sample** = small quantity of soil used for testing; **soil science** = scientific study of the soil, its formation, distribution, etc.; **soil structure** = the arrangement of soil particles in groups or individually; **soil texture** = relative proportions of sand, silt and clay particles in soil; *see also* CHERNOZEM, LOESS, PODZOL, SUBSOIL, TOPSOIL

◇ **soilless gardening** ['sɔɪlləs 'gɑːdənɪŋ] *noun* hydroponics *or* science of growing plants without soil but in sand *or* vermiculite *or* other granular material, using a liquid solution of nutrients to feed them

> COMMENT: soil is a mixture of mineral particles, decayed organic matter and water. Topsoil contains chemical substances which are leached through into the subsoil where they are retained. Soils are classified according to the areas of the world in which they are found or according to the types of minerals they contain or according to the stage of development they have reached

solar ['səʊlə] *adjective* **(a)** referring to the sun; **solar eclipse** = situation when the moon passes between the sun and the earth during the daytime and the shadow of the moon falls across the earth, so cutting off the sun's light; **solar flare** = sudden eruption of bright radiation from the sun which affects radio waves and is associated with sunspot activity; **solar radiation** = rays which are emitted by the sun; **solar system** = arrangement of planets which orbit round the sun **(b)** using the energy of the sun *or* driven by power from the sun; **solar cell** = photoelectric device which converts solar energy into electricity; *see also*

PHOTOVOLTAIC CELL; **solar collector** = SOLAR PANEL; **solar cooker** = cooking stove which directs the rays of the sun to the pot in which food is cooked; **solar dryer** = device for drying crops using the heat of the sun; **solar energy** *or* **solar power** = energy *or* power *or* electricity produced from the radiation of the sun; **solar-generated energy** *or* **power** = energy *or* power generated using radiation from the sun; **solar panel** = device with a dark surface which absorbs the sun's radiation and uses it to heat water

◇ **solarization** [səʊlərаɪ'zeɪʃn] *noun* exposing to the rays of the sun, especially killing pests in the soil by covering the soil with plastic sheets and letting it warm up in the sunshine

◇ **solar-powered** [səʊlə'paʊəd] *adjective* driven by power from the sun; ***a solar-powered steam pump***

COMMENT: the sun emits radiation in the form of ultraviolet rays, visible light and infrared heating rays. Solar energy can be collected by various methods, most often by heating water which is then passed into storage tanks. Although solar power is easy to collect, the problem of storing it is important in order to make sure that the power is available for use during the night or when the sun does not shine

QUOTE: solar power sources and heat pumps may supplement the main energy system, but they do not result in sufficient saving in fuel cost to cover the capital expenditure in this country
Environment Now

solid ['sɒlɪd] **1** *adjective* hard *or* not liquid; ***water turns solid when it freezes;*** **solid food** *or* **solids** = food which is chewed and eaten, not drunk; **solid waste** = waste matter which is hard and not liquid **2** *noun* substance which has volume and a shape (as opposed to a liquid *or* gas); ***the first treatment of sewage involves removing the solids***

◇ **solidification** [sɒlɪdɪfɪ'keɪʃn] *noun* becoming solid

◇ **solidify** [sə'lɪdɪfaɪ] *verb* to become solid; ***carbon dioxide solidifies at low temperatures***

◇ **solid-state** [sɒlɪd'steɪt] *adjective* (electronic device) which works by using the effects of electrical *or* magnetic signals in a solid semiconductor material

QUOTE: more solid-waste material is produced in the USA than in any other country; and most of that ends up in holes in the ground: landfill
Appropriate Technology

solifluction [sɒlɪ'flʌkʃən] *noun* gradual downhill movement of wet soil

solstice ['sɒlstɪs] *noun* one of the two times of year when the sun is at its furthest point north *or* south of the equator; **summer solstice** = 21st June, the longest day in the Northern Hemisphere, when the sun is as its furthest point south of the equator; **winter solstice** = 21st December, the shortest day in the Northern Hemisphere, when the sun is at its furthest point north of the equator

soluble ['sɒljuːbl] *adjective & suffix* meaning which can dissolve in a liquid; **fat-soluble** *or* **water-soluble** = which dissolves in fat *or* water

◇ **solute** ['sɒljuːt] *noun* solid substance which is dissolved in a solvent to make a solution

◇ **solution** [sə'luːʃn] *noun* (i) mixture of water and another chemical substance; (ii) mixture of a solid substance dissolved in a liquid

solum ['səʊləm] *noun* technical word for soil, including both topsoil and subsoil

solvent ['sɒlvənt] *noun* liquid in which a solid substance can be dissolved; ***carbonic acid can act as a solvent for limestone***

somatotropin [səʊmətə'trəʊpɪn] *see* BOVINE SOMATOTROPIN

sonar ['səʊnɑː] *noun* method of finding objects under water by sending out sound waves and detecting returning sound waves reflected by the object

sonde [sɒnd] *noun* device attached to a balloon *or* rocket, for measuring and taking samples of the atmosphere

sonic ['sɒnɪk] *adjective* referring to sound waves; **sonic boom** = loud noise made by the shock waves produced by any object (such as an aircraft *or* a bullet) travelling through the air at *or* faster than the speed of sound

COMMENT: the shock waves can cause objects to resonate so violently that they are damaged. Supersonic aircraft generally fly at speeds greater than the speed of sound only over the sea, to avoid noise nuisance and damage to property

soot [sʊt] *noun* black deposit of fine particles of carbon which rise in the smoke produced by the burning of coal, wood, oil, etc. (It collects on surfaces such as the inside of chimneys); ***legislation to control soot emission***

sorghum ['sɔːgəm] *noun* drought-resistant cereal plant *(Sorghum vulgare)* grown in the semi-arid tropical regions, such as Mexico, Nigeria and Sudan

sound [saʊnd] *noun* noise; **sound insulation** = preventing sound escaping *or* entering; **sound wave** = audible change of pressure which moves through the air

source [sɔːs] *noun* **(a)** substance which produces something; place where something comes from; ***hot rocks are a potential energy source; plants tend to turn towards the source of light*** *or* ***towards the light source*** **(b)** place where a river starts to flow; ***the source of the Nile is in the mountains of Ethiopia***

south [saʊθ] *adjective, adverb & noun* one of the directions on the earth's surface, the direction facing towards the sun at midday and towards the South Pole; ***the wind is blowing from the south; the river flows south into the ocean;*** **South Pole** = point which is furthest south on the earth, through which the lines of longitude run

◇ **southerly** ['sʌðəli] **1** *adjective* to *or* from the south; ***the typhoon is moving in a southerly direction at 25 knots*** **2** *noun* wind which blows from the south

◇ **southern** ['sʌðən] *adjective* in the south; towards the south; ***the herds spend winter on the southern plains;*** **Southern Lights** = Aurora Australis, spectacular illumination of the sky in the Southern Hemisphere caused by ionized particles striking the atmosphere; **Southern Oscillation** = regular cycle by which air is exchanged between the Pacific basin and the Indian Ocean, occurring every two to five years and linked to changes in the sea temperature and to the El Niño effect

sow [səʊ] *verb* to put seeds into soil so that they will germinate and become plants

soya ['sɒjə] *noun Glycine max,* plant which produces edible beans which have a high protein and fat content and very little starch

◇ **soybean** *or* **soya bean** ['sɔɪbiːn *or* 'sɔɪə 'biːn] *noun* bean from a soya plant

COMMENT: soybeans are very rich in protein and, apart from direct human consumption, are used as a livestock feedingstuff and for their oil

QUOTE: soybean is the cheapest source of protein and the beans contain about 40% high quality protein and 20% excellent edible oil. Over 300 food preparations can be made from it
Indian Farming

sp. = SPECIES

space [speɪs] *noun* place *or* empty area between things; **green space** = area of land which has not been built on, containing grass, plants and trees; **open space** = area of land which has no buildings *or* trees on it; **outer space** = area outside the earth's atmosphere in which the sun, stars and planets move

sparse [spɑːs] *adjective* having only a small number of items in a given area; ***a few sparse bushes in the desert; a sparse population of herdsmen***

◇ **sparsely** ['spɑːsli] *adverb* not densely; ***a sparsely populated mountain region***

spathe [speɪð] *noun* large bract, often coloured, around a flower of the lily family

spawn [spɔːn] **1** *noun* **(a)** mass of eggs (of a fish *or* reptile) **(b) mushroom spawn** = spores of the edible mushroom which are sold to be used to propagate mushrooms **2** *verb (of a fish or reptile)* to produce a mass of eggs; **spawning ground** = area of water where fish come each year to produce their eggs

special ['speʃl] *adjective* which refers to one particular thing *or* which is not ordinary; ***the doctor has put him on a special diet***

◇ **specialist** ['speʃəlɪst] *noun* **(a)** person who specializes in a certain branch of study; ***they have called in a contamination specialist to study the problem*** **(b)** organism which only lives on one type of food *or* in a very restricted area

◇ **speciality** [speʃɪ'ælɪti] *noun* particular branch of science which a person studies in detail

◇ **specialization** [speʃəlaɪ'zeɪʃn] *noun* (i) act of studying one particular branch of science in detail; (ii) particular branch of science which a person studies in detail

◇ **specialize in** ['speʃəlaɪz 'ɪn] *verb* to study one particular branch of science in detail; ***he specializes in the study of conifers***

◇ **specialty** ['speʃəlti] *noun US* = SPECIALITY

species ['spiːʃiːz] *noun* division of a genus, a group of living things which can interbreed (NOTE: plural is **species.** Note that when referring to a species, it is usual to abbreviate it to **sp.** after the species name)

◇ **speciation** [spiːsi'eɪʃn] *noun* process of forming a new species

QUOTE: there are about 35 species of sea-horses, all in the genus Hippocampus. Together with several hundred species of pipefish they form the family Syngnathidae
BBC Wildlife

specific [spə'sɪfɪk] *adjective* **(a)** characteristic of something; **specific gravity** = measure of the density of a substance; **specific humidity** = ratio between the amount of water vapour in air and the total mass of the mixture of air and water vapour **(b)** referring to a species; **specific name** = name by which a species is differentiated from other members of the genus (it is the second name in the binomial classification system, the first being the generic name which identifies the genus)

◇ **specificity** [spesɪ'fɪsɪti] *noun* being characteristic of something; ***parasites show specificity in that they live on only a certain limited number of hosts***

specimen ['spesɪmən] *noun* (i) one item out of a group; (ii) small quantity of something given for testing; ***fine specimens of salmon can still be found in the mountain rivers; scientists have taken away soil specimens for analysis***

spectrography [spek'trɒgrəfi] *noun* recording of a spectrum on photographic film

◇ **mass spectrometer** ['mæs spek'trɒmɪtə] *noun* device used in chemical analysis, which separates particles according to their masses

◇ **spectroscope** ['spektrəskəʊp] *noun* instrument used to analyse a spectrum

◇ **spectrum** ['spektrəm] *noun* range of colours (from red to violet) into which white light can be split. (Different substances in solution have different spectra) (NOTE: plural is **spectra**)

spell [spel] *noun* short period when the weather does not change; ***there will be some fine spells over the east of the country; the south has experienced the longest spell of rainy weather since records were first taken***

spent fuel ['spent 'fju:əl] *noun* fuel which has been used in a nuclear reactor but which is still fissile and can be reprocessed; ***tonnes of spent nuclear fuel are sent for reprocessing***

sperm [spɜ:m] *noun* spermatozoon *or* mature male sex cell, which is capable of fertilizing an ovum; **sperm bank** = collection of sperm from donors, kept until needed for artificial insemination; **sperm count** = calculation of the number of sperm in a quantity of semen (NOTE: no plural)

◇ **spermat-** *or* **spermato-** ['spɜ:mətəʊ] *prefix* referring to sperm

◇ **spermatic** [spɜ:'mætɪk] *adjective* referring to sperm

◇ **spermatid** ['spɜ:mətɪd] *noun* immature cell, formed from a spermatocyte, which becomes a spermatozoon

◇ **spermatocyte** [spɜ:'mætəsaɪt] *noun* early stage in the development of a spermatozoon

◇ **spermatogenesis** [spɜ:mætə'dʒenəsɪs] *noun* formation and development of spermatozoa

◇ **spermatogonium** [spɜ:mætə'gəʊniəm] *noun* cell which forms a spermatocyte

◇ **spermatozoon** [spɜ:mætə'zəʊɒn] *noun* sperm, mature male sex cell, which is capable of fertilizing an ovum (NOTE: plural is **spermatozoa**)

◇ **spermicide** ['spɜ:mɪsaɪd] *noun* substance which kills sperm

◇ **sperm whale** ['spɜ:m 'weɪl] *noun* large toothed whale, hunted for various substances which can be extracted from it

COMMENT: a spermatozoon is very small and comprises a head, neck and very long tail. It can swim by moving its tail from side to side

sphagnum ['sfægnəm] *noun* type of moss which grows in bogs; **sphagnum peat** = peaty soil made up of dead sphagnum moss

spider ['spaɪdə] *noun* one of a large group of animals, with two parts to their bodies and eight legs

spike [spaɪk] *noun* tall pointed flower

spill [spɪl] **1** *noun* liquid which has escaped from a container; **oil spill** = escape of oil into the environment (as from a tanker which hits rocks *or* from a ruptured pipeline) **2** *verb* to escape from a container; to overflow; ***he spilled the bottle of herbicide on the grass; about 200,000 barrels of oil spilled into the sea***

◇ **spillage** ['spɪlɪdʒ] *noun* (i) act of escaping from a container; overflowing; (ii) amount of liquid which has escaped from a container; ***spillage of the river into the nearby desert areas***

◇ **spillway** ['spɪlweɪ] *noun* channel down which water can overflow from the reservoir behind a dam

spirochaete ['spaɪərəʊki:t] *noun* bacterium with a spiral shape

spit [spɪt] *noun* long, narrow accumulation of sand or gravel that projects from the shore into the sea

spoil [spɔɪl] **1** *noun* rubbish and waste minerals dug out of a mine; **spoil bank** *or* **spoil heap** = heap of waste soil produced in surface mining and deposited at the side of the worked coal seam **2** *verb* to go bad

◇ **spoilage** ['spɔɪlɪdʒ] *noun* making food bad and inedible (by rotting, damp, etc.)

spongiform ['spʌnʒɪfɔ:m] *see* BSE

spore [spɔ:] *noun* reproductive body of certain plants and bacteria, which can survive in extremely hot *or* cold conditions for a long time

◇ **sporicidal** [spɒrɪ'saɪdəl] *adjective* which kills spores

◇ **sporicide** ['spɒrɪsaɪd] *noun* substance which kills spores

◇ **Sporozoa** [spɔ:rə'zəʊə] *plural noun* type of parasitic Protozoa which includes Plasmodium, the cause of malaria (NOTE: singular is **Sporozoon**)

COMMENT: spores are produced by plants such as ferns or by algae and fungi. They are microscopic and float in the air or water until they find a resting place where they can germinate

spot [spɒt] *noun* particular small piece of land; ***they drove out to a local beauty spot; high levels of radon have been measured at this spot;*** **hot spot =** place where background radiation is particularly high

SPOT = SYSTEME PROBATOIRE D'OBSERVATION DE LA TERRE

sprawl [sprɔːl] *noun* something, such as a plant *or* industrial estate, which is growing *or* has been built so that it stretches out in a disorderly way; **urban sprawl =** mass of houses built without any particular plan in a disorderly way

spray [spreɪ] **1** *noun* **(a)** mass of tiny drops of liquid **(b)** special liquid for spraying onto a plant to prevent insect infestation *or* disease **2** *verb* **(a)** to send out a liquid in a mass of tiny drops; ***they sprayed the room with disinfectant*** **(b)** to send out a special liquid onto a plant to prevent insect infestation *or* disease; ***apple trees must be sprayed twice a year to kill aphids***

◇ **sprayer** ['spreɪə] *noun* machine which forces a liquid through a nozzle under pressure, used to distribute liquid herbicides, fungicides, insecticides and fertilizers

spread [spred] *verb* **(a)** to go out over a large area; ***the locusts spread right across the country*** **(b)** to put manure *or* fertilizer *or* mulch out over a large area

spring [sprɪŋ] *noun* **(a)** place where water comes naturally out of the ground; ***the first settlers found springs of fresh water in the mountains*** **(b)** first season of the year, following winter and before summer, (around April in the Northern Hemisphere) when days become longer and the weather progressively warmer; **spring equinox =** about March 21st, one of the two occasions in the year when the sun crosses the celestial equator and night and day are each twelve hours long; **spring tide =** tide which occurs at the new and full moon when the influence of the sun and moon act together and the difference between high and low water is more than normal; *compare* NEAP TIDE; **spring wheat =** wheat which is sown in the spring and harvested towards the end of the summer (the wheat is softer than winter wheat)

sprout [spraʊt] **1** *noun* little shoot from a plant, with a stem and small leaves **2** *verb (of plant)* to send out new growth

spruce [spruːs] *noun* temperate softwood tree

spur [spɜː] *noun* ridge of land that descends towards a valley floor from higher land above

squall [skwɔːl] *noun* sharp gust of wind

◇ **squally** ['skwɔːli] *adjective* (weather) which is characterized by sharp gusts of wind

Sr *chemical symbol for* strontium

SSSI = SITE OF SPECIAL SCIENTIFIC INTEREST

stable ['steɪbl] *adjective* **(a)** not changing; ***in parts of Southeast Asia temperatures remain stable for most of the year;*** **stable climax =** more or less stable community of plants and animals in equilibrium with its environment, the final stage of an ecological succession; **stable population =** population which remains at the same level, where births and deaths are equal **(b)** (chemical compound) which does not react readily with other chemicals; (substance) which is not radioactive

◇ **stability** [stə'bɪlɪti] *noun* being stable *or* resistance to change

◇ **stabilization lagoon** [steɪbɪlaɪ'zeɪʃn lə'guːn] *noun* pond used to purify sewage by allowing sunlight to fall on the mixture of sewage and water

◇ **stabilize** ['steɪbəlaɪz] *verb* to stop something changing; to stop changing; ***marram grass was planted on the sand dunes to stabilize them***

◇ **stabilizer** ['steɪbəlaɪzə] *noun* **(a)** artificial substance added to processed food to stop the mixture from changing (as in sauces containing water and fat): in the EU emulsifiers and stabilizers have E numbers E322 - E495 **(b)** artificial substance added to plastics to prevent degradation

◇ **stabilizing agent** ['steɪbəlaɪzɪŋ 'eɪdʒənt] *noun* = STABILIZER

stack [stæk] *noun* **(a) (chimney) stack =** very tall chimney, as in a factory, usually containing several flues; ***the use of high stacks in power stations means that pollution is now more widely spread;*** **stack gases =** gases emitted from chimney stacks **(b)** steep-sided pillar of rock which stands in the sea near a cliff

stain [steɪn] **1** *noun* **(a)** coloured mark on a surface **(b)** dye *or* substance used to increase contrast in the colour of a piece of tissue *or* a bacterial sample, etc., before examining it under the microscope **2** *verb* **(a)** to make a coloured mark on a surface; ***peaty water stains the rocks brown*** **(b)** to treat a piece of tissue *or* a bacterial sample, etc., with a dye so as to increase contrast in the colour before examining it under the microscope

◇ **staining** ['steɪnɪŋ] *noun* treating a piece of tissue *or* a bacterial sample, etc., with a dye so as to increase contrast in the colour before examining it under the microscope

stalactite ['stæləktaɪt] *noun* long pointed growth of mineral from the ceiling of a cave, formed by the constant dripping of water which is rich in minerals

◇ **stalagmite** ['stæləgmaɪt] *noun* long pointed growth of mineral from the floor of a cave, formed by the constant dripping of water which is rich in minerals from the tip of a stalactite

QUOTE: the sediments that build up on cave floors and the accumulating layers of stalactites and stalagmites, record what is happening to the climate along similar lines to the annual rings in trees, but going back for many thousands of years
New Scientist

stalk [stɔːk] *noun (of a plant)* (i) main stem which holds the plant upright; (ii) subsidiary stem, branching out from the main stem or attaching a leaf *or* flower *or* fruit

stamen ['steɪmən] *noun* male part of a flower consisting of a stalk (the filament) bearing a container (the anther) in which pollen is produced

stamp out ['stæmp 'aʊt] *verb* to remove completely; ***international organizations have succeeded in stamping out smallpox; the government is trying to stamp out pollution from waste disposal units***

stand [stænd] *noun* group of plants growing together; *(in a forest)* group of standing trees; ***softwoods from the vast conifer stands in Scandinavia could be widely used in place of hardwood***

standard ['stændəd] **1** *adjective* normal; ***it is the standard practice to monitor the pollution levels twice a day*** **2** *noun* something which has been agreed upon and is used to measure other things by; **effluent standard** = amount of sewage which is allowed to be discharged into a river *or* the sea; **emission standard** = amount of an effluent *or* pollutant which is permitted to be released into the environment, such as the amount of sewage which can be discharged into a river *or* the sea; **emission standards for vehicles** = amount of pollutants (hydrocarbons, carbon monoxide and nitrogen oxides) which can be released into the atmosphere by petrol and diesel engines

standing ['stændɪŋ] *adjective* growing upright; **standing crop** = (i) crop (such as corn) which is growing upright in a field; (ii) the numbers and weight of the living vegetation of an area, calculated by weighing the vegetation growing in a sample section; **standing timber** = trees which are growing in a wood, ready to be felled

staphylococcus [stæfɪlə'kɒkəs] *noun* genus of bacteria which grows in a bunch like grapes and causes boils and food poisoning (NOTE: plural is **staphylococci**)

◇ **staphylococcal** [stæfɪlə'kɒkəl] *adjective* (infection, poisoning) caused by staphylococci

COMMENT: staphylococcal infections are treated with antibiotics such as penicillin, or broad-spectrum antibiotics such as tetracycline

staple commodity ['steɪpl kə'mɒdɪti] *noun* basic food or raw material

starch [stɑːtʃ] *noun* substance found in green plants, the usual form in which carbohydrates exist in food, especially in bread, rice and potatoes

◇ **starchy** ['stɑːtʃi] *adjective* (food) which contains a lot of starch

COMMENT: starch is present in common foods and is broken down by the digestive process into forms of sugar

starfish ['stɑːfɪʃ] *noun* one of a group of flat sea animals, characterized by five arms branching from a central body

starvation [stɑː'veɪʃn] *noun* having very little *or* no food; **starvation diet** = diet which contains little nourishment and is not enough to keep a person healthy

◇ **starve** [stɑːv] *verb* to have little *or* no food or nourishment; ***thousands of people were starving in the desert***

stasis ['steɪsɪs] *noun* state when there is no change *or* growth *or* movement

◇ **static** ['stætɪk] *adjective* not changing, not moving, not growing; **static electricity** = electricity which is in a static state as opposed to electricity which is flowing in a current; *see also* ELECTROSTATIC

station ['steɪʃn] *noun* building used for some specific purpose; **field station** = scientific research centre located in the area being researched; **sheep station** = very large sheep farm in Australia

statistics [stə'tɪstɪks] *plural noun* figures relating to measurements taken from samples; ***population statistics show that the birth rate is slowing down***

stator ['steɪtə] *noun* fixed casing of a generator, in which the rotor turns

steam [stiːm] *noun* vapour which comes off boiling water and condenses in the atmosphere; **steam coal** = coal with a lot of sulphur in it, which is suitable for generating steam but not for turning into coke; **steam fog** *or* **steam mist** = fog which forms when a cold air mass moves over a warmer body of water, giving the appearance of steam *or* smoke

stearic acid [sti'ærɪk 'æsɪd] *noun* colourless insoluble waxy acid used for making candles, etc.

steel [stiːl] *noun* hard flexible metal made from iron and carbon; **stainless steel** = steel which does not rust because it contains a large quantity of chromium

◇ **steelworks** [sti:lwɜ:ks] *noun* factory where steel is produced from iron ore

stem [stem] *noun (of a plant)* (i) main stalk which holds the plant upright; (ii) subsidiary stalk, branching out from the main stalk or attaching a leaf *or* flower *or* fruit

steno- ['stenəʊ] *prefix* meaning (i) narrow; (ii) constricted

◇ **stenohaline** [stenəʊ'hælaɪn] *adjective* (i) (organism) which cannot survive variations in salt levels in its environment; (ii) (organism) which cannot survive variations in osmotic pressure of soil water; *compare* EURYHALINE

◇ **stenothermous** [stenəʊ'θɜ:məs] *adjective* (organism) which cannot stand changes of temperature; *compare* EURYTHERMOUS

steppe [step] *noun* wide grassy plain, with no trees, especially in Eurasia (NOTE: the North American equivalent is **prairie**)

sterile ['steraɪl] *adjective* **(a)** free from bacteria *or* microbes *or* infectious organisms **(b)** infertile *or* not able to produce offspring

◇ **sterility** [ste'rɪlɪti] *noun* (i) being free from bacteria *or* microbes *or* infectious organisms; (ii) infertility *or* being unable to produce offspring

◇ **sterilization** [sterɪlaɪ'zeɪʃn] *noun* (i) action of making something free from bacteria *or* microbes (by killing the bacteria); (ii) action of making an organism unable to produce offspring

◇ **sterilize** ['sterɪlaɪz] *verb* **(a)** to make something free from bacteria *or* microbes (by killing the bacteria); ***the soil needs to be sterilized before being used for intensive greenhouse cultivation;*** **sterilized milk** = milk prepared for human consumption by heating in sealed airtight containers to kill all bacteria; *see also* PASTEURELLA **(b)** to make an organism unable to have offspring

Stevenson screen ['sti:vənsən 'skri:n] *noun* shelter that contains meteorological instruments, arranged to give standard readings

stimulate ['stɪmjuleɪt] *verb* to make an organism or organ react *or* respond

◇ **stimulus** ['stɪmjʊləs] *noun* something (such as light *or* heat *or* noise) which makes an organism or organ react *or* respond (NOTE: plural is **stimuli**)

sting [stɪŋ] **1** *noun* **(a)** organ with a sharp point, used by insects to pierce the skin of its victim and inject a toxic substance into the victim's bloodstream **(b)** action of piercing the skin of an insect's victim and injecting a toxic substance into the victim's bloodstream **2** *verb (of an insect)* to pierce the skin of its victim and inject a toxic substance into the victim's bloodstream

> COMMENT: stings by some insects, such as the tsetse fly, can transmit a bacterial infection. Other insects, such as bees, have toxic substances which they pass into the bloodstream of the victim, causing irritating swellings. Some people are particularly allergic to insect stings

stock [stɒk] **1** *noun* **(a)** animals *or* plants which are derived from an ancestor; **stock farming** = breeding livestock for sale; *see also* LIVESTOCK **(b)** plant with roots onto which a piece (the scion) of another plant is grafted; *see also* ROOTSTOCK **(c)** supply of something held for future use; ***stocks of herring are being decimated by overfishing;*** **stock culture** = basic culture of bacteria from which other cultures can be taken **2** *verb* to provide a supply of something for future use; ***he stocked the ponds with a rare breed of fish; a well- stocked garden***

◇ **stockman** ['stɒkmən] *noun* farm worker who looks after animals, especially cattle

◇ **stockpile** ['stɒkpaɪl] *noun* large supply of something held for future use; **nuclear stockpile** = supply of nuclear weapons

stoma ['stəʊmə] *noun* pore in a plant, especially in the leaves (NOTE: plural is **stomata**)

> COMMENT: the stomata are the holes through which a plant takes in carbon dioxide and sends out oxygen

stone [stəʊn] *noun* mineral formation; single piece of this mineral formation; ***the houses are built directly on stone and should withstand earthquakes; the glaciers left heaps of stones as they retreated***

◇ **stoneground** ['stəʊngraʊnd] *adjective* (flour) which has been made by grinding between heavy round stones

◇ **stony** ['stəʊni] *adjective* (ground, etc.) which is covered with stones

store [stɔ:] *verb* to keep something until it is needed; ***whales store energy in the blubber under the skin***

◇ **storage** ['stɔ:rɪdʒ] *noun* keeping something until it is needed; ***the problem of wind power is the question of storage of electricity for use when there is no wind;*** **heat storage** *or* **thermal storage** = keeping heat created during a period of low consumption until a peak period when it is needed

storm [stɔ:m] *noun* violent weather, with wind and rain *or* snow; ***there was a rainstorm during the night; snowstorms swept the northern American states;*** **storm beach** =

accumulation of coarse beach sediments built up above the high-water mark during storms; **storm centre** = low pressure point in the centre of a cyclone; **storm cloud** = dark-coloured cloud in which vigorous activity produces wind with heavy rain *or* snow; **storm drain** = specially wide channel for taking away large amounts of rainwater which fall during tropical storms; **storm sewage** = sewage mixed with storm water after a heavy rainfall; **storm surge** = rush of water in a tide, pushed by the strong winds in a storm; **storm water** = water which falls as rain during a storm and which is cleared by storm-water channels

strain [streɪn] *noun* distinct variety of a species, which will breed true, usually referring to cultivated plants; ***they have developed a new strain of virus-resistant rice***

strait [streɪt] *noun* narrow passage of sea between two larger areas of sea

strata ['strɑːtə] *plural noun see* STRATUM

stratified ['strætɪfaɪd] *adjective* formed of several layers

◇ **stratification** [strætɪfɪ'keɪʃn] *noun* formation of several layers (of sedimentary rocks *or* of water in a lake *or* of air in the atmosphere)

◇ **stratigraphy** [strə'tɪgrəfi] *noun* science of studying rock strata; **glacial stratigraphy** = science of studying layers of polar ice to discover information about climatic conditions when the ice was formed thousands of years ago

◇ **stratocumulus** [strætəʊ'kjuːmjʊləs] *noun* layer of small cumulus clouds, lower than altocumulus, that is, below 3,000m

◇ **stratopause** ['strætəʊpɔːz] *noun* thin layer of the atmosphere between the stratosphere and the mesosphere

◇ **stratosphere** ['strætəʊsfɪə] *noun* layer of the earth's atmosphere, above the troposphere and the tropopause and separated from the mesosphere by the stratopause

◇ **stratospheric** [strætə'sferɪk] *adjective* referring to the stratosphere; ***CFCs are responsible for damage to the ozone in the earth's stratospheric zone***

> COMMENT: the stratosphere rises from about 18km to 50km above the surface of the earth. It is formed of nitrogen (80%), oxygen (18%), ozone, argon and trace gases. The ozone in it forms the ozone layer

stratum ['strɑːtəm] *noun* layer (especially of rock) (NOTE: plural is **strata**)

stratus ['streɪtəs] *noun* type of grey cloud, often producing light rain

straw [strɔː] *noun* dry stems and leaves of cereal crops left after the grains have been removed

> COMMENT: straw can be ploughed back into the soil or is sometimes burned as stubble. It can be cut, compressed into bundles and burned as fuel for heating

stream [striːm] *noun* flow of water *or* air *or* gas; ***a stream of hot gas turns the turbine; mountain streams have cut deep gullies in the hillside;*** **Gulf Stream** = current of warm water in the Atlantic Ocean, which flows north along the east coast of the USA, then crosses the Atlantic Ocean to hit Northern Europe; **jet stream** = (i) wide belt of fast-moving air occurring at the top limit of the troposphere (at about 15 kilometres above the earth's surface); (ii) flow of gases from a jet engine; **stream erosion** = wearing away of earth *or* rock by the effect of a stream; **stream runoff** = rainwater *or* melted snow which flows into streams

◇ **streamflow** ['striːmfləʊ] *noun* amount and speed of water flowing in a stream

streptococcus [streptə'kɒkəs] *noun* genus of bacteria which grows in long chains and causes fevers such as scarlet fever, tonsillitis and rheumatic fever (NOTE: plural is **streptococci**)

◇ **streptococcal** [streptə'kɒkl] *adjective* (infection) caused by streptococci

◇ **streptomycin** [streptə'maɪsɪn] *noun* antibiotic used against many types of infection, especially streptococcal

stress [stres] **1** *noun* condition where an outside influence changes the composition *or* functioning of something; **drought stress** = lack of growth caused by drought; **human-induced stress** = stress in animals caused by humans (such as ringing birds *or* driving cars in deer parks, etc.) **2** *verb* to subject something to stress

> QUOTE: a fatal sequence of stresses may begin with a predisposing stress such as shortage of nutrients. The tree may then be weakened by an inciting stress, such as a severe winter. It is then defenceless against a final contributing stress, such as disease or insect attack, which is the actual cause of death
> **Scientific American**

striation [straɪ'eɪʃn] *noun* mark of a scratch; **glacial striation** = scratches made on rocks by a moving glacier

strip [strɪp] **1** *noun* long thin flat piece; **strip cropping** *or* **strip farming** *or* **strip planting** = method of farming where long thin pieces of land across the contours are planted with different crops in order to reduce soil erosion; **strip cultivation** = method of communal farming where each family has a long thin piece *or* several long thin pieces of land to cultivate **2** *verb* to remove a covering from something; ***spraying with defoliant strips the leaves off all plants;*** **nutrient stripping** = removal of nutrients from sewage to prevent eutrophication of water in reservoirs

◇ **strip mining** ['strɪp 'maɪnɪŋ] *noun* form of mining where the mineral is dug from the surface instead of digging underground to get it; *see also* OPEN-CAST MINING

> QUOTE: a deal allows the company to strip peat from two thirds of the threatened lowland peat bogs
> **Green Magazine**

strontium ['strɒntiəm] *noun* metallic element; **strontium-90** = isotope of strontium which is formed in nuclear reactions and, because it is part of the fallout of nuclear explosions, can enter the food chain, attacking in particular animals' bones (NOTE: chemical symbol is **Sr;** atomic number is **38)**

structure ['strʌktʃə] *noun* **(a)** way in which things are spaced, formed, shaped, etc.; **ecological structure** = spatial and other arrangements of species in an ecosystem; **soil structure** = way in which soil particles are joined together (e.g. loose, firm, etc.) **(b)** something which has been constructed, especially a building

◇ **structural** ['strʌktʃərəl] *adjective* referring to the structure of something; **structural geology** = scientific study of the structure and distribution of the rocks forming the earth's crust

strychnine ['strɪkni:n] *noun* poisonous alkaloid substance made from the seeds of a tropical tree

stubble ['stʌbl] *noun* short stems left in the ground after a crop of wheat *or* other cereal has been cut; **stubble burning** = method of removing dry stubble by burning it before ploughing; **stubble field** = field in which the crop has been harvested and the stubble left in the ground; ***the animals are driven into the stubble fields to graze after the harvest***

> COMMENT: stubble burning has the advantage of removing weed seeds and creating a certain amount of natural fertilizer which can be ploughed into the soil. The disadvantage is that it pollutes the atmosphere with smoke, reducing visibility on roads and releasing large amounts of carbon dioxide. This, together with the possible danger that the fire may get out of control, killing small animals and burning trees and crops, means that it is not recommended as a means of dealing with the stalks of harvested plants

stunt [stʌnt] *verb* to reduce the growth of something; ***the poor mountain soil supports a few stunted trees***

sub- [sʌb] *prefix* meaning (i) less important; (ii) underneath; **subarctic region** = region near the Arctic; **subcloud layer** = air immediately underneath a cloud layer; **sub-saharan area** = area south of the Sahara

◇ **subacute** [sʌbə'kju:t] *adjective* (disease *or* condition) which is between acute and chronic; *compare* ACUTE, CHRONIC

◇ **subclass** ['sʌbklɑ:s] *noun* one of the divisions into which organisms are categorized

◇ **subclimax** ['sʌbklaɪmæks] *noun* stage in the development of a plant community where development stops before climax

◇ **subcontinent** [sʌb'kɒntɪnənt] *noun* very large land mass, attached to a continent; ***the Indian subcontinent***

◇ **subdivision** ['sʌbdɪvɪʒən] *noun* subsidiary category into which organisms may be classified

◇ **subdominant** [sʌb'dɒmɪnənt] *adjective* (species) which is not as important as the dominant species

◇ **subduct** [səb'dʌkt] *verb* to pull something underneath; ***the oceanic crust is being subducted under the continents which surround it***

◇ **subduction** [sʌb'dʌkʃən] *noun* process by which a plate, such as the oceanic crust, is slowly being pulled under another plate

sublimate 1 *noun* ['sʌblɪmət] vapour *or* gas produced during sublimation **2** *verb* ['sʌblɪmeɪt] *(of a solid)* to turn into a vapour *or* gas, without first becoming liquid

◇ **sublimation** [sʌblɪ'meɪʃn] *noun* turning a solid into a vapour *or* gas, without it first becoming liquid

◇ **sublime** [sə'blaɪm] *verb* = SUBLIMATE 2

sublittoral [sʌb'lɪtərəl] *adjective* (zone *or* water) which is further from a shore than the littoral zone; **sublittoral plant** = plant which grows near the sea

submarine [sʌbmə'ri:n] *adjective* beneath the sea; ***corals forming submarine reefs; shellfish collect round warm submarine vents***

submerge [sʌb'mɜ:dʒ] *verb* to cover something with water; ***the coast is dangerous, with submerged rocks near the harbour***

submicroscopic [sʌbmaɪkrə'skɒpɪk] *adjective* too small to be seen with an ordinary microscope

subside [sʌb'saɪd] *verb* **(a)** to go down *or* to become less violent; ***after the rainstorms passed, the flood waters gradually subsided*** **(b)** *(of a piece of ground or a building)* to sink *or* fall to a lower level; ***the office block is subsiding due to the shrinkage of the clay it is built on***

◇ **subsidence** ['sʌbsɪdəns] *noun* **(a)** *(of a piece of ground or a building)* sinking *or* falling to a lower level; ***subsidence caused by the old mine shaft closed the main road*** **(b)** gradual movement downwards of a mass of air; **subsidence inversion** = phenomenon produced when a mass of air gradually sinks and becomes warmer

subsidy ['sʌbsɪdi] *noun* money given by a government *or* organization to help an industry *or* a charity, etc.; ***the car industry had been receiving government subsidies for years***

subsistence [sʌb'sɪstəns] *noun* existing *or* surviving, with very little food; **subsistence farming** = growing just enough crops to feed the farmer and his family and having none left to sell; **subsistence food** = food which is needed to keep a person alive

> QUOTE: pollution from suspended sediments and heavy metals exceeds US Environmental Protection Agency standards and threatens subsistence staples (such as fish and crustaceans)
>
> **Ecologist**

subsoil ['sʌbsɔɪl] *noun* layer of soil under the topsoil, which contains little organic matter and into which chemical substances from the topsoil are leached; **subsoil water** = water held in the subsoil

subspecies ['sʌbspi:ʃi:z] *noun* group of organisms which is part of a species, but which shows slight differences from the main group, with which it can still interbreed

substance ['sʌbstəns] *noun* chemical material; ***toxic substances released into the atmosphere***

substitution [sʌbstɪ'tju:ʃn] *noun* replacing one thing with another; **substitution effect** = the effect on the environment of substituting one form of action for another

◇ **substitute** ['sʌstɪtju:t] **1** *noun* thing used in place of another; **sugar substitutes** = sweeteners (such as saccharin) used in place of sugar **2** *verb* to replace one thing with another; ***farmers have been told to plough up pastureland and substitute woodlots***

substrate ['sʌbstreɪt] *noun* **(a)** substance which is acted on by an enzyme **(b)** matter *or* surface on which an organism lives

◇ **substratum** ['sʌbstrɑ:təm] *noun* layer of rock beneath the topsoil and subsoil (NOTE: plural is **substrata**)

subtropics [sʌb'trɒpɪks] *noun* area between the tropics and the temperate zone

◇ **subtropical** [sʌb'trɒpɪkl] *adjective* referring to the subtropics; ***the islands enjoy a subtropical climate; subtropical plants grow in the sheltered parts of the coast;*** **subtropical high** = area of high pressure normally found in the subtropics

suburb ['sʌbɜ:b] *noun* residential part of a town, away from the centre, but still within the built-up area

◇ **suburban** [sə'bɜ:bən] *adjective* referring to the suburbs

succession [sək'seʃn] *noun* series of stages, one after the other, by which a group of organisms living in a community reaches its final stable state *or* climax; **biotic succession** = changes which take place in a group of organisms under the influence of their changing environment; **primary succession** = ecological community which develops in a place where nothing has lived before (as on cooled lava from a volcano); **secondary succession** = ecological community which develops in a place where a previous community has been removed (as by fire, flooding *or* cutting down of trees)

◇ **successional cropping** [sək'seʃənəl 'krɒpɪŋ] *noun* growing several crops one after the other during the same growing season

◇ **successive** [sək'sesɪv] *adjective* which follows on after something else

succulent ['səkju:lənt] *adjective & noun* (plant, such as a cactus) which has fleshy leaves *or* stems in which it stores water

suffer ['sʌfə] *verb* to experience something painful *or* unpleasant; ***the town suffers from pollution from the nearby chemical works; the whole region is suffering from drought***

sufficiency [sə'fɪʃənsi] *noun* being enough; a large enough amount of something; **self-sufficiency** = ability to provide all that is needed without outside help

suffrutescent *or* **suffruticose** [səfru:'tesənt *or* sə'fru:tɪkəʊz] *adjective* (perennial plant) which has a wooden base to its stem and does not die down to ground level in winter

sugar ['ʃʊgə] *noun* sucrose, sweet substance ($C_{12}H_{22}O_{11}$) obtained from sugar cane; **sugar-free** = (food, drink, diet) which does not contain sugar

sullage ['sʌlɪdʒ] *noun* (i) mud brought down by mountain streams; (ii) liquid waste from a house

sulphate *or US* **sulfate** ['sʌlfeɪt] *noun* salt of sulphuric acid and a metal (SO_4^{2-}); **sulphate of ammonia** = nitrogenous fertilizer available in granular or liquid form; **sulphate of potash** = a fertilizer (potassium sulphate) containing about 50% potash; used by potato growers and market gardeners

◇ **sulphide** *or US* **sulfide** ['sʌlfaɪd] *noun* ion of sulphur (S^{2-}) which is present in chemical compounds and mineral ores

◇ **sulphite** *or US* **sulfite** ['sʌlfaɪt] *noun* salt of sulphuric acid (SO_3^{2-}) which forms part of several chemical compounds which are used in processing paper

◇ **sulphonation** *or US* **sulfonation** [sʌlfə'neɪʃn] *noun* making a substance into sulphonic acid *or* adding sulphonic acid

◇ **sulphonator** [sʌlfə'neɪtə] *noun* apparatus for adding sulphur dioxide to water to remove excess chlorine

◇ **sulphonic acid** *or US* **sulfonic acid** [sʌl'fɒnɪk 'æsɪd] *noun* strong organic acid used in the manufacture of medicines and dyes

◇ **sulphur** *or US* **sulfur** ['sʌlfə] *noun* yellow non-metallic chemical element which is essential to biological life. It is found in volcanoes, in limestone and in some amino acids. It is widely present in fossil fuels which when burnt release sulphur dioxide and sulphur trioxide into the atmosphere; **sulphur cycle** = process by which sulphur flows from the environment, through organisms and back to the environment again; **sulphur dioxide** = unpleasant smelling gas (SO_2) formed when sulphur is burnt with oxygen (as when fuels are burnt); **sulphur oxide** = sulphur dioxide *or* sulphur trioxide, both of which are present in sulphur pollution; **sulphur trioxide** = gas (SO_3) formed when fuels are burnt, usually present in sulphur oxide pollution and forming sulphuric acid when dissolved in water (NOTE: chemical symbol for sulphur is **S;** atomic number is **16.** Note also that words beginning **sulph-** are spelt **sulf-** in US English)

> COMMENT: sulphur dioxide produced from burning coal or oil is an important cause of smog. Today the level of sulphur dioxide has fallen with the reduction in the use of coal and this type of pollution has been replaced by nitrogen dioxide produced by car exhausts

sulphuric acid (H_2SO_4) [sʌl'fjuːrɪk 'æsɪd] *noun* acid formed from sulphur trioxide and water, which forms in the atmosphere and is a nutrient for some types of plant. It is used to make chemical fertilizers and also in the plastics industry

summer ['sʌmə] **1** *noun* season of the year, following spring and before autumn, when the weather is warmest, the sun is highest in the sky and most plants flower and set seed; **summer solstice** = 21st June, the longest day in the Northern Hemisphere, when the sun is as its furthest point south of the equator **2** *verb* to spend the summer in a place; ***the birds summer on the shores of the lake***

◇ **summerwood** ['sʌməwʊd] *noun* wood formed during the later part of the growing season

summit ['sʌmɪt] *noun* **(a)** highest point (on mountain); ***the climber reached the summit of the mountain*** **(b)** meeting between heads of government; *see also* EARTH SUMMIT

sun [sʌn] *noun* very hot star round which the earth and other planets orbit and which gives energy in the form of light and heat

◇ **sunburn** ['sʌnbɜːn] *noun* damage to the skin by excessive exposure to sunlight

◇ **sunburnt** ['sʌnbɜːnt] *adjective* (skin) made brown *or* red by exposure to sunlight

◇ **sunglasses** ['sʌnglɑːsɪz] *plural noun* dark glasses which are worn to protect the eyes from the sun

sunlight ['sʌnlaɪt] *noun* light from the sun; ***he is allergic to strong sunlight***

> COMMENT: sunlight is essential to give the body vitamin D. Excessive exposure to sunlight will not simply turn white skin brown, but may also burn the surface of the skin so badly that the skin dies and pus forms beneath. There is evidence that constant exposure to the sun can cause cancer (or melanoma) of white skin. Depletion of the ozone layer in the atmosphere may increase the incidence of skin cancer

sunrise ['sʌnraɪz] *noun* time when the sun appears above the eastern horizon

◇ **sunset** ['sʌnset] *noun* time when the sun disappears below the western horizon

◇ **sunshine** ['sʌnʃaɪn] *noun* bright light from the sun

sunspot ['sʌnspɒt] *noun* darker patch on the surface of the sun, caused by a stream of gas shooting outwards

> COMMENT: sunspots appear on the surface of the sun in an eleven year cycle. They seem to have an effect on the earth's climate which tends to become more turbulent when the sunspots appear

sunstroke ['sʌnstrəʊk] *noun* serious condition caused by excessive exposure to the sun *or* to hot conditions, where the patient becomes dizzy and has a high body temperature but does not perspire

super- ['suːpə] *prefix* meaning (i) above; (ii) extremely

◇ **supercool** [suːpə'kuːl] *verb* to reduce the temperature of a substance below its normal freezing point without freezing actually occurring; **supercooled fog** = fog which has cooled below freezing point, but still remains liquid

COMMENT: supercooled fog contributes to the phenomenon known as freezing fog, where droplets of fog remain liquid in the air, even though the temperature is below 0°C, but freeze into hoar frost as soon as they touch a surface

superficial [suːpə'fɪʃl] *adjective* on the surface *or* close to the surface

superheated [suːpə'hiːtɪd] *adjective* (steam) heated to a high temperature in a power station

◇ **superheater** [suːpə'hiːtə] *noun* section of a power station boiler, where steam is heated to a higher temperature

COMMENT: steam made by heating water in a boiler is passed through the furnace a second time to heat it again. This superheated steam is produced under very high pressure to turn the first (or high-pressure) turbine in the generating plant

superinsulate [suːpə'ɪnsjuːleɪt] *verb* to equip a building with complete insulation

QUOTE: by investing from two to seven thousand dollars to superinsulate a house, it is possible to reduce annual heating costs to between $20 and $300. The low heating bills result because superinsulated houses store the free heat from people, lighting, electrical appliances and passive solar heat from windows
Scientific American

superorder [suːpə'ɔːdə] *noun* one of the divisions into which organisms are categorized

superorganism [suːpə'ɔːgənɪzm] *noun* community (such as a forest) containing a group of individual organisms

superphosphate [suːpə'fɒsfeɪt] *noun* chemical compound formed from calcium phosphate (basic slag) and sulphuric acid, used as a fertilizer

supersaturated [suːpə'sætʃəreɪtɪd] *adjective* (air) which contains more moisture than that required to saturate it; (solution) which contains more solute than that required to saturate it

supersonic [suːpə'sɒnɪk] *adjective* (aircraft, bullet, etc.) which flies at *or* faster than the speed of sound

supply [sə'plaɪ] *noun* (i) providing something that is needed; (ii) stock of something that is needed; ***when the supply of water runs out; the air force has been dropping supplies for the stranded villagers; supply routes from the port have been cut off;*** **supply and demand** = balance between things produced and required in a society, especially as affecting prices; **electricity supply** = electric power provided for domestic and industrial use; ***the electricity supply is often cut;*** **food supply** = (i) production of food and the way in which it gets to the consumer; (ii) stock of food; **medical supplies** = drugs, bandages, syringes, etc.

support [sə'pɔːt] *verb* to provide what is necessary for organisms to live; ***these wetlands support a natural community of plants, animals and birds***

supra- ['suːprə] *prefix* meaning above *or* over

surface ['sɜːfɪs] *noun* top layer of something; **surface drainage** = removing of surplus water from an area of land by means of ditches and channels; **surface evaporation** = evaporation of water from the surface of a lake, river, etc.; **surface runoff** = flow of rainwater *or* melted snow *or* excess fertilizer from the surface of land into streams and rivers; **surface soil** = = TOPSOIL; **surface tension** = appearance of a film on the surface of a liquid caused by the relationship between the molecules; **surface water** = water (after rain) which lies on the surface of the soil and does not drain into the soil, but flows across the surface as a stream and drains into rivers; **surface wind** = wind which blows across the land surface (as opposed to winds higher in the atmosphere)

surfactant [sɜː'fæktənt] *noun* substance which reduces surface tension

surge [sɜːdʒ] *noun* sudden increase (as in electric current when lightning strikes a building *or* in demand for electricity); ***electricity companies try to reduce the huge surges in demand that occur at certain times of the day;*** **surge arrester** = device to prevent surges of current in an electric system, that can damage equipment

surplus ['sɜːpləs] *noun* extra produce, produce which is more than is needed; ***surplus corn and butter are stored until they can be sold on the world markets; surplus water will flow away in storm drains***

survey ['sɜːveɪ] *noun* (i) investigation *or* inspection; (ii) taking measurements of the height of buildings, mountains, the length of roads, rivers, etc. in order to make a detailed plan *or* map; (iii) document *or* plan *or* map showing the results of an investigation *or* of the measurements taken; **public opinion survey** = asking the general public questions in order to find out what they think about something

survive [sə'vaɪv] *verb* to continue to live; ***they survived a night on the mountain without food; fish can survive for some time in frozen water; the plants survive even the hottest desert temperatures***

◇ **survival** [sə'vaɪvəl] *noun* continuing to live; **survival of the fittest** = natural selection *or* evolution of a species, whereby characteristics which help individual organisms to survive and reproduce are passed on to the offspring and those characteristics which do not help are not passed on; **survival rate** = number of organisms which continue to live; ***the survival rate of newborn babies has begun to fall***

◇ **survivor** [sə'vaɪvə] *noun* organism which continues to live; **there are no survivors** = everyone has died

◇ **survivorship** [sə'vaɪvəʃɪp] *noun* number of individuals of a population surviving at a certain time; **survivorship curve** = graph showing the number of individuals of a population which survive to a certain time

suspend [sə'spend] *verb* to hold particles in a liquid *or* in air; ***an aerosol of suspended particles***

◇ **suspension** ['sə'spenʃn] *noun* liquid with solid particles in it, not settling to the bottom nor floating on the surface

sustain [sə'steɪn] *verb* to keep *or* to support; ***the land is fertile enough to sustain a wide variety of fauna and flora***

◇ **sustainability** [səsteɪnə'bɪlɪti] *noun* ability of a process to leave natural resources undamaged and the environment in good order for future generations

◇ **sustainable** [sə'steɪnəbl] *adjective* (development, process, forestry, etc.) which does not deplete *or* damage natural resources irreparably and which leaves the environment in good order for future generations; ***hardwood from a sustainable source; a conservatory made from sustainable timbers***

QUOTE: rainforest set aside for sustainable logging was being cut indiscriminately
Green Magazine

COMMENT: sustainability or sustainable development have been described as 'meeting the needs of the present without compromising the ability of future generations to meet their own needs', but there is no internationally agreed definition of sustainability

Sv *abbreviation for* sievert

swallow hole ['swɒləʊ 'həʊl] *noun* hole which forms in limestone rock as rainwater drains through it, dissolving minerals in the rock and sometimes forming underground caverns

swamp [swɒmp] *noun* area of permanently wet land and the plants which grow on it; **backswamps** = swampland away from a main watercourse; **mangrove swamp** = swamp covered with mangroves

◇ **swampland** ['swɒmplænd] *noun* area of land covered with swamp

◇ **swampy** ['swɒmpi] *adjective* (land) which is permanently wet and supports a natural community of plants and animals

swarm [swɔːm] **1** *noun* large number of insects (such as bees *or* locusts) travelling as a group **2** *verb (of insects)* to travel as a large group

swash [swɒʃ] *noun* rush of water up a beach from a breaking wave; *compare* BACKWASH

sweetener ['swiːtnə] *noun* artificial substance (such as saccharin) added to food to make it sweet

QUOTE: world sugar stocks do not reflect potential supplies of sweeteners. High sugar prices will lead in due course to an increase in other sweeteners
Farmers Weekly

swidden farming ['swɪdən 'fɑːmɪŋ] *noun* slash and burn *or* form of agriculture where forest is cut down and burnt to create open space for growing crops; the space is abandoned after several crops have been grown and further forest is cut down

swill [swɪl] *noun* waste food from kitchens used for pig feeding

sycamore ['sɪkəmɔː] *noun* *Acer pseudoplatanus,* large hardwood tree of the maple family

symbiosis [sɪmbaɪ'əʊsɪs] *noun* condition where two *or* more organisms exist together and help each other to survive

◇ **symbiont** ['sɪmbaɪɒnt] *noun* one of the organisms living in symbiosis; *compare* COMMENSAL

◇ **symbiotic** [sɪmbaɪ'ɒtɪk] *adjective* referring to symbiosis; living in symbiosis; ***the rainforest has evolved symbiotic mechanisms to recycle minerals;*** **symbiotic relationship** = SYMBIOSIS

◇ **symbiotically** [sɪmbaɪ'ɒtɪkli] *adverb* in symbiosis; ***colonies of shellfish and the parasites that live symbiotically with them***

QUOTE: the mussel filters bacteria from water and also derives nutrition from its symbiotic bacteria like other animals that gain energy from symbionts, the hairy snails can store elemental sulphur
New Scientist

symphile ['sɪmfaɪl] *noun* insect *or* other organism which lives in the nests of social insects such as ants *or* termites, and is fed by them

symptom ['sɪmptəm] *noun* change in the functioning *or* appearance of an organism, which shows that a disease *or* disorder is present

◇ **symptomatic** [sɪmptə'mætɪk] *adjective* which is a symptom; ***the rash is symptomatic of measles***

syn- [sɪn] *prefix* meaning joint *or* fused

syncline ['sɪŋklaɪn] *noun (in rock formation)* concave downwards fold of rock, with the youngest rock on the inside; *compare* ANTICLINE

syndrome ['sɪndrəʊm] *noun* group of symptoms and other changes in the body's functions which, when taken together, show that a particular disease *or* disorder is present

synecology [sɪnɪ'kɒlədʒi] *noun* study of communities of organisms in their environments; *compare* AUTECOLOGY

synergism ['sɪnədʒɪzm] *noun* phenomenon where two substances act more strongly together than they would if they were separate (as when mixtures of pollutants can be more toxic than each pollutant taken separately)

◇ **synergist** ['sɪnədʒɪst] *noun* food additive which increases the effect of another additive

synfuel ['sɪnfju:l] *noun* fuel, similar to those produced from crude oil, but produced from more plentiful resources, such as coal, shale *or* tar

synroc ['sɪnrʊk] *noun* artificial mineral compound, formed of nuclear waste fused into minerals which will never deteriorate

synthesis ['sɪnθəsɪs] *noun* (i) process of combining things to form a whole; (ii) process of producing a compound by a chemical reaction

◇ **synthesize** ['sɪnθəsaɪz] *verb* (i) to combine things to form a whole; (ii) to produce a compound by a chemical reaction; ***essential amino acids cannot be synthesized; the body cannot synthesize essential fatty acids and has to absorb them from food***

◇ **synthetic** [sɪn'θetɪk] *adjective* made artificially by chemical reaction, man-made

◇ **synthetically** [sɪn'θetɪkli] *adverb* (made) artificially by chemical reaction; ***synthetically produced hormones are used in hormone therapy***

synusia [sɪnju:ziə] *noun* group of plants living in the same habitat

system ['sɪstəm] *noun* **(a)** arrangement of things *or* phenomena which act together; **ecosystem** = system which includes all the organisms of an area and the environment in which they live; **frontal system** = series of cold *or* warm fronts linked together; **solar system** = arrangement of planets which orbit around the sun **(b)** arrangement of certain parts of the body so that they work together; **alimentary system** = arrangement of tubes and organs, including the intestine, salivary glands, liver, etc., through which food passes and is digested; **cardiovascular system** = system of organs and blood vessels, including the heart, arteries and veins, in which the blood circulates round the body **(c)** way of classifying something scientifically; **Linnaean system** = method of naming organisms devised by the Swedish scientist, Carolus Linnaeus (1707-1778); *see also* LINNAEAN

◇ **systematic** [sɪstɪ'mætɪk] *adjective* organized in a planned way; part of a system

◇ **systematics** [sɪstɪ'mætɪks] *noun* scientific study of systems, especially of the system of classifying organisms

◇ **systemic** [sɪs'temɪk] *adjective* which affects the whole organism; **systemic fungicide** *or* **insecticide** *or* **pesticide** = fungicide *or* insecticide *or* pesticide which is absorbed into a plant's sap system through its leaves and makes the plant poisonous to fungi *or* insects *or* pests without killing the plant itself; **systemic herbicide** *or* **systemic weedkiller** = herbicide which is absorbed into a plant's sap system through its leaves

Système International *see* SI

Système Probatoire d'Observation de la Terre (SPOT) *noun* French observation satellite transmitting data which give better images of the Earth than those from Landsat

Tt

T *abbreviation for* tera-

Ta *chemical symbol for* tantalum

table ['teɪbl] *noun* level structure; **water table** = top level of water in the ground that occupies spaces in rock *or* soil and is above a layer of impermeable rock; **table mountain** = flat-topped mountain

◇ **tableland** ['teɪbllænd] *noun* area of high flat land

TAC = TOTAL ALLOWABLE CATCH

taiga ['taɪgə] *noun* forested region between the Arctic tundra and the steppe

tailings ['teɪlɪŋz] *plural noun* refuse *or* waste ore from mining operations

tailpipe ['teɪlpaɪp] *noun US* tube at the back of a motor vehicle from which gases produced by burning petrol are sent out into the atmosphere (NOTE: British English is **exhaust pipe**)

tank [tæŋk] *noun* large receptacle for liquid; ***he stopped at the garage to fill his tank with petrol; mobile water tanks have been bringing water to drought-stricken areas;*** **sedimentation tank** = tank in which sewage is allowed to stand so that solid particles can sink to the bottom

◇ **tanker** ['tæŋkə] *noun* (i) large ship for carrying petrol or oil; (ii) truck used to carry liquids such as milk

tantalum ['tæntələm] *noun* rare metal which does not corrode, used to repair damaged bones (NOTE: chemical symbol is **Ta;** atomic number is **73)**

tap [tæp] **1** *noun* pipe with a handle which can be turned to make a liquid *or* gas come out of a container **2** *verb* to remove *or* drain liquid from something; **to tap oil resources** = to bring up oil from the ground; *see also* UNTAPPED

◇ **taproot** ['tæprʊt] *noun* thick main root of a plant which grows straight down into the soil; ***the taproot system has a main root with smaller roots branching off it, as opposed to a fibrous root system which has no main root***

◇ **tap water** ['tæp 'wɔːtə] *noun* mains water which comes out of the tap in a house; ***tapwater often smells of chlorine***

tar [tɑː] *noun* thick black sticky substance derived from coal; **coal tar** = any of several liquids, formed by distillation of coal, used in the pharmaceutical industry and as a wood preservative; **pine tar** = brown *or* black sticky substance derived from pine wood and used in medicines, soap, paint, etc.; **tar oil** = winter wash used to control aphis and scale insects on fruit trees; **tar pit** = natural hole in the earth containing bitumen; **tar sand** = geological formation of sand *or* sandstone containing bitumen, which can be extracted and processed to give oil

tarn [tɑːn] *noun* small lake in a depression high in the hills

tartrazine ['tɑːtrəziːn] *noun* yellow substance (E102) added to food to give it an attractive colour. (Although widely used, tartrazine provokes reactions in hypersensitive people and is banned in some countries)

taungya ['taʊngiə] *noun* tropical agricultural system where arable farming and forestry are practised together

-taxis ['tæksɪs] *suffix* referring to the response of an organism which moves towards *or* away from a stimulus

taxon ['tæksən] *noun* any grouping of a scientific classification of organisms (family, genus, species, etc.) (NOTE: plural is **taxa)**

◇ **taxonomy** [tæks'ɒnəmi] *noun* way of classification of organisms

◇ **taxonomic** *or* **taxonomical** [tæksə'nɒmɪk *or* tæksə'nɒmɪkl] *adjective* referring to taxonomy; ***a taxonomical group or unit***

TBT an antifouling paint

TCDD = TETRACHLORODIBENZOPARADIOXIN highly toxic environmentally persistent by-product of 2,4,5-T

teak [tiːk] *noun* tropical hardwood which is resistant to water

technique [tek'niːk] *noun* way of doing scientific *or* medical work; ***a new technique for dating fossils***

◇ **technical** ['teknɪkl] *adjective* referring to practical *or* scientific work

◇ **technician** [tek'nɪʃn] *noun* person who does practical work in a laboratory *or* scientific institution; ***he is a laboratory technician in a laboratory attached to the research institute***

technology [tek'nɒledʒi] *noun* applying scientific knowledge to industrial processes; ***the technology has to be related to user requirements;*** **alternative technology** = using methods to produce energy which are different and less polluting than the usual ways (i.e.

wind power, tidal power, solar power, as opposed to traditional *or* nuclear power); **appropriate technology** = technology that is suited to the local environment, usually involving skills *or* materials that are easily available locally; **intermediate technology** = technology which is between the advanced electronic technology of industrialized countries and the primitive technology in developing countries; **new technology** = electronic instruments and devices which have recently been developed and are being introduced into industry

◇ **technological** [teknə'lɒdʒɪkl] *adjective* referring to technology; **technological fix** = solution to a problem based on using technology, generally not a really satisfactory solution; **technological revolution** = changing of industrial methods by introducing new technology

◇ **technosphere** ['teknəsfiə] *noun* an environment built *or* modified by humans

> QUOTE: technologies such as improved stoves, locally-designed water pumps and other applications that fall under the category of appropriate technology do not qualify for patents
> **Appropriate Technology**

tectonic [tek'tɒnɪk] **1** *adjective* referring to faults *or* movements in the earth's crust; **tectonic plate** = large area of solid rock in the earth's crust, which floats on the mantle and moves very slowly **2** *noun* **plate tectonics** = theory that the earth's crust is made up of a series of large plates of solid rock which float on the mantle and move very slowly, the places where they meet being subject to earthquakes and volcanic eruptions

> QUOTE: most of the magmatism that creates the Earth's crust occurs in well-defined tectonic zones: mid-oceanic ridges, continental and island arcs, and continental rift zones
> **Nature**

TEL = TETRAETHYL *or* TETRAMETHYL

temperate ['temprət] *adjective* neither very hot nor very cold; **temperate climate** *or* **temperate region** = climate *or* region which is neither very hot in summer nor very cold in winter; **temperate forest** = forest in a temperate region

temperature ['temprətʃə] *noun* measurement of heat in degrees; **critical temperature** = temperature below which a gas will normally become liquid; **global temperature** = temperature over the earth as a whole; **mean temperature** = average temperature; ***the mean temperature for July is 25°;*** **temperature coefficient** = ratio of the speed at which a process develops compared to the speed of the same process at 10°C cooler temperature; **temperature inversion** = atmospheric phenomenon, where cold air is nearer the ground than warm air, making the temperature of the air rise as it gets further from the ground, trapping pollutants between the layers of air; **temperature lapse rate** *see* LAPSE RATE

temporary ['temprəri] *adjective* which is not permanent *or* which does not last a long time; **temporary hardness** = hardness of water caused by carbonates of calcium, which can be removed by boiling the water; **temporary parasite** = parasite which does not live permanently on the host

tender ['tendə] *adjective* (plant) which cannot stand frost

tendon ['tendən] *noun* strip of connective tissue which attaches a muscle to a bone; **tendon sheath** = tube of membrane which covers and protects a membrane

tendril ['tendrɪl] *noun* part of a climbing plant which clings to a surface, so allowing the plant to climb

tera- ['terə] *prefix* meaning one billion (10^{12}); **teragram** = one billion grams

teratogen [te'rætədʒən] *noun* substance which causes birth defects

◇ **teratogenesis** [terætəʊ'dʒenəsɪs] *noun* production of birth defects

◇ **teratogenic** [terætəʊ'dʒenɪk] *adjective* (substance) which causes birth defects

terawatt (TW) ['terəwɒt] *noun* unit of measurement of electric energy, equal to one billion watts

terminal ['tɜːmɪnəl] *adjective* referring to the end; which comes at the end; **terminal moraine** = heap of soil and sand pushed by a glacier and left behind when the glacier melts

termite ['tɜːmaɪt] *noun* insect which looks like a large ant, which lives in colonies and eats cellulose

◇ **termitarium** [tɜːmɪ'teəriəm] *noun* termites' nest, place made by termites to live in, formed in the shape of a hill of hard earth

tern [tɜːn] *noun* species of small gull

terrace ['terəs] **1** *noun* flat strip of land across a sloping hillside, lying level along the contours; **alluvial terrace** *or* **river terrace** = flat plain left when a river cuts more deeply into the bottom of a valley; **continental terrace** = level part of the earth's crust including the continental shelf and the lower-lying areas of the continents themselves; **terrace cultivation** = hill slopes cut to form terraced fields which rise in steps one above the other and are cultivated, often with the aid of irrigation **2** *verb* to build terraces on a mountainside; ***the hills are covered with terraced rice fields;*** **terraced houses** = houses built in a row, all connected together and of a similar style

COMMENT: terracing is widely used to create small flat fields on steeply sloping land, so as to bring more land into productive use, and also to prevent soil erosion

terrain [tə'reɪn] *noun* type of land surface (referring to travelling *or* cultivating); ***buffalo are useful for cultivating land on difficult terrains;*** **all terrain vehicle** = vehicle which can be driven over all types of land surface

terrestrial [tə'restriəl] *adjective* referring to land *or* to the planet earth; **terrestrial animal** = animal which lives on dry land; **terrestrial equator** = imaginary line running round the surface of the earth, at an equal distance from the North and South Poles; **terrestrial magnetism** = magnetic properties of the earth; **terrestrial radiation** = loss of heat from the earth

terricolous [tə'rɪkələs] *adjective* (animal) which lives in *or* on soil

territory ['terɪtri] *noun* area of land occupied and defended by an animal (it may be all or part of the animal's home range); ***a robin will defend its territory by attacking other robins that enter it***

◇ **territorial** [terɪ'tɔːriəl] *adjective* referring to territory; **territorial species** = species which occupies and defends an area of land

◇ **territorialism** *or* **territoriality** [terɪ'tɔːriəlɪzm *or* terɪtɔːri'ælɪti] *noun* phenomenon of having territories

QUOTE: in theory the size of a territory defended by a bird varies according to the abundance of food within the territory
Natural History

QUOTE: the successful territorial expansion of ospreys in Scotland has caused conservationists to predict that the bird of prey will re-establish itself in northern England in the next few years
Independent on Sunday

tertiary ['tɜːʃəri] *adjective* third, coming after secondary and primary; **tertiary consumer** = carnivore which only eats other carnivores; **tertiary industry** = industry which does not produce *or* manufacture, but which offers a service (such as banking, tourism, etc.); **Tertiary Period** = geological period which lasted from about 65 million years ago to 2 million years ago; **tertiary treatment** = advanced stage in the processing of waste

test [test] **1** *noun* examination to see if a sample is healthy *or* if a device is working well; ***laboratory tests showed that the sample was positive; government officials have carried out tests on samples of drinking water; the committee has called for a ban on tests of nuclear weapons;*** **blood test** = laboratory test to find the chemical composition of a patient's blood; **nuclear test** = test on a nuclear weapon; **nuclear test ban** = ban on testing of nuclear weapons **2** *verb* to examine a sample to see if it is healthy *or* a device to see if it is working well; ***they sent the water sample away for testing; the government has resumed the programme of nuclear testing;*** **nuclear testing site** = place where tests are carried out on nuclear weapons

◇ **test tube** ['test 'tjuːb] *noun* small glass tube with a rounded bottom, used in laboratories to hold samples of liquids

testicle *or* **testis** ['testɪkl *or* 'testɪs] *noun* one of two male sex glands (NOTE: plural of **testis** is **testes**)

tetanus ['tetənəs] *noun* infection caused by *Clostridium tetani* in the soil, which affects the spinal cord and causes spasms which occur first in the jaw (also called lockjaw)

COMMENT: people who work on the land or with soil (such as farm workers or construction workers) should be immunized against tetanus

tetrachlorodibenzoparadioxin [tetrəklɔːrəʊdaɪbenzəʊpærədaɪ'ɒksɪn] *noun* dangerous gas made as a by-product of the manufacture of 2,4,5-T. (This was the gas which escaped during the Seveso disaster in 1976)

tetraethyl *or* **tetramethyl** [tetrə'eθɪl *or* tetrə'meθɪl] *noun* **tetraethyl lead** *or* **lead tetraethyl** (**$Pb(C_2H_5)_4$**) = additive added to petrol to prevent knocking

COMMENT: adding these lead compounds to petrol has the effect of releasing lead into the atmosphere as the petrol is burnt in the engine. Unleaded petrol is less polluting

texture ['tekstʃə] *noun* roughness *or* smoothness of a surface or of a substance; **coarse texture** *or* **rough texture** = roughness *or* having very large particles; **fine texture** = smoothness *or* having very small particles; **soil texture** = description of the type of particles that make up a specimen of soil

textured vegetable protein (TVP) ['tekstʃəd 'vedʒɪtəbl 'prəʊtiːn] *noun* substance made from processed soya beans *or* other vegetables, used as a substitute for meat

Th *chemical symbol for* thorium

thallium ['θæliəm] *noun* metallic element, which is poisonous and used in pesticides (NOTE: chemical symbol is **Tl**; atomic number is **81**)

thatch [θætʃ] *verb* to cover a roof with reeds *or* straw; ***reeds provide the best material for thatching***

thaw [θɔ:] **1** *noun* period when the weather becomes warmer after a heavy frost, and ice and snow melt **2** *verb* to melt something which is frozen; ***it is possible that rising atmospheric temperatures will make the polar ice caps thaw and raise the level of the sea***

◇ **thaw water** ['θɔ: 'wɔ:tə] *noun* water from melted snow and ice

theodolite [θi:'ɒdəlaɪt] *noun* instrument formed of a telescope mounted in such a way that it can rotate, used by surveyors for measuring angles

thermal ['θɜ:məl] **1** *adjective* referring to heat; **thermal efficiency** = ability of a process to heat efficiently; **thermal energy** = energy in the form of heat; **thermal pollution** = change in the quality of an environment by increasing its temperature (such as the heat from the cooling towers of a power station *or* from discharge of a coolant); **thermal radiation** = emission of radiant heat; **thermal reactor** = nuclear reactor which heats, using a moderator, as opposed to a breeder reactor; **thermal spring** = hot water running out of the earth continuously; **thermal stratification** = different layers of heat in a body of water; **British Thermal Unit (Btu** *or* **BTU)** = unit of measurement of heat, the amount of heat needed to heat one pound of water one degree Fahrenheit **2** *noun* current of rising hot air in the lower atmosphere

> COMMENT: the BTU is a non-metric unit, and in spite of its name is used in the US rather more than in the UK

thermo- ['θɜ:məʊ] *prefix* referring to (i) heat; (ii) temperature

◇ **thermocline** ['θɜ:məʊklaɪn] *noun* metalimnion, the middle layer of water in a lake, between the epilimnion and the hypolimnion, in which heat decreases rapidly as the depth increases

◇ **thermograph** ['θɜ:məʊgrɑ:f] *noun* instrument which records changes in temperature on a roll of paper

◇ **thermolysis** [θɜ:'mɒlɪsɪs] *noun* reduction in body temperature (as by sweating)

◇ **thermometer** [θə'mɒmɪtə] *noun* instrument for measuring temperature; **maximum-minimum thermometer** = thermometer which shows the highest and lowest temperatures reached since it was last checked, as well as the current temperature

◇ **thermometry** [θer'mɒmɪtri] *noun* science of measuring temperature

◇ **thermonasty** ['θɜ:məʊnæsti] *noun* response of plants to heat

◇ **thermonuclear energy** [θɜ:məʊ'nju:kliə 'enədʒi] *noun* energy produced by fusion of nuclei; **thermonuclear reaction** = fusion of nuclei, which produces huge amounts of energy (used in the hydrogen bomb)

◇ **thermoperiodic** [θɜ:məʊpi:ri'ɒdɪk] *adjective* (organism) which reacts to regular changes in temperature

◇ **thermoperiodicity** *or* **thermoperiodism** [θɜ:məʊpi:riə'dɪsɪti *or* θɜ:məʊ'pi:riədɪzm] *noun* effect on an organism of regular changes in temperature

◇ **thermophilic** [θɜ:məʊ'fɪlɪk] *adjective* (organism) which needs a high temperature to grow, such as the organisms which live in the hot water from thermal springs

◇ **thermoplastic** [θɜ:məʊ'plæstɪk] *adjective* (type of plastic) which can be recycled by heating and cooling

◇ **thermoregulation** [θɜ:məʊregju:'leɪʃn] *noun* control of body temperature (as by sweating, shivering, etc.)

◇ **thermosetting** ['θɜ:məʊsetɪŋ] *adjective* (type of plastic) which is heated while being shaped but which cannot be reheated for recycling

◇ **thermosphere** ['θɜ:məʊsfiə] *noun* zone of the earth's atmosphere above 80km from the surface of the earth, where the temperature increases with the altitude

therophyte ['θerəʊfaɪt] *noun* annual, a plant which grows from a seed in spring, flowers, sets seed and dies within one season, leaving the seed to remain dormant in the soil during the winter

thiamin(e) ['θaɪəmi:n] *noun* Vitamin B_1, vitamin found in yeast, cereals, liver and pork, lack of which can cause beriberi

Third World ['θɜ:d 'wɜ:ld] *noun* the developing countries of Africa, Asia and Latin America; ***a conference of Third-World leaders***

30 per cent club ['θɜ:ti pə'sent 'klʌb] *noun* group of nations who signed a UN agreement to reduce by 1993 emissions of sulphur by 30% as compared to the levels in 1980

thorax ['θɔ:ræks] *noun* (of an insect) middle section of the body, between the head and the abdomen; (of an animal) chest *or* cavity in the top part of the body above the abdomen, containing the diaphragm, heart and lungs, all surrounded by the ribcage

thorium ['θɔ:riəm] *noun* natural radioactive element which decomposes to radioactive radon gas (NOTE: chemical symbol is **Th;** atomic number is **90)**

thorn [θɔːn] *noun* sharp woody point on the stem of a plant; any plant *or* tree which has sharp woody points on its stems *or* branches; **thorn scrub** = area of land covered with bushes and small trees which have sharp woody points on their stems and branches; **thorn woodland** = area of land covered with trees which have sharp woody points on their stems and branches

◇ **thornbush** ['θɔːnbʊʃ] *noun* any shrub *or* bush which has sharp woody points on its stems

threadworm ['θredwɜːm] *noun* pinworm, a thin parasitic worm *Enterobius* which infests the large intestine

threat [θret] *noun* something dangerous which may harm; ***lead pollution is a threat to small babies***

◇ **threaten** ['θretən] *verb* to be a danger to something *or* be able to harm something; ***plant species growing in arid or semi-arid lands are threatened by the expansion of livestock herding; the plan for the motorway threatens to damage the ecology of the wood;*** **threatened species** = species which is not in as much danger of extinction as an endangered species, but which needs protection

Three Mile Island ['θriː 'maɪl 'aɪlənd] nuclear power station in Pennsylvania, USA, where an accident in 1979 almost caused the release of large quantities of radioactive substances

threonine ['θriːəʊniːn] *noun* essential amino acid

threshold ['θreʃhəʊld] *noun* (i) point below which the environment is not harmed by something; (ii) point below which a drug has no effect; (iii) point at which something is strong enough to be sensed by an instrument *or* by a sensory nerve; **audibility threshold** *or* **hearing threshold** = level of sound which a person can just hear; **she has a low hearing threshold** = she cannot hear some sounds which other people can hear; **critical threshold** = point below which something will no longer take place (especially, point at which a species is likely to become extinct); ***the population will rapidly reach a critical threshold if steps are not taken to protect if from extinction;*** **pain threshold** = point at which a person cannot bear pain without crying; **to have a low** *or* **a high pain threshold** = to be unable to bear much pain *or* to be able to bear a lot of pain without crying; **toxic threshold** *or* **toxicity threshold** = point at which a poison starts to have a noticeably harmful effect

thrive [θraɪv] *verb (of animal or plant)* to do well *or* to live and grow strongly; ***these plants thrive in very cold environments***

thrombin ['θrɒmbɪn] *noun* substance which coagulates blood

◇ **thromboplastin** ['θrɒmbəʊplæstɪn] *noun* substance which converts prothrombin into thrombin

throw [θrəʊ] *noun* amount of movement up *or* down of rocks at a fault line

thrush [θrʌʃ] *noun* infection of the mouth *or* vagina with the bacterium *Candida albicans*

thrust [θrʌst] *noun* force in the crust of the earth which squeezes and so produces folds; **thrust fault** = fault in which the upper layers of rock have been pushed forward over the lower layers

thunder ['θʌndə] *noun* loud sound generated by lightning in the atmosphere; ***the storm was accompanied by thunder and lightning***

◇ **thunderstorm** ['θʌndəstɔːm] *noun* storm with rain, thunder and lightning

> COMMENT: the distance of a storm from the person hearing the thunder can be calculated by counting the number of seconds between the lightning flash and the sound of the thunder. Divide this figure by three and you have the distance in kilometres to the centre of the storm

thyroid ['θaɪrɔɪd] **1** *adjective* referring to the thyroid gland; **thyroid hormone** = hormone produced by the thyroid gland **2** *noun & adjective* **thyroid (gland)** = endocrine gland in the neck

> COMMENT: the thyroid gland is activated by the pituitary gland and produces thyroxine, a hormone which regulates the body's metabolism. The thyroid gland needs a supply of iodine in order to produce thyroxine

◇ **thyroxine** [θaɪ'rɒksɪn] *noun* hormone produced by the thyroid gland which regulates the body's metabolism and conversion of food into heat

Ti *chemical symbol for* titanium

tick [tɪk] *noun* tiny parasite of the order Acarida, which sucks blood from the skin; **tick fever** = infectious disease transmitted by bites from ticks

tide [taɪd] *noun* regular rising and falling of the sea, in a twice-daily rhythm; **high tide** *or* **low tide** = points when the level of the sea is at its highest *or* lowest; **neap tide** = tide which occurs at the first and last quarters of the moon, when the difference between high and low water is less than normal; **spring tide** = tide which occurs at the new and full moon when the influence of the sun and moon act together and the difference between high and low water is more than normal; **red tide** = phenomenon where the sea becomes red, caused by

Rhodophyta, a type of very small algae *or* phytoplankton; **tide range** = difference between high water and low water

◇ **tidal** ['taɪdəl] *adjective* referring to the tide; **tidal current** = flow of water into *or* out of a bay, harbour *or* estuary; **tidal energy** *or* **tidal power** = energy obtained from the movement of the tides; electricity produced by turbines driven by the force of the tides; **tidal marsh** = marsh in which the water level rises and falls twice daily; **tidal power plant** *or* **tidal power station** = installation where electricity is produced by turbines driven by the force of the tides; **tidal range** = difference between high water and low water; **tidal river** = river in which the water level rises and falls twice daily; **tidal wave** = tsunami *or* wave caused by an earthquake under the sea, which moves rapidly across the surface of the sea and becomes very large when it hits the shore. Note that it, in fact, has no connection with the tides; **tidal zone** = area where the water level rises and falls twice daily

> COMMENT: the tides are caused by the gravitational pull of the moon, taken together with the centrifugal force of the earth as it rotates, which make the water on the surface of the earth move to high peaks at opposite sides of the earth and low troughs halfway between. Twice each month, at the new and full moon, the sun, moon and the earth are directly aligned, giving the highest gravitational pull and causing the spring tides

tilapia [tɪ'læpiə] *noun* tropical white fish, very suitable for growing in fish farms

till [tɪl] **1** *noun* boulder clay *or* clay soil mixed with rocks of different sizes, found in glacial deposits **2** *verb* to plough *or* to dig *or* to harrow the soil, to make it ready for the cultivation of crops

◇ **tillage** ['tɪlɪdʒ] *noun* husbandry *or* action of tilling the soil

tilth [tɪlθ] *noun* good light crumbling soil, suitable for growing plants; tilled soil; ***work the soil into a fine tilth before sowing seeds***

timber ['tɪmbə] *noun* trees which have been *or* which are to be felled and cut into logs

◇ **timberline** ['tɪmbəlaɪn] *noun* (i) line at a certain altitude, above which trees will not grow; (ii) line in the Northern *or* Southern hemisphere, north *or* south of which trees will not grow; ***the slopes above the timberline were covered with boulders, rocks and pebbles***

tin [tɪn] *noun* metallic element, used especially to form alloys (NOTE: chemical symbol is **Sn**; atomic number is **50**)

◇ **tin (can)** [tɪn 'kæn] *noun* metal container for food *or* drink, made of iron with a lining of tin *or* made entirely of aluminium (NOTE: US English is **can**)

tip [tɪp] **1** *noun* place where rubbish is thrown away; ***the siting of the council refuse tip has caused a lot of controversy*** **2** *verb* to throw away rubbish; **controlled tipping** = disposal of waste in special landfill sites, as opposed to throwing it away anywhere; *see also* FLY-TIPPING (NOTE: the US English is **sanitary landfill**)

tissue ['tɪʃu:] *noun* material made of cells, of which the parts of an animal's *or* plant's body are formed; ***most of an animal's body is made up of soft tissue, with the exception of the bones and cartilage;*** **adipose tissue** = body fat *or* tissue where the cells contain fat which replaces the normal fibrous tissue when too much food is eaten; **connective tissue** = tissue which forms the main part of bones and cartilage, ligaments and tendons, in which a large amount of fibrous material surrounds the tissue cells; **tissue culture** = live tissue grown in a culture in a laboratory; also a method of plant propagation which reproduces clones of the original plant on media containing plant hormones

titanium [taɪ'teɪniəm] *noun* light metallic element which does not corrode, used in aircraft manufacture (NOTE: chemical symbol is **Ti**; atomic number is **22**)

titration [taɪ'treɪʃn] *noun* process of measuring the strength of a solution

◇ **titre** ['taɪtə] *noun* measurement of the strength of a solution

Tl *chemical symbol for* thallium

toadstool ['təʊdstu:l] *noun* general name for a poisonous fungus which looks like a mushroom

tobacco [tə'bækəʊ] *noun* leaves of a plant which are dried and smoked, either in a pipe *or* as cigarettes *or* cigars

> COMMENT: tobacco contains nicotine, which is an addictive stimulant. This is why it is difficult for a person who smokes a lot of cigarettes to give up the habit. Nicotine can enter the bloodstream and cause poisoning; tobacco smoking also causes cancer, especially of the lungs and throat. Nicotine is a poisonous alkaloid obtained mainly from tobacco and is used in agriculture as an insecticide

toe = TONNES OF OIL EQUIVALENT

tolerance ['tɒlərəns] *noun* ability of the body to accept something *or* not to react to something; ***he has been taking the drug for so long that he has developed a tolerance to it;*** **drug tolerance** = condition where a drug has been given to a patient for so long that his body no longer reacts to it and the dosage has to be increased; **tolerance dose** = amount of radiation which can be given without harm in radiotherapy

◇ **tolerant** ['tɒlərənt] *adjective & suffix* which accepts something *or* does not react to something; ***a salt-tolerant plant***

◇ **tolerate** ['tɒləreɪt] *verb* to accept something *or* not to react to something

◇ **toleration** [tɒlə'reɪʃn] *noun* accepting something *or* not reacting to something; **toleration level** = level below which an organism will accept something *or* not react to something; ***waste hot water from power stations can kill freshwater fish by raising the water temperature above toleration levels***

ton [tʌn] *noun* unit of measurement of weight; *GB* **(long) ton** = unit of measurement of weight (= 1,016 kilograms); **(metric) ton** = tonne *or* unit of measurement of weight (= 1,000 kilograms); *US* **(short) ton** = unit of measurement of weight (= 907 kilograms)

◇ **tonne** [tʌn] *noun* metric ton (= 1,000 kilograms); **tonnes of oil equivalent (toe)** = unit of measurement of the energy content of a fuel, calculated by comparing its heat energy with that of oil

QUOTE: at present there are millions of tonnes of fly-tipped material which local authorities cannot afford to have removed
London Environment Alert!

QUOTE: Saudi Arabia has become not only self-sufficient in wheat, but also exports substantial quantities. Total production of wheat rose to 2.6m tonnes in 1987, 3m tonnes in 1988, and an estimated 3.2m tonnes in 1989. This compares with the annual Saudi consumption of about 800,000 tonnes
Middle East Agribusiness

toothed [tu:θt] *adjective* (animal) with teeth; **toothed whales** = Odontoceti *or* whales, including sperm whales, killer whales, porpoises and dolphins, which have teeth

topical ['tɒpɪkl] *adjective* referring to one particular part of the body; **topical drug** = drug which is applied to one part of the body only

◇ **topically** ['tɒpɪkli] *adverb* (applied) to one part of the body only; ***the drug is applied topically***

topography [tə'pɒgrəfi] *noun* study of the physical features of a geographical area

◇ **topographic(al)** [tɒpəʊ'græfɪkl] *adjective* referring to topography

topotype ['tɒpəʊtaɪp] *noun* population which has become different from other populations of a species because of adaptation to local geographical features

topset bed ['tɒpset 'bed] *noun* layer of fine-grained sediment in a delta

topsoil ['tɒpsɔɪl] *noun* top layer of soil, often containing organic material, disturbed in farming, from which chemical substances are leached into the subsoil below

tor [tɔ:] *noun* pile of blocks or rounded rocks found on summits and hillsides

tornado [tɔ:'neɪdəʊ] *noun* violent storm with a column of rapidly turning air at the centre of an area of very low pressure, giving very high winds and causing damage to buildings

COMMENT: tornadoes are formed by rising air currents associated with large cumulonimbus clouds. They rotate anticlockwise in the Northern Hemisphere and clockwise in the Southern. Passing over the sea they pick up water and become waterspouts

torrent ['tɒrənt] *noun* violent rapidly flowing stream of water *or* lava; violently heavy rainfall

◇ **torrential** [tə'renʃl] *noun* like a torrent; ***the storm brought a torrential downpour of rain***

total ['təʊtəl] *adjective* complete *or* covering everything; ***the total world population of the animal is no more than four or five hundred pairs;*** **total allowable catch (TAC)** = maximum amount of fish which can be caught over a period of time if their numbers are to be sustainable by reproduction; **total eclipse** = eclipse when the whole of the sun *or* moon is hidden

town [taʊn] *noun* collection of buildings *or* urban area, smaller than a city and larger than a village; **boom town** = town where the population and number of buildings is growing very quickly; town where business is thriving; ***the oilmen have turned this once sleepy port into a boom town;*** **town hall** = building in a town where the administrative offices of the town are situated and where the local council has its meetings; **town planner** = person who supervises the design of a town *or* the way the streets and buildings in a town are laid out and developed; **town planning** = designing a town *or* city, including the way the streets and buildings are laid out and developed

◇ **townscape** ['taʊnskeɪp] *noun* appearance of a town, the way in which the streets and buildings are laid out

◇ **township** ['taʊnʃɪp] *noun (in the US and Canada)* administrative area which is a subdivision of a county; *(in South Africa)* **Black township** *or* **White township** = area of a large town *or* city inhabited by Blacks *or* Whites

tox- *or* **toxo-** ['tɒksəʊ] *prefix* meaning poison

◇ **toxic** ['tɒksɪk] *adjective* poisonous *or* harmful; **toxic agent** = substance which is poisonous *or* harmful; **toxic substance** = substance, such as lead, which is poisonous *or* harmful to living organisms; **toxic waste** = waste which is poisonous *or* harmful

◇ **toxicity** [tɒk'sɪsɪti] *noun* level to which a substance is poisonous *or* harmful; amount of poisonous *or* harmful material in a substance; ***scientists are measuring the toxicity of car exhaust fumes; certain gases are dangerous because of their toxicity to humans;*** **acute toxicity** = level of concentration of a toxic substance which makes people seriously ill *or* or can even cause death; **chronic toxicity** = exposure to harmful levels of a toxic substance over a long period of time

◇ **toxico-** ['tɒksɪkəʊ] *prefix* meaning poison

◇ **toxicological** [tɒksɪkə'lɒdʒɪkl] *adjective* referring to toxicology; ***irradiated food presents no toxicological hazard to humans***

◇ **toxicologist** [tɒksɪ'kɒlədʒɪst] *noun* scientist who specializes in the study of poisons

◇ **toxicology** [tɒksɪ'kɒlədʒɪ] *noun* scientific study of poisons and their effects on the human body

◇ **toxicosis** [tɒksɪ'kəʊsɪs] *noun* poisoning

◇ **toxin** ['tɒksɪn] *noun* poisonous substance from a plant *or* animal

◇ **toxoid** ['tɒksɔɪd] *noun* toxin which has been treated and is no longer poisonous, but which can still provoke the formation of antibodies

> COMMENT: toxoids are used as vaccines and are injected into a patient to give immunity against a disease

TPO = TREE PRESERVATION ORDER

trace [treɪs] *noun* very small amount; ***there are traces of radioactivity in the blood sample;*** **trace element** = element which is essential to organic growth but only in very small quantities; **trace gases** = gases (such as xenon *or* helium) which exist in the atmosphere in very small quantities; **trace metal** = metal which is essential to organic growth but only in very small quantities

> COMMENT: plants require traces of copper, iron, manganese, zinc; human beings require the trace elements chromium, cobalt, copper, magnesium, manganese, molybdenum, selenium and zinc

◇ **tracer** ['treɪsə] *noun* foreign substance, usually radioactive, that can be attached to material so that the course of that material (for example, round the body) can be followed

tract [trækt] *noun* **(a)** part of an internal organ *or* system; ***an infection of the upper respiratory tract;*** **digestive tract** = arrangement of tubes and organs, including the intestine, salivary glands, liver, etc., through which food passes and is digested **(b)** large area, especially of land; ***large tracts of forest have been destroyed by fire***

tractor ['træktə] *noun* heavy vehicle, with large wheels, which provides the main source of power in modern farms

> COMMENT: The general purpose tractor does most of the work on arable and livestock farms and may be either two- or four-wheel drive. Lighter tractors, usually two-wheel drive models, are used by market gardeners. More powerful four-wheel drive tractors are needed for ploughing and heavy cultivation

trade wind ['treɪd 'wɪnd] *noun* wind which blows towards the equator, from the northeast in the Northern Hemisphere and from the southeast in the Southern Hemisphere; **trade-wind cumulus** = cumulus clouds usually associated with the trade winds

traditional [trə'dɪʃnəl] *adjective* which has always been done in the same way; ***traditional technologies exist which meet basic subsistence needs; the traditional system of agriculture has been revolutionized by the application of modern technology***

traffic ['træfɪk] *noun* vehicles coming and going on a road, planes in the air, ships at sea, etc.; **traffic-calming measure** = means of reducing the number and speed of motor vehicles using a road (such as building a ring road, imposing speed limits, etc.); **traffic congestion** *or* **traffic jam** = blockage caused by too many motor vehicles on a road

trail [treɪl] *noun* (i) path *or* track; (ii) mark *or* scent left by an animal; **nature trail** = path through the countryside with signs to draw attention to important and interesting features

trait [treɪ] *noun* characteristic which is particular to an organism

trans- [trænz] *prefix* meaning through *or* across

◇ **transalpine** [trænz'ælpaɪn] *adjective* on *or* to the other side of the Alps; crossing the Alps; ***all transalpine roads are closed***

◇ **transboundary pollution** *or* **transfrontier pollution** [trænz'baʊndri *or* trænzfrʌn'tiə pə'lu:ʃn] *noun* airborne *or* waterborne pollution produced in one country, which crosses to another

transect ['trænsekt] *noun* line used in ecological surveys to provide a way of measuring and showing the distribution of organisms

transfer ['trænsfə] **1** *noun* passing from one place to another; **heat transfer** = passing heat from one medium *or* substance to another; **population transfer** = movement of people *or* organisms from one place to another; **technology transfer** = passing technology from a developed country to a developing one **2** *verb* to pass from one place to another

> QUOTE: we have warned against making the assumption that technology can be transferred from one country to another without testing and adaptation
>
> **Appropriate Technology**

transformation [trænsfə'meɪʃn] *noun* change in structure *or* appearance

transhumance [træns'hju:məns] *noun* practice of moving flocks and herds up to summer pastures and bringing them down to the valley again in winter

transition [træn'zɪʃn] *noun* change from one state to another; **demographic transition** = pattern of change of population growth from high birth and death rates to low birth and death rates; **transition phase** = period when something is changing from one state to another

translocate [trænslə'keɪt] *verb* to move substances through the tissues of a plant; **translocated herbicide** = herbicide which kills a plant after being absorbed through its leaves

◊ **translocation** [trænslə'keɪʃn] *noun* movement of substances through the tissues of a plant

transmit [trænz'mɪt] *verb* to pass on (a disease) to another animal or plant; ***some diseases are transmitted by insect bites***

transpire [træn'spaɪə] *verb (of plant)* to lose water through stomata; ***in tropical rainforests, up to 75% of rainfall will evaporate and transpire into the atmosphere***

◊ **transpiration** [trænspɪ'reɪʃn] *noun* loss of water from a plant through its stomata

> COMMENT: transpiration accounts for a large amount of water vapour in the atmosphere. A tropical rainforest will transpire more water per square kilometre than is evaporated from the same area of sea. Clearance of forest has the effect of reducing transpiration, with the accompanying change in climate: less rain, leading to eventual desertification

transplant 1 *noun* ['trænsplɑ:nt] **(a)** taking a growing plant from one place and planting it in the soil in another place **(b)** taking an organ *or* tissue from one person and putting it into *or* onto another **2** *verb* [træns'plɑ:nt] **(a)** to take a growing plant from one place and plant it in the soil in another place **(b)** to take an organ *or* tissue from one person and put it into *or* onto another

transport ['trænspɔ:t] *noun* system of motor vehicles, trains, aircraft, trams, boats, etc. for moving people and goods around; **public transport** = system of buses, trains, aircraft, trams, boats, etc. which all people may use to travel on; **road transport** = moving goods and people around in motor vehicles; **transport system** = arrangement for moving people and goods around

◊ **transportation** [trænzpɔ:'teɪʃn] *noun* = TRANSPORT

transuranic element [trænzju'rænɪk 'elɪmənt] *noun* artificial radioactive element which is beyond uranium in the periodic table

> COMMENT: the transuranic elements have higher atomic numbers than uranium (atomic number 92) and, apart from neptunium and plutonium, do not occur naturally but are formed from uranium in nuclear reactions (such as americium, atomic number 95)

trash [træʃ] *noun US* rubbish *or* household waste

◊ **trash can** ['træʃ 'kæn] *noun US* container into which household rubbish is placed to be collected by municipal refuse collectors (NOTE: British English is **dustbin)**

treat [tri:t] *verb* to process sewage

◊ **treatment** ['tri:tmənt] *noun* processing of sewage; **biological treatment** = processing sewage by allowing bacteria to break up the organic matter; **chemical treatment** = processing sewage by adding chemical substances to it; **mechanical treatment** = processing sewage by mechanical means, such as agitating, stirring, etc.; **sewage treatment plant** = place where sewage from houses and other buildings is brought for processing; **waste water treatment plant** = place where liquid waste is processed

tree [tri:] *noun* very large plant with a wooden stem; ***ecologists are worried that the cutting of forest trees in many parts of the world will bring about desertification and change global climate patterns; tree ring records from high altitudes show sharply reduced annual growth;*** **tree cover** = number of trees growing on a certain area of land; **tree fern** = very large fern found in Australasia, which grows like a tree with a single thick stem; **tree preservation order (TPO)** = order from a local government department which stops a tree from being felled; **tree ring** = ring of new wood formed each year in the trunk of a tree and which can easily be seen when the tree is cut down; **tree savanna(h)** = dry grass-covered plain with some trees

◇ **treeline** ['tri:laɪn] *noun* (i) line at a certain altitude, above which trees will not grow; (ii) line in the Northern *or* Southern hemisphere, north *or* south of which trees will not grow; ***the slopes above the treeline were covered with boulders, rocks and pebbles***

> QUOTE: about 10 per cent of Britain is covered by trees
> **Times**

tremor ['tremə] *noun* slight movement; **earth tremor** = slight earthquake

trench [trenʃ] *noun* **(a)** long narrow hole in the ground; **drainage trench** = long hole cut in the ground to allow water to run away **(b)** **oceanic trench** = long deep valley in the floor of the ocean, where two tectonic plates meet, usually associated with volcanic activity

trend [trend] *noun* gradual development; ***there is a trend towards organic farming and away from using chemical fertilizers***

> QUOTE: computers enable ecologists to study their recorded observations and indicate links between data showing trends which can be statistically explained
> **Environment Now**

tributary ['trɪbjʊtəri] *noun* stream *or* river flowing into a larger river

trichlorophenoxyacetic acid [traɪklɔ:rəʊfenɒksɪə'setɪk 'æsɪd] *noun* 2,4,5-T *or* herbicide which forms dioxin as a by-product during the manufacturing process. (This was the gas which escaped during the Seveso disaster in 1976)

trickle ['trɪkl] *verb (of liquid)* to flow gently; **trickle system** = irrigation system where water is brought to the base of each plant and drips slowly into the soil; **trickling filter** = filter bed through which liquid sewage is passed to purify it

trigger ['trɪgə] *verb* to start something happening; ***it is not known what triggered the avalanche; the plague of locusts was triggered by unusually warm weather***

trioxide [traɪ'ɒksaɪd] *noun* any oxide containing three oxygen atoms per molecule; **sulphur trioxide** = gas (SO_3) formed when fuels are burnt, usually present in sulphur oxide pollution and forming sulphuric acid when dissolved in water

tritium ['trɪtiəm] *noun* rare isotope of hydrogen

trivial ['trɪviəl] *adjective* not important

> QUOTE: the radioactivity in the waste will become trivial after hundreds of years
> **New Scientist**

troph(o)- ['trɒfəʊ] *prefix* referring to food *or* nutrition

◇ **trophic** ['trɒfɪk] *adjective* referring to nutrition; **trophic chain** = food chain *or* series of organisms which pass energy from one to another as each provides food for the next. The first organism in the food chain is the producer and the rest are consumers; **trophic level** = one of the levels in a food chain; **trophic structure** = structure of an ecosystem, shown by food chains and food webs; *see also* DYSTROPHIC, EUTROPHIC, MESOTROPHIC, OLIGOTROPHIC

> COMMENT: the three trophic levels are: producers (organisms, such as plants, which take energy from the sun or the environment and convert it into matter); primary consumers (organisms, such as herbivores, which eat producers); secondary consumers (organisms, such as carnivores, which eat other consumers)

-trophy ['trɒfi] *suffix* meaning (i) nourishment; (ii) development of an organ; **dystrophy** = wasting of an organ *or* muscle *or* tissue due to lack of nutrients in that part of he body

-tropic ['trɒpɪk] *suffix* meaning (i) turning towards; (ii) which influences; **heliotropic plant** = plant which grows *or* turns towards the light

tropic ['trɒpɪk] *noun* **Tropic of Cancer** = parallel running round the earth at latitude 23°28N; **Tropic of Capricorn** = parallel running round the earth at latitude 23°28S; **the tropics** = region between the Tropic of Cancer and the Tropic of Capricorn *or* tropical countries *or* hot areas of the world; ***he lives in the tropics; disease which is endemic in the tropics***

◇ **tropical** ['trɒpɪkl] *adjective* referring to the tropics; ***the disease is carried by a tropical insect;*** **tropical air** = mass of air which originates in the tropics; **tropical climate** = type of climate found in the tropics and characterized by very high temperatures; **tropical disease** = disease which is found in tropical countries, such as malaria, dengue, Lassa fever; **tropical medicine** = branch of medicine which deals with tropical diseases; **tropical rainforest** = biome where almost constant rain and high temperature permit plants to grow throughout the year; **tropical storm** = violent storm occurring in the tropics; *see also* CYCLONE, HURRICANE, TYPHOON; **tropical zone** = region between the Tropic of Cancer and the Tropic of Capricorn

-tropism ['trɒpɪzm] *suffix* meaning the action of turning towards a stimulus; **phototropism** = turning *or* growing towards a stimulus of light, found in certain plants

(NOTE: opposed to **-nasty** which refers to a response without turning)

tropopause ['trɒpəpɔ:z] *noun* layer of the atmosphere between the troposphere and the stratosphere

troposphere ['trɒpəsfiə] *noun* lower layer of the earth's atmosphere, immediately above the ground, rising to about 18km above the earth's surface at the equator and 8km at the North and South Poles

◇ **tropospheric** [trɒpə'sferɪk] *adjective* referring to the troposphere

COMMENT: the troposphere is formed largely of nitrogen (78%) and oxygen (21%), plus argon and some trace gases. The temperature in the troposphere falls about 6.5°C per thousand metres of altitude; the temperature at 16km being about minus 55°C. The troposphere contains almost all the atmosphere's water vapour

trough [trɒf] *noun* **(a)** low point; ***the graph shows the peaks and troughs of pollution over the seasons*** **(b)** long narrow area of low pressure with cold air in it, leading away from the centre of a depression (NOTE: the opposite is **ridge**)

true [tru:] *adjective* correct; **to breed true** = to reproduce all the characteristics of the species; ***F_1 hybrids do not breed true;*** **true north** = direction along any line of longitude towards the North Pole (as opposed to magnetic north)

trunk [trʌŋk] *noun* **(a)** main part of a tree **(b)** **trunk road** = main road, especially one that may be used by large trucks; **trunk sewer** = main sewer pipe into which smaller sewers flow

trypanosome [[trɪ'pænəsəʊm] *noun* parasite which causes sleeping sickness, transmitted by the tsetse fly

◇ **trypanosomiasis** [trɪpænəsəʊ'maɪəsɪs] *noun* sleeping sickness, a serious disease attacking both livestock and humans, spread by the tsetse fly

tryptophan ['trɪptəʊfæn] *noun* essential amino acid

tsetse fly ['tsetsi: 'flaɪ] *noun* African insect which passes trypanosomes into the bloodstream of humans and other animals, causing sleeping sickness

tsunami [tsu'nɑ:mi] *noun* tidal wave *or* wave caused by an earthquake under the sea, which moves rapidly across the surface of the sea and becomes very large when it hits the shore (NOTE: although commonly called a tidal wave it, in fact, has no connection with the tides)

tsutsugamushi disease [tsu:tsəgə'mu:ʃi dɪ'zi:z] *noun* scrub typhus, a severe form of typhus found in Southeast Asia, caused by the Rickettsia bacterium and transmitted to humans by mites

tuber ['tju:bə] *noun* fat part of an underground stem *or* root, which holds nutrients and which has buds from which shoots develop; ***a potato is the tuber of a potato plant***

◇ **tuberous** ['tju:bərəs] *adjective* (i) like a tuber; (ii) (plant) which grows from a tuber

tufa ['tju:fə] *noun* form of calcareous deposit found near hot springs

tundra ['tʌndrə] *noun* cold treeless arctic and alpine region which may be covered with low shrubs, grasses, mosses and lichens

turbid ['tɜ:bɪd] *adjective* (liquid) which is cloudy because of particles suspended in it

◇ **turbidity** [tɜ:'bɪdɪti] *noun (of liquid)* cloudiness

turbine ['tɜ:baɪn] *noun* mechanical device which converts moving liquid *or* steam *or* air into energy by turning a generator; **gas turbine** = internal combustion engine where expanding gases from combustion chambers drive a turbine; a rotary compressor driven by the turbine sucks in the air used for combustion; **steam turbine** *or* **wind turbine** = turbine driven by steam *or* wind

COMMENT: water turbines create electricity from water power. Water is channelled from a reservoir through pipes which turn the vanes of the turbine, which then turn the rotor of the generator. The main types of water turbine are: axial-flow turbine, with blades like those on a ship's propeller, rotating horizontally; the impulse turbine, where jets of water are directed at bucket-shaped blades which catch the water; reaction turbine, where the blades on the turbine adjust to the angle at which the jets of water hit them. Pumped storage turbines act as generators when water pressure is high, but become water pumps when pressure is low, pumping water back into the reservoir

turbulence ['tɜ:bjʊləns] *noun* rushing *or* uneven movement of air currents

◇ **turbulent** ['tɜ:bjʊlənt] *adjective* rushing *or* uneven (air current)

TVP = TEXTURED VEGETABLE PROTEIN

TW = TERAWATT

twin [twɪn] *noun* one of two babies *or* animals born at the same time from two ova fertilized at the same time *or* from one ovum; *see also* DIZYGOTIC, MONOZYGOTIC

twister ['twɪstə] *noun US (informal)* tornado

twitcher ['twɪtʃə] *noun (informal)* birdwatcher who will travel anywhere to see a rare bird

2,4-D ['tu:fɔ: 'di:] selective translocated hormone herbicide especially effective against broad-leaved weeds growing in cereals

2,4,5-T ['tu:fɔ:faɪv 'ti:] *noun* trichlorophenoxyacetic acid, a herbicide which forms dioxin as a by-product during the manufacturing process. (This was the gas which escaped during the Seveso disaster in 1976)

typhoid fever *or* **typhoid** ['taɪfɔɪd] *noun* infection of the intestine caused by *Salmonella* in food or water, which causes fever and diarrhoea and may be fatal

typhoon ['taɪfu:n] *noun* name for a tropical cyclone in the Far East

typhus ['taɪfəs] *noun* one of several fevers caused by the Rickettsia bacterium, making the patient very weak; *see also* SCRUB TYPHUS

tyramine ['taɪrəmi:n] *noun* enzyme found in cheese, beans, tinned fish, red wine and yeast extract, which can cause high blood pressure if found in excessive quantities in the brain

tyrosine ['taɪrəʊsi:n] *noun* amino acid in protein which is a component of thyroxine

Uu

U *chemical symbol for* uranium

UKAEA = UNITED KINGDOM ATOMIC ENERGY AUTHORITY

ultra- ['ʌltrə] *prefix* meaning (i) further than; (ii) extremely

◇ **ultrabasic** [ʌltrə'beɪsɪk] *adjective* (rock) which has less silica and more magnesium than basic rock

◇ **ultramicroscopic** [ʌltrəmaɪkrə'skɒpɪk] *adjective* too small to be seen with an ordinary microscope

◇ **ultrananoplankton** [ʌltənænəʊ'plæŋktən] *noun* very small plankton

◇ **ultraplankton** [ʌltrə'plæŋktən] *noun* very small plankton

◇ **ultrasonic** [ʌltrə'sɒnɪk] *adjective* referring to ultrasound

◇ **ultrasonics** [ʌltrə'sɒnɪks] *noun* study of ultrasound

◇ **ultrasound** *or* **ultrasonic waves** ['ʌltrəsaʊnd *or* ʌltrə'sɒnɪk 'weɪvz] *noun* very high-frequency sound waves; ***the nature of the tissue may be made clear on ultrasound examination***

> COMMENT: the very high-frequency waves of ultrasound can be used to detect objects. Bats and dolphins use ultrasonic waves to find their way and ultrasound can be used in medicine to locate and record organs *or* growths inside the body (in a similar way to the use of X-rays) by recording the differences in echoes sent back from different tissues

ultraviolet rays (UV rays) [ʌltrə'vaɪələt 'reɪz] *noun* short invisible rays, beyond the violet end of the colour spectrum, which form the tanning and burning element in sunlight; **ultraviolet lamp** = lamp which gives off ultraviolet rays which tan the skin, help the skin produce Vitamin D and kill bacteria

> COMMENT: UV rays form part of the high-energy radiation which the earth receives from the sun. UV rays are classified as UVA and UVB rays. UVB rays form only a small part of radiation from the sun but they are dangerous and can cause skin cancer if a person is exposed to them for long periods. The effect of UVB rays is reduced by the ozone layer in the stratosphere

UN = UNITED NATIONS

un- [ʌn] *prefix* meaning reversing an action *or* state

◇ **unblock** [ʌn'blɒk] *verb* to remove something which is blocking

◇ **unboiled** [ʌn'bɔɪld] *adjective* which has not been boiled; ***in some areas, it is dangerous to drink unboiled water***

UNCED = UNITED NATIONS CONFERENCE ON ENVIRONMENT AND DEVELOPMENT

unconfined [ʌnkən'faɪnd] *adjective* (aquifer, ground water) of which the upper surface is at ground level

◇ **uncontaminated** [ʌnkən'tæmɪneɪtɪd] *adjective* which has not been contaminated

◇ **uncontrollable** [ʌnkən'trəʊləbl] *adjective* which cannot be controlled

◇ **uncontrolled** [ʌnkən'trəʊld] *adjective* which is not controlled; **uncontrolled dumping** = throwing away waste anywhere; **uncontrolled dumpsite** = place where waste is left on the ground and not buried in a hole; **uncontrolled fire** = fire which has ignited accidentally

◇ **uncoordinated** [ʌnkəʊ'ɔːdɪneɪtɪd] *adjective* not joined together *or* not working together

◇ **uncultivated** [ʌn'kʌltɪveɪtɪd] *adjective* (land) which is not cultivated

under- ['ʌndə] *prefix* meaning less than *or* not as strong as; **undernourished** = having too little food; **underproduction** = producing less than normal

◇ **underdeveloped country** [ʌndədɪ'veləpd 'kʌntri] *noun* country which has not been industrialized

underground [ʌndə'graʊnd] *adjective & adverb* beneath the surface of the ground; ***foxes live in underground holes; an underground nuclear test was carried out at the desert test site; worms live all their life underground; if power cables were placed underground they would be less of an eyesore;*** **underground water** = water in porous rocks underground

undergrowth ['ʌndəgrəʊθ] *noun* shrubs and other plants growing under large trees

undershot wheel [ʌndə'ʃɒt 'wiːl] *noun* type of waterwheel where the wheel rests in the flow of water which passes underneath it and makes it turn. (It is not as efficient as the overshot wheel where the water falls on the wheel from above)

understorey ['ʌndəstɔːri] *noun* lowest layer of small trees and shrubs in a wood, below the canopy

> QUOTE: beech is mainly in the South Island. It is a fairly open forest, though its understorey is often a complex association of young beech, ferns, shrubs and smaller tree species
>
> **Ecologist**

underwood ['ʌndəwʊd] *noun* small trees in a wood, below the canopy

UNEP = UNITED NATIONS ENVIRONMENT PROGRAMME

Ungulata *or* **ungulates** [ʌŋgjuː'lɑːtə *or* 'ʌʊŋgjuːleɪts] *noun* grazing animals, such as cattle, which have hooves

uni- ['juːni] *prefix* meaning one

◇ **unicellular** [juːnɪ'seljʊlə] *adjective* (organism) formed of one cell

UNICEF ['juːnɪsef] = UNITED NATIONS CHILDREN'S FUND

unit ['juːnɪt] *noun* (i) single part (as of a series of numbers); (ii) basic standard of measurement; **SI units** = international system of units for measuring physical properties (such as weight *or* speed *or* light, etc.); ***lumen is the SI unit of light per second***

United Kingdom Atomic Energy Authority (UKAEA) *noun* official organization in the United Kingdom responsible for all aspects of atomic energy, both commercially and in the research field

United Nations (UN) [juː'naɪtɪd 'neɪʃnz] *noun* international organization, formed in 1945 to promote international cooperation and peace

◇ **United Nations Children's Fund (UNICEF)** *noun* agency of the United Nations providing money for education and health programmes aimed at children and mothers

◇ **United Nations Conference on Environment and Development (UNCED)** *noun* meeting held in Rio de Janeiro, Brazil, in June 1992, to discuss climate change, biological diversity and many other environmental problems; *see also* EARTH SUMMIT

◇ **United Nations Environment Programme (UNEP)** *noun* programme set up by the United Nations in 1972 to monitor all developments in the global environment, to promote good management of the global environment and to safeguard the global environment for future generations

◇ **United Nations Organization (UNO)** *noun* = UNITED NATIONS

unleaded petrol [ʌn'ledɪd 'petrəl] *noun* petrol with a low octane rating, which has no lead additives (such as tetraethyl lead) and therefore creates less lead pollution in the atmosphere

unlined [ʌn'laɪnd] *adjective* (landfill site) with no lining, so that waste liquids can leak out into the surrounding soil

unnatural [ʌn'nætʃrəl] *adjective* not normal; contrary to nature

unneutralized [ʌn'njuːtrəlaɪzd] *adjective* which has not been made neutral

◇ **unpasteurized** [ʌn'pɑːstʃəraɪzd] *adjective* which has not been pasteurized; ***unpasteurized milk can carry bacilli***

◇ **unpolluted** [ʌnpə'luːtɪd] *adjective* which has not been polluted; ***unpolluted atmosphere in the mountain areas***

◇ **unsaturated fat** [ʌn'sætʃəreɪtɪd 'fæt] *noun* fat which does not have a large amount of hydrogen and so can be broken down more easily; *see also* FAT, SATURATED

◇ **unsettled** [ʌn'setld] *adjective* (weather) which changes frequently from rainy to fine and back again

◇ **unsightly** [ʌn'saɪtli] *adjective* not pleasant to look at; ***the company is proposing to run a line of unsightly pylons across the moors***

◇ **unspoilt** [ʌn'spɔɪlt] *adjective* (landscape) which has not been ruined by development; ***the highland region is still unspoilt; the conservancy council is hoping to preserve the area of unspoilt woodland***

◇ **unstable** [ʌn'steɪbl] *adjective* not stable *or* which may change easily; **unstable air mass** = mass of air in which a sample of wet air, in rising, cools less rapidly than the surrounding air and thus continues to rise until ultimately condensation and precipitation of the water content occur

◇ **unsterilized** [ʌn'sterɪlaɪzd] *adjective* which has not been sterilized; ***the milk was sold in unsterilized bottles***

◇ **unsustainable** [ʌnsəs'teɪnəbl] *adjective* (development, process, etc.) which depletes *or* damages natural resources irreparably and which does not leave the environment in good order for future generations

◇ **untapped** [ʌn'tæpt] *adjective* which has not yet been tapped; ***untapped mineral resources***

◇ **untreated** [ʌn'tri:tɪd] *adjective* which has not been treated; ***untreated sewage leaked into the river***

> QUOTE: some 99.8% of logging is unsustainable
> **Green Magazine**

UNO ['ju:nəʊ] = UNITED NATIONS ORGANIZATION

updraught ['ʌpdrɑ:ft] *noun* rising air current (usually of warm air)

upfreezing ['ʌpfri:zɪŋ] *noun US* phenomenon where loose rocks rise to the surface of the soil in spring, caused by the freezing of the soil in winter

upland ['ʌplænd] *noun* area of high land; ***the uplands*** *or* ***upland areas have different ecosystems from the lowlands*** (NOTE: opposite is **lowland**)

> QUOTE: fallout from the Chernobyl accident contaminated upland areas with radioactive caesium
> **New Scientist**

upstream [ʌp'stri:m] *adverb & adjective* towards the source of a river; ***the river is contaminated for several miles upstream from the estuary; pollution has spread into the lake upstream of the waterfall; downstream communities have not yet been affected***

upwelling [ʌp'welɪŋ] *noun* process by which warmer surface water in the sea is drawn away from the shore and replaced by colder water from beneath the surface

uranium [ju'reɪniəm] *noun* natural radioactive metallic element which is an essential fuel for nuclear power (NOTE: chemical symbol is **U**; atomic number is **92**)

> COMMENT: three uranium isotopes are found in ores: uranium-234, uranium-235 and uranium-238. Of these, U-235 is the only fissionable isotope occurring in nature and so is an essential fuel for reactors

urban ['ɜ:bən] *adjective* referring to towns; **urban decay** *or* **urban erosion** = condition where part of a city *or* town becomes old *or* dirty *or* ruined, because businesses and wealthy families have moved away from it; **urban fringe** = land at the edge of a city *or* town; **urban redevelopment** *or* **urban renewal** = rebuilding old parts of a city *or* town to create new houses, new shops, offices and factories; **urban sprawl** = mass of houses built without any particular plan in a disorderly way

◇ **urbanization** [ɜ:bənaɪ'zeɪʃn] *noun* movement of people from the countryside to the city, from small settlements to larger ones; making the countryside more like a city *or* town, with buildings and industries

urea [ju:'riə] *noun* substance produced in the liver from excess amino acids and excreted by the kidneys into the urine

uric acid ['ju:rɪk 'æsɪd] *noun* chemical compound which is formed from nitrogen in waste products from an animal's body

urine ['jʊərɪn] *noun* liquid containing uric acid, secreted as waste from an animal's body

user ['ju:zə] *noun* person who uses something; **road user** = person who drives on the road

utility [ju:'tɪləti] *noun* company that organizes an essential public service, such as providing electricity, gas *or* public transport; **public utility** = essential service such as electricity, gas, water, telephone, railway, etc. which is available to all people in general

UV ['ju:'vi:] = ULTRAVIOLET

Vv

vaccinate ['væksɪneɪt] *verb* to give a person a vaccine which provides immunity to a specific disease; ***she was vaccinated against smallpox as a child*** (NOTE: you vaccinate someone **against** a disease)

◇ **vaccination** [væksɪ'neɪʃn] *noun* action of vaccinating

◇ **vaccine** ['væksiːn] *noun* substance which contains the germs of a disease and which provides immunity to that disease when someone is vaccinated with it; ***the hospital is waiting for a new batch of vaccine to come from the laboratory; new vaccines are being developed all the time***

◇ **vaccinia** [væk'sɪniə] *noun* cowpox *or* infectious disease of cattle, used to prepare vaccine to vaccinate against smallpox (NOTE: Originally the words **vaccine** and **vaccination** applied only to smallpox immunization, but they are now used for immunization against any disease)

> COMMENT: a vaccine contains antigens (the germs of the disease), sometimes alive and sometimes dead, and this is introduced into the patient so that his body will develop immunity to the disease. The antigens provoke the body to produce antibodies, some of which remain in the bloodstream for a very long time and react against the same antigens if they enter the body naturally at a later date when the patient is exposed to the disease. Vaccination is mainly given against cholera, diphtheria, rabies, smallpox, tuberculosis and typhoid fever

vadose [væ'dəʊz] *adjective* (area) which lies between the surface of the ground and the water table

vagrant ['veɪgrənt] *noun* bird which only visits a country occasionally

vale [veɪl] *noun* valley

valine ['veɪliːn] *noun* essential amino acid

valley ['væli] *noun* long low area, usually with a river at the bottom, between hills *or* mountains; ***fog forms in the valleys at night;*** **valley bog** = peat bog which forms in the damp bottom of a valley; **valley glacier** = large mass of ice which moves down a valley from above the snowline towards the sea

value ['væljuː] *noun* quantity shown as a number; **calorific value** *or* **energy value** = number of calories which a certain amount of a substance (such as a certain food) contains; ***the tin of beans has a calorific value of 250 calories;*** **food value** = amount of energy produced by a certain amount of a certain food; **pH value** = number which indicates how acid *or* alkaline a solution is

vapour ['veɪpə] *noun* particles of liquid suspended in air and visible as clouds, smoke, etc.; **water vapour** = particles of water suspended in air; **vapour concentration** = ratio of the mass of a vaporized substance in a given quantity of air to the amount of air; **vapour trail** *or* **condensation trail** = white streak in the sky left by an aircraft flying at high altitude and caused by condensation and freezing of components of its exhaust gases, mainly water

◇ **vaporize** ['veɪpəraɪz] *verb* to turn a liquid into a vapour (either by heating *or* by blowing it forcibly into air)

◇ **vaporizer** ['veɪpəraɪzə] *noun* instrument which sprays liquid in the form of very small drops like mist; *(in medicine)* device which warms a liquid to which medicinal oil has been added so that it provides a vapour which a patient can inhale

variable ['veəriəbl] **1** *adjective* which changes *or* which is not constant; ***the weather forecast is for variable winds*** **2** *noun* quantity *or* quality which changes

◇ **variant** ['veəriənt] *noun* specimen of a plant *or* animal which is different from the normal type

◇ **variation** [veəri'eɪʃn] *noun* change from one thing *or* level to another; ***there is a noticeable variation in temperature in the desert regions; the chart shows the variations in atmospheric pressure over a period of six months;*** **climatic variation** = change from one type of climate to another

variety [və'raɪəti] *noun* type of organism, especially a cultivated plant

vector ['vektə] *noun* insect *or* animal which carries a disease *or* parasite and can pass it to other organisms; ***the tsetse fly is a vector of sleeping sickness***

veer [vɪə] *verb (of wind)* to change direction, clockwise in the Northern Hemisphere and anticlockwise in the Southern Hemisphere (NOTE: opposite is **back**)

vegan ['viːgən] *noun & adjective* strict vegetarian *or* (person) who eats only vegetables and fruit and no animal products like milk, fish, eggs *or* meat

vegetable ['vedʒɪtəbl] *noun* plant grown for food, not usually sweet; ***green vegetables are***

a source of dietary fibre; **the vegetable kingdom** = the plant kingdom *or* category of all organisms classed as plants; **vegetable oil** = oil obtained from a plant and its seeds, low in saturated fat

◇ **vegetarian** [vedʒɪ'teəriən] *noun & adjective* (person) who does not eat meat; ***he is on a vegetarian diet***

vegetation [vedʒɪ'teɪʃn] *noun* (i) plants in general; (ii) plants which are found in a particular area; ***the vegetation was destroyed by fire; very little vegetation is found in the Arctic regions; he is studying the vegetation of the island***

◇ **vegetative** ['vedʒətətɪv] *adjective* referring to vegetation; ***the loss of vegetative cover with the destruction of forest, increases the accumulation of carbon dioxide in the atmosphere;*** **vegetative propagation** *or* **vegetative reproduction** = reproduction of plants by taking cuttings *or* by grafting, not by seed

vehicle ['vi:ɪkl] *noun* machine for moving people *or* goods, e.g. a car, truck, train, boat, etc.; **goods vehicle** = truck *or* van for moving objects; **heavy goods vehicle** = large truck for moving objects

veil [veɪl] *noun* thin layer of cloud *or* mist

vein [veɪn] *noun* **(a)** layer of valuable mineral found in rock; ***they have struck a vein of gold*** **(b)** blood vessel which takes deoxygenated blood containing waste carbon dioxide from the tissues back to the heart **(c)** thin tube which forms part of the structure of a leaf

vent [vent] *noun* hole through which gases *or* lava escape from a volcano; **hydrothermal vent** = place on the ocean floor where hot water and gas flow out of the earth's crust

ventilate ['ventɪleɪt] *verb* to provide fresh air; ***the mine is ventilated with air pumped down from the surface***

◇ **ventilation** [ventɪ'leɪʃn] *noun* providing fresh air *or* allowing air to circulate

◇ **ventilator** ['ventɪleɪtə] *noun* device which circulates fresh air into a room *or* building

ventral ['ventrəl] *adjective* referring to (i) the abdomen; (ii) the underneath of a plant's *or* an animal's body; (iii) to the front of the human body; ***the fish has two ventral fins*** (NOTE: opposite is **dorsal)**

venturi effect [ven'tjʊri ɪ'fekt] *noun* rapid flow of a liquid *or* a gas as it passes through a narrower channel

vermicide ['vɜ:mɪsaɪd] *noun* substance which kills worms

vermiculite [vɜ:'mɪkju:laɪt] *noun* substance which is used in soilless gardening

> COMMENT: vermiculite occurs naturally as a form of silica and is capable of retaining moisture. It is processed into small pieces and used instead of soil in horticulture

vermin ['vɜ:mɪn] *noun* organisms which are looked upon as pests by some people; *see also* PEST

vermis ['vɜ:mɪs] *noun* central part of the cerebellum

vernacular [və'nækjʊlə] *adjective & noun* (language) spoken in a certain area; **vernacular building** = building built in the local style, which is different from the style in other parts of the country; **vernacular material** = building material found in the local area

vernal ['vɜ:nəl] *adjective* referring to the spring; **vernal equinox** = about March 21st, one of the two occasions in the year when the sun crosses the celestial equator and night and day are each twelve hours long

◇ **vernalization** [vɜ:nəlaɪ'zeɪʃn] *noun* making a seed germinate early by refrigerating it for a time

vertebra ['vɜ:tɪbrə] *noun* one of the ring-shaped bones which link together to form the backbone (NOTE: plural is **vertebrae)**

◇ **vertebral** ['vɜ:tɪbrəl] *adjective* referring to the vertebrae

◇ **vertebrate** ['vɜ:tɪbrət] *adjective & noun* (animal) which has a backbone (NOTE: opposite is **invertebrate)**

vessel ['vesəl] *noun* container, especially a container for nuclear fuel *or* radioactive waste; **pressure vessel** = container that houses the core, coolant and moderator in a nuclear reactor

vestigial [ves'tɪdʒiəl] *adjective* which exists in a simple and reduced form; ***some snakes still have vestigial legs***

vibrate [vaɪ'breɪt] *verb* to shake repeatedly; ***the passing traffic makes the foundations of the bridge vibrate***

◇ **vibration** [vaɪ'breɪʃn] *noun* act of vibrating; ***vibrations caused by aircraft can shatter windows; the vibrations from traffic have weakened the foundations of the house***

victim ['vɪktɪm] *noun* person *or* animal that is injured *or* harmed *or* attacked

vigour ['vɪgə] *noun* strength; **hybrid vigour** = increase in size *or* rate of growth *or* fertility *or* resistance to disease found in offspring of a cross between two species

◇ **vigorous** ['vɪgərəs] *adjective* which grows strongly; ***plants put out vigorous shoots in warm damp atmosphere***

vinyl ['vaɪnəl] *noun* plastic that is not biodegradable; **vinyl chloride** = chemical compound of chlorine and ethylene, used as a refrigerant and in the making of PVC

virement ['vaɪəmənt] *noun* transfer of money from one account to another *or* from one budget to another

virgin ['vɜːdʒɪn] *adjective* untouched by humans *or* in its natural state; ***virgin rainforest is being cleared at the rate of 1,000 hectares per month;*** **virgin land** = land which has never been cultivated; **virgin ore** = ore as it is extracted from the ground, before it has been processed

virus ['vaɪrəs] *noun* tiny cell composed largely of genetic material and protein, which can only develop in other cells and often destroys them; ***scientists have isolated a new flu virus; shingles is caused by the same virus as chickenpox***

◇ **viral** ['vaɪrəl] *adjective* (disease) caused by a virus

> COMMENT: many common diseases such as measles or the common cold are caused by viruses; viral diseases cannot be treated with antibiotics (which only destroy bacteria)

viscous *or* **viscid** ['vɪskəs *or* 'vɪsɪd] *adjective* (liquid) which is thick, sticky and slow-moving

◇ **viscosity** [vɪs'kɒsɪti] *noun* state of a liquid which is thick, sticky and slow-moving

visible ['vɪzɪbl] *adjective* which can be seen; ***everywhere there were visible signs of the effects of acid rain***

◇ **visibility** [vɪzə'bɪləti] *noun* ability to see objects in certain conditions; ***in fog, the visibility is less than 1,000 metres***

visitant *or* **visitor** ['vɪzɪtənt *or* 'vɪzɪtə] *noun* migrant bird which comes to a region regularly; ***the flycatcher is a summer visitor to Britain***

vital ['vaɪtəl] *adjective* most important for life; ***vital nutrients are leached from the topsoil; oxygen is vital to the human system;*** **vital organs** = the most important organs in the body (such as the heart, lungs, brain) without which a human being cannot live; **vital statistics** = official statistics relating to the population of a place (such as the percentage of live births per thousand, the incidence of a certain disease, the numbers of births and deaths)

vitamin ['vɪtəmɪn] *noun* essential nutrient usually needed by animals in minute quantities for growth and health; **vitamin deficiency** = lack of necessary vitamins; ***he is suffering from Vitamin A deficiency; Vitamin C deficiency causes scurvy***

viviparous [vɪ'vɪpərəs] *adjective* (animal) which bears live young (such as mammals, as opposed to birds and reptiles which lay eggs and are oviparous)

◇ **vivisection** [vɪvɪ'sekʃn] *noun* dissecting a living animal as an experiment

volatile ['vɒlətaɪl] *adjective* (liquid) which turns into gas at normal room temperature; **volatile oil** = concentrated oil from a scented plant used in cosmetics and as an antiseptic

volcano [vɒl'keɪnəʊ] *noun* mountain surrounding a hole in the earth's crust, formed of solidified molten rock sent up from the interior of the earth

◇ **volcanic** [vɒl'kænɪk] *adjective* referring to volcanoes; **volcanic activity** = earthquakes, eruptions, lava flows, smoke emissions, etc. which show that a volcano is not extinct; **volcanic ash** = ash and small pieces of lava and rock which are thrown up by an erupting volcano; **volcanic dust** = fine ash thrown up by an erupting volcano; **volcanic rock** = rock formed from lava

> COMMENT: volcanoes occur along faults in the earth's surface and exist in well-known chains. Some are extinct, but others erupt relatively frequently. Some are always active, in that they emit sulphurous gases and smoke, without actually erupting. Volcanic eruptions are a major source of atmospheric pollution, in particular of sulphur dioxide. Very large eruptions (such as that of Tambora in Indonesia in 1815) cause a mass of dust to enter the atmosphere, which has a noticeable effect on the world's climate. (The Tambora eruption caused the 'year without a summer' in 1816)

> QUOTE: subduction-zone volcanoes make up about 400 of the world's 500 active volcanoes and are particularly prone to violent eruptions
> **New Scientist**

volume ['vɒljuːm] *noun* **(a)** amount of space that a substance occupies **(b)** loudness of a sound

-vore *or* **-vorous** [vɔː *or* vərəs] *suffix* meaning which eats; **carnivore** = animal which eats meat; **herbivorous** = which eats plants

vortex ['vɔːteks] *noun* flow of a liquid in a whirlpool *or* of a gas in a whirlwind; ***the most destructive winds are in the vortex, where the rotation of the whirlwind produces very high wind speeds;*** **circumpolar vortex** *or* **polar vortex** = circular movement of air around one of the poles

vulcanism ['vʌlkənɪzm] *noun* movement of magma or molten rock onto or towards the Earth's surface

vulnerable ['vʌlnrəbl] *adjective* (species) which is likely to become endangered unless protective measures are taken

vulture ['vʌltʃə] *noun* large bird of prey which feeds on carrion

Ww

wader ['weɪdə] *noun* bird which walks in shallow water and feeds on organisms *or* plants found in the water

wadi ['wɒdi] *noun* gully with a stream at the bottom, found in the desert regions of North Africa; *compare* ARROYO

Waldsterben ['væltʃteəbən] *noun* (German word meaning 'the dying of trees') forest dieback, a disease affecting pine trees, where the pine needles turn yellow

COMMENT: there are many different theories about the cause of forest dieback: sulphur dioxide, nitrogen oxides, ozone are all possible causes, as are acid rain or acidification of the soil which makes the trees weak, and prevents them taking up nutrients from the soil

QUOTE: since 1980, many forests in the eastern US and parts of Europe have suffered a loss of vitality - a loss that could not be linked to any of the familiar causes, such as insects, disease or direct poisoning by a specific air or water pollutant. The most dramatic reports have come from Germany, where scientists, stunned by the extent and speed of the decline, have called it Waldsterben
Scientific American

Wallace's line ['wɒlɪsɪs 'laɪn] *noun* line dividing the Australasian biogeographical region from the Southeast Asian region

warble fly ['wɔːbl 'flaɪ] *noun* parasitic fly, whose larvae infest cattle

warfarin ['wɔːfərɪn] *noun* substance used to poison rats

warm [wɔːm] **1** *adjective* quite hot *or* pleasantly hot; ***these plants grow fast in the warm season;*** **warm front** = movement of a mass of warm air which displaces a mass of cold air and gives rain **2** *verb* to make something hotter; ***the greenhouse effect has the result of warming the general atmospheric temperature; at the poles, warming would be two or three times the global average if the greenhouse effect makes the earth's temperature rise;*** **global warming** = gradual rise in temperature over the whole of the earth's surface, caused by the greenhouse effect; **sudden warming** = rapid rise in the temperature of the stratosphere, which occurs at the beginning of spring

◇ **warm-blooded** ['wɔːm 'blʌdɪd] *adjective* (animal, such as a mammal) which has warm blood, as opposed to reptiles *or* fish; *see also* HOMOIOTHERM

wash out ['wɒʃ 'aʊt] *verb* to remove (a mineral) by running water; ***most minerals are washed out of the soil during heavy rains***

◇ **washout** ['wɒʃaʊt] *noun* process where raindrops form normally in the atmosphere and collect pollutant particles as they fall; *compare* RAINOUT

waste [weɪst] **1** *adjective* useless *or* which has no use; ***waste products are dumped in the sea; waste matter is excreted by the body in the faeces or urine;*** **waste ground** = area of land which is not used for any purpose; **waste product** = substance which is not needed in a process **2** *noun* rubbish *or* material which is not needed; **high-level waste** = waste which is very hot, emits strong radiation and is very difficult to dispose of safely; **household waste** = rubbish from houses; **indeterminate waste** = waste which is more radioactive than low-level waste, but not as dangerous as high-level waste; **industrial waste** = rubbish from industrial processes; **low-level waste** = waste which is only slightly radioactive and does not cause problems for disposal; **nuclear waste** = radioactive waste from a nuclear reactor (including spent rods and coolant); **radioactive waste** = used radioactive materials produced by nuclear power stations, industrial plants, hospitals, etc.; **pathological waste** = waste (such as waste from a hospital) which may contain pathogens and which could cause disease; **sewage waste** = waste water and other refuse such as faeces, carried away in sewers; **waste disposal** = getting rid of waste; **waste disposal unit** = device which fits into the plughole of a kitchen sink and which grinds up household waste so that it can be flushed away; **waste dump** = place where waste is thrown away; **waste management** = action of controlling and processing waste; **waste processing** *or* **waste treatment** = treating waste material to make it reusable *or* so it may be disposed of safely; **waste processing plant** *or* **waste treatment plant** = place where waste material is treated to make it reusable *or* so it may be disposed of safely; **waste site** = place where waste is thrown away; **waste sorting** = separating waste into different materials, such as glass, metal, paper, plastic, etc. **3** *verb* to use more than is needed

◇ **wastage** ['weɪstɪdʒ] *noun* act of wasting; amount wasted; ***there is an enormous wastage of mineral resources***

◇ **wasteland** ['weɪstlænd] *noun* area of land which is no longer usable for agriculture or for any other purpose; ***overgrazing has produced wastelands in Central Africa***

◇ **wastepaper** ['weɪstpeɪpə] *noun* paper which has been thrown away after use

◇ **wastewater** ['weɪstwɔːtə] *noun* water as part of effluent *or* sewage, especially from industrial processes; ***there is considerable interest in the anaerobic treatment of industrial wastewaters; wastewater will add small but significant quantities of heavy metals to the aquatic environment;*** **wastewater treatment** = processing wastewater to make it reusable *or* so it may be disposed of safely

> COMMENT: high-level radioactive waste is potentially dangerous and needs special disposal techniques. It is sealed in special containers, sometimes in glass, and is sometimes disposed by dumping at sea

water ['wɔːtə] **1** *noun* common liquid (H_2O) which forms rain, rivers, the sea, etc., and which makes up a large part of the bodies of organisms; **drinking water** = water which is safe to drink; **heavy water** = deuterium oxide (D_2O) *or* water containing deuterium instead of the hydrogen atom, used as a coolant or moderator in certain types of nuclear reactor such as the CANDU; **light water** = ordinary water (as opposed to heavy water *or* deuterium oxide) used as a coolant in some types of power station; **runoff water** = rainwater *or* melted snow which flows into rivers and streams; **water balance** = (i) state where the water lost in a an area by evaporation *or* by runoff is replaced by water received in the form of rain; (ii) state where the water lost by the body (in urine *or* perspiration, etc.) is balanced by water absorbed from food and drink; **water closet** = toilet in which the excreta are flushed away by running water into the sewage system; **water meadow** = grassy field near a river, which is subject to flooding; **water mill** = mill driven by water; **water plant** = plant which grows in water; **water pollution** = polluting of the sea, rivers, lakes, canals; **water quality** = how good water is to drink; ***the authorities are always trying to improve water quality;*** **water-salt balance** = state where the water in the soil balances the amount of salts in the soil; **water softener** = device attached to the water supply to remove nitrates *or* calcium from the water; **water vapour** = particles of water suspended in air **2** *verb* to give water to (a plant)

> COMMENT: water is essential to plant and animal life. Since the human body is formed of more than 50% water, a normal adult needs to drink about 2.5 litres (5 pints) of fluid each day. Water pollution can take many forms: the most common are discharges from industrial processes, household sewage and the runoff of chemicals used in agriculture

waterborne ['wɔːtəbɔːn] *adjective* (disease) carried in water; ***after the floods diarrhoea, dysentery and other waterborne diseases spread rapidly***

◇ **watercourse** ['wɔːtəkɔːs] *noun* stream *or* river *or* canal

◇ **waterfall** ['wɔːtəfɔːl] *noun* place in a river where water falls over a steep vertical drop

◇ **waterfowl** ['wɔːtəfaʊl] *noun* birds which live on water (such as ducks)

◇ **waterhole** ['wɔːtəhəʊl] *noun* (i) place in a desert *or* elsewhere where water rises naturally to the surface; (ii) similar place, created by boring holes in the ground; ***in the evening, the animals gather round the waterholes to drink***

◇ **waterlogged** ['wɔːtəlɒgd] *adjective* (soil) which is saturated with water and so cannot keep oxygen between its particles. (Hence most plants cannot grow in waterlogged soil)

◇ **waterproof** ['wɔːtəpruːf] *adjective* which will not let water through

◇ **watershed** ['wɔːtəʃed] *noun* line between the headstreams of river systems, dividing one catchment area from another

◇ **waterside** ['wɔːtəsaɪd] *adjective* (plant) which grows next to a river *or* lake, etc.

◇ **water-soluble** [wɔːtə'sɒljuːbl] *adjective* which can dissolve in water

waterspout ['wɔːtəspaʊt] *noun* phenomenon caused when a rapidly turning column of air forms over water, sucking the water up into the column

> COMMENT: waterspouts form in summer weather as air rises rapidly from the warm surface of the sea. Waterspouts turn cyclonically (i.e. anticlockwise in the Northern Hemisphere). When the waterspout moves onto dry land, it loses momentum and falls as saltwater rain. The rain sometimes contains fish which have been sucked up into the spout

water table ['wɔːtə 'teɪbl] *noun* top level of water in the ground that occupies spaces in rock *or* soil and is above a layer of impermeable rock

◇ **waterway** ['wɔːtəweɪ] *noun* river *or* canal used as channel for moving around

◇ **waterwheel** ['wɔːtəwiːl] *noun* wheel with wooden steps *or* buckets, which is turned by the flow of water against it and itself turns machinery, such as a mill *or* an electric generator; *see also* OVERSHOT, UNDERSHOT

◇ **waterworks** ['wɔːtəwɜːks] *noun* plant for treating and purifying water before it is pumped into pipes for distribution to houses, factories, schools, etc.

wave [weɪv] *noun* **(a)** pressure change *or* signal that rises and falls as it travels through a medium; **sound wave** = audible change of pressure which moves through the air **(b)** mass of water moving across the surface of a body of water, rising higher than the surrounding water as it moves; **wave power** *or* **wave energy** = energy obtained from the movement of the waves; electricity produced by turbines driven by the force of the waves

◇ **wavelength** ['weɪvleŋθ] *noun* distance between two similar points in adjacent waves

> COMMENT: in harnessing wave power, the movement of waves on the surface of the sea is used to make large floats move up and down. These act as pumps which pump a continuous flow of water to turn a turbine

◇ **wave refraction** ['weɪv rɪ'frækʃn] *noun* tendency of wave crests to turn from their original direction and become more parallel to the shore as they move into shallower water

waymarking ['weɪmɑːkɪŋ] *noun* indicating the direction of a public path by special signs

> QUOTE: the Trust has improved and waymarked 24 kilometres of footpath
> **Environment Now**

WCED = WORLD COMMISSION ON ENVIRONMENT AND DEVELOPMENT

wealth [welθ] *noun* large amount of money *or* valuable possessions, such as jewellery *or* mineral resources, etc.

◇ **wealthy** ['welθi] *adjective* (person *or* nation) which has a large amount of money *or* valuable possessions

weather ['weðə] **1** *noun* daily atmospheric conditions (e.g. sunshine, wind, precipitation) in a certain area, as opposed to climate which is the average weather for the area; **weather centre** = office which analyses weather reports and forecasts the weather; **weather chart** *or* **weather map** = chart *or* map showing the state of the weather at a particular moment *or* changes which are expected to happen in the weather in the near future; **weather forecast** = description of what the weather will be for a period in the future; **weather forecasting** = scientific study of weather conditions and patterns, which allows you to describe what the weather will be for a period in the future; **weather report** = written *or* spoken statement describing what the weather has been like recently *or* what it is like at the moment *or* what it will be for a period in the future; **weather ship** = ship which takes readings of weather conditions and relays them back to a central station; **weather station** = place where weather is recorded **2** *verb* to change the state of soil *or* rock through the influence of the weather (rain, sun, frost, wind, etc.) or by chemical pollutants present in the rain or the atmosphere

◇ **weathering** ['weðərɪŋ] *noun* changing the state of soil *or* rock through the influence of the weather (rain, sun, frost, wind, etc.) or by chemical pollutants present in the rain or the atmosphere

web [web] *noun* threads secreted by a spider in the form of a net; **food web** = series of food chains which are linked together in an ecosystem

◇ **webbed** [webd] *adjective* with skin between the toes; ***ducks and other aquatic birds have webbed feet***

weed [wiːd] *noun* wild plant which grows in cultivated land and which the farmer *or* gardener tries to remove

◇ **weedkiller** ['wiːdkɪlə] *noun* substance used to kill weeds; **selective weedkiller** = weedkiller which is supposed to kill only certain plants and not others

> COMMENT: although weedkillers are widely used in farming and horticulture, they can have serious side effects. They may blow away from the area where they are being sprayed; excess weedkiller enters the soil and is leached away by rain into watercourses, killing water plants

well [wel] *noun* hole dug in the ground to the level of the water table, from which water can be extracted either by a pump *or* by a bucket; **artesian well** = well bored into a confined aquifer

◇ **wellhead** ['welhed] *noun* top of a well

Wentworth-Udden scale ['wentwɜːθ 'ʌdən 'skeɪl] *noun* scale for measuring and describing the size of grains of minerals

> COMMENT: the scale runs from the largest size, the boulder, down to the finest grain, or clay. The approximate diameters of each grain are: boulder (up to 256mm); cobble (above 64mm); pebble (between 4 and 64mm); gravel (between 2 and 4mm); sand (between 0.06 and 2mm); silt and clay are the finest sizes

west [west] *adjective, adverb & noun* one of the directions on the earth's surface, the direction facing towards the setting sun; ***the wind is blowing from the west; the river flows west into the ocean;*** **west wind** = wind which blows from the west

◇ **westerly** ['westəli] **1** *adjective* to *or* from the west; ***the ship was proceeding in a westerly direction; a westerly airstream covers the country*** **2** *noun* wind which blows from the west

◇ **western** ['westən] *adjective* in the west; towards the west; ***the herds spend winter on the western plains***

wet [wet] *adjective* not dry *or* covered in liquid; **wet bulb hygrometer** = thermometer where the bulb is kept wet, used with a dry bulb thermometer to give a humidity reading; **wet season** = time of year in some countries when a lot of rain falls (as opposed to the dry season)

◇ **wetland** ['wetlænd] *noun* area of land which is often covered by water *or* which is very marshy; ***every effort has been made to halt the draining of the wetlands; there are two important wetland sites in the area***

whale [weɪl] *noun* one of various types of very large mammals living in the sea

◇ **whaler** *or* **whaling boat** ['weɪlə *or* 'weɪlɪŋ bəʊt] *noun* boat which is specially equipped for catching whales

◇ **whaling** ['weɪlɪŋ] *noun* catching whales to use as food *or* for their oil, etc.; **whaling fleet** = fleet of boats specially equipped for catching whales

> COMMENT: whales are the largest mammals still in existence. There are two groups of whales: the toothed and the baleen whales. Baleen whales have no teeth and feed by sucking in large quantities of water which they then force out again through their baleen, which is a series of fine plates like a comb hanging down from the upper jaw. The baleen acts like a sieve and traps any plankton and krill which are in the water. The toothed whales have teeth and eat fish; they include the sperm whale, the killer whale and porpoises and dolphins. Whales are caught mainly for their oils, though also in some cases for food. Some species of whale have become extinct because of overexploitation and the population of many of the existing species is dangerously low. Commercial whaling is severely restricted

wheat [wi:t] *noun Triticum* sp., the most important of the cereals, grown in the temperate regions; **spring wheat** = varieties of wheat which are sown in spring and harvested towards the end of the summer; **winter wheat** = varieties of wheat sown in the autumn or early winter months and harvested the following summer

◇ **wheatgerm** ['wi:tdʒɜ:m] *noun* central part of the wheat seed, which contains valuable nutrients

◇ **wheatmeal** ['wi:tmi:l] *noun* brown flour with a large amount of bran, but not as much as wholemeal

whirlpool ['wɜ:lpu:l] *noun* rapidly turning eddy of water

◇ **whirlwind** ['wɜ:lwɪnd] *noun* column of rapidly turning air at the centre of an area of very low pressure. (Over water it becomes a waterspout and over desert a dust devil)

wholefood ['həʊlfu:d] *noun* naturally *or* organically grown food, which has not been processed; ***a wholefood diet is healthier than eating processed foods***

◇ **wholegrain** ['həʊlgreɪn] *noun* food (such as rice) of which the whole of the seed is eaten

◇ **wholemeal** ['həʊlmi:l] *noun* flour which has had nothing removed or added to it and contains a large proportion of the original wheat seed, including the bran; **wholemeal bread** *or US* **wholewheat bread** = bread made from wholemeal flour

wild [waɪld] *adjective* not domesticated; **wild rice** = species of grass which is found naturally in North America and which is similar to rice

◇ **wildlife** ['waɪldlaɪf] *noun* wild animals and birds; ***plantations of conifers are poorer for wildlife than mixed or deciduous woodlands; the effects of the open-cast mining scheme would be disastrous on wildlife, particularly on moorland birds***

wilderness ['wɪldənəs] *noun* area of wild uncultivated land, usually far from human habitation, but sometimes referring to undeveloped land in an urban area; **wilderness area** = area of undeveloped land, such as a forest, which is set aside and protected as a national park, etc.

> QUOTE: the recent successes of Berlin's conservationists in preserving some of the city's valuable wilderness areas have shown that it may be possible to maintain a park-like environment despite the pressures of urban growth
>
> **Environment**

willow ['wɪləʊ] *noun* strong temperate hardwood (genus *Salix*), often used for coppicing or pollarding

wind [wɪnd] *noun* air which moves in the lower atmosphere; ***the weather station has instruments to measure the speed of the wind;*** **wind-driven** *or* **wind-powered** = (machine *or* turbine) that is powered by the wind; **wind erosion** = erosion of soil *or* rock by wind; **wind farm** *or* **wind park** = group of large windmills, built to harness the wind to produce electricity; **wind pollination** = pollination of flowers when the pollen is blown by the wind; **wind power** = power generated by using wind to drive a machine (as in a windmill) *or* to drive a turbine which creates electricity; **wind pump** = pump driven by the wind, which raises water out of the ground; **wind rose** = chart showing the direction of the prevalent winds in an area; **wind turbine** = turbine driven by wind

◇ **wind chill factor** ['wɪnd 'tʃɪl 'fæktə] *noun* way of calculating the risk of exposure in cold weather by adding the speed of the wind to the number of degrees of temperature below zero

> QUOTE: with wind farms popping up all over the world, wind power's future looks bright. The only problem could be environmental impact - locating hundreds of machines in windy areas can lead to conflict. Even though wind turbines do not generate waste or pollutants, care must be taken to avoid noise pollution
>
> **Environment Now**

> QUOTE: a proposal to site one of Britain's first wind parks on an exposed area of high fell has angered the government's advisers on environmental protection
>
> **New Scientist**

windmill ['wɪndmɪl] *noun* construction with sails which are turned by the wind, so driving a machine; *see also* PANEMONE

> COMMENT: windmills were originally built to grind corn or to pump water from marshes. Large modern windmills are used to harness the wind to produce electricity

windrow ['wɪndrəʊ] *noun* cut stalks of a crop, gathered into a row to be dried by the wind

windward ['wɪndwəd] *adjective* exposed to the wind; ***the trees provide shelter on the windward side of the house***

winter ['wɪntə] **1** *noun* last season of the year, following autumn and before spring, when the weather is coldest, the days are short, most plants do not flower *or* produce new shoots and some animals hibernate; **nuclear winter** = period which scientists believe will follow after a nuclear war, when there would be no warmth and light because dust particles would obscure the sun and most life would be affected by radiation; **winter solstice** = 21st December, the shortest day in the Northern Hemisphere, when the sun is at its furthest point north of the equator; **winter wheat** = varieties of wheat that are sown in the autumn or early winter and harvested the following summer **2** *verb* to spend the winter in a place

◇ **winterbourne** ['wɪntəbɔːn] *noun* stream which flows only in the wetter part of the year, usually in winter

> QUOTE: the birds wintered in the tall grass prairies of Southern Louisiana
>
> **Birder's World**

wither ['wɪðə] *verb (of plants, leaves, flowers)* to shrivel and die

wolds [wəʊldz] *noun* belts of upland country, on chalk or limestone

wood [wʊd] *noun* **(a)** (i) hard tissue which forms the main body of a tree; (ii) construction material that comes from trees; **wood alcohol** = methanol *or* alcohol manufactured from waste wood, which can be used as a fuel *or* solvent; *see also* HARDWOOD, SOFTWOOD **(b)** large number of trees growing together; **oak wood** *or* **beech wood** number of oak trees *or* beech trees growing together

◇ **-wood** [wʊd] *suffix* referring to wood; ***tropical hardwoods; sandalwood has an aromatic scent; an elm- wood stool;*** **fuelwood** = wood which is grown to be used as a fuel; **pulpwood** = softwood used for making paper

◇ **woodfuel** ['wʊdfjuːəl] *noun* wood which is used as fuel

◇ **woodlands** *or* **woods** ['wʊdləndz *or* wʊdz] *noun* land covered with trees with clear spaces between them (i.e. not as dense as a forest); **ancient woodland** = wooded area which has been covered with trees for many hundreds of years

◇ **woodlot** ['wʊdlɒt] *noun* small area of land planted with trees

◇ **wood pulp** ['wʊd 'pʌlp] *noun* softwood which has been pulverized into small fibres and mixed with water, used to make paper

> QUOTE: three-quarters of our breeding land birds, half of our butterflies and one sixth of our flowers need woodland habitats
>
> **Sunday Times**

woodstove *or* **wood-burning stove** ['wʊdstəʊv *or* 'wʊdbɜːnɪŋ 'stəʊv] *noun* heating device which burns wood

> COMMENT: although woodstoves are apparently contributing to a better environment by not burning fossil fuels, by using renewable resources and by not releasing dangerous pollutants into the atmosphere, they do in fact burn vast quantities of wood, and can contribute to deforestation unless the wood is coppiced

work [wɜːk] *noun* energy used when something is forced to move

◇ **work out** ['wɜːk 'aʊt] *verb* to use up something completely; ***the coal mine was worked out years ago***

◇ **worker** ['wɜːkə] *noun* sterile female bee which forages for food

◊ **workings** ['wɜːkɪŋz] *plural noun* underground tunnels in a mine

world [wɜːld] *noun* the planet earth; people who live on the earth; ***a map of the world; to sail round the world***

◊ **World Commission on Environment and Development (WCED)** *noun* official body set up by the United Nations in 1983 to formulate a global programme for sustainable development. The Commission's report, published in 1987, made recommendations on pollution control, conservation measures, economic cooperation, etc.

◊ **World Wide Fund for Nature (WWF)** ['wɜːld 'waɪd fʌnd fə 'neɪtʃə] *noun* international organization, set up in 1961, to protect endangered species of wildlife and their habitats, and now also involved with projects to control pollution and promote policies of sustainable development

◊ **worldwide** [wɜːld'waɪd] *adjective & adverb* concerning *or* covering the whole world; all over the world; ***the worldwide energy crisis; we sell our products worldwide***

worm [wɜːm] *noun* earthworm, an invertebrate animal with a long thin body living in large numbers in the soil

> COMMENT: earthworms provide a useful service by aerating the soil as they tunnel. They also eat organic matter and help increase the soil's fertility. It is believed that they also secrete a hormone which encourages rooting by plants

WWF = WORLD WIDE FUND FOR NATURE

Xx Yy Zz

X-chromosome ['ekskrəʊməsəʊm] *noun* sex chromosome; *see also* Y-CHROMOSOME *and comment at* SEX

> COMMENT: every person has a series of pairs of chromosomes, one of which is always an X chromosome; a normal female has one pair of XX chromosomes, while a male has one XY pair. Haemophilia is a disorder linked to the X chromosome

Xe *chemical symbol for* xenon

xeno- ['zenəʊ] *prefix* meaning different

◊ **xenobiotics** [zenəʊbaɪ'ɒtɪks] *noun* chemical compounds that are foreign to an organism

xenon ['ziːnɒn] *noun* inert gas, traces of which are found in the atmosphere (NOTE: chemical symbol is **Xe**; atomic number is **54**)

xeric ['zɪərɪk] *adjective* dry (environment); *compare* HYDRIC

xero- ['zɪərəʊ] *prefix* meaning dry

◊ **xeromorphic** [zɪərəʊ'mɔːfɪk] *adjective* (plant) which can prevent water loss from its stems during hot weather

◊ **xerophilous** [zɪ'rɒfɪləs] *adjective* (plant) which lives in very dry conditions

◊ **xerophyte** ['zɪərəfaɪt] *noun* plant which is adapted to living in very dry conditions

◊ **xerosere** ['zɪərəʊsiːə] *noun* succession of communities growing in very dry conditions

◊ **xerothermic** [zɪərəʊ'θɜːmɪk] *adjective* (organism) which is adapted to living in very dry conditions

X-ray ['eksreɪ] **1** *noun* **(a)** ray with a very short wavelength, which is invisible, can go through soft tissue and register on a photographic plate **(b)** photograph taken using X-rays **2** *verb* to take a photograph using X-rays

> COMMENT: because X-rays go through soft tissue, it is sometimes necessary to make internal organs opaque so that they will show up on the film. In the case of stomach X-rays, patients take a barium meal before being photographed; in other cases, such as kidney X-rays, radioactive substances are injected into the bloodstream or into the organ itself. X-rays are used not only in radiography for diagnosis but as a treatment in radiotherapy. Exposure to X-rays, either as a patient being treated, or as a radiographer, can cause radiation sickness

xylem ['zaɪləm] *noun* wood *or* solid tissue in a plant which takes water from the roots to the rest of the plant and is formed of dead cells (as opposed to phloem which is formed of living cells)

xylophagous [zaɪ'lɒfəgəs] *adjective* wood-eating

xylophilous [zaɪ'lɒfɪləs] *adjective* preferring to grow on wood

yaws [jɔːz] *plural noun* tropical disease caused by the spirochaete *Treponema pertenue*

COMMENT: symptoms include fever with raspberry-like swellings on the skin, followed in later stages by bone deformation

Y-chromosome ['waɪkrəʊməsəʊm] *noun* male chromosome; *see also* X-CHROMOSOME *and comment at* SEX

COMMENT: the Y chromosome has male characteristics and does not form part of the female genetic structure. A normal male has an XY pair of chromosomes

yeast [jiːst] *noun* fungus used in the fermentation of alcohol and in making bread

COMMENT: yeast is a good source of Vitamin B and can be taken in dried form in tablets

yellowing ['jeləʊɪŋ] *noun* condition where the leaves of plants turn yellow, caused by lack of light; **lethal yellowing (LY)** = disease which attacks and kills coconut palms

yield [jiːld] **1** *noun* quantity of a crop *or* a product produced from a plant *or* from an area of land; ***the normal yield is 2 tonnes per hectare; the green revolution increased rice yields in parts of Asia*** **2** *verb* to produce a crop *or* a product; ***the rice can yield up to 2 tonnes per hectare; the oil deposits may yield 100,000 barrels a month***

QUOTE: research has shown what must have been known to stockmen for years, that kindness to cows increases their milk yield, yet the pharmaceutical industry is determined to increase milk production by injecting cows with the hormone bovine somatotropin (BST)

Ecologist

yolk [jəʊk] *noun* yellow central part of an egg; **yolk sac** = membrane which encloses the yolk in embryo fish, reptiles and birds

zero population growth ['ziːrəʊ pɒpjuː'leɪʃn grəʊθ] *noun* state when the numbers of births and deaths in a population are equal and so the size of the population remains the same

zinc [zɪŋk] *noun* white metallic trace element, essential to biological life, used in alloys and as a protective coating for steel (NOTE: chemical symbol is **Zn**; atomic number is **30**)

zone [zəʊn] *noun* area of land *or* sea *or* of the atmosphere; **equatorial zone** = area near the equator, mostly with a very hot and humid climate, except for land at high altitudes as in South America; **pedestrian zone** = area in a town where motor vehicles are not allowed and where people may only go on foot

◇ **zonal** ['zəʊnəl] *adjective* referring to a zone; **zonal airstream** = stream of air blowing from west to east in the upper atmosphere of the Northern Hemisphere

◇ **zoning** ['zəʊnɪŋ] *noun* order by a government *or* local council that an area of land shall be used only for a certain type of building *or* for a certain purpose, e.g. agricultural, industrial, recreational, residential, etc.

zoo- [zəʊ] *prefix* meaning animal

◇ **zoogeographical** [zəʊədʒiːə'græfɪkl] *adjective* referring to animals and geography; **zoogeographical region** = large area of the world where the fauna is different from that in other areas; *see also* BIOGEOGRAPHICAL REGION

◇ **zoologist** [zuː'ɒlədʒɪst] *noun* scientist who specializes in zoology

◇ **zoology** [zuː'ɒlədʒi] *noun* scientific study of animals

◇ **zoophyte** ['zəʊəfaɪt] *noun* animal which looks like a plant

◇ **zooplankton** [zəʊə'plæŋktən] *noun* microscopic living animals which live and drift in water; ***some zooplankton eat algae***

zygote ['zaɪgəʊt] *noun* fertilized ovum, the first stage of development of an embryo

SUPPLEMENT

OUTLINE CLASSIFICATION OF LIVING ORGANISMS

Recent classifications include five kingdoms:

Kingdom Monera: typically single-cell organisms without (unlike the other kingdoms) a membrane surrounding the cell nucleus.

Kingdom Protista: single-cell organisms, and single-cell organisms living in colonies.

Kingdom Fungi: multicell organisms which absorb nutrients and have many nuclei per cell.

Kingdom Plantae: multicell organisms which photosynthesize food with photosynthetic pigments.

Kingdom Animalia: multicell organisms which ingest food and have no photosynthetic pigments. Estimates of the number of species differ from one authority to another depending on how the species are classified. Estimates are likely to change because so many speces have not yet been discovered or catalogued. This is particularly true of phylum Arthropoda, which contains more species than the total of all other animal species and may number in the millions. Most of these live in tropical rainforests and will disappear without being discovered as the forests are cut down.

phylum	*representative common name*	*estimated number of known species*
KINGDOM MONERA		
Cyanophyta	blue-green algae	2,500
Myxobacteriae	gliding bacteria	-
Schizophyta	true bacteria	3,000
Actinomycota	branching bacteria	-
Spirochaetae	spirochetes	-
KINGDOM PROTISTA		
Euglenophyta	euglenoids	450
Xanthophyta	yellow-green algae	400
Chrysophyta	golden-brown algae	10,000
Pyrrophyta	dinoflagellates	1,100
Hyphochytridiomycota	hyphochytrids	-
Plasmodiophoromycota	plasmodiophores	-
Sporozoa	sporozoans	-
Cnidosporidia	cnidosporidians	-
Zoomastigina	animal flagellates	-
Sarcodina	rhizopods	-
Ciliophora	ciliates	-
KINGDOM FUNGI		
Myxomycota	slime moulds	450
Acrasiomycota	slime moulds	25
Labyrinthulomycota	slime moulds	15
Oomycota	water moulds	200
Chytridomycota	chytrids	-
Zygomycota	black moulds	300
Ascomycota	sac fungi	30,000
Basidiomycota	mushrooms	15,000

KINGDOM PLANTAE

division	*common name*	*estimated number of known species*
Rhodophyta	red algae	4,000
Phaeophyta	brown algae	1,500
Chlorophyta	green algae	10,000
Charophyta	stoneworts	-
Bryophyta	mosses	25,000
Lycopodiophyta	clubmosses	900
Equisetophyta	horsetails	-
Polypodiophyta	ferns	9,000
Pinophyta	conifers	640
Magnoliophyta	flowering plants	250,000

KINGDOM ANIMALIA

phylum	*common name*	*estimated number of known species*
Mesozoa	mesozoans	-
Porifera	sponges	4,000
Cnidaria	jellyfish	10,000
Ctenophora	comb jellies	100
Phatyhelminthes	flatworms	14,000
Nemertea	proboscis worms	550
Nematoda	roundworms	8,000
Acanthocephala	hook-headed worms	100
Nematomorpha	horsehair worms	200
Rotifera	rotifers	1,200
Gastrotricha	gastrotrichs	100
Entoprocta	bryozoa	60
Ectoprocta	colonial bryozoa	5,000
Brachiopoda	lamp shells	200
Phoronida	phoronids	10
Annelida	worms	10,000
Onychophora	onychophors	-
Arthropoda	insects	800,000
Mollusca	snails	80,000
Brachiata	beard worms	-
Chaetognatha	arrow worms	30
Echinodermata	starfish	6,000
Hemichordata	acorn worms	100
Chordata	vertebrates	70,000

GEOLOGICAL TIME SCALE AND RADIATION OF SOME FLORA AND FAUNA

era	*period*	*epoch*	*millions of years ago*	*radiation of some flora and fauna*
Cenozoic	Quaternary	Holocene	0.01	
		Pleistocene	2	*Homo (man)*
	Tertiary	Pliocene	5	
		Miocene	23	
		Oligocene	37	
		Eocene	55	
		Palaeocene	65	
Mezozoic	Cretaceous		136	
				flowering plants
	Jurassic		190	birds
	Triassic		225	mammals
Palaeozoic	Permian		280	
	Carboniferous		345	reptiles
	Devonian		395	
				conifers and amphibians
	Silurian		440	
				land plants
	Ordovician		500	
				fish
	Cambrian		570	
Precambrian	(comprising 87% of geological time)		3300	invertebrates first known fossils
			4600	origin of the earth

ESTIMATED HUMAN POPULATION GROWTH SINCE 8000 B.C.

year	*millions*
8000BC	6
1 AD	255
1000	255
1250	400
1500	460
1600	600
1700	680
1800	950
1900	1600
1920	1860
1930	2070
1940	2295
1950	2500
1960	3050
1970	3700
1980	4400
1985	4800
1990	5290
1994	5700
Projected	
2000	6000
2025	8200
2090	10500

The human population increased extremely slowly for the first 200,000 years of its existence. It began to accelerate with the arrival of agriculture, about 10,000 years ago. It reached one billion by about 1800, two billion by about 1930, three billion by about 1960, four billion by about 1970 and five billion by about 1987. At present the world population is expanding by about 86 million a year and is expected to double in size within the next hundred years. An ideal population size is one that can be sustained indefinitely without harming its environment.

ENDANGERED SPECIES

The following species are all categorized by the International Union for Conservation of Nature and Natural Resources as 'endangered'. This category of threat means that the species has a 20% probability of extinction within 20 years or 10 generations, whichever is longer. When available, information about numbers is given. A fuller list (of 4,629 threatened species) can be found in the 1994 IUCN Red List of Threatened Animals and more details in the various IUCN Data Books.

MAMMALS

Black rhinoceros (*Diceros bicornis*). Sub-Saharan Africa. About 2,000 remain.

[The **great one-horned rhino** (*Rhinoceros unicornis*) survives in protected areas of northeast India and Nepal, where it numbers about 1,960. The smaller **Javan one-horned rhino** (*Rhinoceros sondaicus*) numbers about 50-60 in Java. Only 500 of the **Sumatran rhino** (*Dicerorhinus sumatrensis*) survive in Indonesia, Malaysia, Thailand and Vietnam.]

Blue Whale (*Balaenoptera musculus*). All Oceans. Less than 3,000 surviving.

Giant Panda (*Ailurupoda melanoleuca*). China. 1,000-2,000 surviving.

Golden Bamboo Lemur (*Hapalemur aureus*). Madagascar. 200-400 surviving.

Golden Rumped Lion Tamarin [monkey] (*Leontopithecus chrysopygus*). Brazil. About 100 surviving.

Indus River Dolphin (*Platanista minor).* Pakistan. 500 surviving.

Mediterranean Monk Seal (*Monachus monachus*). Mediterranean. Less than 500 surviving.

Mountain Gorilla (*Gorilla gorilla*). Rwanda; Uganda; Zaire. Fewer than 500 surviving.

Pygmy Hippopotamus (*Hexaprotodon liberiensis*). West Africa. Less than 100 surviving.

Spanish Lynx (*Lynx pardinus*). Portugal; Spain. Less than 1,000 surviving.

Tiger (*Panthera tigris*). Asia. 30-80 South China tigers; 150-200 Siberian tigers; 650 Sumatran tigers; 1,000-1,700 Indo-Chinese tigers; 3,350-4,700 Bengal tigers surviving.

Wild Yak (*Bos mutus*). China; India; Nepal. Only a few hundred surviving.

Woolly Spider Monkey (*Brachyteles arachnoides*). Brazil. 200-300 surviving.

BIRDS

Bermuda Petrel (*Pterodroma cahow*) Bermuda. About 50 pairs surviving.

Black Stilt (*Himantopus novaezeelandia*). New Zealand. 80 surviving of the world's rarest wading bird.

Gurney's Pitta (*Pitta gurneyi*). Thailand. 25-35 pairs surviving.

Imperial Woodpecker (*Campephilus imperialis*). Mexico. Only sporadic sightings of this woodpecker, the largest in the world.

Ivory-billed Woodpecker (*Campephilus principalis*). Cuba. Remnant population in Cuba; almost extinct in the USA.

Madagascar Fish Eagle (*Haliaetus vociferoides*). Madagascar. 50 pairs surviving.

Spanish Imperial Eagle (*Aquila adalberti*). Portugal; Spain.

White Cockatoo (*Cacatua alba*). Indonesia.

White-winged Guan (*Penelope albipennis*). Peru. 100 surviving.

REPTILES AND AMPHIBIANS

Desert Tortoise (*Gopherus agassizii*). Mexico; USA. 100,000 surviving.

Galapagos Giant Tortoise (*Geochelone elephantopus*). Galapagos Island.

Kemp's Ridley Sea Turtle (*Caretta caretta*). Mexico; North America. 400-600 nesting females (males do not come ashore).

Komodo Dragon (*Varanus komodoensis*). Indonesia.

FISH

Baltic Sturgeon (*Acipenser sturio*). Black Sea. 300-1,000 surviving.

Eastern Freshwater Cod (*Maccullochella ikei*). New South Wales, Australia.

Moduc Sucker (*Catostomus microps*). California, USA. About 1,300 surviving.

Swan Galaxias (*Galaxias fontanus*). Tasmania.

Totoaba Seatrout (*Totoaba macdonaldi*). Gulf of California. Now confined to the extreme northern end of the gulf.

INVERTEBRATES OTHER THAN INSECTS

Birdwing Pearly Mussel (*Conradilla caelata*). Tennessee; Virginia, USA.

Louisiana Pearlshell (*Margaritifera hembeli*). Louisiana, USA.

Tooth Cave Spider (*Leptoneta (Neoleptoneta) myopica*). Texas, USA.

Tulotoma Snail (*Tulotoma magnifica*). Alabama, USA.

INSECTS

Colombia River Tiger Beetle (*Cicindela columbia*). Idaho, USA.

Homerus Swallowtail (*Papilio homerus*). Jamaica.

Otway Stonefly (*Eusthenia nothofogi*). Australia.

Pygmy Hog Sucking Louse (*Haematopinus oliveri*). India. Thought to number only a few hundred.

St Helena Giant Earwig (*Labidura herculeana*). St Helena, South Atlantic Ocean.

PLANTS

Baishan Fir (*Abies beshanzuensis*). China. Only five specimens known to remain on Baishanzu mountain in Zhejiang Province.

PLANTS contd.

Bigcone Pinyon Tree (*Pinus maximartinezii*). Mexico. 1,000-10,000 trees surviving.

Brazilian rosewood (*Dalbergia nigra*). Brazil. Atlantic coastal forests where this rosewood is found have declined dramatically.

Golden Gladiolus (*Gladiolus aureus*). South Africa. About 20 plants surviving.

Green Pitcher Plant (*Sarracenia oreophila*). Alabama; Georgia, USA. About 26 colonies exist in the wild.

Shirhoy Lily (*Lilium mackliniae*). India. A few scattered populations survive in Manipur.

Adiantum asarifolium. Mauritius. Very few specimens left of this fern.

RECENT MAJOR NATURAL DISASTERS

In their book *Natural Disasters: Acts of God or acts of Man?* Anders Wijkman and Lloyd Timberlake point out that there are over 3,000 deaths per disaster in low-income countries, compared with 500 in high-income countries. The major factors in disasters in the Third World are poverty, environmental degradation and rapid population growth.

July 1979-May 1981
Drought. *Hebei province, China.* Worst for sixty years. Much of area normally one of the most fertile in China, became a virtual desert. 14,000,000 peasants left dependent on government grain rations. It was estimated 2,000,000 children were suffering from malnutrition.

10 October 1980
Earthquakes. *El Asnam, Algeria.* 7.5 and 6.5 on Richter scale. 3,600 dead. 8,252 injured. 300,000 homeless. The town had been destroyed in 1954 by another earthquake which killed 1,000. It was rebuilt by the French with supposedly 'quake-proof' structures. These were the worst affected by the 1980 quake.

late August-early September 1980
Hurricane David. *Dominican Republic.* 20 killed. 60,000 homeless. Banana and grapefruit crops destroyed.

8 May 1980
Volcanic eruption. *Mount St Helens, Washington State, USA.* 2,000 people evacuated. Large quantities of volcanic ash blanketed communities to a depth of several feet in parts of Washington, Idaho and Montana.

3 November 1980
Earthquake. *South Italy.* 6.8 on Richter scale. Aftershocks continued for 10 days. 200 towns and villages affected: Naples, Salerno, Avellino worst affected. 4,500 dead. 8,000 injured. 350,000 homeless.

9-14 July 1981
Floods. *Sechuan province, China.* 1,385 dead. 14,109 injured. Affected 10 million people in 135 of the province's 212 counties, threatened the Gezhouba dam and destroyed more than 1,000,000 acres of crops.

late 1982-1983
Floods. *Eastern Bolivia.* 40,500 hectares (100,000 acres) of prime agricultural land destroyed. At the same time in the high valleys of Bolivia and in southern Peru drought conditions devastated the potato crop affecting about 2 million people.

3 December 1982
Earthquake. *North Yemen, Dhamar region.* 6.0 on Richter scale. 1,588 killed. 1,604 injured. 200,000 homeless.

February 1983 -July 1984
Drought. *Northeast Brazil. (Maranhao to northern part of Minas Gerais).* Fifth year of drought. 850,000 sq kms affected. By mid-1982 70% of population suffering from malnutrition.

24-25 May 1985
Cyclone and tsunami. *S.E. Bangladesh.* 11,000 killed. 250,000 homeless.

13 November 1985
Volcanic eruption. *Armero, Colombia.* The Nevado del Ruiz volcano erupted leaving 22,000 dead 4,000 injured and 20,000 homeless. The village of Armero was totally engulfed by a torrent of mud when La Lagunilla river burst its banks. 11,000 hectares of agricultural land ruined.

19 September 1985
Earthquake. *Mexico City, Mexico.* 8.1 on Richter scale. More than 7,000 dead. 30,000 injured. 10,000 seriously. 100,000 homeless.

21-22 August 1986
Toxic gases following volcanic eruption. *Lake Nyos, Cameroon.* Toxic gases (a mixture of carbon dioxide and hydrogen sulphide) were released from volcanic Lake Nyos, near the Nigerian border, killing 1,746 and injuring 3,000, though exact numbers of casualties will probably never be known. In Nyos village only 2 out of a population of 700 survived.

10 October 1986
Earthquake. *San Salvador, El Salvador* 7.5 on Richter scale. 1,500 killed. 300,000 homeless. On the outskirts of the city a landslide buried 100 people.

March 1987
Earthquake. *Ecuador.* 7.3 on Richter scale. Followed by 1,300 aftershocks. 2-4,000 killed. 50,000 homeless. Mudslides entombed whole villages.

August-September 1987
Floods. *Bangladesh.* 70 killed (a very low estimate). 24,000,000 homeless and without food (out o a total population of 100,000,000). 2,000,000 homes and 2,600 kms of road destroyed. 4,300,000 acres of land devastated.

September 1987
Heavy rains and mudslides. *Medellin, Columbia.* Shanty town of Villa Tina was buried by mud. 355 people dead. 200 seriously injured. 2,100 homeless.

February 1988
Mudslides. *Rio de Janeiro, Brazil.* In a shanty town in Rio after weeks of torrential rain. 300 killed 735 seriously injured. 11,000 homeless.

August-September 1988
Floods. *Bangladesh.* Widespread flooding on a similar scale to the floods of 1987.

9 September 1988
Hurricane Gilbert. *Jamaica.* Over 200 mph winds. Left a trail of destruction across Jamaica, Grand Cayman, the Yucatan Peninsula, Mexico, Texas. At least 240 killed, most of those in Yucatan.

Mid-1988
Drought. *USA.* Farming lands were turned into disaster areas and world food production fell.

6 November 1988
Earthquake. *China.* 7.6 on Richter scale. 900 people killed.

7 December 1988
Earthquake. *Armenia.* 6.7 on Richter scale. 55,000 dead. 250,000 left homeless.

28 April 1989
Tornado. *Bangladesh.* 1,000 people killed and many villages reduced to rubble.

23 September 1989
Hurricane. *Caribbean/USA.* Hurricane Hugo caused widespread devastation in the north-eastern Caribbean and south-eastern USA. Winds reached up to 140 mph. Guadeloupe and Leeward Islands: 6 killed, 10,000 homeless and many crops destroyed. Montserrat: 10 killed, hospital and houses

destroyed and airport runway blocked. Puerto Rico: 25 killed, 100,000 homeless. Dominica: much of banana crop destroyed. South Carolina: 5 killed, houses damaged, services disrupted.

17 October 1989
Earthquake. *San Francisco, USA.* 6.9 on Richter scale. About 100 people killed, mostly when the elevated section of Interstate 880 highway collapsed crushing people in their cars.

27 February 1990
Mudslide. *Peru.* The village of San Miguel de Rio Mayo, 800 km north of Lima, was buried. 200 people killed or missing.

29 May 1990
Earthquake. *Peru.* 5.8 on Richter scale. About 200 people killed, village flattened by landslides.

1991
Epidemic. *Latin America.* Cholera broke out in January in Peru and spread to Brazil, Colombia, Chile and Ecuador. WHO estimates 42,000 people could die unless steps are taken to combat malnutrition and lack of basic sanitary and medical facilities.
Famine. Still critical in Sudan, Ethiopia, Somalia, Mozambique and Angola, all countries in which armed conflict exacerbated food distribution problems.

2 February 1991
Earthquake. *Afghanistan.* 6.5-6.8 on Richter scale. 1,000 killed, many more injured. Resulting floods killed 200 and left 3.000 people homeless.

Mid-March 1991
Floods and mudslides. *Malawi.* 516 people dead or missing. 40-50,000 homeless.

5-6 April 1991
Earthquake. *Peru.* Three earthquakes. the worst of which was 6.9 on the Richter scale. 35 killed. 750 injured.

22 April 1991
Earthquake. *Costa Rica/Panama.* 7.5 on Richter scale. About 80 people killed, 800 injured, thousands of homes destroyed.

29-30 April 1991
Cyclone. *Bangladesh.* Winds up to 145 mph. 139,000 killed and thousands more threatened by epidemic. Up to 10,000,000 people homeless, 4,000,000 at risk from starvation. Toll would have been worse but for 300 cyclone shelters built under Cyclone Preparedness Programme set up by Red Crescent.

30 April 1991
Earthquake. *Georgia, Russia.* 7.2 on Richter scale. 100 people killed, several villages totally destroyed.

9-15 June 1991
Volcano. *Philippines.* After lying dormant for 600 years Mount Pinatubo suffered several eruptions, some shooting volcanic ash and rock 19 miles into the air. At least 343 people were killed and 100-200,000 made homeless. 600,000 Filipinos lost their means of livelihood.

3 December 1991
Earthquake. *Romania.* 5.7 on Richter scale. 1,700 people made homeless but no one killed.

23 December 1991
Floods. *Texas, USA.* 13.3 inches of rain fell in six days. At least 14 people were killed.

Late December 1991
Cyclone. *Hanoi, Vietnam.* 220 people killed.

1-2 February 1992
S.E. Turkey. Avalanches smothered remote villages killing at least 167 people, most of them soldiers.

August 1992
Hurricane. *Bahamas, and Florida and Louisiana, USA.* Winds up to 264 km/hr. Hurricane Andrew caused extensive damage to property, estimated at $20,000 million in Florida. 38 people killed. Up to 200,000 people homeless.

September 1992
Flooding. *Pakistan.* Brought on by continuous torrential rains. Engulfed thousands of villages and inundated vast acres of cropland especially in Punjab, the worst affected province. More than 2,000 people killed. At least 3,000,000 homeless.

1 September 1992
Tidal wave. *Nicaragua.* Coastal settlements devastated by tidal wave up to 15 metres high, triggered by earthquake beneath Pacific Ocean. Over 200 people dead. 4,200 homeless.

12 October 1992
Earthquake. *Cairo, Egypt.* 5.9 on Richter scale. 552 people killed. About 3,000 injured. Hundreds made homeless.

December 1992
Mudslide. *Bolivia.* Town of Caranavi engulfed. Homes pushed into Tipuani River. Around 350 people killed. 135 injured.

12 December 1992
Earthquake. *Nusa Tenggara Province, Indonesia.* 6.8 on Richter scale. As many as 2,000 people killed. Created a 24-metre-high tidal wave which devastated the coastal town of Maumere, killing more than 1,000 people and damaging a third of the buildings.

February 1993
Floods. *Yemen & Iran.* Widespread damage. At least 500 people killed. More than 1,500 people displaced.

March 1993
Severe storms. *USA.* Heavy snow and floods caused widespread destruction from the Florida Keys through the Deep South to northern New England. 50 tornadoes hit Florida. At least 163 people killed.

14 March 1993
Earthquake. *Erzincan, Turkey.* 6.8 on Richter scale. More than 500 people killed. Over 1,000 injured

April 1993
Floods and landslides. *Colombia.* At least 59 people killed in north-west area. Thousands homeless.

9 May 1993
Landslide. *Ecuador.* Over 200 people killed in a village in south Ecuador as 15,000 tonnes buried the community.

May 1993
Floods and mudslides. *Tadjikistan.* At least 200 people killed.

June 1993
Floods. *Bangladesh.* About 100 people killed. More than 1 million displaced. Worst affected area was Sylhet district.

15 June 1993
Mudslide. Kabul, Afghanistan.

115 people reported killed.

July 1993
Torrential rains and mudslides. *India, Nepal, Bangladesh.* Worst hit areas of India in Assam and West Bengal. More than 3,700 people killed. Millions made homeless.

July 1993
Flooding. *USA.* Midwest affected as Mississippi-Missouri river system broke its banks following heavy rain. 40,000 sq km flooded. At least 45 dead. 30,000 homes ruined.

12 July 1993
Earthquake. *Japan.* 7.8 on Richter scale. Ensuing tidal waves struck parts of Hokkaido, Honshu and devastated coastal regions of Okushiri. Triggered landslides and fires. Over 150 people killed.

10 August 1993
Tropical storm. *Venezuela.* Torrential rain and high winds caused mudslides in shanty towns above Caracas. At least 150 killed. More than 400 injured.

30 September 1993
Earthquake. *India.* 6.5 on Richter scale. Devastation in states of Maharashtra, Karnataka and Andhra Pradesh. Over 9,000 people killed. Umarga and Khilari worst hit towns. More than 30 villages flattened.

17 January 1994
Earthquake. *Los Angeles, USA.* 57 people killed, thousands injured and about 25,000 homeless. Damage estimated at costing up to $20,000 million.

16 February 1994
Earthquake. *Indonesia.* Struck southern Sumatran province of Lampung, killing 200 people and injuring 2,700.

2 May 1994
Cyclone. *Bangladesh.* 165 people officially reported killed as cyclone hit south-eastern coast.

7 June 1994
Earthquake. *Colombia.* 250 people killed as earthquake triggered mudslide in south-western region. 11,000 homeless.

18 August 1994
Earthquake. *Algeria.* 5.6 on Richter scale. 149 people killed in western region.

January 1995
Flooding. *NW Europe.* More than 300,000 people fled their homes in Belgium, France, Germany and The Netherlands. Immediate natural cause of flood was heavy rain and snow-melt.

17 January 1995
Earthquake. *Kobe, Japan.* 7.2 on Richter scale. About 5,000 people killed. 250,000 homeless. Widespread damage estimated at $50,000 million.

28 May 1995
Earthquake *Sakhalin Island, Russia* 7.5 on Richter scale. Estimated 2,000 people killed. Many thousands homeless

RECENT MAJOR MAN-MADE DISASTERS

10 July 1976
Seveso, Italy. Release of dioxin (TCDD - tetrachlorodibenzo-p-dioxin) poisonous gas from factory producing trichlorophenol. Within a few days 80,000 domestic fowl and half the pig population had died. A sharp rise in percentage of deformed births.

March 1978
off coast of Brittany, France. Oil tanker Amoco Cadiz, runs aground. 220,000 tons of oil spilt. Over 1 million sea birds killed as well as other marine life.

1978
Love Canal, near Niagara Falls, USA. This abandoned canal became a dumping ground for chemical wastes (including polychlorinated biphenyls, dioxin and pesticides) produced in the 1940s and 1950s. The site was filled in and given to the City of Niagara Falls. Housing was built on it. In 1978 leakage of toxic chemicals began to be detected in the basements of houses, A very high incidence of chromosomal damage was found amongst residents. The area was completely evacuated. It is not known what the long-term effects will be.

28 March 1979
Nuclear meltdown at *Three Mile Island, Pennsylvania, USA.* Mechanical and electrical failures were compounded by errors made by operators at the plant. It is not known how many deaths may have resulted.

June 1979
Gulf of Mexico. Oil rig out of control. Oil slick 640 miles long, with over 500,000 tons of oil spilt.

July 1979
off coast of Tobago, West Indies. Two supertankers, Atlantic Express and Aegean Captain, collide. 236,000 tons of oil spilt. Countless birds and marine life killed.

2 December 1984
Bhopal, India. Gas leak from pesticide plant. At least 2,000 killed and 220,000 treated for various ailments.

26 April 1986
Chernobyl, Ukraine. Roof of nuclear power station blown off, a hydrogen explosion then showered radioactive debris 1,500 metres into the air. Operational errors blamed. All Eastern Bloc countries, much of Scandinavia, all of Western Europe except Spain and Portugal are known to have been affected in varying degrees. At least 30 people at Chernobyl died almost immediately from burns or acute radiation sickness. The number of people finally affected will never be known.

1 November 1986
River Rhine. Following a fire at a chemical plant over 30 tons nf pesticides, fungicides and other chemicals were washed into the river by firemen dousing the flames. Mercury was the main chemical involved. For almost two days none of the governments along the Rhine knew the true nature of the chemicals flowing down Europe's largest waterway. 200 miles of Upper Rhine practically dead and it is estimated that it will take 10-30 years to restore it to life.

September 1987
Goiana, Brazil. Radioactive caesium chloride powder from a cylinder stolen from a hospital radiotherapy unit and sold for scrap found on a rubbish dump. It was handled casually by children

and others ignorant of its dangers. Only four people died but this does not take into account longer-term effects from radioactive pollution of area.

March 1989

Alaska, USA. The tanker Exxon Valdez ran aground off Alaska causing massive damage to the coastline, wildlife and habitat.

19 August 1989

Liverpool, Great Britain. 156 tonnes of crude oil emptied into the Mersey estuary from a Shell oil pipeline causing a 10-mile slick, killing at least 300 birds and affecting 2,000 others.

12 September 1990

Kazakhstan. An explosion and fire at a factory making nuclear fuels at Ulba in East Kazakhstan contaminated a large area with highly toxic compounds of beryllian metal. The health of up to 120,000 people potentially at risk.

24 September 1990

Bangkok, Thailand. 58 people died and 100 were injured when a truck carrying two tanks of liquefied petroleum gas crashed and exploded in the centre of the capital.

Early 1991

Persian Gulf. During the Gulf War Iraqi forces allegedly released 5-10,000,000 barrels of oil into the Gulf creating a vast oil slick affecting 400 miles of coast and endangering both the marine environment and the livelihoods of local fishermen.

Kuwait. 600 oilwells in Kuwait set on fire by Iraqi troops causing massive air pollution and contamination of agricultural land and water supplies especially in the fertile, well-populated Tigris and Euphrates valleys in Iraq 'Black rain' damaged crops in Iran, Pakistan. Bulgaria and Afghanistan. The final well fire was capped in late 1991.

11 April 1991

Genoa, Italy. A Cypriot-registered tanker Haven exploded during a routine pumping operation in the Ligurian Sea off Genoa. The resulting oil slick caused pollution along the Mediterranean coast westwards towards Nice. The 40 km slick was broken up by gale-force winds which diminished damage potential.

13 June 1991

Chile. The government ordered 40% of all cars off the streets of the capital. Santiago, one of the most polluted in the world. At least 1,465 children had been treated for breathing problems since 10 June.

12 July 1991

Washington State, USA. After a collision with a Chinese ship, a Japanese fish-processing ship sank and discharged 100,000 gallons of oil into the sea creating an oil slick threatening the Olympia National Park. Hundreds of birds were killed and rare species such as the bald eagle, peregrine falcon and sea otter endangered.

14 August 1991

California, USA. A tanker rail car spilled 19,500 gallons of metam sodium pesticide into the Sacramento River. It drifted 45 miles downstream, killing wild life. 200 people were admitted to hospital suffering from fume inhalation and skin irritation.

5 December 1991

Russia. The new environment minister, Victor Danilov-Daniliyan, admitted that large parts of Russian territory had been made uninhabitable for decades to come by nuclear accidents and waste.

contd./

8 January 1992
Athens. Greece. When smog in the capital exceeded emergency levels, the government banned all cars from the centre, industrial production was cut and central heating was turned off in public buildings except hospitals.

January 1992
Turkey & Greece. Hundreds of dead or dying dolphins were washed up on beaches, the victims of a virus linked with water pollution. The rare monk seal, of which only about 300 remain, is also at risk.

3 December 1992
Spain. A Greek-owned oil tanker ran aground and exploded off the north-western Spanish port of La Coruna, spilling an estimated 70,000 tonnes of oil into the sea, with serious ecological consequences. Strong winds and heavy seas helped disperse much of the oil.

January 1993
World. NASA research showed ozone levels lower than ever before. Levels were measured up to 14% below normal in the mid-latitudes of the northern hemisphere. In the second half of 1992, ozone levels were 2-3% lower than in any previous year. Global thinning of the ozone layer means a widespread threat to human health.

January 1993
Malaysia. A supertanker fully laden with almost 2 million barrels of crude oil collided with another tanker and started spilling oil into the sea off the northern tip of Sumatra.

5 January 1993
Shetland Islands, UK. The oil tanker Braer, carrying 84,000 tonnes of light crude oil, was driven ashore. Less damage than originally feared as hurricane winds were effective in breaking down and dispersing the spill. Salmon farms, fisheries and wildlife adversely affected.

6 April 1993
Tomsk, Russia. Explosion at the Tomsk-7 nuclear reprocessing plant released a cloud of radioactive gas over western Siberia. Reported to have contaminated 120 sq km of forest, although official reports claimed the amount of plutonium released was negligible.

Summer 1993
Europe Toxic green algae (first noticed in 1984 off the coast of Monaco) choked parts of the Mediterranean. Clumps of the weed, which grows following the pollution of the sea by chemicals that fertilize the water, stretched over 4.2 sq km and there were colonies as far apart as Majorca and Livorno, on the Italian coast. Holidaymakers in France, Spain and Italy were warned not to uproot it.

February 1994
Komi Republic, Russia. Oil spill polluted wide areas of tundra. Estimates of the leak from the 47-km-long pipeline ranged from 14,000 to 200,000 barrels. It was initially contained by an earth dam but this was breached by heavy rains in August.

Although these disasters seem (and indeed are) catastrophic, more pollution and human mortality occurs on a smaller but cumulatively greater scale each year. For instance, spills from oil tanker disasters are less than the cumulative total of small leaks, discharges and minor spills which occur throughout the year from such sources as tankers washing their empty holds with sea water, oil washed from cars and lorries into sewers, and pollution caused by routine handling of oil in refineries and garages. Sudden large-scale disasters do make an instant impact on us, and bring to our attention the way in which we are increasingly putting ourselves and our environment at risk.

I want to order/Please send me details of:

Mail or fax to: 1 Cambridge Road, Teddington, TW11 8DT, UK (fax 0181 943 1673)

English		
Accounting	0-948549-27-0	☐
Agriculture, 2nd ed	0-948549-78-5	☐
American Business	0-948549-11-4	☐
Automobile Eng.	0-948549-66-1	☐
Banking & Finance	0-948549-12-2	☐
Business, 2nd ed	0-948549-51-3	☐
Computing, 2nd ed.	0-948549-44-0	☐
Vocabulary for Computing	0-948549-58-0	☐
Ecology & Environment, 3rd	0-948549-74-2	☐
Goverment & Politics	0-948549-05-X	☐
Hotels, Tourism & Catering	0-948549-40-8	☐
Information Technology	0-948549-03-3	☐
Law, 2nd ed	0-948549-33-5	☐
Library & Information Mgmt	0-948549-68-8	☐
Marketing	0-948549-08-4	☐
Medicine, 2nd ed	0-948549-36-X	☐
Multimedia	0-948549-69-6	☐
Personnel Management,2e	0-948549-79-3	☐
Printing & Publishing	0-948549-09-2	☐
Science & Technology	0-948549-67-X	☐

English-French/French-English		
Business, 2nd ed	0-948549-64-5	☐
Computing, 2nd ed	0-948549-65-3	☐
Ecology & Environment	0-948549-29-7	☐

English-Swedish/Swedish-English		
Business (hb)	0-948549-14-9	☐
Computing/IT (hb)	0-948549-16-5	☐
Law (hb)	0-948549-15-7	☐
Medicine (hb)	0-948549-23-8	☐

English-German/German-English		
Agriculture (hb)	0-948549-25-4	☐
Banking & Finance (hb)	0-948549-35-1	☐
Business new edn. (hb)	0-948549-50-5	☐
Computing/IT (hb)	0-948549-20-3	☐
Ecology (hb)	0-948549-21-1	☐
Law (hb)	0-948549-18-1	☐
Marketing (hb)	0-948549-22-X	☐
Medicine (hb)	0-948549-26-2	☐
Print/Publishing (hb)	0-948549-19-X	☐

English-Spanish/Spanish-English		
Business (hardback)	0-948549-30-0	☐

English-Chinese		
Business (hb)	0-948549-63-7	☐

Name: . . . .

Address: . . . .

. . . . Postcode: . . . .